# Wave Optics

LECTURES IN OPTICS

**Volume 3**

By the Author

*Lectures in Optics, Vol. 1, Introduction to Optics*

*Lectures in Optics, Vol. 2, Geometrical Optics*

*Lectures in Optics, Vol. 3, Wave Optics*

*Lectures in Optics, Vol. 4, Visual Optics*

*Lectures in Optics, Vol. 5, Ocular Imaging*

# Wave Optics

LECTURES IN OPTICS

**Volume 3**

**George Asimellis**

**SPIE PRESS**
Bellingham, Washington USA

Library of Congress Cataloging-in-Publication Data

Names: Asimellis, George, 1966- author.
Title: Wave optics / George Asimellis.
Description: Bellingham, Washington, USA : SPIE Press, [2020] | Series: Lectures in optics ; vol. 3 | Includes index.
Identifiers: LCCN 2019001143| ISBN 9781510622630 (softcover) | ISBN 1510622632 (softcover) | ISBN 9781510622647 (pdf) | ISBN 1510622640 (pdf)
Subjects: LCSH: Wave theory of light. | Light. | Polarization (Light) | Dispersion. | Interference (Light)
Classification: LCC QC403 .A85 2020 | DDC 535/.13--dc23
LC record available at https://lccn.loc.gov/2019001143

Published by
SPIE
P.O. Box 10
Bellingham, Washington 98227-0010 USA
Phone: +1 360.676.3290
Fax: +1 360.647.1445
Email: books@spie.org
Web: http://spie.org



Printed in the United States of America.
First Printing.
For updates to this book, visit http://spie.org and type "PM296" in the search field.

SPIE.

COVER IMAGE:

MACRO PHOTO OF DIFFRACTION EFFECTS AND MISTY DROPLETS OVER A BLUE RAY DISK.

IMAGE CREATION: EFSTRATIOS I. KAPETANAS. FACEBOOK.COM/PHOTOSTRATOSKAPETANAS/

*Enalithos (Ena-, ένας = One & -lithos, λίθος = Stone) encounters the Greek philosopher Socrates, the questioner of everything and everyone. Athens, Greece, 400 BC (© www.fiami.ch).*

Cartoon illustrations pertaining to Einstein's virtual colloquium with Greek philosophers are part of a series on the history of science entitled 'The Lives of Einstein,' published by www.fiami.ch. Enalithos later becomes Alberto Unasso (when meeting Galileo), and then Albert Singlestone (when meeting Isaac Newton), and finally, Albert Einstein.

Kind permission has been granted by Fiami (www.fiami.ch) to use some of these comic strip illustrations in this book.

# TABLE OF CONTENTS

# Foreword

The study of optics has had an enormous impact on modern life in so many different ways that we tend to take it for granted. Take for example telecommunications, medicine, entertainment, and the arts. Here is a text that introduces readers to various important behaviors of light so that they can understand how optics came to be such an integral part of our existence. It is accessible to anyone who has a basic background in mathematics.

I had the privilege of advising Professor Asimellis during his PhD studies at Tufts University. I have always enjoyed discussing optics with him and am very happy to see that he has written a textbook quite unlike any of the others.

The writing is colloquial and includes historical references and philosophical insights that shed light on the thought processes that went into making the underlying discoveries. We learn from the past to advance to the future. It takes some effort to understand how something invisible enables us to see, and how something you cannot hold has such a powerful influence on our lives.

The author appreciates these paradoxes and has done an excellent job helping us to go from understanding nothing to catching enough of a glimpse of the truth to be able to begin to make contributions ourselves. This textbook covers in its entirety the essential topics of wave and physical optics on a level suitable for most college and engineering curriculae. There are many other excellent texts that go more thoroughly into the theory and application of optics, but for a general introduction that encourages the reader to have the confidence to wade in more deeply, one need go no further than this monograph.

**Mark Cronin-Golomb, PhD**
Professor, Department of Biomedical Engineering, Tufts University
Medford, Massachusetts
June 2020

# Preface

Wave optics...geometrical optics—Are they that different? At first glance, perhaps yes. They appear to be almost unrelated. The physical properties of light primarily influence wave optics, while the natural rectilinear propagation and the simple laws of reflection and refraction appear to be the main laws that govern geometrical optics.

Yet, upon diving into the details, one comes to realize that, while the location and size of an image are governed by simple geometrical laws, the fine details of an image, such as its resolution, are governed by the physical properties described by wave optics.

Light is ultimately a wave phenomenon. Hence, it is natural that a volume of this *Lectures in Optics* series should be devoted to presenting a view of optics deriving from electric field oscillations and waves. Wave optics concerns the nature of light, especially, its vector nature, its interaction with matter, the complexity of the refractive index, the interference of light with light, and realization of the infinitesimal wavelets that explain diffraction effects. Finally, wave optics and physical optics, as well as quantum optics, merge to form the principles of lasers.

The origins of this textbook can be traced back to the Laboratory Optics course that the author had the honor of teaching at the Department of Physics, Aristotle University of Thessaloniki, Greece. What grew out of that course text is an attempt to provide a modernized textbook based on updated lecture notes and the narrative flow of classroom instruction. Readers are expected to be knowledgeable of college-level mathematics, including algebra, trigonometry, linear algebra, ordinary differential equations, and partial differential equations.

A certain familiarity with vector notation and advanced calculus will be helpful; however, the derivation of certain results is outside the scope of this book and is not emphasized. This book covers the essentials needed for any college-level Wave Optics curriculum in Physics and Engineering departments, as well as Optometric professional programs, and will be useful to those seeking a bottom-up textbook that foregoes a formal style and presents an attractive and updated perspective.

**George Asimellis, PhD**
Pikeville, Kentucky
June 2020

## Acknowledgments

I would like to thank the following colleagues for their helpful suggestions, encouragement, and recommendations:

**Mark Cronin-Golomb**, PhD
Professor, Department of Biomedical Engineering, Tufts University

**Robert W. Arts**, PhD
Professor of Education & Physics, University of Pikeville

**Frank Spors**, EurOptom, MS, PhD, FAAO
Associate Professor, College of Optometry, Western University of Health Sciences

I would also like to thank the following members of the Frank M. Allara Library, University of Pikeville for their contributions in providing grammatical and English-language corrections:

Karen Evans, Director of Library Services

Edna Fugate, University Archivist and Reference Librarian

Jerusha Shipstead, Reference and Instruction Librarian

Katherine Williams, Instructional Design Librarian & Coordinator of Instruction

Most of the artwork in this book was created by the author; any figure that was not created by the author is attributed to the source provided in the caption.

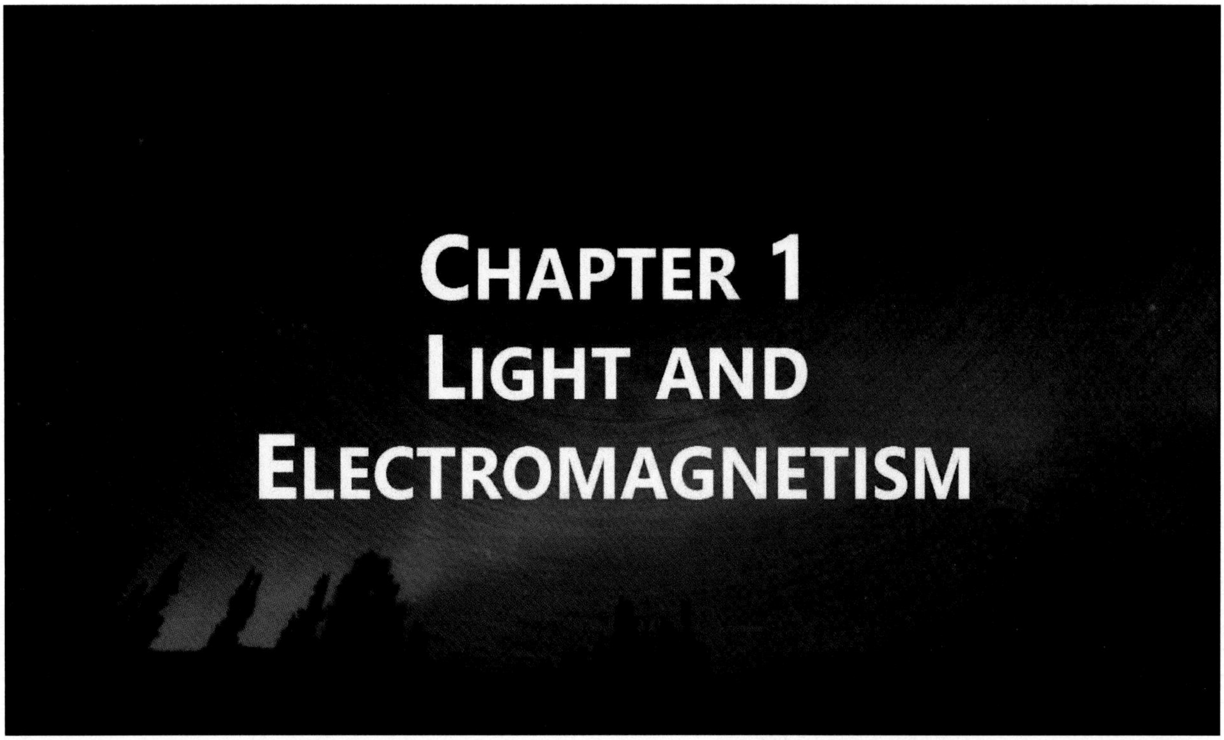

# CHAPTER 1 LIGHT AND ELECTROMAGNETISM

**Light** is the physical entity that the human organ of vision, the eye, is sensitive to. For centuries, the quest to understand this entity has been a challenge for which the answers vary depending on the way we comprehend and interact with light.

In the field of optics, light is studied using the effects of its propagation and interaction with matter, as well as of its creation and detection. Optics interprets a wide variety of simple effects, such as the reflection of a mirror or the refraction from a lens, as well as more-complex effects, such as the function of human vision or the operation of a digital photography camera.

Whether we are enjoying the sun's warmth or utilizing the energy produced by a photovoltaic collector, we understand that light is associated with energy transfer. We know from the Greek poet Homer that the Achaeans used light signals to deliver the news of their Trojan War victory to those back home in Greece. If we are talking on the phone or browsing the Internet, we are doing so by the transfer of electronic signals, as nowadays, copper has generally been replaced by optical fiber in such applications. Light, properly modulated, can carry information.

Today, optics is considered a very broad-spectrum scientific and technological field dealing with effects and applications in which light carries energy or information. Such applications are, for example, optical telecommunications, where light pulses from a semiconducting laser propagate through optical fibers carrying information, optical signal processing applications in optical memories, and medical imaging.

# 1.1 THE NATURE OF LIGHT

On a microscopic scale, in the world of infinitesimal dimensions, natural effects can be quite different from what we might expect based on our everyday experiences. An atom, for example, in first approximation, may appear like a tiny solar system (as presented in detail in § 6.1), with the planets (electrons) orbiting around the sun (nucleus). In a more accurate approximation, atoms are like a dust cloud or fog. But the exact depiction of an atom is still unknown. Any attempt to describe nature in human terms is fascinating as well as challenging, perhaps even more so today, since our current understanding of the physical world has advanced so much—or at least we think it has. 'Nature likes to hide,' as Heraclitus of Ephesus once said.

*Figure 1-1: The knowledge we receive from our senses is misleading. The 'insolent youth' is Enalithos (© www.fiami.ch).*

Any attempt to understand light presents a similar challenge. Over the years, our quest to comprehend the concept of light has had to overcome preconceptions and mysticisms, as well as theological doctrine. Light represents, or is associated with, the very Divine itself in most (if not all) religions.

Science, however, is not driven by doctrine. The scientific evaluation of the question of light has to be objective. We, scientists, do not just 'believe in' something. We support evidence-based suggestions. We propose models that are subject to verification until proven accurate or else rejected. Observation and experiment are the ultimate tools for such an evaluation.

Light, which is perhaps the first entity that we sense at the very beginning of life and that renders our world visible and therefore detectable, is itself invisible, cautiously hiding its nature. What we see with our eyes or detect with sensors is not light itself (either as a stream of photons or an electromagnetic disturbance) but the result of light interaction with the sensory organ of vision, the eye.

*Figure 1-2: Galileo is excited to meet Alberto Unasso and exchange ideas about scientific beliefs (© www.fiami.ch).*

The sense of vision is affected by light's specific properties, such as its frequency and type of photons, as well as its luminous intensity, which is proportional to the photon population. It is interesting to ask the question, If we had a different ability to 'see' light, how differently would we understand light? It is possible that the development of our models about light would have been vastly different in this case.

### 1.1.1 Early Theories

Greek philosophers presented quite a number of (indeed, contradicting) theories on light. Lucretius, in his *De Rerum Natura* (*On the Nature of Things*), stated one of the first concepts supporting the corpuscular nature of light and associated light with the heat emitted by the sun. Evaluating his proposal after so many centuries, we may find that he was quite close to our current concept of light! Hero of Alexandria (*Ἥρων ὁ Ἀλεξανδρεύς*) studied the effect of reflection and is credited with the first statement suggesting that light obeys the principle of propagating via the shortest path.

The first descriptive study of optical effects was presented by Johannes Kepler, who was renowned for his planetary motion laws presented in his book *Dioptrice* (1611). In the framework of his planetary observations, Kepler successfully described image formation and magnification, the telescope (he is credited with inventing the astronomical telescope), and total internal reflection.

Around the same time, Isaac Newton, in his book *Opticks, or A Treatise of the Reflections, Refractions, Inflections and Colours of Light* (1704), and Christiaan Huygens in his book *Traité de la Lumière* (1690), presented two seemingly opposing theories about the nature of light. According to Newton, light is a stream of tiny particles, called corpuscles. Without providing specifics of their characteristics, he described them as having vector attributes such as momentum, speed, and energy. According to Huygens, light is composed of wave pulses (wavelets), and light propagation can be predicted by the wavefronts formed by these wavelets.

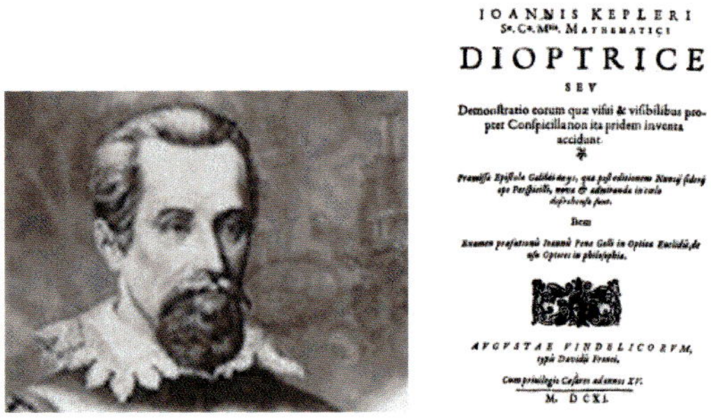


*Figure 1-3: Johannes Kepler (1571–1630).*

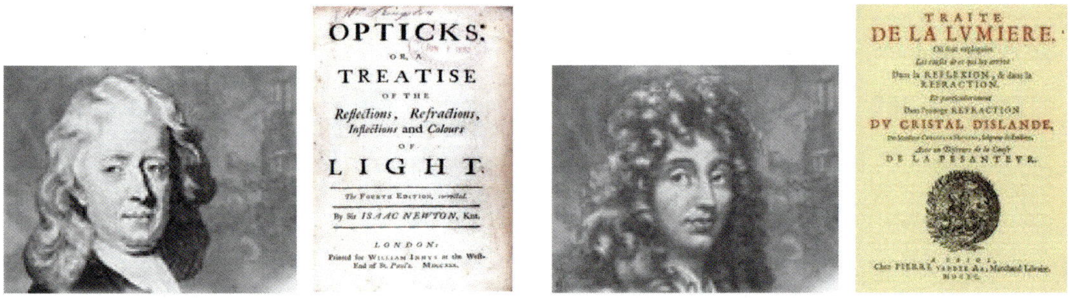


*Figure 1-4: Sir Isaac Newton (1642–1727) and Christiaan Huygens (1629–1695).*

None of these seemingly different concepts regarding the nature of light prevailed over the others. This is because each concept provided equally acceptable explanations for the then-known optical phenomena such as reflection, refraction, lens imaging, prism function, etc. For example, we know that when light passes from one medium to another, it changes speed; this is refraction. In addition, if light forms an angle with respect to the normal to the interface, it bends. This effect can be satisfactorily explained by both the corpuscular theory and the wave theory.

The corpuscular theory predicted a ray vector, whose tangent along the interface is preserved. It is interesting that the current corpuscular theory, photonics, verifies this theory with a momentum vector, the wavevector. Photonics explains that if the ray vector turns, the new path can be found.

The wave theory can also present a very plausible explanation for refraction. We imagine a barrel rolling on the street at a fixed speed (in magnitude and direction). The barrel moves along a straight path, which means that it does not turn. If, even momentarily, the barrel rolls onto grass, then, because this portion of the path corresponds to a slower speed, the barrel turns to roll in a new direction.

This is what happens in a wave propagating in air when it encounters along its path another medium, such as a piece of glass. In air, propagation follows a straight path, just as the barrel does on the street. Because the propagation speed in glass is lower during the time when the wavefront transcends the interface, its various parts have different speeds—the part in glass has less speed than the part still in air. Thus, as the various points of the wavefront pass from one medium to the other, the wave changes direction.

There is a similar agreement in the laws of reflection and, generally, other effects described by geometrical optics. However, effects such as interference (Chapter 4) and diffraction (Chapter 5) can be satisfactorily described only in terms of wave optics—the early corpuscular theory could not offer a satisfactory interpretation. Classical particles simply add up, and it is not possible to observe interference between two of such particles, for example, one canceling the presence of the other! If a classical particle encounters an obstacle along its path, this obstacle will allow either transmission exactly across it, or no transmission at all; diffraction effects are not observed around it.

## 1.1.2 The Wave Nature of Light

The experiment that proved the wave nature of light beyond any doubt was performed by Thomas Young in 1801. In this experiment, light forms interference fringes (see § 4.2.1), which is a characteristic wave effect. Two coherent light sources may form alternating bright and dark areas, called fringes. Bright areas appear where the two wave perturbations 'add up' by arriving at the same point in phase (simultaneously as crests, for example) and interfering constructively. Dark areas appear where the perturbations arrive at the same point with opposite phases (one as a crest and the other as a trough) and cancel each other in a negative (destructive) interference.

Thus, it became scientifically acceptable to assert that light is a wave (propagating in a hypothetic elastic medium, the ether), even without the knowledge of (1) which type of disturbance is associated with the wave, (2) what ether actually is, and (3) how a light ray can also behave like a wave.

Soon after Michael Faraday's experiment in 1845, it became known that magnetic fields interact with light by rotating the light's polarization plane. A few years later, James Clerk

Maxwell predicted the existence of electromagnetic waves while solving his equations. The speed associated with these waves was the same as the speed of light, which was already known; in 1862, Léon Foucault measured the light speed in air at 298,000 ± 500 km/s. With appreciation for this calculated speed of light, Maxwell proposed in 1870 that light is an electromagnetic wave!

Maxwell's prediction was eventually verified 17 years later, when Heinrich Rudolf Hertz, attempting to resolve whether 'the electric force propagates with infinite speed,' according to Wilhelm Eduard Weber, or 'behaves as a wave,' according to Maxwell, managed to prove experimentally that electromagnetic waves do exist. Thus, electromagnetic waves were no longer just an inspiration, a theoretical prediction, or a solution to specific equations. They were actually emitted in Hertz's lab, where a source [an inductor–capacitor (LC) circuit] radiated a signal that was detected by a receiver oscillating at the same frequency. Hertz proved that what was exchanged, in addition to being electromagnetic, was indeed a wave; he observed reflection, refraction, and even interference in the form of standing waves. Based on these observations, he calculated the propagation speed. By measuring the distance between two standing waves, Hertz derived the wavelength; then, multiplying the wavelength by the known frequency, he calculated the speed to be ≈ 300,000 km/s, which equals the speed of light.

*Figure 1-5: Michael Faraday (1791–1867) and James Clerk Maxwell (1831–1879).*

Soon thereafter, two new effects, the spectral distribution of black-body radiation density (discussed in § 1.4.3) and the photoelectric effect (§ 1.4.5), raised some objections that the electromagnetic theory could not satisfactorily explain. Attempts to interpret these effects led to a major revolution in modern physics. The first catalysts for this advancement were Max Planck's idea that energy is quantized, and the then-unknown amateur scientist Albert Einstein's statement made more than a century ago that light is composed of particles, called photons, which are light quanta, and the energy of each photon is proportional to its frequency.

Thus, the 'old' corpuscular theory returned to prominence under the mantle of quantum physics, this time not to completely replace the wave theory, but to complement it. Exactly how light may behave is not absolutely granted—it can appear as a wave or a particle. The modern corpuscular theory agrees, in many cases, with the wave theory for a large photon population.

*Table 1-1: Comparison of the corpuscular and wave theories of light.*

| Corpuscular Theory | Wave Theory |
|---|---|
| Particles that belong to various colors have the same speed in vacuum but different speeds in some media, e.g., glass, water, air. | Waves with wavelengths that belong to different colors have the same speed in vacuum but different speeds in other media, e.g., glass, water, air. |
| White light is composed of particles that belong to different colors. A prism separates the colors with refraction. | White light is a composition of waves of different wavelengths that correspond to different colors. A prism separates the colors with refraction. |
| Reflection and refraction are interpreted with the laws of motion. | Reflection and refraction are interpreted with wave propagation laws. |
| The photon spin state and its direction of momentum determine the polarization plane. | The direction of oscillation of the electric field and the Poynting vector determine the polarization plane. |
| ☠ Problem! Light passing through an aperture forms a geometrical shadow, so the corpuscular theory does not explain diffraction. | ✓ A wave passing through a large aperture forms a geometrical shadow, but when it passes through a small aperture, we notice diffraction effects. |
| ☠ Problem! Interference <u>is not explained</u> by the corpuscular theory. | ✓ Waves that arrive in phase interfere constructively, while waves that arrive in the opposite phase interfere destructively. |
| ✓ The photoelectric effect is in full accordance with the photon model. | ☠ Problem! The photoelectric effect <u>is not</u> explained by wave theory. |

## 1.1.3 Wave Characteristics

The notion of a wave is employed in the physical sciences to describe the process of a propagating disturbance, which can be expressed by a temporal and spatial variation of a physical quantity. For a wave to exist, there needs to be a physical quantity that is subject to a disturbance; there has to be a source that disrupts something. This disruption propagates away from the source. There must also be a mechanism of interaction from one point to another and (although not always) from a point to a medium. In other words, what happens at point B is influenced by what happens (i.e., the type and magnitude of disruption) at point A. The events taking place at points A and B are linked by the physical laws (the mechanism) that explain what happens and why.

Gravitational waves and electromagnetic waves, for example, do not require a medium, as they can propagate in vacuum, while mechanical waves do require a medium. In a wave, there is energy and momentum transfer with no associated mass (of the medium material) transfer; the medium within which the wave propagates—if it exists—does not move macroscopically. Finally, a wave can be detected by a receiver, a sensor, or a detector that responds to the perturbation and receives (not necessarily all of) the energy and momentum carried by the wave.

A familiar wave is seen in sports stadiums when cheering spectators throw their hands into the air, prompting the adjacent spectators to do the same. Gradually, this hand movement propagates through the crowd.

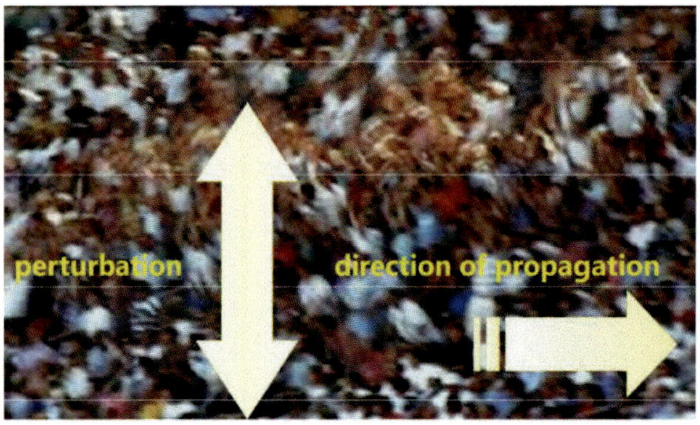


*Figure 1-6: Cheering spectators at an athletic event.*

The disruption here is the hand movement propagating from spectator to spectator. We observe that, while the disturbance is propagating, no spectators actually move from their seats. When the wave ceases, every sports fan is in the same position as before the wave.

In this specific example, because the disorder—the hand movement—is perpendicular to the rows of stands (which is the direction of propagation), it is a **transverse wave**. In a transverse wave, the disturbance is normal (perpendicular) to the direction of propagation. The sea is made of such waves. The disruption in this case is the vertical movement of surface points from their equilibrium position, and the wave propagates horizontally along the surface.

If the sports fans were waving from right to left, which means parallel to the rows of stands, then we would have a **longitudinal wave**. In a longitudinal wave, the disturbance is parallel to the direction of propagation. This is what happens if we compress a spring along its axis. Such is the case in sound, described mathematically by a longitudinal density disturbance along the direction of propagation. Often enough, this very physical disorder causes, in turn, disturbances to other physical quantities. For example, in sound waves, the disturbed physical entity is density, and this density disturbance also induces pressure or temperature disturbances. Sound can be described as a longitudinal disturbance of either pressure or temperature that propagates (hence, it is a wave).

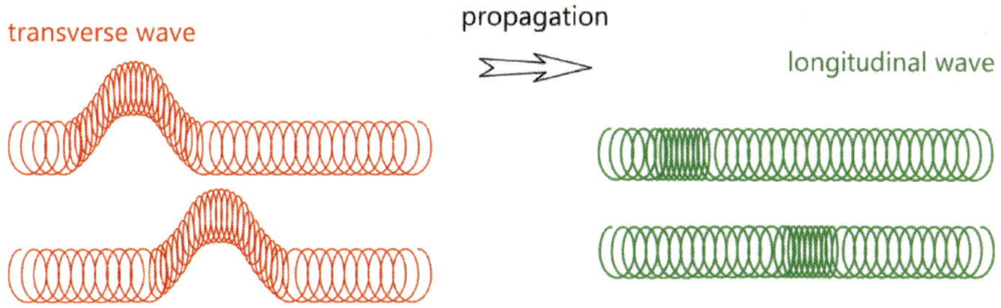


*Figure 1-7: Transverse and longitudinal waves in a spring.*

The physical quantity disturbed in a wave can be one-dimensional (1-D) or three-dimensional (3-D), vectorial or numeric, and is described by a function in space and time. If the physical size of the disturbance is $y$ and the disturbance propagates along the –$z$ axis, then the function is expressed as $y(z, t)$, where $t$ is the variable describing time.

The shape or instant of the –$y$ disturbance does not change for a sufficiently large distance (temporal or spatial) in relation to the duration of the disturbance. We can follow a disturbance as it propagates over time as well as over space; consider two different instances—in time (from $t$ to $t+\delta t$) and in space (from $z$ to $z+\delta z$).

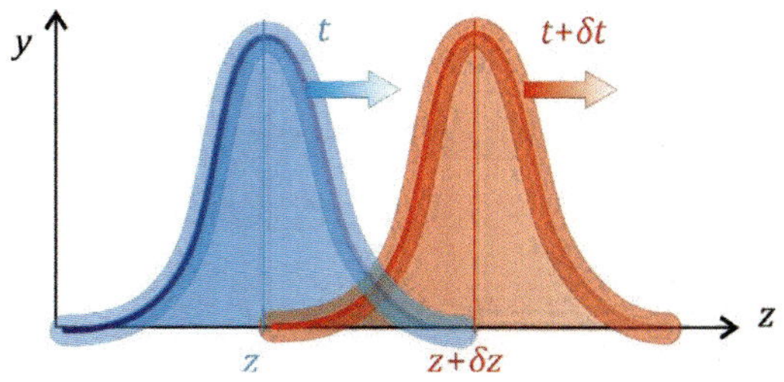


*Figure 1-8: One-dimensional wave disturbance propagating along the +z direction.*

For the disturbance to propagate along the –$z$ axis, the physical variable $y(z, t)$ that describes this disturbance must obey the **wave equation**:

Wave Equation: $$\frac{\partial^2 y(z,t)}{\partial z^2} - \frac{1}{u^2}\frac{\partial^2 y(z,t)}{\partial t^2} = 0 \tag{1.1}$$

One of its solutions has the form: $$y = y(z, t) = y(u \cdot t - z) \tag{1.2}$$

The wave equation [Eq. (1.1)] can describe a wide variety of waves. It applies to transverse mechanical waves on a string, as well as to longitudinal sound waves in air, or elastic waves in a solid medium. It can also describe voltage and current density in a wire, and electromagnetic waves in vacuum or in air, such as light or radio waves propagating through a multitude of optical media. The physical quantity $u$ always has dimensions of speed and is proportional to frequency × wavelength ($u = \nu \cdot \lambda$); however, depending on the nature of the wave and the medium in which it propagates, it has different expressions. For example, for waves in a string, the expression of the quantity $u$ is $(T/\rho)^{1/2}$, where $T$ is the mechanical tension, and $\rho$ is the linear density. The ratio $T/\rho$ expresses the square of the propagation speed in such waves.

The solutions $y(z, t)$ produced by Eq. (1.1) are termed **wave functions**. A wave function $y$ may combine $z$ and $t$ in the form of $u \cdot t - z$, or in any other equivalent expression, for example, $(u \cdot t - z)^2$, but <u>not</u> in the form of, for example, $(u \cdot t^2 - z)$ or $(u^2 \cdot t + z)$. The solutions must have factors carrying the same order for both the $u \cdot t$ and the $z$ terms. A wave function thus describes the propagation of a disturbance of the physical quantity $y$ with speed $u$ along the direction $+z$.

It is common to consider the disturbance as an oscillation (i.e., a periodic change) and, moreover, as a simple harmonic oscillation described by a simple trigonometric function. In this case, the wave source is a harmonic oscillator.

Why is this view widely accepted? The answer is two-fold. First of all, using trigonometry helps with the math, and secondly, as will be discussed in § 1.1.5, any periodic change can be decoded in a series of simple harmonic oscillations. For this reason, we extensively use simple harmonic wave expressions in the mathematical models.

### 1.1.4 The Electromagnetic Wave

A **field** is a distribution of physical quantity values in space and time. Fields can be scalar or vectorial, depending on the nature of the physical quantity. For example, on a weather map, the temperature map presents the distribution of temperature values over an area and (possibly) its progression over time. This is a scalar field because temperature has no direction, only magnitude. A surface wind velocity is described by assigning a vector to each point on a map.

Each vector represents the speed and direction of the movement of air at that point. This is a vector field. Fields can be static (time-stationary) or they may progress over time.

Some fields are dynamic, meaning that forces can be exerted. For example, a positive charge creates a static electric field. Another positive or negative charge in the vicinity may be repelled or attracted; i.e., an electric force is exerted on a charge (positive or negative) by the electric field created by the first charge. Thus, the positive charge is the source of a static electric field that can be visualized with electric field dynamic lines originating from the source charge and extending to infinity [Figure 1-9 (left)]. A dynamic line traces a path (straight or curved line) that follows the direction of the vector field. The tangent to the dynamic line is parallel to the field at that point. The density of the dynamic lines is proportional to the strength of the field.

*Figure 1-9: The density of the dynamic electric field lines around a source charge. The greater the charge, the denser the dynamic lines. Left and center show a positive charge, and right shows a negative charge.*

The magnitude of the electric field is fixed over time (a static field), and its value is determined by the distribution of electric charges. The law that describes this field is Coulomb's law, named after the French physicist Charles Augustin de Coulomb, who formulated it.

*Figure 1-10: Electric field near two charges, resulting from the interaction between the two fields.*

A constant electric current flowing in a conductor generates a static magnetic field that exerts forces on a magnet or another conduit. The magnitude of this static field is fixed over time. The relationship between the current and the magnetic field it produces is described by a relationship known in the literature as the Biot–Savart law.

If the field source (the charged sphere) begins to oscillate (accelerate), then it not only disrupts the temporally constant electric field around it, but also creates a time-dependent magnetic field. This change in the magnetic field also causes a change in the electric field. A pulsating electric field generates a pulsating magnetic field, and vice versa. Thus, the **electromagnetic field** is formed from interdependent, nonstatic electric and magnetic fields.

Feynman: It is harder to understand the electromagnetic field than to understand invisible angels; the former requires a vivid imagination of a myriad of complex waves. *The Strange Theory of Light and Matter* (1985).

What kind of wave is light? The answer to this question was proposed in the mid-19th century: Light is an electromagnetic wave. The task to describe an electromagnetic wave is a challenging one. Physicists like to include the properties known as electric charge and magnetic moment in their description of the notion of fields. A field is stronger if the force exerted on a unit charge is greater.

We therefore describe electromagnetic waves as perturbations of the electric and magnetic fields—oscillations that are mutually dependent in perpendicular planes.

Feynman: I see some kind of vague showy, wiggling lines—here and there an E and a B written on them somehow, and perhaps some of the lines have arrows on them—an arrow here or there which disapears when I look too closely at it. When I talk about the fields swishing through space, I have a terrible confusion between the symbols I use to describe the objects and the objects themselves. I cannot really make a picture that is even nearly like the true waves. So if you have difficulty making such a picture, you should not be worried that your difficulty is unusual.

*The Feynman Lectures on Physics, Vol. II*

Consider again the pulsating charge. In a stationary charge, the dynamic lines are also straight. The question is: What indications are there that the generating charge of these dynamic lines is, indeed, oscillating? Assuming that we are 3000 miles away, can we sense (and if so, how) the 'message' that this charge oscillates? Maxwell's equations show that the electric field lines are no longer straight, but rather meander; i.e., the electric field at any point fluctuates and flips periodically in what is called propagating perturbation. The same happens in the magnetic field.

There are therefore two interdependent perturbations of the electric and magnetic fields. A change in one causes a change in the other. Both fields are disrupted perpendicularly to each other and perpendicularly to the direction of propagation. The two together are, in fact, the same entity, a **transverse electromagnetic** (TEM) **wave**.

A wave has a very specific propagation speed, even if the oscillating charge that created it stops vibrating. This speed is none other than the speed of light. Thus, the 'message' from the charge oscillation travels at the speed of light and reaches the observer. The wave equation for the electric field $\mathbf{E}$ and for the magnetic field $\mathbf{H}$ describes the propagation in vacuum for any electromagnetic perturbation:

$$\nabla^2\mathbf{E} - \varepsilon_o\mu_o\frac{\partial^2}{\partial t^2}\mathbf{E} = 0 \quad \text{and} \quad \nabla^2\mathbf{H} - \varepsilon_o\mu_o\frac{\partial^2}{\partial t^2}\mathbf{H} = 0 \qquad (1.3)$$

Comparing this equation to Eq. (1.1), the speed of the electromagnetic wave in vacuum is

$$c = \frac{1}{\sqrt{\varepsilon_o\mu_o}} = \frac{1}{\sqrt{\left(4\pi\times10^{-7}\text{N}/\text{s}^2\text{C}^{-2}\right)\times\left(8.8542\times10^{-12}\,\text{C}^2\text{N}^{-1}\text{m}^{-2}\right)}} = 2.9979\times10^8\ \text{m/s} \qquad (1.4)$$

where $\varepsilon_0$ is the electric permeability (also referred to as permittivity), and $\mu_0$ is the magnetic permeability of vacuum (or free space). This holds for all electromagnetic waves!

The propagation speed of light in vacuum $c$ is considered a global constant, the only constant (perhaps in the entire field of physics, including Einstein's theory of special relativity) that is greater than any other speed and is unaffected by the system of reference. Initially, the constant $c$ derived from the Weber constant; today $c$ is attributed to the Latin *celeritas*, for speed.

Einstein: You see, wire telegraph is a kind of a very, very long cat. You pull its tail in New York and its head is meowing in Los Angeles. Radio operates exactly the same way: you send signals here, they receive them there. The only difference is that there is no cat.

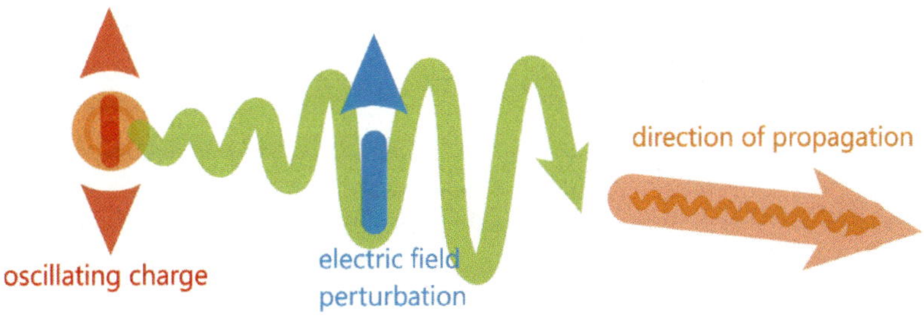


*Figure 1-11: The dynamic lines from the electric field are disturbed by the oscillation from the charge that generates the field.*

A simple representation of an electromagnetic wave is a set of harmonically oscillating electric and magnetic fields. Such a wave may be produced by a harmonically oscillating pair of equal and opposite-sign charges (electric dipoles) oscillating with angular frequency $\omega$. Far away from this dipole, the mathematical representations of this wave take the following form:

$$\mathbf{E} = (1/r)\cdot\cos(\omega\cdot t - \mathbf{k}\cdot\mathbf{r} + \varphi_o) \quad \text{and} \quad \mathbf{H} = (1/r)\cdot\cos(\omega\cdot t - \mathbf{k}\cdot\mathbf{r} + \varphi_o) \tag{1.5}$$

This is a simple **harmonic, spherical wave**. It is harmonic because it is described by a simple trigonometric function, and it is spherical because its amplitude (see below) is dependent on the parameter $1/r$, the reciprocal of the distance $r$ from the source.

Note ⏱: Vectors are denoted in bold: For example, $\mathbf{k}$ and $\mathbf{r}$ indicate the wavevector and the directional vector, respectively; $\mathbf{k}\cdot\mathbf{r}$ indicates the dot product of these two vectors. Scalar variables are indicated by italics: For example, $\omega$ is the angular frequency, $t$ is the time, and $\varphi_o$ is the initial phase. (These variables will be introduced shortly.) If a vector variable is not indicated in bold, then it is its magnitude that is being considered, appearing in italics. For example, $E$ indicates the magnitude of the electric field vector.

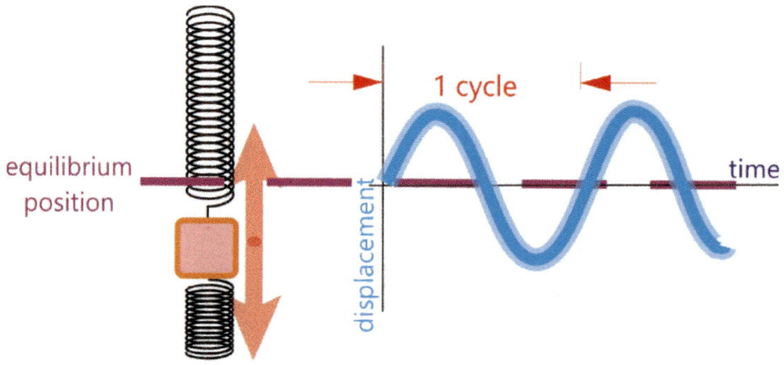


*Figure 1-12: A harmonic oscillator.*

Farther away from the source, we omit the amplitude dependence on distance, and the relationships in Eq. (1.5) take the following forms:

$$\mathbf{E} = \mathbf{E}_o\cdot\cos(\omega\cdot t - \mathbf{k}\cdot\mathbf{r} + \varphi_o) \quad \text{and} \quad \mathbf{H} = \mathbf{H}_o\cdot\cos(\omega\cdot t - \mathbf{k}\cdot\mathbf{r} + \varphi_o) \tag{1.6}$$

thus forming a simple **harmonic, plane wave**. Here vectors $\mathbf{E}_o$ and $\mathbf{H}_o$ are the amplitudes of the field vectors $\mathbf{E}$ and $\mathbf{H}$. The wave is plane because it has a fixed amplitude, which does not depend on location. We also use the following complex expressions:

$$\mathbf{E} = \mathrm{Re}\,\{\mathbf{E}_o\cdot\exp[i(\omega\cdot t - \mathbf{k}\cdot\mathbf{r} + \varphi_o)]\} \quad \text{and} \quad \mathbf{H} = \mathrm{Re}\,\{\mathbf{H}_o\cdot\exp[i(\omega\cdot t - \mathbf{k}\cdot\mathbf{r} + \varphi_o)]\} \tag{1.7}$$

which are particularly useful for mathematical processing, such as for the the vector sum or superposition. Their real part, appearing as Re{ }, expresses the respective physical quantity.

Like every wave, electromagnetic waves have the following attributes:

- **Frequency** $\nu$, which is defined as the number of oscillations in the unit of time. The frequency of a given wave is fixed and always equals the frequency of its source. Frequency has units of hertz (1 Hz = 1 cycle per second). For example, a bus that arrives at a bus stop every 10 minutes will arrive 6 times within one hour. The frequency is 6 cycles per hour.
- **Period** $T$, which is defined as the duration of an oscillation. In the bus example, the period is 10 minutes.
- **Angular frequency** $\omega$, which is the 'cycle' frequency, defined as $\omega = 2\pi \cdot \nu$. The angular frequency expresses the rate of phase change in time.
- **Wavelength** $\lambda$, which is defined as the distance between two consecutive disturbance crests along the wave's propagation. Wavelength has dimensions of length; for visible light, we use micrometers (1 µm = $10^6$ m) and nanometers (1 nm = $10^9$ m). Wavelength and frequency determine the <u>speed of the wave</u> in that medium: $u = \nu \cdot \lambda$.

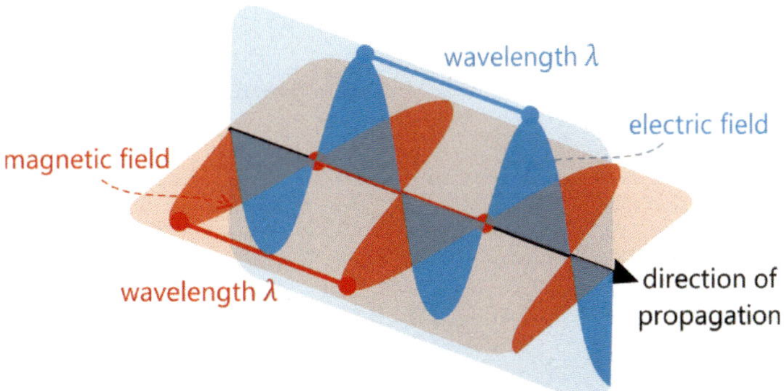


*Figure 1-13: Light as an electromagnetic wave consists of mutually perpendicular electric and magnetic fields oscillations. A wavelength can be visualized as the distance between two consecutive crests.*

- **Phase** $\varphi$, the internal clock of the disturbance. Over the length of a wavelength, the phase changes from 0° to 360° or from 0 to $2\pi$ rad. In the harmonic expressions of Eq. (1.6), the common phase in the disturbance is the argument in the trigonometric function:

$$\varphi = \omega \cdot t - \mathbf{k} \cdot \mathbf{r} + \varphi_0 \tag{1.8}$$

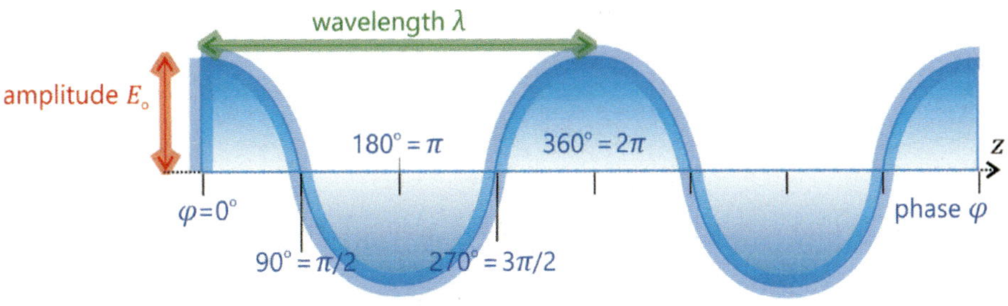


*Figure 1-14: Amplitude, wavelength, and phase in a harmonic wave.*

- **Wavevector k**, which expresses the rate of phase change in space along the direction of propagation. For this reason, a wavevector's magnitude is the spatial frequency. A wavevector has dimensions of inverse length. It is a vector, whose magnitude $k$ is

$$k = 2\pi/\lambda \tag{1.9}$$

- **Field**, the respective value of the disturbance, either the electric field **E** or the magnetic field **H**. These fields are vectorial, which means that they have direction in addition to magnitude and phase. In the electromagnetic field, these field vectors are always perpendicular to each other and perpendicular to the direction of propagation as well.
- **Amplitude**, representing the maximum values of the electric field $\mathbf{E}_o$ and magnetic field $\mathbf{H}_o$.
- **Luminous intensity** $I$, often simply called intensity, which is the physical quantity that is directly detectable. The relationship between the intensity and the electric field is

$$I = c\varepsilon_o \cdot \langle \mathbf{E} \cdot \mathbf{E} \rangle = \sqrt{(\varepsilon_o/\mu_o)} \cdot \langle \mathbf{E} \cdot \mathbf{E} \rangle \tag{1.10}$$

where $\langle\,\rangle$ denotes the time-average of the enclosed quantity. For a harmonic wave, this becomes

$$I = c\varepsilon_o \cdot \langle \mathbf{E} \cdot \mathbf{E} \rangle = \tfrac{1}{2}\, c\varepsilon_o\, E_o^2 \tag{1.11}$$

The Poynting vector **S** (units cd/m$^2$) expresses the flow of electromagnetic energy. Its direction of propagation is parallel to that of the wave. The magnitude of luminous intensity is the time-averaged Poynting vector value. Three vectors, the electric field **E**, the magnetic field **H**, and the direction of propagation **S** form a Cartesian axis system.

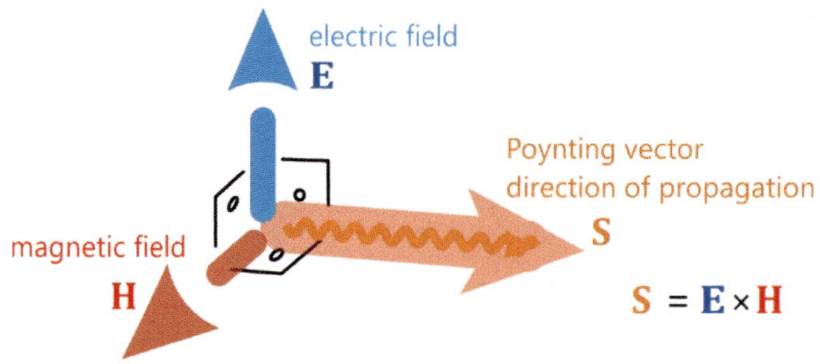


*Figure 1-15: Poynting vector, the direction of energy propagation in an electromagnetic field.*

### *1.1.4.1 Maxwell's Equations*

James Clerk Maxwell systematically described electricity's and magnetism's mutual dependence. With simple equations, he combined four vector fields: the electric field **E**, the magnetic field **H**, the electric displacement $\mathbf{D}$ (= $\varepsilon_o\mathbf{E}+\mathbf{P}$), and the magnetic induction $\mathbf{B}$ (= $\mu_o\mathbf{H}$), as well as two source descriptors, the density of static charges $\rho$ and the current density **j**.

The interrelation of the space and time derivatives of these vectors is described by the following laws:

**1. Gauss's law for electricity** relates the distribution of electric charge to the resulting electric field. It states that the electric flux out of a closed surface equals the enclosed charge divided by the permittivity. The dot-product space derivative $\nabla\cdot$ form of this law is

Gauss's Law for Electricity: $$\nabla\cdot\mathbf{E} = \rho/\varepsilon_o \tag{1.12}$$

**2. Gauss's law for magnetism** states that the magnetic flux out of a closed surface is zero, since the magnetic sources are always dipole sources:

Gauss's Law for Magnetism: $$\nabla\cdot\mathbf{B} = 0 \tag{1.13}$$

**3. Faraday's law** describes the interaction between the magnetic field and an electric circuit that produces an electromotive force, known as electromagnetic induction. This force depends on the rate of change of the magnetic flux. The cross-product derivative (curl, $\nabla\times$ ) form is

Faraday's Law: $$\nabla\times\mathbf{E} = -\frac{\partial\mathbf{B}}{\partial t} \tag{1.14}$$

**4.** The **Ampère–Maxwell law**, which is an extension of Ampère's original law, Ampère's circuital law. This law relates the magnetic field to the electric currents and the rate of change of the electric field that produces them:

Ampère–Maxwell Law: $$\nabla\times\mathbf{B} = \mu_o\mathbf{j}+\mu_o\varepsilon_o\frac{\partial\mathbf{E}}{\partial t} \tag{1.15}$$

These laws appear either in a differential form, as shown above, or in an integral form. The two forms are equivalent and are related by the Kelvin–Stokes theorem. Adding to the confusion, these laws appear differently when using SI units compared to when using CGS units, another system used in electromagnetism. Other forms use the magnetic field $\mathbf{B} = \mu_o\mathbf{H}$ and the electric displacement $\mathbf{D} = \varepsilon_o\mathbf{E}+\mathbf{P}$, where $\mathbf{P}$ is the electric polarization.

The wave equation presented in Eq. (1.3) is derived by taking the curl of one field (**E**) and the time derivative of the other field (**H**). It is derived using space and time differentials and vector identities.[1]

[1] Pearson JM. A Theory of Waves. Allyn and Bacon, Boston, 1966.

Knowing the electric or magnetic field at a specific point in space and time, with these equations, it is possible to predict what will happen at other points in space and time. The theory of electromagnetism is one of the most developed theories today, and the majority of observed natural phenomena are electromagnetic.

What we call the electromagnetic spectrum includes a very wide range of waves, extending from radio waves all the way to X rays and gamma rays. These waves differ only in terms of their frequency. Light is just a small part of the electromagnetic spectrum, located approximately in the middle, and it stands out because it is perceived by the human eye in different colors.

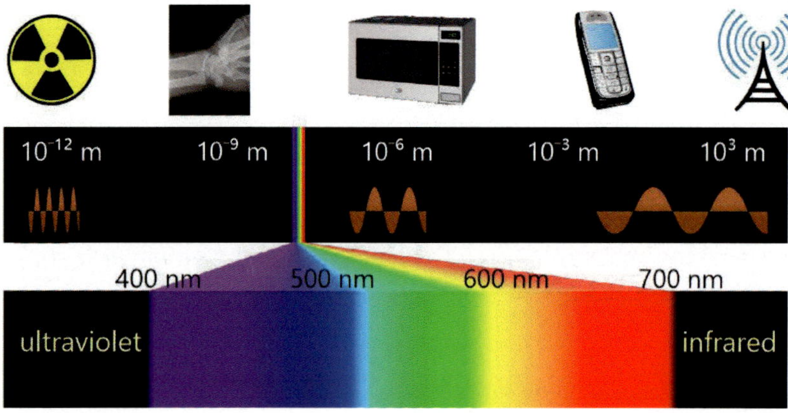


*Figure 1-16: Visible light as part of the electromagnetic spectrum. It is perceived by the human eye in the colors of the rainbow, each of which corresponds to a specific wavelength.*

Specifically, visible light has frequencies that ranges from $4.3\times10^{14}$ Hz for the red (wavelength 0.7 μm = 700 nm) to $5.7\times10^{14}$ Hz ($\lambda$ = 0.4 μm = 400 nm) for the blue. We realize that the frequency associated with visible light can be billions of hertz!

**Electromagnetic Wave**

- The fundamental disturbance is associated with small oscillations in the electric and magnetic fields.
- The electromagnetic wave may have many forms, from radio waves to X rays, and the difference lies in their frequency.

In the immediate vicinity of visible light, there is the ultraviolet (UV) with $\lambda_{UV} < 0.4$ μm and the infrared (IR) with $\lambda_{IR} > 0.7$ μm. Ultraviolet is often referred to as black light. Although it cannot be perceived by the photoreceptors of the human eye, UV light can potentially cause damage to the eye.

Like any electromagnetic wave, light propagates in vacuum at the speed of ... light: $c = \nu \cdot \lambda = 3\times10^8$ m/s. Light travels from the sun to the earth, a distance of 150,000,000 km, in 8 min, 20 s (500 s total), while it travels from the moon to the earth, a distance of 384,000 km, in 1.3 s. The distance from Washington, D.C. to Albany, New York (500 km / 311 miles) is covered in just 1 ms!

Even within the visible region of the spectrum, not all parts are the same, and the different parts are perceived as different colors. Color is nothing more than the property by which human vision distinguishes a very small part of visible light (or a combination of parts) that has a very specific frequency/wavelength.

Example ☞: In relation to the visible, are the ultraviolet frequencies: (a) lower, (b) higher, or (c) the same?

The answer is (b), higher. Ultraviolet extends to the area of the electromagnetic spectrum with shorter wavelengths compared to the visible. Ultraviolet frequencies are higher than visible frequencies. The product of the frequencies is constant and equal to the speed of light in the medium.

A source whose spectral content has several color components may form what is perceived as being white or nearly white light (white light source). This light can be analyzed for its chromatic components by a prism.

We emphasize here that color-specific radiation refers to radiation that can be correlated to a specific wavelength. An average color, as perceived by the human sensory organ of sight, may be composed of several monochromatic parts.

### 1.1.5 Realistic Waves and the Harmonic Wave

Figure 1-13 and the relationships in Eq. (1.6) present the simplest expression of a plane electromagnetic wave. This wave has a fixed amplitude and is absolutely monochromatic, which means that it has a specific wavelength $\lambda$, or equivalently, it has exactly one specific frequency $\nu$. The wave extends infinitely: Values for the field magnitudes $E$ and $H$ exist for every value of $t$. We start to suspect that this is too good to be true...

The relationships in Eq. (1.5) are the simplest expressions of a spherical electromagnetic wave, which is absolutely monochromatic, too. Its amplitude reduces inversely with $r$, so its intensity reduces with the inverse of $r^2$, but its total energy (flow) is preserved, since the area of a sphere increases with $r^2$. In distances far away from the source, the reduction of the field amplitude with the distance $r$ may be ignored, so the plane wave can be expressed as a generalized spherical wave with an infinite radius of curvature.

These harmonic wave expressions are, of course, mathematical idealizations. In reality, light is emitted in the form of a burst of pulses with a much shorter time duration and does not necessarily have this form; i.e., it might not be harmonic, plane, monochromatic, etc. For example, it is not even possible to have only one frequency value; there is always a spectral range $\delta\nu$, so the light's color is a superposition of many neighboring colors/frequencies. This pulse has a finite time duration $\approx 1/\delta\nu$, which is inversely proportional to the spectral range.

So why do we use these ideal harmonic waves? The answer is that they exemplify the simplest representations for which we can develop an initial mathematical formulation and then include additional parameters. Furthermore, although it does not exist on its own in nature, an ideal harmonic wave exists at the heart of a very real wave! Any wave pulse can be analyzed via a Fourier analysis to a series of nonlimited ideal harmonic components.

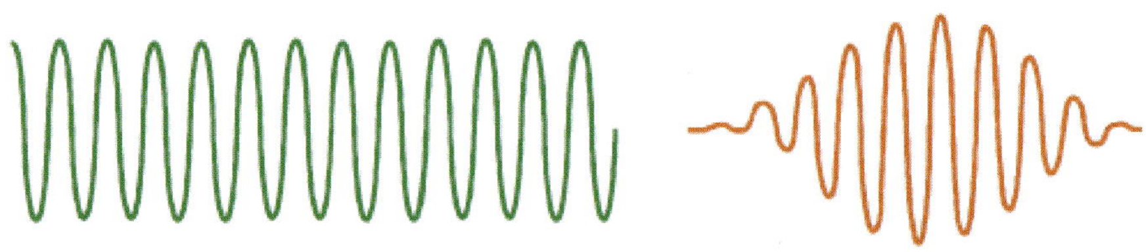

*Figure 1-17: Representation in time of (left) a harmonic wave of a single frequency and (right) a short pulse resulting from a summation of harmonic waves with different frequencies.*

## 1.2 RAYS AND WAVEFRONTS

What is an optical ray? It is nothing but an abstract notion that we draw on paper or plot on a computer screen. In general, the notion of a ray is applicable in every wave but is primarily used in light. A ray is an idealized model in which light is visualized as an infinitesimally thin pencil of light that defines a specific path of propagation.

Another idealized model is the wavefront, which is defined as the surface that joins points that share the same phase, for example, the crests or troughs of localized wave oscillations. Two consecutive wavefronts are separated by a distance equal to the wavelength:

Wavefront Definition: $$\mathbf{k}\cdot\mathbf{r} = \text{constant} \tag{1.16}$$

The wavevector, which expresses the rate of phase propagation in space, is perpendicular to the wavefronts at every point. In a harmonic wave we define the **phase velocity** $\mathbf{u}_{\text{phase}}$ as a vector along the direction of the wavevector **k**:

$$\mathbf{r} = (\omega \cdot t) \cdot (\mathbf{k}/k) \Rightarrow \mathbf{r}/t = \mathbf{u}_{\text{phase}} = \omega \cdot (\mathbf{k}/k); \quad u_{\text{phase}} = \omega/k \tag{1.17}$$

If we travel at a speed $\mathbf{u}_{\text{phase}}$, we can be 'riding' on the exact same phase as that of the disturbance, such as a crest. This vector is called the phase velocity because it is the propagation speed of the phase of the wave perturbation. Unless otherwise noted, the magnitude $u$ of the wave velocity is simply the magnitude $u_{\text{phase}}$ of the phase velocity. The greater the wavevector projection along a direction, the denser the equiphasic surfaces that depict the phase change density. In summary:

- The phase propagates with the phase velocity whose magnitude is $\omega/k$.
- The phase propagates along the direction of the wavevector.

Rays in isotropic media are always perpendicular to the wavefronts. Any wave can be described equally well using either the ray or the wavefront model.

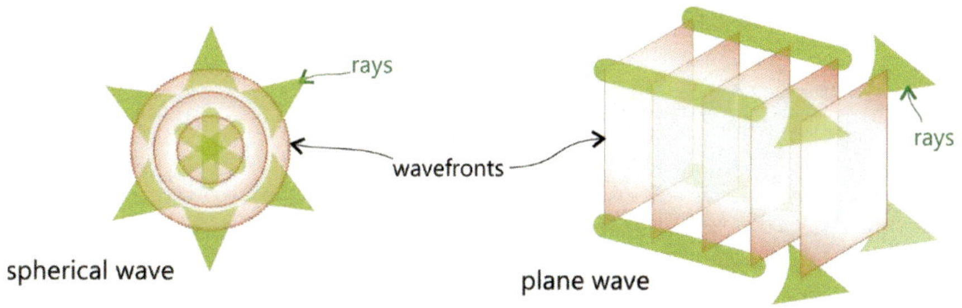


*Figure 1-18: Rays are perpendicular to the wavefronts and can be drawn much more easily than wavefronts, which require several parallel planes.*

In Figure 1-18 (left), the waves are emitted by a point source and propagate uniformly in all directions. The wavefronts are spherical, forming a succession of concentric spheres about the source. This is a **spherical wave**. In Figure 1-19 (right), the source is so far away that we can consider it to be located at so-called optical infinity. The wavefronts form flat surfaces, planes that are parallel to each other; the rays are also parallel, forming a collimated pencil of rays. The beam is called **collimated**, and the wave is called a **plane wave**.

*Figure 1-19: Wavefronts and rays: (left) a spherical wave and (right) a plane wave.*

The rays in Figure 1-19 (left) depict the entire picture, including all of the circles from a point source. Often, we only see a part of such a wave—a subset of diverging rays, whose extrapolations appear to originate from the source. The wave is still considered spherical or circular, and the wavefronts correspond to portions of surfaces with an increasingly larger radius.

**Rays of Light:**

- are straight lines and are always normal to the wavefronts,
- correspond to the direction of propagation,
- their paths follow the laws of reflection and refraction.

Spherical wavefronts can be either **diverging** or **converging**. Diverging rays originate from a unique source point or aperture and propagate away from each other without crossing. The wavefronts form successive, expanding spheres originating from the source.

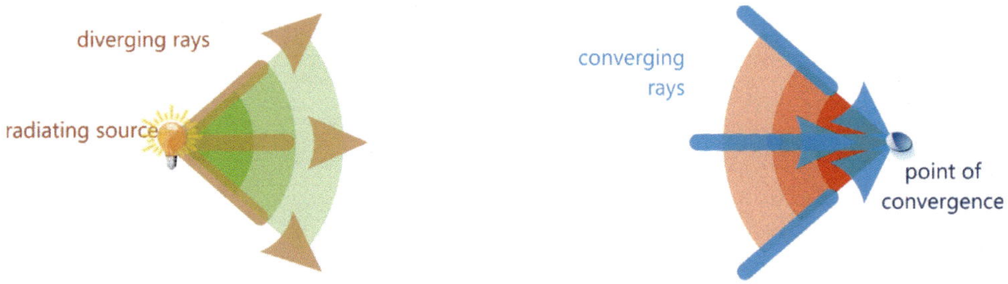


*Figure 1-20: (left) Diverging and (right) converging rays and their corresponding wavefronts.*

Converging rays are all directed toward a unique focus point, and their corresponding converging wavefronts form spheres contracting toward that point. This may be the result of various phenomena such as reflection from a concave mirror or refraction by a converging lens. In diverging wavefronts, the radius of curvature increases as the wave propagates, while in converging wavefronts, the radius of curvature decreases as the wave propagates.

Wavefronts are quite suitable for describing the effects associated with the wave nature of light. The rules were set by the **Huygens–Fresnel principle**, which postulates that light is formed by many elementary wavelets, and its propagation is predicted by the tangent of all these elementary wavelets, which forms the wavefront. This principle was initially called simply Huygens' principle, named after Christiaan Huygens, who pioneered the wave theory of light in 1670. His work was later rediscovered after the eventual prominence of the wave theory of light by the work of Fresnel and others. When applied to the propagation of light waves, the Huygens–Fresnel principle states the following:

*Every point on a wavefront can be considered a source of secondary spherical wavelets that spread out in the forward direction, traveling at the speed of light. The new wavefront is the tangential surface to all secondary wavelets.*

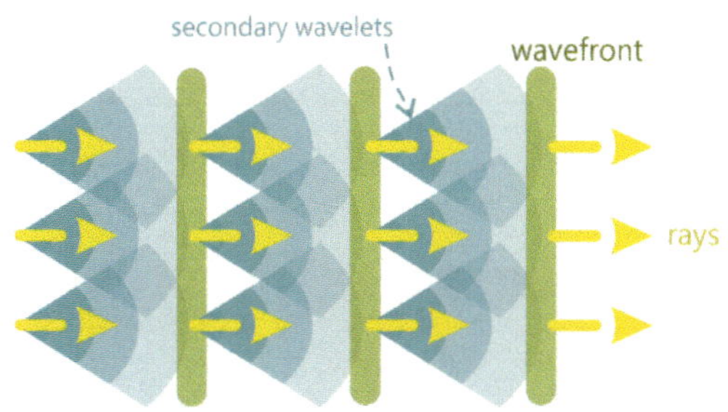


*Figure 1-21: Illustration of the Huygens–Fresnel principle.*

We can ignore the wave nature of light if the wavelength is small compared to the dimensions of the aperture in an optical system. However, if the cross-section (for example, its diameter) of an obstacle or an opening has dimensions on the order of a few millimeters, which is comparable to the optical-range wavelength (dimensions less than 100 $\lambda$), then the approximation is no longer valid and we cannot ignore the wave nature of light, which manifests as diffraction effects (Chapter 5). In this case, the wave nature of light may dominate, and certain assumptions based in geometrical optics are no longer valid.

According to geometrical optics, rays passing by an obstacle or through a small aperture do not diverge and continue on to cast a geometric shadow. According to wave optics, if the wave encounters a small opening or an obstacle, this opening or obstacle is considered to be a source of secondary waves. In this case, the shadow does not conform to a geometrical projection but rather to a diffraction pattern.

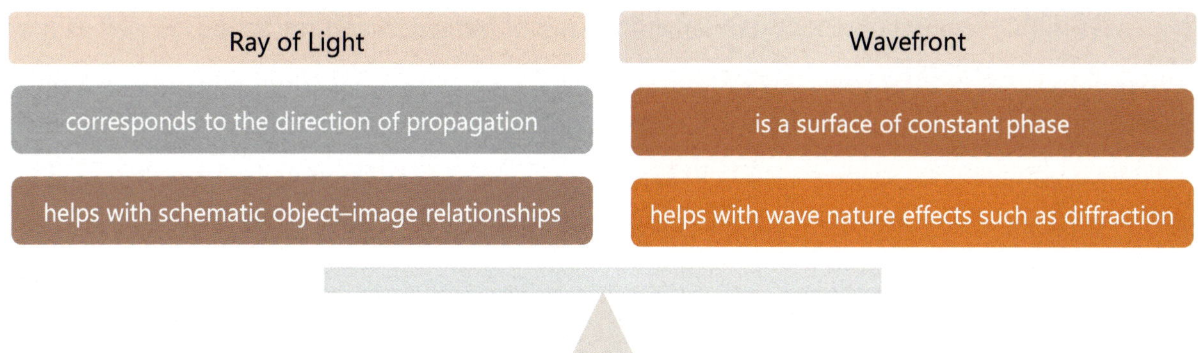

# 1.3 PROPAGATION OF LIGHT

## 1.3.1 Light is Always 'in a Hurry'...

In addition to understanding the exact nature of light, the understanding of light propagation has also attracted much attention and effort. A general principle, the **principle of least time**, or equivalently, the principle of minimum optical path, explains all phenomena relating to the propagation of light. This principle states that, of all possible paths that light may follow from one point to another, it follows the path that corresponds to the least time. So, we see that light is always 'in a hurry'!

Although credited to the French mathematician Pierre de Fermat, the Greek philosopher Hero of Alexandria had discovered that principle much earlier. In Hero's version, the principle covered only the reflective part of light propagation. Fermat (in 1657) expanded the principle to include the refractive part of light propagation.

Principle of Least Time, or Fermat's Principle (1657):

Of all possible paths from one point to another, light follows the path that takes the least time.

From this general principle, we derive the laws of reflection and refraction, as well as the **principle of reversibility**, which states that light will follow exactly the same path if its direction of travel is reversed. These principles and laws form the constitution of geometrical optics. The principle of reversibility is similar to the principle of least action, which applies in classical mechanics and is in full agreement with both the particle and the wave nature of light.

*Figure 1-22: Examples of rectilinear propagation of light in air. (left) Laser beams on an optical bench; (right) straight-line shadows falling on a living room floor on a warm winter day.*

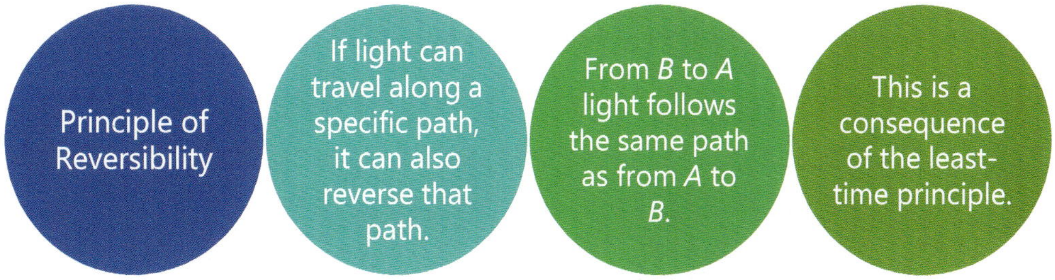


So, we may now think that it is justified to accept that light travels in straight lines. This, however, is not so certain. In fact, if we apply the rule that governs light propagation, the shortest-time path, we conclude that light travels in straight lines. There are some conditions, however. One is that there is 'nothing' between the two points. Nothing means vacuum only. What if there is, for example, air or a piece of glass? The straight-line rule also applies if the medium in which light propagates is the same; i.e., its speed is the same at all points. Such a medium is **homogeneous**. If the speed of light is the same regardless of its direction of propagation, then the medium is **isotropic**.

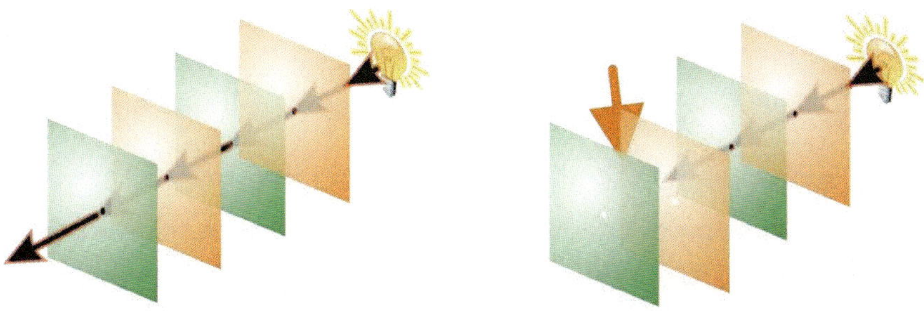

*Figure 1-23: Rectilinear propagation of light in which (left) the pinholes are aligned so the ray goes through, and (right) the pinholes are not aligned so the ray is interrupted.*

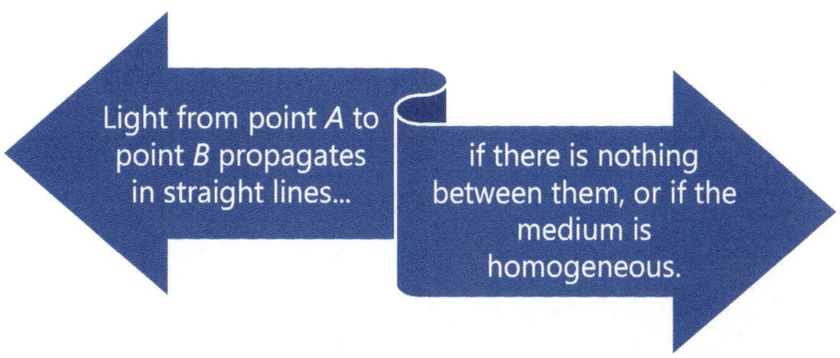


To conclude, if there is 'nothing' between two points, then there is no reason for the rays to change their direction. Light propagation is **rectilinear**; in other words, the rays continue on their straight path. The same applies if the medium through which light propagates (such as water, air, or glass) is homogeneous: The path taken by light is a straight line. Light from a lamp is spread evenly in all directions, and each ray follows a straight path.

*Figure 1-24: Alberto Unasso experiences the principle of reversibility of light propagation (© www.fiami.ch).*

## 1.3.2 Index of Refraction

What about the speed of light within different optical media? Is it the same as in vacuum?

Water, glass, and other optical media, despite being transparent, slow down light. In other words, in different transparent media, light travels at different speeds and, specifically, at lower speeds than in vacuum. If we divide the magnitude of the speed of light in vacuum $c$ by the speed in the medium $u$ (the magnitude of the phase velocity), we obtain a number that is greater than 1.0, which is the **index of refraction** $n$ (or **refractive index**) for that medium:

Refractive Index: $$n = c / u \quad (1.18)$$

The greater the value of the refractive index in a specific medium, the slower the light propagates. In vacuum, the refractive index has its lowest value, 1.0; in any other medium, the refractive index is greater than 1.0. For example, in water $n \approx 1.33$, in various glasses $n$ ranges from 1.4 to 1.9, and in diamond $n \approx 2.4$. The highest known refractive-index value is for the element germanium (Ge, $n \approx 4$), but this is the case only in the infrared range.

The refractive index is a dimensionless quantity (just a number) because it is a fraction of speed over speed. The value of the refractive index depends on the material properties of the medium. Environmental properties such as temperature or pressure (if gaseous) also affect the refractive index value.

*Table 1-2: Values of the refractive index for various media.*

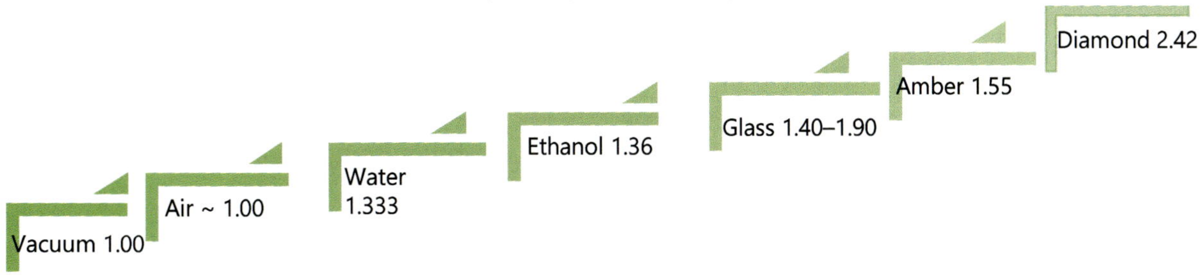


Glass has a wide range of refractive index values, spanning from 1.4 to 1.9. Essentially, glass is a molten, noncrystalline form of silicon dioxide ($SiO_2$), common sea sand, also referred to as amorphous silica and glassy silica.[2] It is often mixed with alkali-lime silicates containing about 10% potassium oxide ($K_2O$). These are the crown-type of silica glasses. Flints are silica glasses that contain lead oxide (PbO) and are typically heavier; the name appears to have originated from an early manufacturing process that involved pulverization of silica flints (= hard stones). As alternatives to silica glasses, other materials can be barium oxide (BaO), boron oxide ($B_2O_5$), phosphorus pentoxide ($P_2O_5$), germanium oxide ($GeO_2$), and rutile ($TiO_2$).

The main classes of glass are **crown glass**, **flint glass**, and rare earth glass with an even greater refractive index. Most important to remember, however, is that the refractive index depends on the wavelength (Figure 1-25): In general, the refractive index decreases with an increase in wavelength.

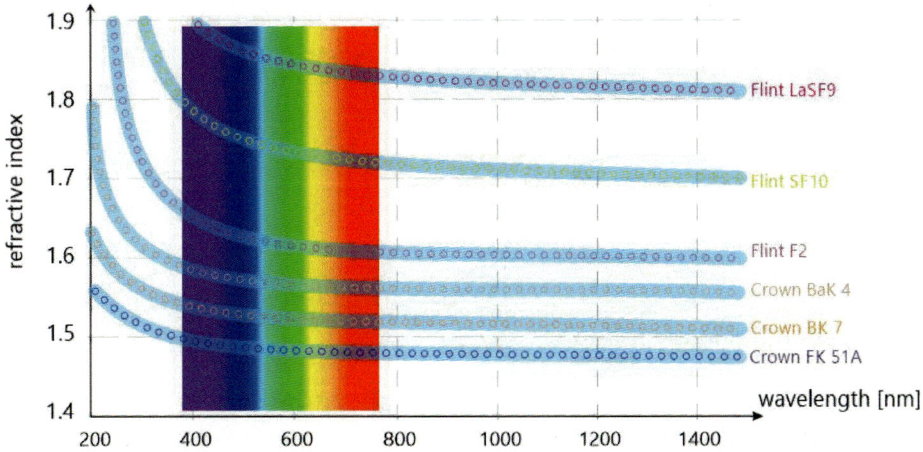


*Figure 1-25: Values of the refractive index for various glasses.*

When we mention a value for the refractive index, we must define the wavelength. Values quoted for the refractive index when no wavelength is mentioned refer to the yellow sodium line ($\lambda$ = 589 nm) and, more recently, to the yellow helium d-line ($\lambda$ = 587.6 nm).

[2] Hart G. The nomenclature of silica. Am Mineral. 1927; 12:383–95.

*Figure 1-26: The rainbow is a manifestation of the dependence of the refractive index on the wavelength. Albert Singlestone is on his way to meet Isaac Newton (© www.fiami.ch).*

The fact that when light propagates into a different medium (for example, from air to water) it changes speed means that either the frequency or the wavelength must change, since their product, which is the propagation speed $u = \nu \cdot \lambda$, changes. But because frequency is a characteristic of the source, it does not change when the wave propagation becomes more difficult (propagation speed slows). Therefore, it is the wavelength in the specific material that has to be reduced by a factor equal to the refractive index $n$:

Wavelength in Medium with $n$: $$\lambda_n = \lambda_o/n \tag{1.19}$$

We now consider the propagation of light in an optical medium (glass). Assume that the duration of observation is $t_{AX}$ (where the subscript AX stands for from point $A$ to point $X$). Then light travels a distance equal to $L_{AX} = u \cdot t_{AX}$. If, at the same time, light is propagating in vacuum, the distance would have to be $L_{o\,AX} = c \cdot t_{AX}$. This distance is the **optical path length** (OPL). Comparing the two distances, we find that

$$\frac{L_o}{L} = \frac{c \cdot t_{AX}}{u \cdot t_{AX}} \Rightarrow \frac{L_o}{L} = \frac{c}{u} = n \Rightarrow L_o = n \cdot L \tag{1.20}$$

In the simplest case of a homogeneous medium, the optical path length is the product of the length of the path traveled and the index of refraction:

Optical Path Length: $$L_{oAX} = c \cdot t_{AX} = n \cdot L_{AX} \tag{1.21}$$

Obviously, this distance in vacuum is the largest possible because in glass (and more generally, any medium other than vacuum) light has a slower speed. The optical path length is, in other words, a measure of the propagation 'difficulty' between two points along a given curve. OPL has units of length and is proportional (× $c$) to the time needed for light to propagate from point $A$ to point $X$.

Another interesting notion is **optical density**, which essentially expresses the value of the refractive index in a medium. A medium that has a higher refractive index is optically denser. For example, water with $n = 1.33$ is optically denser than air with $n \approx 1$, but less optically dense than glass with $n \approx 1.5$.

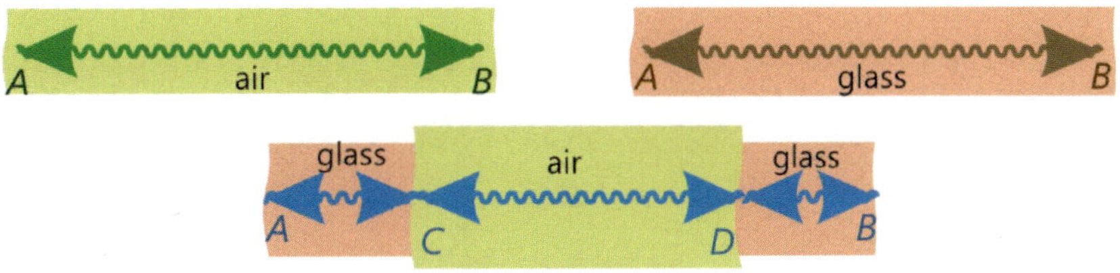


*Figure 1-27: Calculation of optical path length.*

Example ☞: In Figure 1-27, in the first case, points $A$ and $B$ are separated by 10 m, and we consider air with $n = 1.0$ as the medium between the two points. The optical path length ($L_{o\,AB}$) is simply the distance between these two points. In the second case, where the entire segment $AB$ is within glass with $n \approx 1.5$, the optical path length is $L_{o\,AB} = 1.5 \cdot AB = 15$ m.

If there are different media along a path (case 3), the optical path length is found by multiplying each segment by the refractive index and adding the products:

$L_{o\,AB} = 1.5 \cdot AC + 1.0 \cdot CD + 1.5 \cdot DB = 1.5 \cdot 2.5\text{m} + 1.0 \cdot 5\text{m} + 1.5 \cdot 2.5\text{m} = 12.5$ m.

### Optical Path Length (OPL)

- OPL may be considered as the distance light travels multiplied by the 'degree of difficulty' in that medium, which is the refractive index.
- OPL is always greater than (equal only in vacuum) the actual path length traveled by a ray.

# 1.4 FROM PARTICLES TO WAVES TO PHOTONS...

## 1.4.1 Isaac Newton's Initial Particle Theory

When admiring the colors in an abalone shell, a rainbow, or a soap bubble, we might realize that light is made up of many colors. We can observe the analysis of white light—the splitting of white light into seven colors—experimentally through a prism. A band of the iris colors (*ίριδα*) will appear. If we pass each color again, but separately, through a second prism, there will be no more analysis. Thus, we realize that these iris colors have a fundamental nature.

Newton: Are not rays of light very small bodies emitted from shining substances?

This was an observation first noted by Isaac Newton. Newton realized that light is a mix of colors and offered his own theory for the nature of light: It is a flow of tiny particles, which he called corpuscles. Different colors correspond to different particles. Each of the colors in the rainbow is part of the white light. The prism color separation does nothing more than simply separate the different particles according to the color to which they correspond.

*Figure 1-28: Isaac Newton displays to Albert Singlestone the miracles of light (© www.fiami.ch).*

### 1.4.2 Challenges to the Classical Wave Theory

This initial corpuscular theory, however, could not offer a satisfactory explanation for interference and diffraction effects. The explanation, the wave theory of light, emerged in the early $19^{th}$ century. The proposition was proven by Young's interference experiment (presented in § 4.2.1). For a number of years, the nature of these light waves was unknown, as the nature of the wave oscillations remained a mystery.

Later in the $19^{th}$ century, Maxwell predicted, and Hertz proved, that electromagnetic waves result from perturbations of the electric and magnetic fields. For a while, all optics effects seemed to have been explained. The electromagnetic wave theory of light properly interpreted all known optics effects, including emission, propagation, interaction with various media, interference, and diffraction. The theory of electromagnetism was solidly founded and there was no doubt about its validity.

This did not last long, however. Just as Newton's corpuscular theory collapsed when investigators were presented with the challenges of interference and diffraction, the classical light wave theory was 'deposed' when it could not produce a satisfactory interpretation of two seemingly nonconforming effects, the spectral distribution of radiation density in a black body and the photoelectric effect.

The classical wave (electromagnetic) theory of light simply would not provide adequate explanations for these effects. It was thought that it was simply a matter time until the proper interpretation would be found, of course, within the framework of electromagnetism. In the worst case, the explanations would be notable, but they are just exceptions. Alas!

### 1.4.3 Black-Body Radiation

What is this black body after all? How can a black material be capable of both absorbing and emitting radiation? Let's clarify that any body that has the ability to absorb radiation can equally well re-emit it.

According to Gustav Kirchhoff, a black body has the ability to entirely absorb the electromagnetic radiation incident upon it. Despite its name, it does not always appear black because it emits radiation; therefore, depending on its temperature, it may have a color. Thus, a black body appears black when only absorption is concerned. Since by definition it is fully absorbing, it appears black. However brilliant it may appear when emitting radiation, our sun may be described as a black body!

A black body can be represented by a small aperture cavity with absorbing walls [Figure 1-29 (left)]. Any radiation entering the cavity through the aperture is eventually absorbed after some reflections and does not escape; radiation can exit the cavity via the aperture.

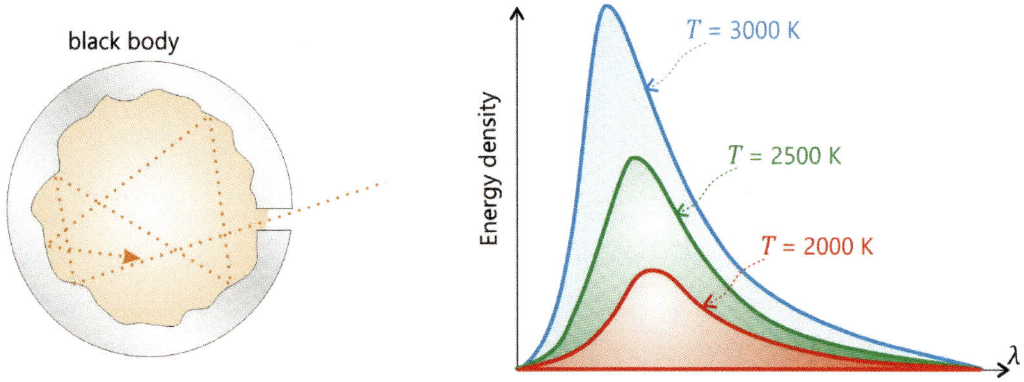


*Figure 1-29: (left) Black-body schematic representation and (right) black-body distribution of spectral density.*

The characteristics of the emitted radiation are reported as a distribution of spectral density, or the fraction of radiated power per unit of surface within a specific wavelength (or frequency) range. Such a graph is presented in Figure 1-29 (right). The area under the spectral distribution curve expresses the total energy emitted per unit of time normal to the unit surface (radiant intensity). Here are the specifics of black-body radiation:

(1) A warmer body emits more radiation, depending only on the absolute temperature $T$:

$$I = \sigma \cdot T^4 \tag{1.22}$$

where $\sigma$ is the Stefan–Boltzmann constant: $\sigma = 5.670367 \times 10^{-8}\ \mathrm{kg \cdot s^{-3} \cdot K^{-4}}$

(2) For a fixed temperature, the distribution decreases for longer $\lambda$ (lower frequencies) but, more importantly, it decreases drastically for much shorter $\lambda$ (higher frequencies). The maximum density at the peak of the curve ($\lambda_{MAX}$) shifts to progressively shorter wavelengths as the temperature increases. The curve's peak determines the black-body 'color' and relates to the absolute temperature via the empirical law derived by Wilhelm Wien, known as Wien's displacement law:

$$\lambda_{MAX} \cdot T = 29.8978 \times 10^6\ \text{Å} \cdot \mathrm{K} \tag{1.23}$$

which is ≈ 30 million if we express the wavelength in angstroms (Å) and the temperature in K. The hotter the black body, the smaller the emission maximum wavelength. The sun's surface temperature is ≈ 5500 K (5200 °C), with radiation peaking at the yellow/green $\lambda_{MAX} = 5250$ Å = 0.525 µm. It's no surprise that the human eye is most sensitive to this yellow!

The above characterizations can be explained by a simple observation. A blacksmith from the good-old days would put a piece of iron in the fire. As the iron was heated, it would become progressively red, then yellow, then blue. The same phenomena might be noticed in a legacy electric heater. The filament or the glowing iron is a radiating black body; as it is heated, the electrons begin to collide with the lattice, causing oscillations, and these in turn cause electromagnetic emission, which is the light we see and the heat we feel. We can therefore agree that a black body can have a certain color!

The glowing iron color changes as the temperature increases because the emission peak gradually shifts from red to yellow, and then to blue. Any glowing light source can be likened to a black body radiating at such a temperature to emit the same color. This is the **color temperature** of the source (or the equivalent color temperature for sources with discrete, not continuous, spectra). The value of the color temperature is given in Kelvin (K).

These specifics of black-body radiation cannot be interpreted using the classical wave theory. Specifically, the disagreement is in the short wavelengths/high frequencies (the left side of the graph in Figure 1-30). There is good agreement between the experimental results and the classical wave theory only at long wavelengths/low frequencies.

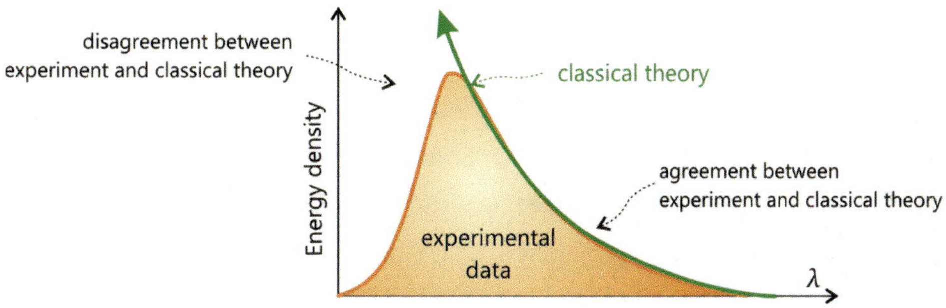


*Figure 1-30: The classical EM theory agrees with experimental data only at long wavelengths.*

The radiated energy results from oscillation (resonant) modes, which are standing waves within the cavity; this energy is proportional to the mode density, which can be expressed by a term that is proportional to the square of the frequency ($8\pi\nu^2/c^3$).

Classical wave theory assumes a continuous radiation distribution, the Boltzmann distribution (discussed in § 6.1.2), named after the Austrian physicist Ludwig Boltzmann. The form utilizes the Boltzmann constant $k_B$, a physical constant with dimensions of energy per degree of temperature, having a value of $1.38064852\times10^{-23}$ J·K$^{-1}$ :

$$f(\varepsilon) = \exp\left(-\frac{\varepsilon}{k_B T}\right) \tag{1.24}$$

Integrating Eq. (1.24) with respect to $\varepsilon$, we find an equal mean energy per mode that is independent of the frequency:

$$E_{\text{average, classical}} = \int \varepsilon \cdot f(\varepsilon) d\varepsilon = \int \varepsilon \cdot \exp\left(-\frac{\varepsilon}{k_B T}\right) d\varepsilon = k_B T \tag{1.25}$$

This distribution is multiplied by the oscillation mode density to produce the spectral distribution of radiated energy, which predicts a parabolic increase in distribution with frequency:

Rayleigh–Jeans Approximation:

$$I_{\text{classical}}(T) = \frac{8\pi \nu^2}{c^3} \cdot k_B T \tag{1.26}$$

This is the classical approximation credited to John William Strutt (Lord Rayleigh) and James Hopwood Jeans. According to this approximation, the spectral distribution curve for a specific temperature increases parabolically with the frequency, since it is proportional to $\nu^2$. In other words, for high frequencies (ultraviolet), we should observe a drastic increase in density energy, termed the ultraviolet catastrophe or the Rayleigh–Jeans catastrophe. However, this has never been observed. Nature appears to have found a way to avoid this apparent apocalyptic call.

### 1.4.4 The Particle Theory: The Revenant

An explanation of black-body radiation density distribution was proposed in 1900 by Karl Ernst Ludwig Max Planck, who presented a new idea: Energy is emitted not in a continuous fashion, but in wave packets having specific, minimum-energy quantities, or **quanta of energy**.[3]

Thus, it was understood that energy cannot be indefinitely sliced into infinitesimal quantities, but only into discrete quantities. The smallest of such quantities relates to frequency $\nu$:

$$\varepsilon = h \cdot \nu \tag{1.27}$$

where $h$ is Planck's constant: $h = 6.626 \times 10^{-34}$ J·s

The energy quanta fall in the order of magnitude of an electron volt (eV), where 1 eV is the energy of an electron accelerated by an electric potential of 1 V (volt):

Electron Volt:

$$1 \text{ eV} = 1.6 \times 10^{-19} \text{ J} \tag{1.28}$$

This unit is quite convenient for expressing energies on the atomic scale, as well as photon energies in the visible, which range from 1.8 eV (the red) to 3.1 eV (the violet); the

[3] Karl Max Planck. On the law of distribution of energy in the normal spectrum. Annalen der Physik. 1901; 4:553.

median green at 530 nm has a photon energy of 2.34 eV. As photon energies decrease, wavelengths shift to longer ones: The thermal energy of a vibrating molecule may be on the order of 0.04 eV, and in X rays, photon energies can be up to 200,000 eV. A photon with energy of 1 eV belongs to the infrared region of 1240 nm. Thus, a mnemonic rule that associates the energy in electron volts with the wavelength is

Electron Volt and Wavelength: $\varepsilon\ [\text{eV}] = 1240/\lambda\ [\text{nm}]$ (1.29)

Planck explained this rule very simply, as follows: Energy radiates/exists only in multiple integers of the energy block $h \cdot v$. The lower the frequency, the smaller this block of energy, regardless of the photon population. At this part of the spectrum, classical theory and quantum theory agree. However, for high frequencies/short wavelengths, the energy block $h \cdot v$ is much larger and therefore not so easily emitted. This version reflects our new understanding of energy, which agrees with Nature but does not agree with classical wave theory.

For high frequencies/short wavelengths, the radiation spectral density is much lower than what classical wave theory predicts. The classical wave theory's destructive prediction of a substantial increase in radiation density with an increase in frequency (the ultraviolet catastrophe) does not exist! The new expression of the spectral distribution of radiated energy becomes

$$I_{\text{quantum}}(T) = \frac{8\pi v^2}{c^3} \times \frac{1}{\exp\left(\frac{hv}{k_B T}\right) - 1} \tag{1.30}$$

For low frequencies/long wavelengths, the asymptotic behavior of Eq. (1.30) coincides with the classical wave theory prediction [Eq. (1.26)]. The difference between the two is noted at short wavelengths or high frequencies: The smallest energy block, $h \cdot v$, is no longer too small, and the quantum nature of energy is evident.

Applying the suggestion that energy cannot be split infinitely but is rather split into blocks with energy proportional to frequency, Planck properly interpreted the distribution of black-body spectral density. The theory that energy has some particle characteristics was such a breakthrough that even Planck himself did not fully grasp it. He thought that the quanta proposal was just (another) mathematical tool that helped to explain an effect, not the foundation of a major revolution in physics!

Thus, it was the study of black-body radiation that brought both the concept of energy quantization to our current understanding of physics and the Nobel Prize in Physics to Max Planck (in 1918).

## 1.4.5 The Photoelectric Effect

The attempt to interpret another 'disobeying' effect provided the next step in developing the foundation of the new theory. This noncompliant force is the **photoelectric effect.** Briefly, if a clean metal sheet (preferably, an alkali element) is illuminated, electrons can be emitted from it.

Observations relating to the photoelectric effect that contradict the classical physics theory as understood so far are the following:

- For every material, there is a wavelength threshold, beyond which there is no emission. Light with a long wavelength cannot extract electrons, regardless of its energy. For example, if potassium is illuminated with 700 nm wavelength light (the upper limit of the visible range), then, regardless of the intensity, there is no electron emission.

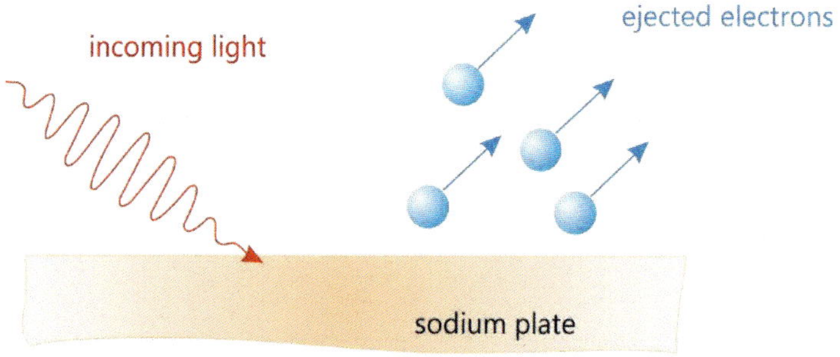


*Figure 1-31: Ejection of electrons from an alkali material (such as sodium): the photoelectric effect.*

- Electrons are emitted instantly, with no delay. In other words, no electron has to 'wait' to accumulate energy. It is either instantly ejected or not ejected at all.
- Light with short wavelengths, such as the blue, even in dim illumination, can cause electrons to be ejected. The electron kinetic energies emitted in this case are significantly greater than those emitted from longer-wavelength light, such as the yellow.
- An increase in luminous intensity is related to an increase in the number of ejected electrons but is not related to the ejected electrons' kinetic energy. The speed of the ejected electrons increases only with a higher frequency, not with a higher intensity (brighter light).
- The electron kinetic energy is in a linear relation to the light frequency: $E = \alpha \times \nu + \beta$. The linear slope $\alpha$ is the same for all frequencies and materials. The intersect point $\beta$ corresponds to a negative number that is different for every material. It appears thus that the entity interacting with the electrons has an energy that is proportional to the frequency.

Sunburn is an effect that behaves similarly to the photoelectric effect. A threshold energy > 3.5 eV is required for the breakdown of an organic material. This energy corresponds to the UV. So, while we may be enjoying the sunshine, caution is required for proper skin protection; in particular, a UV-absorbing sunscreen is needed.

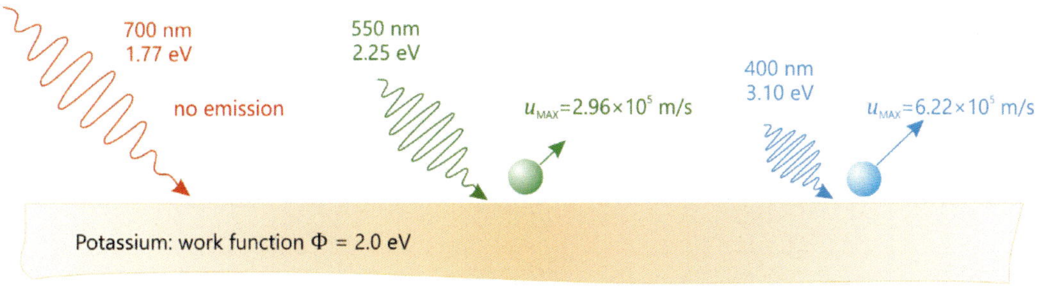


*Figure 1-32: Photoelectric effect and photon energy.*

Here, too, classical wave theory cannot offer an explanation. Classical light is a continuous wave whose energy is simply proportional to the square of its amplitude. With brighter light, more energy is incident on the metal. Thus, the electron receives more energy and should be ejected off the metal with a higher speed, regardless of the frequency (color) of the light. However, this is not what was observed.

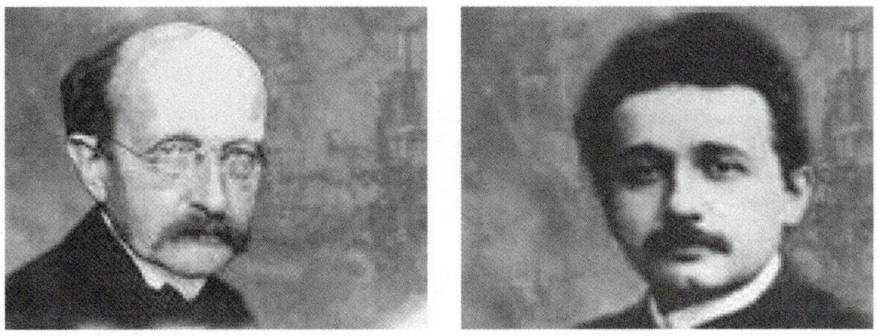

*Figure 1-33: Karl Ernst Ludwig Max Planck (1858–1947) and Albert Einstein (1879–1955).*

Albert Einstein applied Planck's energy quanta to light. He proposed that light is energy made of energy quanta; thus, light is not a continuous wave, but is composed of many indivisible 'packets' or **photons**. Light of a specific frequency $\nu$, and therefore of a specific color, comes in the same energy quanta $\varepsilon = h \cdot \nu$. In other words, light is a 'shower' of photons, in which it is difficult to discern each individual photon. The slope $\alpha$ in the linear relationship corresponds directly to Planck's constant. Brighter light means more photons. Blue light means photons of higher energy. Because energy is not continuous but comes in blocks, atoms absorb and emit energy not in a continuous fashion but with specific elementary quantities.

Light–electron interaction is an interaction between independent particles and involves a complete energy exchange. The photon gives the electron a specific amount of energy (= $h \cdot \nu$), from which, if the work function Φ (the energy threshold required to discharge electrons from the lattice field) is subtracted, the difference is the electron kinetic energy. Thus, each electron is ejected from the metal at exactly the same speed if it is illuminated with exactly the same light frequency, independent of the incident intensity.

With the increase in intensity, the photon population increases, so more electrons are ejected. The energy that a photon carries differs for the various colors (frequencies) and thus changes the kinetic energy of the emitted electrons. If the light frequency drops below a threshold, the photon energy is not enough to produce the effect. The 'exit work' Φ is increased, and no emission occurs at all, regardless of the photon population (light intensity). The energy of the emitted electrons is simply not enough.

*Figure 1-34: Einstein's contribution to modern optics dates back to 1905 (© www.fiami.ch).*

Under this new theory, light cannot be split infinitesimally, but only down to certain minimum-energy 'packets,' called photons.

Photons are bosons, which are elementary particles that follow Bose–Einstein statistics. Bosons make up one of the two classes of particles, the other being fermions (the electron is the most well-known fermion). Photons tend to be random and unpredictable; however, they also tend to cluster together. This is due to an important characteristic of bosons: Their statistics do not restrict the number of bosons that can occupy the same quantum state. Thus, bosons may form a photonic state, which is many look-alike photons having similar properties. This is a light-ray beam!

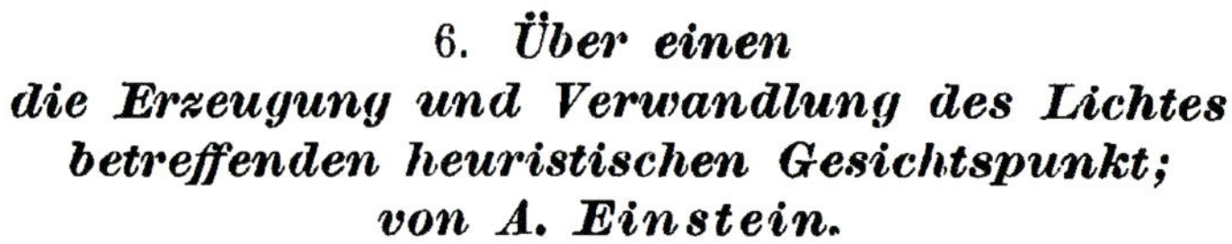
6. ***Über einen***
***die Erzeugung und Verwandlung des Lichtes***
***betreffenden heuristischen Gesichtspunkt;***
***von A. Einstein.***

*Figure 1-35: The Nobel Prize in Physics (1921) was awarded to Albert Einstein for his contributions to theoretical physics and especially for his discovery of the law of the photoelectric effect.*

*Figure 1-36: Isaac Newton gives 'permission' to Albert Singlestone to improve his corpuscular theory of light. This took place some centuries later, in 1905 (© www.fiami.ch).*

## 1.4.6 The Wave Nature of the Photon

The concept of a photon is entirely different from that of an electromagnetic wave. Before the discovery of the photon, we were content with absolute wave characteristics and mathematical idealizations. With the 'second coming' of the corpuscular theory, the very framework of physics changed. The photon and the electromagnetic wave were not viewed as mutually exclusive; instead, they could both be used to explain the physical world. The photon offered a new dimension to the definition of a wave. The German physicist Werner Heisenberg furthered the concepts of quantum mechanics with his suggestion that light quanta have wave properties.

As is the case for every particle, the photon has an associated wave property. A large population of identical photons is a photonic state that corresponds to a plane, monochromatic, polarized electromagnetic wave. The wave's direction of propagation is the direction of photon momentum; the color of the wave (frequency) is the same as the color of the photons, and the polarization state of the wave is associated with the spin state of the photon.

Macroscopically, the wave attributes and photon attributes blend together, becoming undistinguishable. The photon corpuscular character may be noted only for high frequencies (photon energy). A photonic state is, of course, an idealized one. Even a strictly monochromatic wave is not composed of absolutely identical photons. Common white light is composed of many colors and is not polarized, either. White light is considered to be a superposition of many photonic states.

The photon energy $\varepsilon$ is related to the wave frequency ($\hbar = h/2\pi$):

$$\varepsilon = h \cdot \nu = \hbar \cdot \omega \tag{1.31}$$

The photon momentum $\mathbf{p}$ is related to the wavevector $\mathbf{k}$:

$$\mathbf{p} = \hbar \cdot \mathbf{k} \tag{1.32}$$

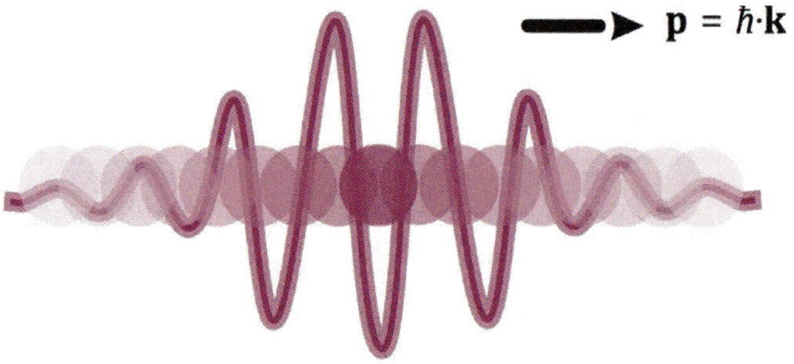


*Figure 1-37: Photon momentum.*

The above relationships describe the dual nature of the photon, which connects wave characteristics (frequency $\nu$ and wavelength $\lambda$) and particle attributes, such as energy $\varepsilon$ and momentum $\mathbf{p}$. The photon has properties that some may say make it both a particle and a wave, and that others might describe as being like neither a particle nor a wave.

The concept of a photon provides an explanation for the effects of light propagation (reflection, refraction, diffraction), as well as its interaction with matter (photoelectric effect, absorption, emission). For some effects, we do not have to consider photons at all, as the wave nature is adequate. There are cases, however, in which the concept of a photon is necessary to interpret light behavior. Thus, today we have a relative certainty as to how light behaves: Depending on the scale of measurement and the instrument of observation, sometimes the wave nature is convenient for explaining light's behavior, and at other times the particle nature is necessary.

### 1.4.7 Photon Propagation

Because we know how a wave propagates, we know (to a sufficient degree) how a photonic state propagates, too. The principle of the least optical path (as presented in § 1.3.1) offers very good guidance. However, if we try to expand this principle to the world of the infinitely small, we must abandon our common sense. We cannot 'tag' a photon and force it to provide a specific location and speed reference as it moves. In fact, quantum electrodynamics predicts that a photon propagating from point *A* to point *B* can follow whatever path it wants, all paths being equally probable.

This resembles the uncertainty relating to the motion of an electron inside an atom: At any time, the electron is likely to have any speed and location. Just as in the case with a photon, we cannot determine whether it will pass by a point or not; we can only state the likelihood (probability) of this occurring. If a photon is directed toward a sensor, all we know is the probability that the event is recorded. This relates to the **quantum efficiency** of the sensor, which can be, for example, only 20%: This means that of the five photons that hit the sensor, only one is logged. Which one exactly, we may not know.

Feynman: We can only consider probabilities in the world of such small dimensions.

What can be measured, or predicted, is the probability of a photon passing from one place to another. This can be represented by a probability vector, called **Feynman's vector**, after the great American physicist Richard Phillips Feynman. This vector can be described by a complex number that has both magnitude and phase. Its magnitude is the probability of the event, and its phase is determined by the time of travel.

Two of such vectors have equal magnitudes if they correspond to equiprobable events. The probability of two consecutive independent events occurring is calculated by adding independent vectors and then squaring their sum; this is how the principle of linear superposition is explained in Feynman's quantum electrodynamics (QED). This extremely simple principle can explain both interference and the principle of the least optical path (see § 4.2.6).

*Figure 1-38: Richard Phillips Feynman (image from the California Institute of Technology Institute Archives used with permission).*

Just as light has both a wave nature and a particle nature, so does the photon: It is both a wave and a particle.

→ The two natures of light do not contradict one another.

→ These natures coexist, and sometimes it serves us to employ the wave nature and sometimes the particle nature!

# 1.5 Light Sources

## 1.5.1 Let There Be Light...

... and God said, "Let there be light," and there was light. God saw that the light was good ...

... καὶ εἶπεν ὁ Θεός· γενηθήτω φῶς· καὶ ἐγένετο φῶς. Καὶ εἶδεν ὁ Θεὸς τὸ φῶς, ὅτι καλόν ...

Genesis 1:3-4.

The sun, the stars, a lightning bolt, a firefly—these are all objects that emit light. For this reason, they are called **sources**. The process they use is called **light emission**, which is essentially no different from broadcasting radio waves from a radio station antenna. What is different is the wavelength.

Light emission, as well as its opposite, light absorption, results from interaction between light and matter. During emission, a form of energy is converted into light; during absorption, the energy of a photon is converted into another form of energy.

Light emitted by a filament in an incandescent lamp is an example of electricity being converted to light. By applying a voltage across the filament, electrons are accelerated, picking up kinetic energy. These electrons collide with the filament metal (ionic) lattice, causing it in turn to oscillate. This oscillation causes the ions to emit light. In the ionic lattice, there are many possible combinations of 'springs' that maintain the ions in their lattice position, so there is a very large number of possible oscillation modes. This is why the emitted light has a continuous spectrum: It contains all of the frequencies within a wide range.

However, light emission by a low-pressure gas discharge lamp is completely different. When a voltage is applied, the ions inside the tube accelerate toward the anode or cathode, depending on their charge. Along their path, the ions collide with neutral gas atoms, causing excitations, which are electron transitions to specific, higher energy states. With or without some intermediate transitions, the result is photon emission at specific frequencies (wavelengths) that match the allowed transitions of electrons to lower energy states for the given element/molecule. The spectrum of this light is no longer continuous, but is now discrete (linear), as it contains only specific colors.

## 1.5.2 Light–Matter Interactions

Light (in general, electromagnetic radiation) and matter interact continuously. Their interactions can be classified macroscopically as refraction, reflection, scattering, and absorption.

Reflection and refraction describe effects of light encountering a surface such as an air–glass interface. The reflected part of the beam returns to the original medium (in this case, air). The refracted part of the beam propagates in the second medium (in this case, glass).

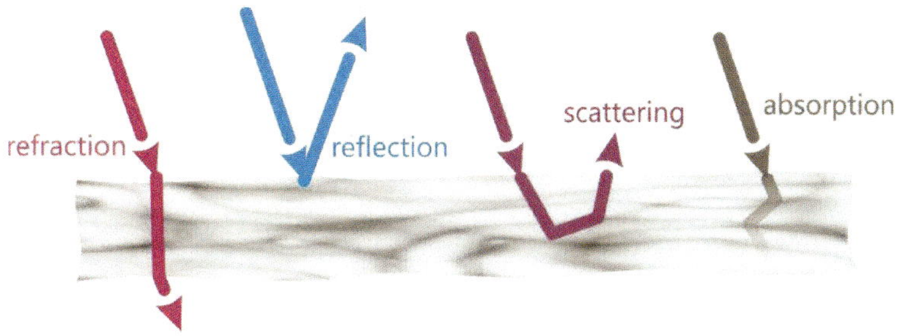


*Figure 1-39: Interactions between light and matter.*

Scattering and absorption are light–matter interactions that result in changes to the light's direction of propagation or its energy. In **scattering**, light exits the medium in a random direction without transforming into any other form of energy. An example is sunlight scattering from white clouds.[4] In **absorption**, light energy is converted to another energy form (heat), which is what happens when the sun warms the planet.

Note: As discussed in § 2.5.1, scattering is responsible not only for white clouds, but also for the blue sky. There are several different manifestations of scattering. Mie scattering, associated with relatively large scattering particles such as water droplets, produces spectrally uniform (white) and mostly forward/backward scatter. Rayleigh scattering, associated with relatively small scattering particles such as gas molecules (e.g., nitrogen or oxygen), produces stronger scattering at right angles with respect to the direction of propagation compared to other angles. Rayleigh scattering is much stronger for the blue (short wavelengths) than the red (long wavelengths). Therefore, the sky is mostly blue at right angles to the sun.

Microscopically, on the atomic/molecular scale, there are only two light–matter interactions: scattering (in which a photon collides with an atom or molecule and bounces) and absorption (in which the photon energy is such that the atomic/molecular system 'takes' the

[4] The sky is blue, again, due to scattering. This is Rayleigh scattering, an effect associated with no change in the re-emitted (scattered) wavelength. This effect involves particles that are physically smaller than the incident wavelength.

photon's energy, which dissipates in the form of heat). Both reflection and refraction can be explained as a macroscopic manifestation of organized forms of scattering from the surface of the material.

The basic aspect of light–matter interaction is the ability of a photon to excite an electron, meaning its ability to induce the transition of an electron to a state of higher energy.

*Figure 1-40: Albert Einstein is as excited as an electron and as tired as the ether (© www.fiami.ch).*

Therefore, the result of the interaction between light and matter depends primarily on the frequency $\nu$. The photon energy $h\cdot\nu$ is compared to the available energy transitions, which are dependent on the atomic energy level differences $E_2 - E_1$.

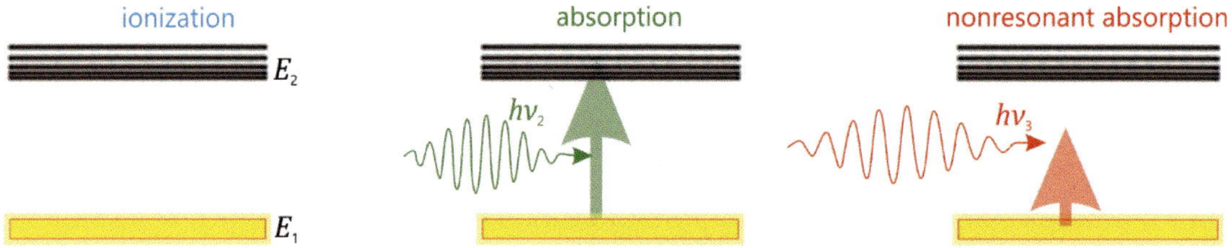


*Figure 1-41: Photon absorption depends on the energy structure of the material.*

The energies of a light photon that corresponds to the visible frequencies range from 1.7 to 3.5 eV, which are comparable to the energies associated with transitions of electrons between any two molecular or atomic energy states. Thus, the main mechanism of photon absorption corresponding to the visible radiation involves the elevation of an electron to a higher energy state (stimulation).

If the energy of the photon $h\cdot\nu$ matches the atomic state differences, $h\cdot\nu = E_2 - E_1$, then the photon's frequency coincides with the resonance frequencies of the medium and there is **resonant** (quantum) **absorption**. The electron, by absorbing energy $h\cdot\nu$, shifts from a lower energy state $E_1$ to a higher energy state $E_2$ (see also § 6.1.3.3).

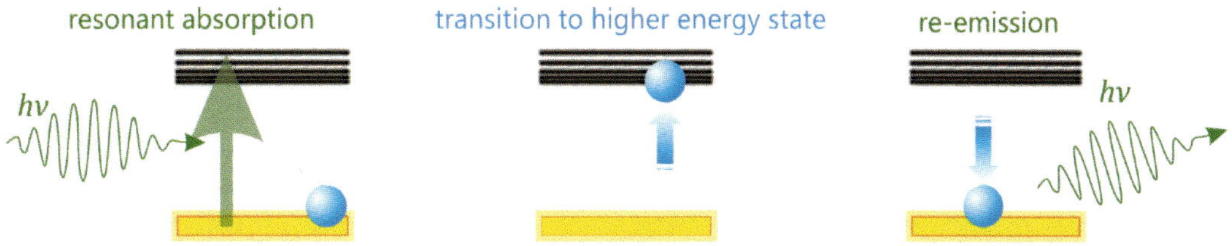


*Figure 1-42: Mechanism of quantum absorption and spontaneous emission.*

The electron will not stay in an excited state for very long before various de-excitation mechanisms come into play. The two most prominent of these mechanisms are **spontaneous emission**, in which a photon is emitted (Figure 1-42), and **stimulated emission**, where the incident photon encounters an already excited state (Figure 1-43), in which case two identical photons are emitted simultaneously. Both mechanisms emit photons having the same frequency as the initially incident photon (see also § 6.1.3.2).

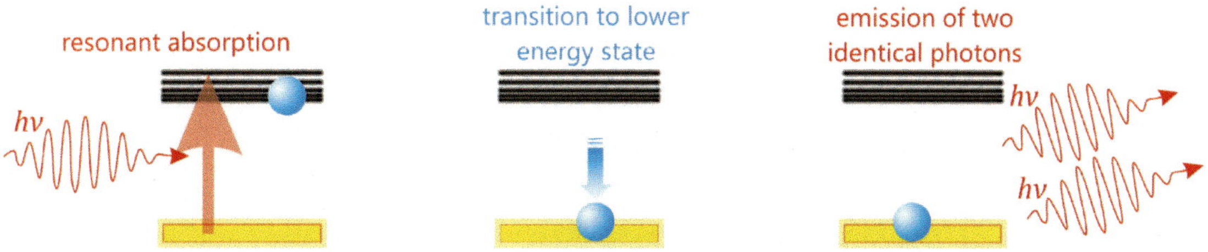


*Figure 1-43: Mechanism of stimulated emission.*

Other de-excitation mechanisms include a transition to intermediate, degenerate states, with subsequent emission of photons of less energy (longer wavelength). The associated effects in this case are fluorescence and phosphorescence.

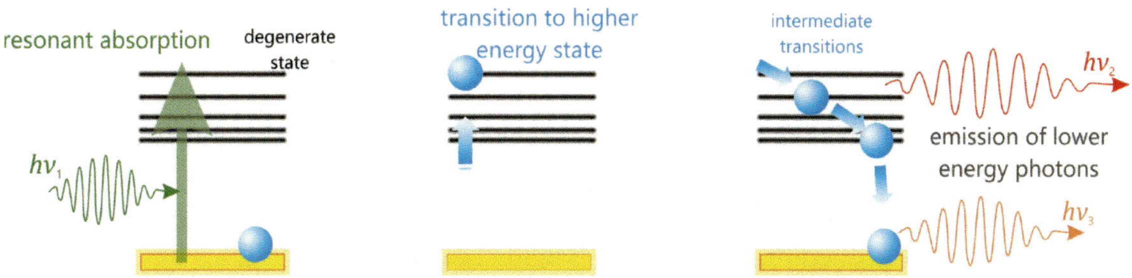


*Figure 1-44: Quantum absorption and fluorescence.*

**Fluorescence** results from transitions between states of the same multiplicity (such as from simple to simple), while phosphorescence results from transitions between states of different multiplicities. The reaction time for fluorescence is characteristically different from that of phosphorescence. If emission is almost immediate ($< 10^{-8}$ s), fluorescence occurs, while if the emission is not immediate (> $10^{-8}$ s), the effect is called luminescence. If ruby is illuminated with green light (0.550 μm), there is fluorescence. Ruby then re-emits red light at 0.693 μm; this property of ruby was the basis of the first laser (as discussed in § 6.4.2.5).

When charged particles are blown toward the earth by the solar wind, they are largely deflected by the earth's magnetic field and are trapped, becoming concentrated over either pole. Some of these particles enter the atmosphere and collide with high-altitude atmospheric particles. These collisions result in fluorescence that we perceive as auroral light. The variations in color are due to the different types of colliding gas molecules. Green is the most common auroral color, produced by oxygen molecules at a height of about 60 miles above the earth. The rarer red auroral light is produced at heights of up to 200 miles. Nitrogen produces blue or purple auroral light.

*Figure 1-45: (left) Northern lights (aurora borealis) and (right) southern lights (aurora australis). Both are manifestations of fluorescence. (Left image by Manda Maggs used with permission; right image by NASA.)*

If the frequency of the incident wave does not coincide with any of the resonance frequencies (i.e., is nonresonant, $h \cdot \nu \neq E_2 - E_1$), then there is no quantum absorption. We distinguish the following two cases:

- If $h \cdot \nu < E_2 - E_1$, then forced bipolar oscillation occurs. This is a form of classical, nonresonant **absorption**. The electric field of the incident wave exerts forces that lead to an oscillation of the electronic cloud. This oscillation is considered a small, first-class disturbance of the state of equilibrium. It may lead to lattice vibrations, which are macroscopically observed as thermal absorption—the material heats up. In other cases, particularly in rare, thin media, the oscillating dipole simply re-emits energy in the form of a photon of the same frequency to directions perpendicular to the oscillation plane. This is the phenomenon of **elastic scattering**, which is employed analytically in the interpretation of polarization effects in nature (presented in detail in § 2.5).

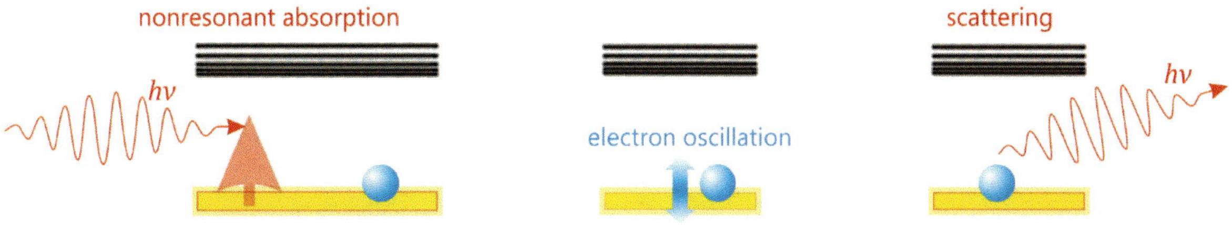


*Figure 1-46: Mechanism of elastic scattering.*

- If $h \cdot \nu > E_2 - E_1$, then the absorbed photon energy exceeds the necessary amount of energy for an electron to ascend to a higher level, possibly leading to electron extraction. This is called photoionization, which is a form of a photoelectric effect. The principle of the conservation of energy stipulates that the kinetic energy of the electron is the difference between the absorbed photon energy and the work function = $\Phi \times$ the electric charge $e$, which is the difference (change) in potential electric energy.

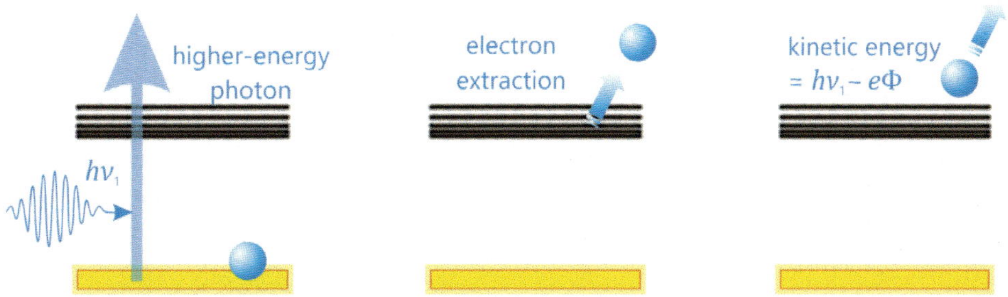


*Figure 1-47: Mechanism of photoionization.*

# 1.6 LIGHT AND ELECTROMAGNETISM QUIZ

1) Which of the following cannot be represented by energy carried in the form of an electromagnetic wave (select two)?
   a) heat from the sun
   b) lighting
   c) thunder
   d) the northern lights
   e) an earthquake tremor

2) The particle theory of light can offer a satisfactory (even if not always the prevailing) explanation of all except which one of these effects/properties?
   a) the speed of light in various media
   b) the composition of light as several colors
   c) polarization properties
   d) diffraction properties
   e) the photoelectric effect

3) The classical electromagnetic theory of light is in good agreement with the experimental results of the photoelectric effect in what areas of the spectrum?
   a) very low energy, long wavelength
   b) high energy, short wavelength
   c) visible light range
   d) ultraviolet light range

4) Which one of the following differences between elastic mechanical waves (such as sound) and electromagnetic waves (such as light) is true?
   a) The propagation speed in mechanical waves is dependent on the medium in which they traverse, while the propagation speed in electromagnetic waves is independent of the propagating medium.
   b) In mechanical waves, there is an associated material mass transfer, while in electromagnetic waves, there is no mass transfer.
   c) Mechanical waves can be longitudinal or transverse, whereas electromagnetic waves are only transverse.
   d) Mechanical waves do not produce interference effects, whereas electromagnetic waves produce interference effects.
   e) Simple harmonic oscillation can be implemented only in mechanical waves, not in electromagnetic waves.

5) As a wave propagates away from its source, which is oscillating at frequency $\nu$, it encounters two different media A & B. The propagation speed in medium A ($u_A$) is greater than that in medium B ($u_B$). The wave frequency(ies) and wavelength(s) in the two media are linked by what relationships (two correct answers)?
   a) $\nu_A = \nu_B = \nu$
   b) $\nu_A > \nu_B$
   c) $\nu_A < \nu_B$
   d) $\lambda_A = \lambda_B$
   e) $\lambda_A > \lambda_B$
   f) $\lambda_A < \lambda_B$

6) Which of the following two combinations of speed ($u$), time ($t$), and directional distance ($z$) are acceptable combinations in a wave equation (two correct answers)?
   a) $u \cdot t - z$
   b) $u \cdot t^2 - z$
   c) $u^2 \cdot t - z$
   d) $(u \cdot t - z)^2$

7) In a simple, plane harmonic wave, over the course of a half wavelength, the phase increases by ...
   a) 0°
   b) 90°
   c) 180°
   d) 270°
   e) 360°

8) In a simple, plane harmonic wave, over the course of a half wavelength, the wave amplitude changes by ...
   a) no change
   b) half, same sign
   c) half, opposite sign
   d) full, opposite sign

9) In a simple, plane harmonic wave, over the course of a half wavelength, the field vector changes by ...
   a) no change
   b) half, same sign
   c) half, opposite sign
   d) full, opposite sign

10) In a simple, plane harmonic wave representation of the electric field, the amplitude is $E_o$ = 4 N/C. The value of the electric field varies from ...
    a) 0 to 8 N/C
    b) 0 to 4 N/C
    c) –4 to 4 N/C
    d) –16 to 16 N/C

11) As a wave propagates, its phase increases by 4π. The optical path length traveled equals ...
    a) one-fourth of a wavelength
    b) one-half of a wavelength
    c) one wavelength
    d) two wavelengths

12) The visible part of the electromagnetic spectrum ranges from the violet ($\lambda \approx$ 400 nm) to the red ($\lambda \approx$ 700 nm). In comparison, the infrared part of the electromagnetic spectrum has ...
    a) longer wavelengths, higher frequencies
    b) longer wavelengths, lower frequencies
    c) shorter wavelengths, higher frequencies
    d) shorter wavelengths, lower frequencies

13) Rays are perpendicular to wavefronts in (select all that apply) ...
    a) all waves
    b) spherical waves
    c) plane waves
    d) converging waves

14) The units of the index of refraction are ...
    a) units of speed
    b) units of length
    c) no units
    d) units of reciprocal length

15) The value of the index of refraction is ...
    a) at least 1.0
    b) at most 1.0
    c) between 0.5 and 1.5
    d) between 0 and 2.0

16) The refractive index for the violet blue in a flint F2 glass type is about 1.65, while for the red it is about 1.61. This is ...
    a) an exception that is encountered only in this type of flint glass
    b) a typical manifestation of blue having higher refractive index $n$ than red in most media
    c) an unusually large difference in values of the refractive index
    d) an unusually small difference in the values of refractive index

17) Medium A has refractive index $n_A$ = 1.5, while medium B has refractive index $n_B$ = 1.8 (recall that the speed of light in vacuum is $3\times10^8$ m/s). It can be inferred that the speed of light (two correct) ...
    a) in medium A is $2\times10^8$ m/s
    b) in medium A is $4.5\times10^8$ m/s
    c) in medium B is $1.66\times10^8$ m/s
    d) in medium B is $5.4\times10^8$ m/s

18) Green light of vacuum wavelength $\lambda$ = 500.0 nm enters medium A ($n_A$ = 1.5). The wavelength within the medium is ...
    a) 333.3 nm
    b) 500.0 nm
    c) 501.5 nm
    d) 750.0 nm

19) Between points $K$ and $L$ there is air. The optical path length between these two points is 2.0 m. What would be the optical path length if between points $K$ and $L$ we insert a block of glass of medium B ($n_B$ = 1.8) that covers the entirety of the length between $K$ and $L$?
    a) 1.11 m
    b) 2.0 m
    c) 3.6 m
    d) 3.8 m

20) Back to Q 19. In the space between $K$ and $L$ we insert a 1.0 m thick block of glass of medium A ($n_A$ = 1.5). What would be the optical path length between points $K$ and L?
    a) 1.5 m
    b) 2.0 m
    c) 2.5 m
    d) 3.0 m

21) A radiowave has a wavelength in the vicinity of 1 m. What is this wave's frequency (in hertz)?
    a) $3.0\times10^4$ Hz
    b) $3.0\times10^6$ Hz
    c) $3.0\times10^8$ Hz
    d) $3.0\times10^{12}$ Hz

22) A microwave oven operates at 24.5 GHz (1 GHz = $10^9$ Hz). What is the operating wavelength of this device?
    a) 1.2 cm
    b) 12 cm

c) 120 cm
d) 1200 cm

23) A photon with an energy of 1 eV belongs to what part of the EM spectrum?
   a) microwave
   b) infrared
   c) visible
   d) ultraviolet

24) Which one of the following energies most likely belongs to a photon in the ultraviolet region?
   a) 0.35 eV
   b) 1.0 eV
   c) 2.5 eV
   d) 3.5 eV

25) The wavelength of a 2 eV photon is ...
   a) 2 nm
   b) 310 nm
   c) 620 nm
   d) 1240 nm

26) As a black body steadily increases its temperature, the peak radiation wavelength ...
   a) shifts to a longer wavelength
   b) remains the same (no shift)
   c) shifts to a shorter wavelenth

27) The Sun's surface temperature is 5200° C. Based on Wien's displacement law, its peak emission wavelength is ...
   a) 520 nm
   b) 525 nm
   c) 550 nm
   d) 570 nm

28) A hotter-than-the-Sun star has a surface temperature of 5250 K. Again, based on Wien's displacement law, its peak emission wavelength is ...
   a) 520 nm
   b) 525 nm
   c) 550 nm
   d) 570 nm

29) A photon of what eV energy is most likely to be capable of discharging electrons from zinc (Zn) metal, whose work function is $\Phi \approx 4.3$ eV?
   a) 3.3 eV
   b) 3.4 eV
   c) 3.8 eV
   d) 4.4 eV

30) Platinum (Pt) has a work function of $\Phi = 6.35$ eV. We shine light of 190 nm on a platinum sheet. The discharged electrons ...
   a) There are no discharged electrons.
   b) have kinetic energy of about 6.52 eV
   c) have kinetic energy of about 6.35 eV
   d) have kinetic energy of about 0.17 eV

31) A photon of 243 nm can barely extract an electron out of a gold metal sheet. What is gold's work function $\Phi$?
   a) 243 eV
   b) 5.1 eV
   c) 2.43 eV
   d) 0.51 eV

32) Skin tissue sunburn occurs when skin is exposed to UV radiation of wavelengths shorter than 350 nm. What is the work function $\Phi$ [eV]?
   a) 350 eV
   b) 35 eV
   c) 3.5 eV
   d) 0.35 eV

33) A resonant photon absorption occurs when the photon energy is _________ the energy difference between the two atomic energy levels.
   a) less than
   b) equal to
   c) greater than

34) A nonresonant photon absorption occurs when the photon energy is _________ the energy difference between the two atomic energy levels.
   a) less than
   b) equal to
   c) greater than

35) In a ruby fluorescence, the excitation wavelength is green (550 nm). The emitted light has a wavelength that is ...
   a) shorter than 550 nm
   b) equal to 550 nm
   c) longer than 550 nm

# 1.7 LIGHT AND ELECTROMAGNETISM SUMMARY

## Nature of Light

Light properties can be described by considering either the corpuscular nature or the wave nature of light. The wave theory of light is used to explain the phenomena of polarization, interference, and diffraction. It also explains phenomena associated with the formation of images by lenses, prisms, and mirrors (which is the realm of geometrical optics).

The nature of light is dual; the photon, an elementary particle, represents a quantum of light (or some other electromagnetic radiation) that carries a specific amount of energy that is proportional to its frequency. The photon has both wave and particle properties. The energy-particle (quantum) theory of light is essential in explaining light–matter interaction. Under this theory, energy quanta are emitted by light sources and absorbed by atoms or molecules in the medium.

## History of Light Theories

The two early theories of light were those presented by Huygens (light is a wave) and Newton (light is a particle). The notable experiment by Thomas Young in 1801 that demonstrated interference proved the wave nature of light. Later on, the electromagnetic theory established that light is electromagnetic waves, being composed of mutually dependent perturbations of the electric and the magnetic fields. Maxwell's equations provide a formalization of a set of the known properties of the electromagnetic field and therefore of light.

In their early forms, these theories explained some optical phenomena (such as rectilinear propagation, reflection, and refraction) equally well, but failed to agree on other effects. The particle theory failed to explain diffraction and interference, while, later, it was discovered that the wave theory could not provide sufficient explanation for effects such as black-body radiation and the photoelectric effect.

The breakthrough was the establishment of quantum optics. Light comes in a stream of quanta, or photons, whose energy $E$ is described by the equation $E = h \cdot \nu$, where $\nu$ is the frequency of the light wave, and $h$ is Planck's constant. Photon energies are reported in the unit of electron volts (eV), where 1 eV is the amount of energy gained (or lost) by the charge of a single electron when it moves across an electric potential difference of 1 V. 1 eV is about 1.6 $\times 10^{-19}$ J (joules).

The photon energies that correspond to the visible range fall between 1.65 eV (the red) and 3.1 eV (the violet). A photon of wavelength 193 nm has an energy of 6.4 eV; it belongs to the UV range of the EM spectrum. This energy is sufficient to induce damage to living cells (an example of which is sunburns).

## Wave Properties of Light

As a wave, light has a wavelength $\lambda$ and a frequency $\nu$. Visible light is in the narrow portion of the electromagnetic spectrum with wavelengths between 400 nm (0.4 μm) and 700 nm (0.7 μm). In the immediate vicinity of visible light, we have the shorter-wavelength ultraviolet (UV) with $\lambda_{UV} < 0.4$ μm and the longer-wavelength infrared (IR) with $\lambda_{IR} > 0.7$ μm.

Phase $\varphi$ is a wave property that describes the internal clock of the disturbance. Over the length of a wavelength, the phase changes from 0° to 360° or from 0 to $2\pi$ rad.

Just like any electromagnetic wave, light propagates in vacuum at the speed of light: $c = \nu \cdot \lambda = 3 \times 10^8$ m/s. In any other medium, light travels at a slower speed $u$. The ratio of the speed of light in vacuum to the speed of light in the medium is the index of refraction $n$. The index of refraction is a number (which later, in Chapter 3, we will revise) that takes values greater than 1.0. When light enters a medium of refractive index $n$, its wavelength $\lambda_n$ is reduced in comparison to the wavelength in vacuum $\lambda_o$: $\lambda_n = \lambda_o / n$.

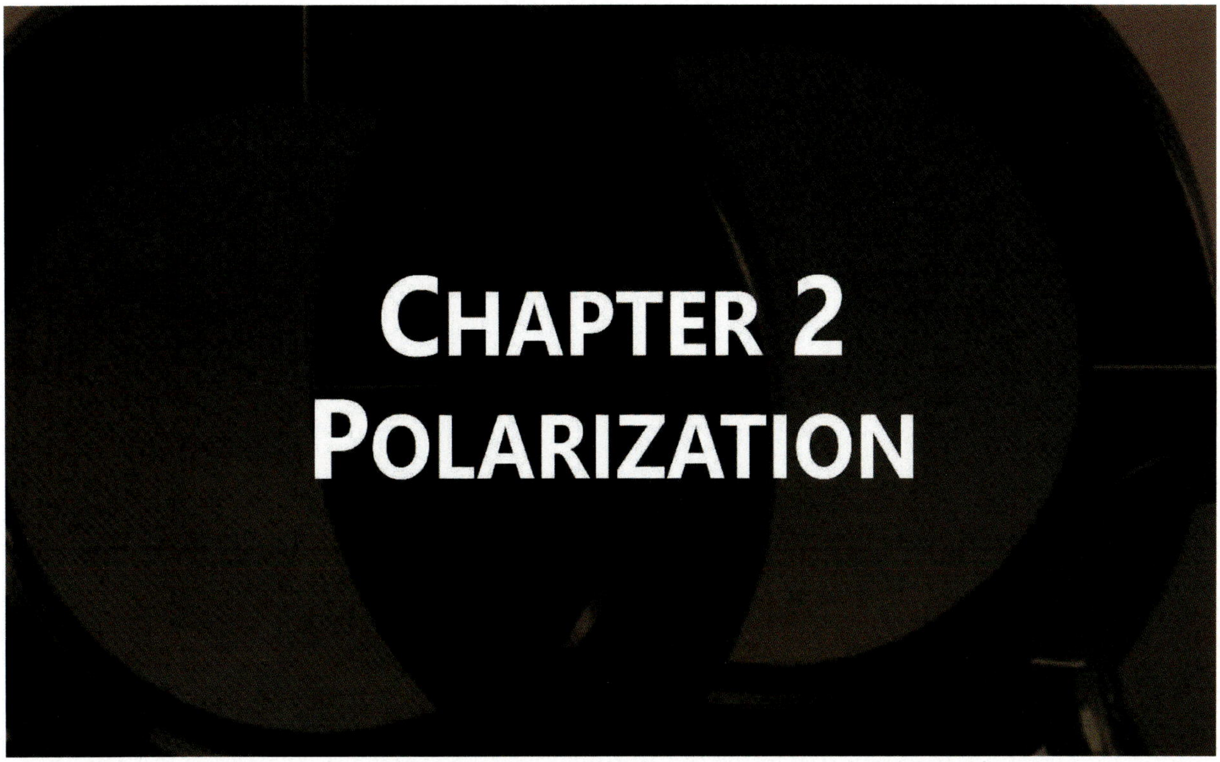

# CHAPTER 2
# POLARIZATION

## 2.1 LIGHT IS A TRANSVERSE WAVE

Light sensation results from an interaction between light, whose specifics are color (frequency) and intensity (photon population), and a sensory organ. What is recorded corresponds to a time-average of the light intensity, meaning that we lose some information pertaining to the vector nature of light, which is described by polarization.

In geometrical optics, it is not necessary to invoke the vector nature of light, which describes the orientation of the electric field associated with light. In addition, we are not concerned with the possible distinction between the corpuscular (photon particle) and the wave nature of light because both aspects offer essentially the same interpretations for most imaging phenomena. In interference and diffraction, the wave nature prevails, while in other effects, such as the photoelectric effect, the photon nature of light provides suitable answers.

Polarization can be described by either the wave nature or the photonic nature of light. The electromagnetic wave description uses transverse electric and magnetic field disturbances. In this case, polarization is determined by the orientation stability of the electric field vector. Using photon 'language,' polarization is associated with the stability of the spin of the photonic state. Traditionally, we explain polarization using the electromagnetic field vector description.

Polarization effects are responsible for a wide range of phenomena, such as the blue color of the sky, image formation in a liquid crystal display, and birefringence (§ 2.6.1) in certain anisotropic crystals.

The human eye cannot distinguish polarized light from nonpolarized light. To do so, we need to put on polarized sunglasses! There are insects whose sensory organs of vision are capable of making this distinction. Apparently, nature provides another class of color perception. Despite the fact that our eyes cannot follow the vector of the electric field as it oscillates over time and propagates in space, we may still wonder how we would sense light if we had that unique detection ability—perhaps quite differently from the way we do!

## 2.1.1 The Transverse Vector Nature of Light

Light can be described as a transverse, directional electromagnetic vector wave. The disturbance involves two mutually dependent vector fields, the electric field $\mathbf{E}$ and the magnetic field $\mathbf{H}$. The dependence of these two vector fields is described by the laws of electromagnetism (§ 1.1.4.1). The transverse nature means that these two vectors oscillate on a plane surface perpendicular to the direction of propagation. If we are looking directly at a ray, the fields are oscillating up and down, right to left, and anything between these directions, but not back and forth along the direction of propagation. The projection of the vectors $\mathbf{E}$ and $\mathbf{H}$ along the direction of propagation is always zero. Mathematically, this is described by the dot product of the vectors $\mathbf{E}$ (or $\mathbf{H}$) and $\mathbf{r}$, where $\mathbf{r}$ is the direction of propagation:

Transverse Vector Nature of Light: $$\mathbf{E}\cdot\mathbf{r} = 0 \quad \text{and} \quad \mathbf{H}\cdot\mathbf{r} = 0 \tag{2.1}$$

The dot product, appearing as (·) between two vectors, is the product of their magnitudes times the cosine of their relative angle. For nonzero magnitudes, their dot product is zero if the cosine of their relative angle is zero, which means that they are perpendicular (the directions of the two vectors form a 90° angle). Thus, Eq. (2.1) is a statement that $\mathbf{E}$ and $\mathbf{H}$ are perpendicular to the direction of propagation $\mathbf{r}$.

A simple wave expression for the electric field [Eq (1.6)] component of the wave is

$$\mathbf{E} = \mathbf{E}_o\cdot\cos(\omega\cdot t - \mathbf{k}\cdot\mathbf{r} + \varphi_o) \tag{2.2}$$

The detected quantity is the **intensity**, which is proportional to the square of the electric field according to the relationship in Eq. (1.11). It is a scalar, not a vector, quantity; i.e., there is no direction, only magnitude:

$$I = c\varepsilon_o\langle\mathbf{E}\cdot\mathbf{E}\rangle \propto E_o^{\,2} \tag{2.3}$$

Expressions (2.2) and (2.3) correspond to an ideal harmonic wave (presented in § 1.1.4) with a fixed amplitude magnitude and a fixed initial phase $\varphi_o$. The wave has a flat wavefront and has no beginning and no end. This harmonic wave is an idealized, yet convenient, model—by itself, such a wave does not exist in nature; however, this model is very useful. It is useful because an actual light wave consists of many harmonic waves with different frequencies, phases, amplitudes, oscillation field directions, etc.

In this particular wave, the vector of the electric field **E** oscillates parallel to a virtual line, the axis, which is the direction of the amplitude vector $\mathbf{E}_o$. As we will soon see (§ 2.2), this wave is, in fact, linearly polarized. The same applies to the magnetic field: Vector **H** also oscillates along a specific axis. These two axes are perpendicular to each other, forming an orthogonal system that also includes the direction of propagation. Because the two fields **E** and **H** are dependent on each other, we will only use the electric field; of course, the same conclusions about the state of polarization will apply if we describe only the magnetic field.

A wave with a known direction of propagation can be formed by a superposition of many ideal harmonic waves. These waves can vary in frequency ($\omega_1$, ... $\omega_N$), amplitude magnitude ($E_{o1}$, ... $E_{oN}$), initial phase ($\varphi_{o1}$, ... $\varphi_{oN}$), and wavevector magnitude, but still maintain the same direction of propagation. The wave is transverse, a property that restricts the orientation of the field oscillation planes; the vector of the electric field is always on a plane perpendicular to the direction of propagation, and it can oscillate in any direction, as long as it is on such a plane.

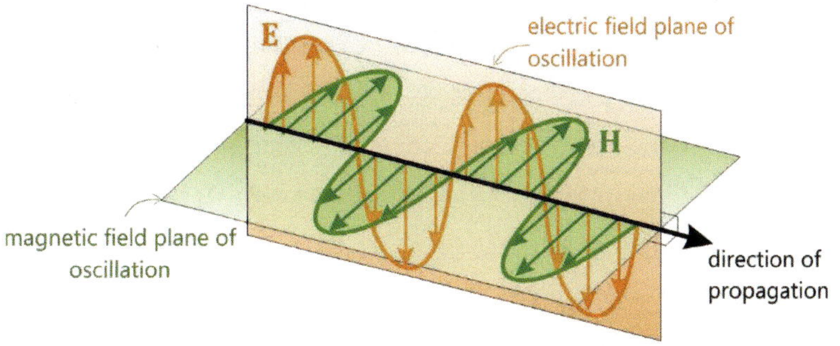


*Figure 2-1: Simple harmonic electromagnetic wave showing the electric field and the magnetic field.*

We can visualize such a wave using only the electric field (Figure 2-2). For simplicity, we assume just one wavelength and the same initial phase (unlike a real case) among all components. Such a wave may be the result of a rotation of the wave depicted in Figure 2-1; the rotation is restricted to the plane that is perpendicular to the direction of propagation. There could be an infinite number of such electric field vector orientations. The electric field can vary over time in a completely random way, both in magnitude and in the direction of oscillation, as long as this direction is restricted to a plane that is perpendicular to the fixed direction of propagation.

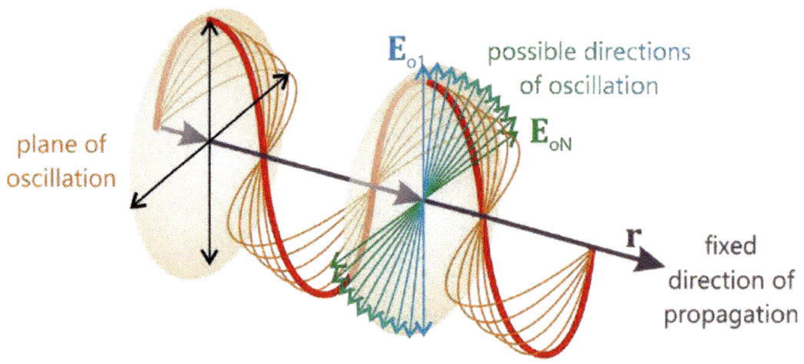


*Figure 2-2: The electric field of an electromagnetic wave has an infinite number of possible orientations, under one constraint: They all have to be perpendicular to the direction of propagation.*

This is **unpolarized** or **natural** light. The term 'natural' applies even to artificial light, such as the monochromatic light from a laser source.

**Unpolarized (natural) light**

Unpolarized, or natural, light exists when the electric field oscillates randomly.

This occurs along a plane perpendicular to the direction of propagation.

Consider light emitted by a source. Whatever the emission mechanism, a pulse or wave packet is emitted (a photon, if we use the particle aspect). This short pulse has a random initial electric field orientation. It could have been produced by grid oscillations in an incandescent lamp, or from spontaneous emission in a gas lamp. Even if it originates from stimulated emission in a laser (§ 6.1.3.2), where there is control of the initial phase and frequency, this light can be unpolarized (natural) because there are several possible modes of oscillation (see § 6.2.2.2) and therefore a very large number of electric field orientations. In unpolarized light, we cannot determine the specific axis of electric field oscillation.

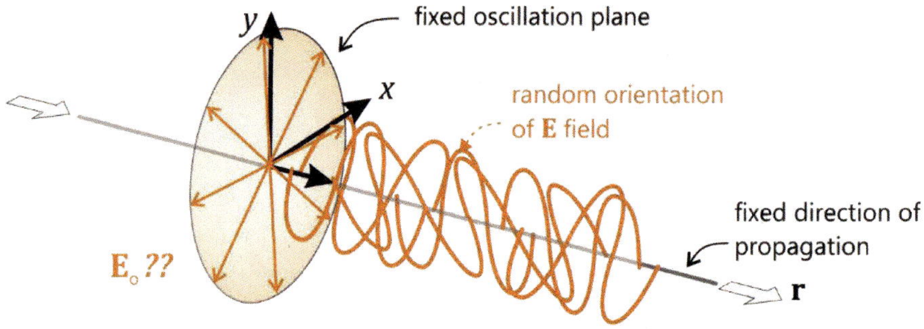


*Figure 2-3: In natural light, the electric field oscillates on a plane that is perpendicular to the direction of propagation; however, on this plane, we cannot identify the oscillation axis where the amplitude is aligned.*

If we could record several screenshots of the electric field, we would see the **E** vector vibrating on a plane perpendicular to the direction of propagation (denoted by **r** in Figure 2-3), but with no specific orientation. We cannot locate a specific oscillation orientation because the orientation changes rapidly and randomly—again, we emphasize—along the plane that is perpendicular to the direction of propagation. Another rule, deriving from exactly the random nature of this orientation, is that at any given time and point, equal amounts of the vector magnitude $E$ are projected along either the $-x$ coordinate or the $-y$ coordinate axis.

## 2.2 LINEARLY (PLANE-) POLARIZED LIGHT

If the oscillating electric field oscillates along only one direction (one axis) such that the orientation of the **E** vector is fixed at all points in space and time, then the wave is called **linearly** (or **plane**-) **polarized**.

**Plane-polarized light**

Linearly, or plane-, polarized light exists when the electric field oscillates on only one specific axis.

This axis is perpendicular to the direction of propagation because the wave is transverse.

This fixed oscillation direction of the electric field is the **polarization axis**. The plane defined by the direction of propagation and the polarization axis is the **polarization plane**.

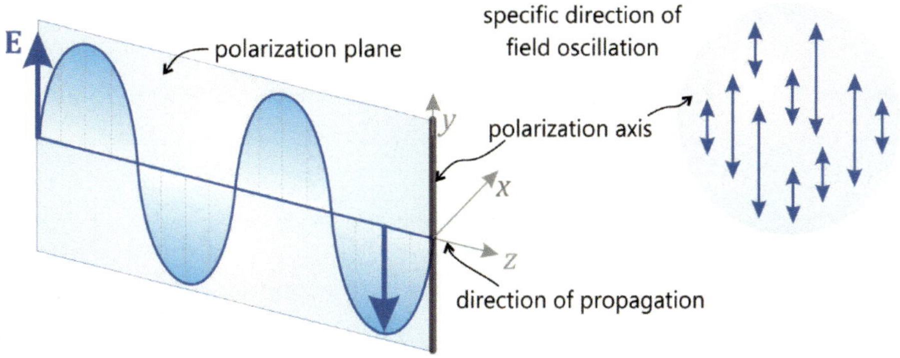


*Figure 2-4: Simple case of plane-polarized light. The wave propagates along the –z axis; the polarization axis is –y, and the polarization plane is y–z.*

The expression for the electric field associated with the wave depicted in Figure 2-4 is

$$\mathbf{E} = E_o \cdot \mathbf{y} \cdot \cos(\omega \cdot t - k \cdot z + \varphi_o) \tag{2.4}$$

This wave has a fixed amplitude vector $\mathbf{E}_{oy}$ with a specific direction of propagation –$z$. It is linearly polarized because the electric field oscillates along the –$y$ axis, the axis of polarization. The polarization plane is $y$–$z$. Here $\mathbf{y}$ indicates the unit vector along the –$y$ axis. Another simple case is

$$\mathbf{E} = E_o \cdot \mathbf{x} \cdot \cos(\omega \cdot t - k \cdot z + \varphi_o) \tag{2.5}$$

which is depicted in Figure 2-5. This wave also has a fixed amplitude vector $\mathbf{E}_{ox}$ with a specific direction of propagation –$z$, and it is linearly polarized because the electric field oscillates only along the –$x$ axis, which is the axis of polarization. Here $\mathbf{x}$ indicates the unit vector along the –$x$ axis. The only difference, in this case, is that the axis of polarization is –$x$ and the polarization plane is $x$–$z$.

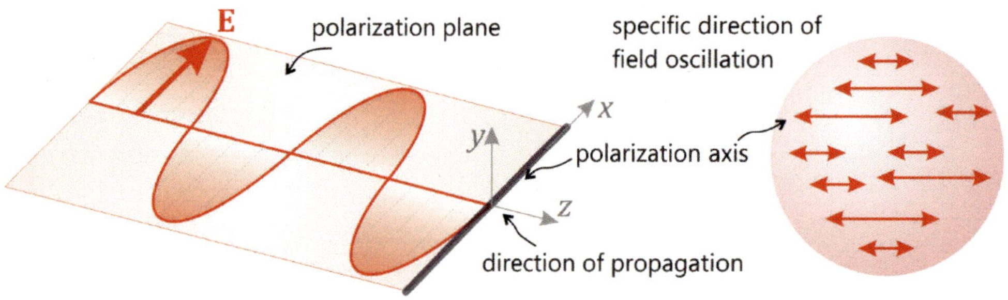


*Figure 2-5: Simple case of plane-polarized light. The wave propagates along the –z axis, the polarization axis is –x, and the polarization plane is x–z.*

In the general case of a linearly polarized wave, the polarization axis may not coincide with either –$x$ or –$y$. The wave may be a combination of both –$x$ and –$y$ components. The general expression of the electric field for linearly polarized light propagating along the –$z$ axis is therefore

$$\mathbf{E} = (E_{ox} \cdot \mathbf{x} + E_{oy} \cdot \mathbf{y}) \cdot \cos(\omega \cdot t - k \cdot z + \varphi_o) \tag{2.6}$$

which suggests that the electric field can have –$x$ and –$y$ components, but no –$z$ component.

Such a wave can be derived from two mutually perpendicular (orthogonal), linearly polarized waves, which can be expressed as

$$\mathbf{E}_x = E_{ox} \cdot \mathbf{x} \cdot \cos(\omega \cdot t - k \cdot z + \varphi_o) = (E_o \cdot \cos\alpha) \cdot \mathbf{x} \cdot \cos(\omega \cdot t - k \cdot z + \varphi_o) \text{ and}$$

$$\mathbf{E}_y = E_{oy} \cdot \mathbf{y} \cdot \cos(\omega \cdot t - k \cdot z + \varphi_o) = (E_o \cdot \sin\alpha) \cdot \mathbf{y} \cdot \cos(\omega \cdot t - k \cdot z + \varphi_o) \tag{2.7}$$

These waves must have the same initial phase ($\varphi_{ox} = \varphi_{oy}$); therefore, $\delta\varphi = (\varphi_{oy} - \varphi_{ox}) = 0$, or, generally, $\varphi_{ox} = \varphi_{oy} \pm m\pi$, so $\delta\varphi = \pm m\pi$, where $m$ is an integer. Here an angle (tilt) $\alpha$ is formed between the polarization axis and the $-y$ axis, and is defined as

$$\tan\alpha = \frac{E_{ox}}{E_{oy}} \tag{2.8}$$

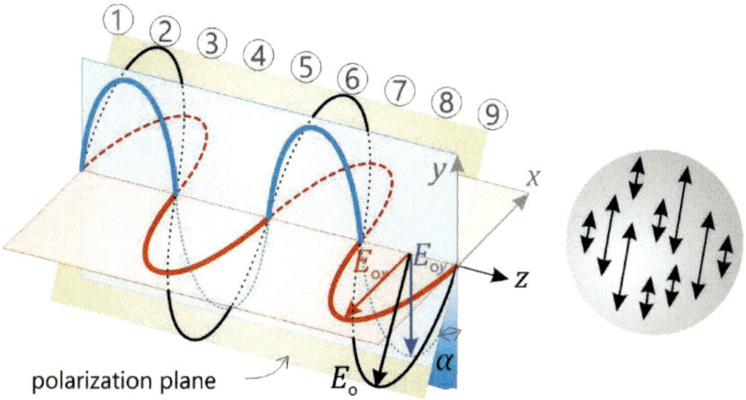


*Figure 2-6: Synthesis of linearly polarized light from two orthogonal components with a zero phase difference.*

The composition of the two vectors is illustrated in Figure 2-6 and produces a linearly polarized wave. Note that the electric field oscillates only along a specific axis, indicated by the red vector.

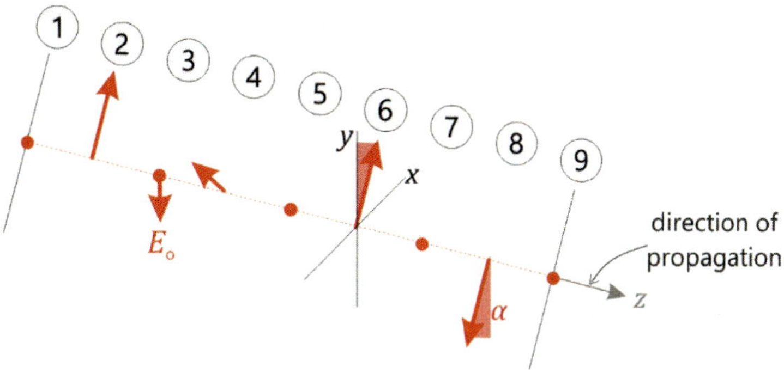


*Figure 2-7: The electric field vector resulting from the synthesis of the waves shown in Figure 2-6 has a specific direction (axis) of oscillation, forming an angle $\alpha$ with the $-y$ axis.*

Example ☞:

A ray of natural light and a ray of linearly polarized light propagate along the same direction $-z$. They have the same color and equal intensity.

(i) For linearly polarized light, the electric field may have $\mathbf{E}_x$ and $\mathbf{E}_y$ components, but no $\mathbf{E}_z$ component. True or False?

(ii) For linearly polarized light, the electric field may have $\mathbf{E}_y$ and $\mathbf{E}_z$ components, but no $\mathbf{E}_z$ component. True or False?
(iii) For natural (unpolarized) light, the electric field may have $\mathbf{E}_x$ and $\mathbf{E}_y$ components, but no $\mathbf{E}_z$ component. True or False? [Bonus question: What is different about example (i)?]

Both beams propagate along the $-z$ axis. Let us not forget that light is a transverse wave, so the directions of oscillations from the electric field are always perpendicular to $-z$, on either $-x$ or $-y$, or any combination of both. Therefore, (i) is true, and (ii) is false. There cannot be a $-z$ component.
Since the 'natural' beam propagates along $-z$, the electric field may lie along either $-x$ or $-y$, or any combination, but no $-z$ component is allowed. Therefore, (iii) is true. What is different about (i) is that, whatever orientation the electric field in polarized light has, that orientation is fixed, while for natural light, it varies randomly.

We emphasize that the necessary condition under which a linearly polarized wave can be formed is that the two components must have a fixed phase difference $\delta\varphi = (\varphi_{oy} - \varphi_{ox})$ equal to 0 or a multiple integer of $\pi$ rad. The odd multiples of $\pi$ simply indicate an inverted oscillation direction of the corresponding component. The amplitudes of the two components are not required to be equal. In each case, the fixed oscillation direction of the electric field is along the hypotenuse of the triangle composed of $\mathbf{E}_x$ and $\mathbf{E}_y$.

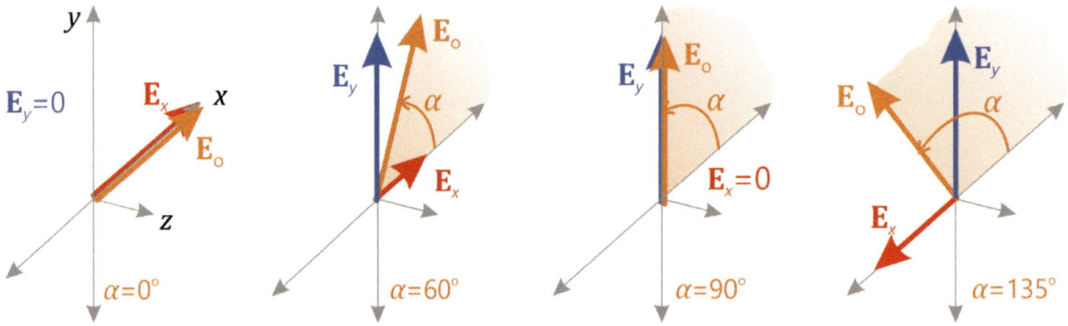


*Figure 2-8: Analysis of linearly polarized light into two components along the $-x$ and $-y$ axes. The wave propagates along the $-z$ axis.*

If the amplitude magnitudes are equal, the resulting amplitude forms an angle $\alpha = \pm 45°$ with the axes. If the amplitude magnitudes are not equal, the resulting amplitude and the $-x$ axis form an angle (tilt) of any value.

The relationships in Eqs. (2.6) and (2.7) work either way: Just as the two linearly polarized waves were composed to produce a polarized wave in Eq. (2.6), we can decompose a linearly polarized wave into two linearly polarized components in select orthogonal axes. These two components propagate along the same direction and have the same frequency and wavelength as the original wave. The phase difference between them is fixed in time and space, and equals 0 or a multiple integer of $\pi$ rad.

We can also consider natural light composed of two perpendicularly polarized waves with equal amplitude magnitudes, whose phase difference changes rapidly and randomly. The requirement of equal amplitude magnitude is necessary because there is no preferred direction of oscillation, so, in any given reference system, the two components (projections) of the electric field are equal. The random and rapidly changing orientation of the electric field is a result of the random and rapidly changing phase difference between the two components.

## 2.2.1 Partially Polarized Light

A state of absolute polarization is ideal; however, in reality, linearly polarized light is never fully polarized. Electric fields can be almost parallel, but not exactly. Equivalently, there may be a preferred direction of oscillation of the electric field, but this might not be the only direction of oscillation. There is thus always a mix of natural (unpolarized) and polarized light; this is the case of **partially polarized** light. The total intensity $I_{\text{TOT}}$ is the sum of an unpolarized [nonpolarized (NP)] component $I_{\text{NP}}$ and a linearly polarized component $I_{\text{P}}$.

The **degree of** (partial) **polarization** $p$ is defined as the ratio of the intensity of the linearly polarized component to the total intensity:

Degree of Polarization:
$$p = \frac{I_{\text{P (polarized)}}}{I_{\text{TOT (total)}}} = \frac{I_{\text{P}}}{I_{\text{P}} + I_{\text{NP}}} \tag{2.9}$$

The degree of polarization is a dimensionless physical quantity with values ranging from 0.0 to 1.0. The value of $p = 0.0$ corresponds to the completely unpolarized state. If light is fully linearly polarized, the degree of polarization is $p = 1.0$. In the case of partial polarization, $0.0 < p < 1.0$. In § 2.3.2 we present a way to measure the degree of polarization.

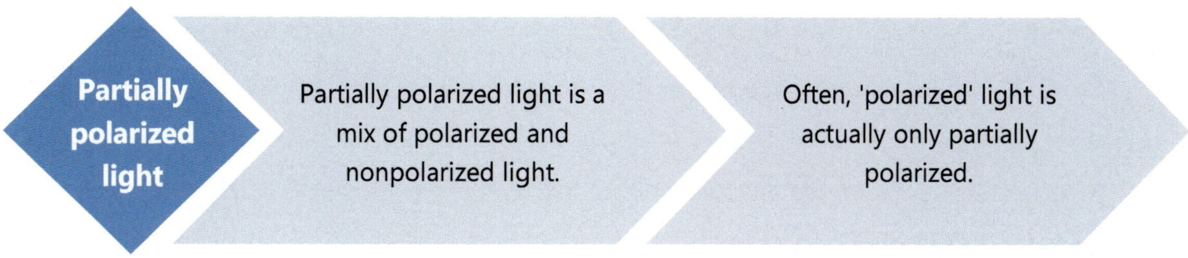

# 2.3 FROM UNPOLARIZED TO POLARIZED LIGHT

## 2.3.1 Creation of Linearly Polarized Light

We cannot restrict the electric field to oscillate along a select, specific direction. However, it is relatively easy to allow propagation of the portion of light in which the electric field oscillates along a particular direction. Thus, we can create linearly polarized light. This can be achieved with a linear polarizer (see § 2.3.2.1). Consider a fictional system of parallel bars that allow only oscillations parallel to the direction of the gaps between the bars. Oscillations perpendicular to the gaps are completely cut off (attenuated). This direction denotes the **polarization axis**.

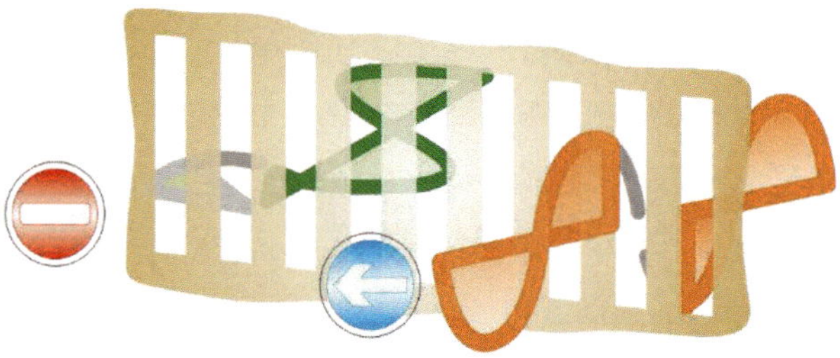

*Figure 2-9: A linear polarizer allows for (left) complete cut off and (right) complete crossing of oscillations.*

If light with a random electric field encounters a linear polarizer, the transmitted light has only one specific direction of oscillation—parallel to the axis of the polarizer. Light oscillating along a perpendicular direction is eliminated. Thus, the transmitted light is linearly polarized with the polarization plane formed by the axis of the polarizer and the direction of propagation.

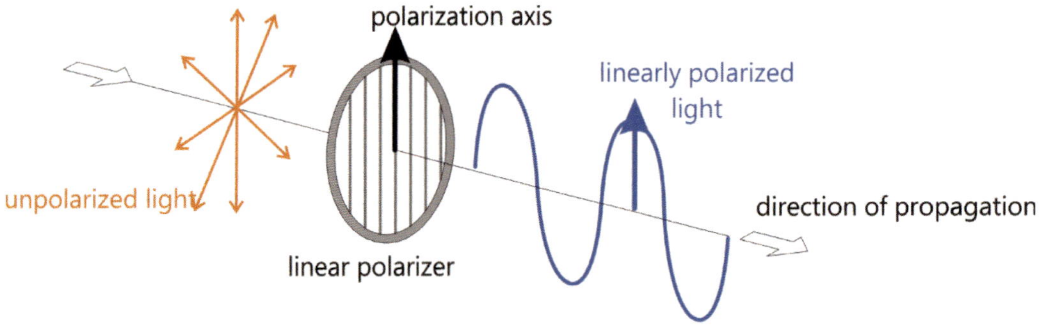


*Figure 2-10: Conversion from unpolarized light to linearly polarized light.*

If we rotate the polarization axis on the perpendicular plane by angle $\vartheta$, then the polarization axis of the transmitted light is parallel to that angle. The polarization plane (Figure 2-11) follows the rotation of the linear polarizer.

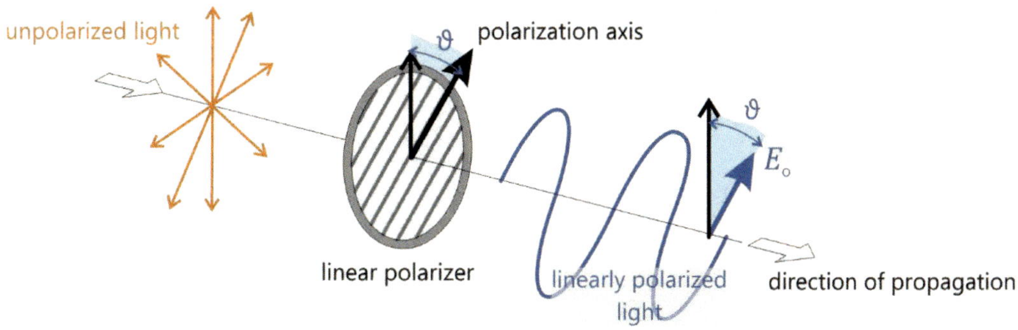


*Figure 2-11: Rotation of the polarization plane resulting from a linear polarizer rotation.*

## 2.3.2 Detection of Linearly Polarized Light

How can we determine if the light transmitted through a polarizer is indeed linearly polarized? The difficulty in detecting polarized light is that the eye, as well as any light detector, does not record the electric field **E**, but rather the associated intensity $I$. It is impossible to record the electric field, which is a vector that oscillates with a period on the order of magnitude of $10^{-14}$ s.

We therefore use an additional linear polarizer, called an **analyzer**, which is placed after the first polarizer. Light leaving the first polarizer is incident on the analyzer. The intensity of the light exiting the analyzer depends on the state of polarization of the light entering the analyzer.

First Case: The polarizers are parallel to each other. Let's set their axes oriented vertically.

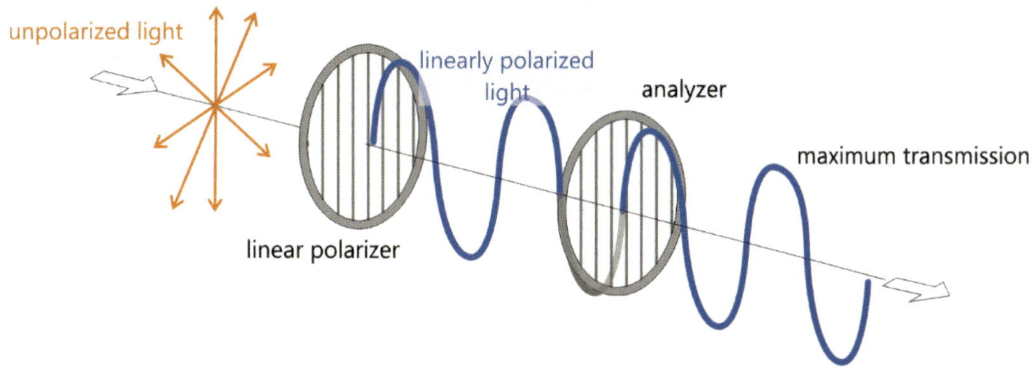


*Figure 2-12: Detection of linearly polarized light under the condition of maximum detected intensity. The polarization plane and the analyzer axis are parallel.*

Light becomes vertically linearly polarized when passing through this linear polarizer, regardless of the initial polarization state, even if it was unpolarized. This polarized light is incident on the analyzer and passes through without further loss; i.e., it is already polarized in full compatibility with the second polarizer, being both vertical. There is therefore maximum detected intensity.

Second Case: The polarizers are crossed, which means that they form a right (90°) angle.

Let's set the first axis oriented vertically and the second oriented horizontally. Light passing through the first polarizer becomes vertically linearly polarized. This polarized light is incident on the analyzer and does not pass at all.

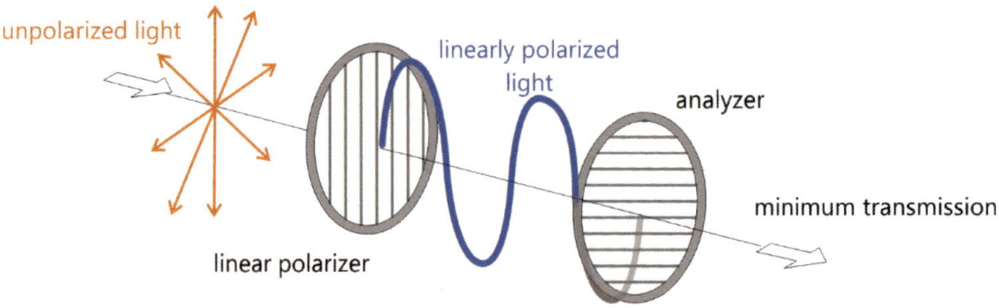


*Figure 2-13: Detection of linear polarized light under the condition of minimum detected intensity. The polarization plane and the analyzer axis are perpendicular.*

We now rotate the analyzer a full circle. When the analyzer is parallel to the polarizer, it allows complete transmission of the linearly polarized light. When the analyzer is at a right angle to the polarizer, there is complete attenuation. The four positions alternate every 90°. For $\vartheta = 0°$ or 180°, there is complete attenuation, and for $\vartheta = 90°$ or 270°, there is no attenuation.

Third case: The polarizers form a random angle $\vartheta$.

The transmitted intensity depends on the relative angle $\vartheta$ formed by the two polarizers. The analyzer allows transmission only of the electric field component (projection) that is parallel to its axis. The incident beam has an amplitude of magnitude $E_{\mathrm{oI}}$, and the transmitted beam has an amplitude of magnitude $E_{\mathrm{oT}}$. These two magnitudes are related according to

$$E_{\mathrm{oT}} = E_{\mathrm{oI}} \cdot \cos\vartheta \tag{2.10}$$

The detected intensity is proportional to the square of the electric field [see Eq. (2.3)]:

$$I_{\mathrm{T}} = (E_{\mathrm{oT}})^2 = (E_{\mathrm{oI}} \cdot \cos\vartheta)^2 = E^2_{\mathrm{oI}} \cdot \cos^2\vartheta = I_{\mathrm{oI}} \cdot \cos^2\vartheta \tag{2.11}$$

This is Malus' law, stated in 1809 by the French physicist and mathematician Étienne-Louis Malus:

Malus: The light intensity passing through two successive polarizers is proportional to the cosine squared of the relative angle of the two polarization axes.

Malus' law is fully compatible with the conditions for $\vartheta = 0$ or $\pi$ (0° or 180°), resulting in maximum transmission ($I_T = I_I$); for $\vartheta = \pi/2$ or $3\pi/2$ (90° or 270°), the intensities are zero ($I_T = 0$), resulting in complete attenuation (minimum detected intensity).

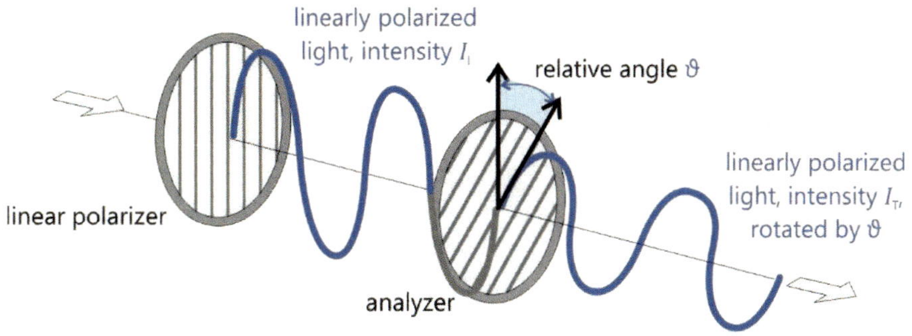


*Figure 2-14: Polarizer and analyzer oriented according to a relative angle $\vartheta$.*

If the polarizer is fixed in place and the analyzer is rotated (Figure 2-14), we observe periodic variations in the transmitted beam intensity. The variation in relative intensity (the ratio of the detected intensity to the maximum detected intensity) is proportional to $\cos^2\vartheta$, according to Malus' law.

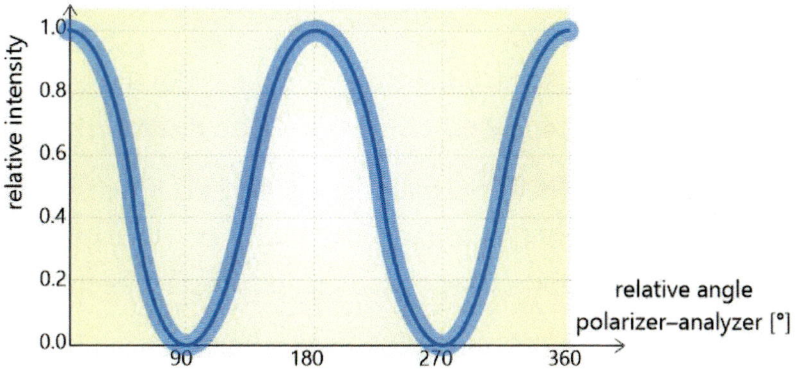


*Figure 2-15: Simulation of the variation in the relative intensity of linearly polarized light as the analyzer is rotated.*

Example ☞: Natural (unpolarized) light falls upon two cascaded polarizers. Their relative angle is $\vartheta = 45°$. What percentage of the initial light exits the second polarizer?

Assume that 100 (arbitrary) units of light intensity are incident on polarizer ❶. Light leaving this polarizer is linearly polarized and has 50 units of light intensity. The axis of this polarization is parallel to that of the polarizer, which we set at 0°. This light is incident on polarizer ❷, whose axis forms an angle of 45°. Light exiting this polarizer is linearly polarized at 45°. It has $50 \cdot \cos^2(45°) = 50 \cdot (0.707)^2 = 50 \cdot 0.5 = 25$ units. Therefore, 25% of the light that was initially incident on the first polarizer exits the second polarizer.

Example☞: Light leaving a linear polarizer whose axis is vertical is incident on an analyzer whose axis forms an angle $\vartheta$ = 30° with the horizontal. What percentage of the light exits the analyzer?

Assume that 100 units of light intensity are leaving the polarizer. This light is vertically linearly polarized and is incident on the analyzer whose axis forms a relative angle of 60° with the vertical. Light leaving this analyzer is linearly polarized at 60° with the vertical and has $100 \cdot \cos^2(60°) = 100 \cdot 0.5^2 = 100 \cdot 0.25 = 25$ units. Therefore, 25% of the light passes through the analyzer.

What happens if natural light, rather than polarized light, is incident on a polarizer–analyzer? At any given moment, the electric field vector forms an angle $\vartheta$ with the axis of the polarizer. Malus' law predicts that the transmitted intensity is proportional to $\cos^2\vartheta$. Because $\vartheta$ is random and rapidly changing, what we record is the time average of $\cos^2\vartheta$, which is simply 0.5. To prove this, we use the trigonometric relationship $\cos^2\vartheta = ½ + ½ \cos(2\vartheta)$:

$$\left\langle \cos^2 \vartheta \right\rangle = \frac{1}{2} + \frac{1}{2}\left\langle \cos \vartheta \right\rangle = \frac{1}{2} + 0 = \frac{1}{2} \tag{2.12}$$

Thus, if natural light passes through a polarizer, the intensity leaving the polarizer is fixed, equal to ½ of the initial light for any orientation (angle $\vartheta$):

Detection of Unpolarized Light: $$I_{MAX} = I_{MIN} = \frac{1}{2} I_{TOT} \tag{2.13}$$

This is one way of detecting natural light, but it is not absolute. The same behavior is displayed by circularly polarized light (see § 2.4); § 2.4.3 presents a way to distinguish between the two.

Complete attenuation (zero detected intensity) occurs only if the two polarizers are ideal, or if completely linearly polarized light is incident on an analyzer. For partially polarized light, the minimum detected light does not correspond to complete attenuation.

Suppose now that partially polarized light is incident on the analyzer with the linearly polarized component along the 0° axis. An analyzer at $\vartheta$ = 0° or 180° allows transmission of only ½ of the natural light ($I_\Phi$) and the entire linearly polarized component.

When the analyzer is at $\vartheta$ = 90° or 270°, (again) only ½ of the natural, unpolarized light ($I_{NP}$) is allowed to be transmitted—this occurs for any $\vartheta$—and none of the polarized component is allowed. The maximum and minimum detected light intensity can be expressed as

Detection of Partially Polarized Light: $$I_{MAX} = \frac{1}{2} I_{NP} + I_P \quad \text{and} \quad I_{MIN} = \frac{1}{2} I_{NP} \tag{2.14}$$

rearranged as $$I_{NP} = 2 \cdot I_{MIN} \quad \text{and} \quad I_P = I_{MAX} - I_{MIN} \tag{2.15}$$

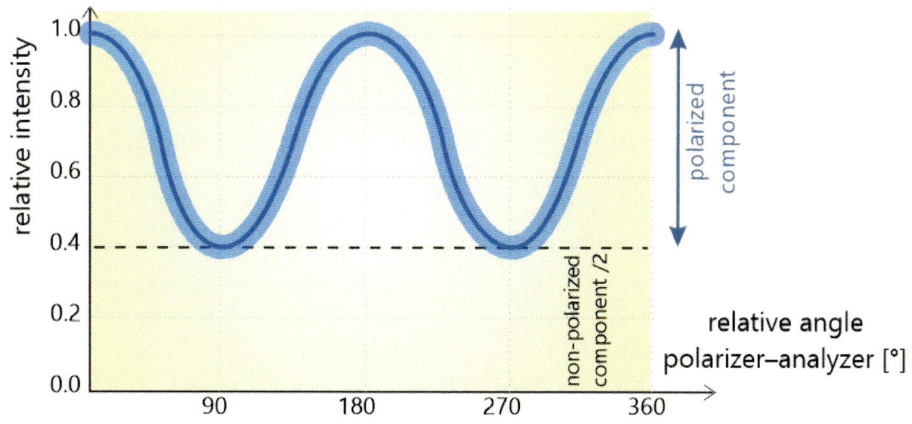


*Figure 2-16: Variation in the relative intensity of partially polarized light as the analyzer is rotated.*

Example ☞: We can calculate the degree of polarization $p$ from the data shown in Figure 2-16. We use the maximum ($I_{MAX}$ = 1.0) and the minimum ($I_{MIN}$ = 0.4) observed intensities:

$I_{NP} = 2 \cdot I_{MIN} = 2 \cdot 0.4 = 0.8$ and $I_P = I_{MAX} - I_{MIN} = 1.0 - 0.4 = 0.6$

Therefore, the degree of polarization [see Eq. (2.9)] is: $p = I_P / I_{TOT} = I_P / (I_P + I_{NP}) = 0.6/(0.6+0.8) \approx 0.43$.

**To detect linear polarization, we place an analyzer in front of the beam.**

We observe the intensity of the light transmitted through the analyzer as we rotate the analyzer.

- If fluctuations are observed, the light is linearly polarized.
  - Polarization is complete if there is zero minimum transmission at two opposite locations.
  - Polarization is partial if the minimum transmission is not zero.
  - The axis of polarization is ∥ to the direction of maximum transmission and ⊥ to the direction of minimum transmission (darkness).
- If no fluctuations are observed, the light is either unpolarized or circularly polarized.

Practice with partial polarization:

Maximum $I_{MAX}$ = 1.0; minimum $I_{MIN}$ = 1.0 ⇒ $I_{NP}$ = 2.0, $I_P$ = 0.0, $p$ = 0.0 (completely nonpolarized).

Maximum $I_{MAX}$ = 1.0; minimum $I_{MIN}$ = 0.75 ⇒ $I_{NP}$ = 1.50, $I_P$ = 0.25, $p$ = 0.14 (partially polarized).

Maximum $I_{MAX}$ = 1.0; minimum $I_{MIN}$ = 0.50 ⇒ $I_{NP}$ = 1.00, $I_P$ = 0.50, $p$ = 0.33 (partially polarized).

Maximum $I_{MAX}$ = 1.0; minimum $I_{MIN}$ = 0.00 ⇒ $I_{NP}$ = 0.00, $I_P$ = 1.00, $p$ = 1.00 (completely linearly polarized).

### 2.3.2.1 The Linear Polarizer

Polarizing materials, known as Polaroid™, use polymeric membranes invented by Edwin H. Land in 1930. These materials contain iodoquinine sulfate microcrystals specifically oriented to selectively absorb the components of a particular oscillation direction of the electric field, which corresponds to the attenuation axis. Light with an electric field oscillation perpendicular to this axis passes with no loss at all—this is the transmission axis of the polarizer. This anisotropic optical absorption property is called **dichroism**.

Tourmaline, a material that exhibits dichroism, was the original Polaroid sheet material. Today, Polaroid sheets contain poly(vinyl alcohol) (PVA) macromolecules, which absorb the electric field parallel to them. Certainly, there is neither absolute absorption nor ultimate transmission. The ratio of the intensity transmitted parallel to the polarizer axis to the intensity transmitted perpendicular to the polarizer axis is the **extinction coefficient**. A high-quality polarizer will have an extinction coefficient that is as large as possible. The term **polarizer**, proposed by A. Francis Hallimond in his work on the polarizing microscope, is generally used for any device that creates linearly polarized light. Polarizing sunglasses are actually filters with built-in linear polarizers whose axes are typically vertical.

*Figure 2-17: Crossed linear polarizers. Near-zero light passes through the area where the polarizers overlap at right angles.*

It is interesting to observe partially polarized light through such a polarizer. The sky appears more blue (see § 2.5.1), and reflection from a glass (see § 2.5.2.1) or water surface is eliminated under specific angles. Perhaps of more interest is the complete dimming of the screens of our mobile phones or of a liquid crystal display, such as the LCD screens in modern televisions and computers, because light from such screens is fully linearly polarized (see § 2.6.3).

Polarizing sunglasses (shades) incorporate a set of polarizing filters overlaid on the regular shade filter/lens material. The filters are oriented such that they block the horizontally polarized light component since this is the dominant linear polarization of the natural light reflecting at an angle close to Brewster's angle (to be discussed in § 2.5.2).

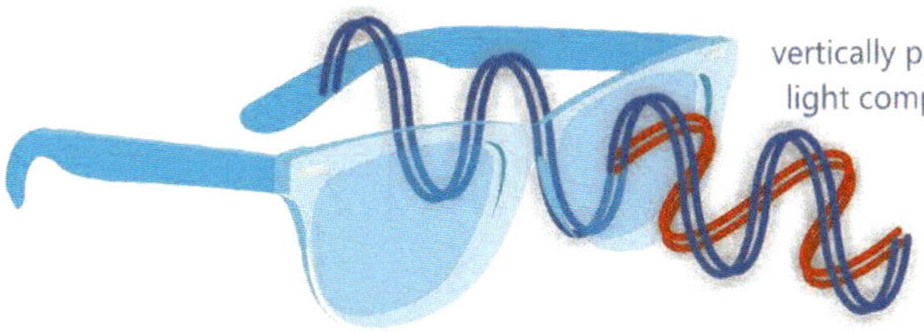


*Figure 2-18: Polarized sunglasses.*

Polarized stereograms are used in the evaluation of stereopsis, the perception of depth produced by the reception in the brain of slightly different visual stimuli from each eye. A polarized stereogram consists of two images encoded for viewing via crossed polarizers. Through a spectacle set of crossed polarizers, a stereogram appears as one image to one eye and as another image to the other eye.

A vectograph is a polarized stereogram, in which one-half of a chart is seen by one eye and the other half is seen by the other eye, while some lines, letters, or numbers are seen binocularly to lock fusion. The vectograph is useful for balancing ocular refraction and to detect ocular suppression and fixation disparity. The Titmus stereo test (Figure 2-19) consists of various vectographs, including one with a stereoscopic pattern representing a fly, to establish gross stereopsis.

*Figure 2-19: Use of polarizing reading glasses for viewing the Titmus stereo test.*

Photography is another application of polarization, in which polarizers are placed as threaded filters in front of the camera lens. Much like polarizing sunglasses, these filters act as analyzers for the incoming polarized light. Thus, they are able to nearly attenuate the light's polarized components. For example, as will be discussed in § 2.5.1, the sky blue that is at right angles to the light from the sun (not the entire sky) is strongly polarized. Therefore, the attenuation of the bright blue renders a deep dark blue with dramatic contrast, particularly in a sky strewn with white clouds. Additionally, polarizing filters can eliminate undesired reflections, such as those from a glass plate or water surface, again, because at a certain reflection angle (Brewster's angle § 2.5.2), the reflected light is strongly polarized. A polarizer crossed with that light can eliminate reflections completely.

Such filters drop the exposure by a stop.[5] This is because they absorb, on average, 50% of the natural light. Polarizing filters should be avoided when they do not offer any advantage, such as when the sky is overcast, or in low-light and indoor photography.

*Figure 2-20: Polarizing filters used in photography improve the contrast of a blue sky.*

Modern single-lens reflex (SLR) cameras use built-in photometry to evaluate the exposure. A portion of the available image is directed to a photometer via a prism, which polarizes this part of the image. A linear polarizing lens can affect this exposure metering, which is why circular polarizing filters (presented in § 2.4.2) are recommended instead. These filters consist of a linear polarizer and a $\lambda/4$ retardation plate. They retain all of the advantages of a linear polarizer. Since the light directed to the photometer is circularly polarized, any possible polarization by the internal prism does not affect the exposure metering.

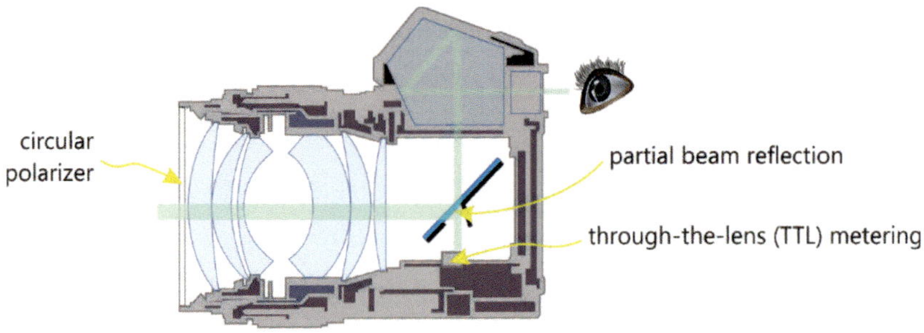


*Figure 2-21: Polarizing filters in a single-lens reflex (SLR) camera.*

[5] *Introduction to Optics* § 6.2.3 Light Control in a Photography Camera.

# 2.4 CIRCULARLY POLARIZED LIGHT

Circularly polarized light is not easily encountered in nature. For certain insects (scarab beetles), the light reflected off their exoskeleton is fully left-circularly polarized if the incident light is linearly polarized.

## 2.4.1 The Components of Circularly Polarized Light

Despite its name, circularly polarized light can result from the addition of two orthogonal, linearly polarized waves with equal amplitude magnitudes ($E_{ox} = E_{oy} = E_o$) and a fixed phase difference $\pm\pi/2$, i.e., $\varphi_{ox} = 0$ and $\varphi_{oy} = \pm\pi/2$. In a very simple case, the two orthogonal waves can be

$$\mathbf{E}_x = E_{ox}\cdot\mathbf{x}\cdot\cos(\omega\cdot t - k\cdot z) \quad \text{and} \quad \mathbf{E}_y = E_{oy}\cdot\mathbf{y}\cdot\cos(\omega\cdot t - k\cdot z \pm\pi/2) = \mp E_{oy}\cdot\mathbf{y}\cdot\sin(\omega\cdot t - k\cdot z) \quad (2.16)$$

These waves propagate in such a way that one is at its maximum when the other is at its minimum: $E_x = E_o$ and $E_y = 0$ for $\omega\cdot t - k\cdot z = 0$, and $E_x = 0$ and $E_y = E_o$ for $\omega\cdot t - k\cdot z = \pi/2$.

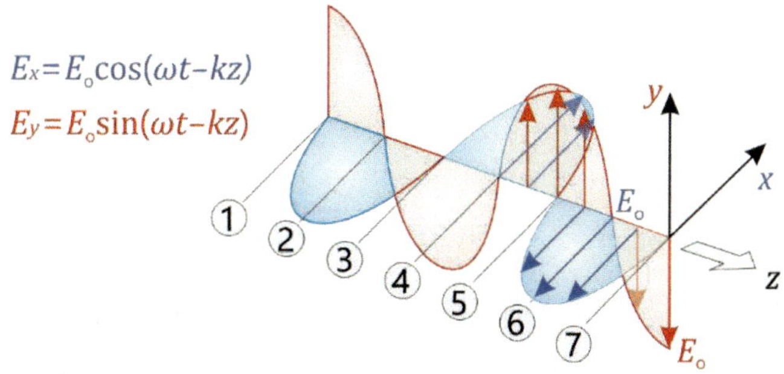


*Figure 2-22: Perpendicular linearly polarized waves with equal amplitude magnitudes and a phase difference between the two electric field vectors of $\pi/2$.*

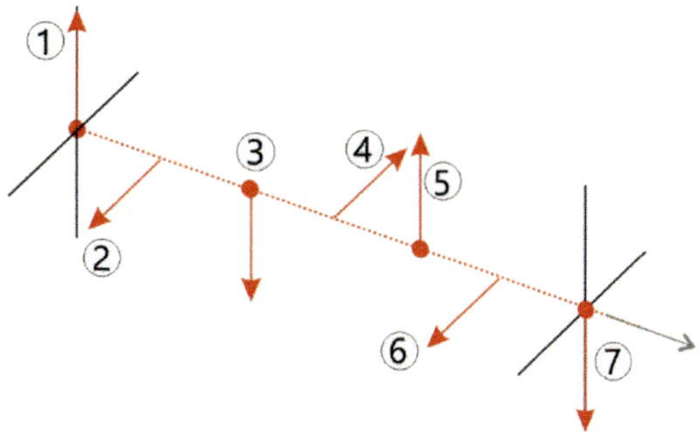


*Figure 2-23: Time development of the resultant electric field vector in circularly polarized light.*

At any given time $t$ and any point $z$, the electric field resulting from the vector sum is

$$\mathbf{E} = \mathbf{E}_x + \mathbf{E}_y = E_{ox}\cdot\mathbf{x}\cdot\cos(\omega\cdot t - k\cdot z) \mp E_{oy}\cdot\mathbf{y}\cdot\sin(\omega\cdot t - k\cdot z) \quad (2.17)$$

The electric field vector is perpendicular to the direction of propagation ($-z$) and has a fixed magnitude $E_o$. Its tip runs the circumference of a circle with radius $E_o$, rotating with angular velocity $\omega$. This is **circularly polarized light**.

**Circlarly polarized light**

In circularly polarized light, the electric field circumscribes a circle on a plane perpendicular to its direction of propagation.

It may result from a combination of two orthogonal, linearly polarized waves with equal amplitudes and a phase difference of $\pm\pi/2$.

The angle (slope) $\alpha$ that the combined vector of the electric field (Figure 2-24) forms with the $-x$ axis at any point $z$ is

$$\tan\alpha = E_y / E_x = \mp \tan(\omega \cdot t - k \cdot z) \quad \text{or} \quad \alpha = \mp (\omega \cdot t - k \cdot z) \qquad (2.18)$$

At a given point along the axis of propagation (simplest case, $z = 0$), where the $-y$ component phase difference is $+\pi/2$, slope $\alpha$ decreases steadily with time (by $\omega \cdot t$). Therefore, the vector of the electric field rotates in a clockwise fashion. This is **right-circularly polarized** (RCP) light.

In the case where the $y$-component phase difference is $-\pi/2$, slope $\alpha$ increases steadily with time. Therefore, the vector of the electric field rotates in a counterclockwise fashion. This is **left-circularly polarized** (LCP) light [Figure 2-24 (left)].

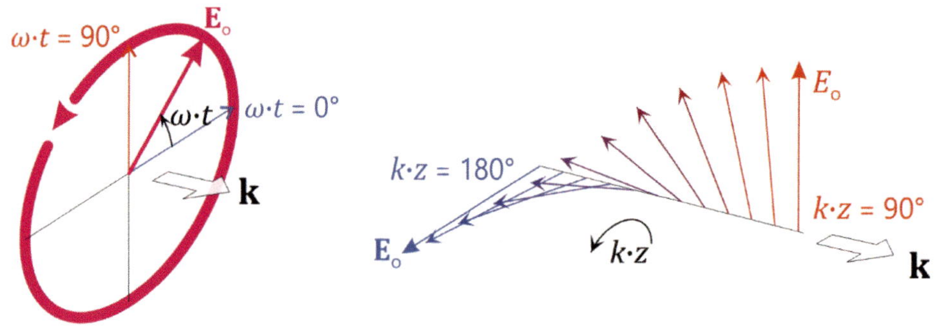


*Figure 2-24: Evolution of the electric field vector* **E** *for left-circularly polarized light (left) over time (z is fixed) and (right) over space (t is fixed).*

For observation at a given point in time (simplest case, $t = 0$), we draw a snapshot of LCP light. If we plot the vector for the same time at different points along the $-z$ axis, the value of the angle formed by the vector $\mathbf{E}$ decreases steadily with the distance [by $k \cdot z = (2\pi/\lambda) \cdot z$]. Thus, the tip of vector **E** circumscribes a clockwise helix, with a step inversely proportional to the wavelength [(Figure 2-24 (right)]. For RCP light, the corresponding helix is left-handed.

What if the conditions were relaxed a bit? Let's look at the general case of two orthogonal, linearly polarized waves, with unequal amplitude magnitudes ($E_{ox} \neq E_{oy}$) and for which the phase difference $\varphi$ can have any value. The composition of such an electric field vector results in an ellipse [Figure 2-25 (right)] over a plane that is perpendicular to the direction of propagation.

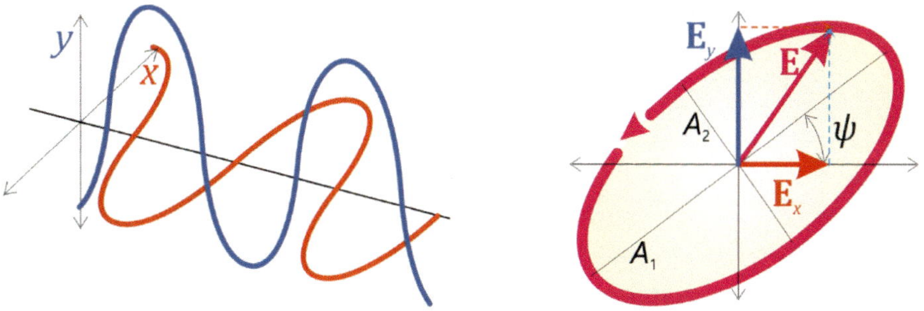


*Figure 2-25: Electric field for left (counterclockwise) elliptically polarized light.*

The ellipse's major axis forms an angle $\psi$ ($0 \leq y < \pi$) with the $-x$ axis, which is determined by the following:

$$\psi = \frac{1}{2}\tan^{-1}\left[\frac{2 \cdot E_{ox} \cdot E_{ox}}{E_{ox}^2 - E_{oy}^2}(\cos\varphi)\right] \tag{2.19}$$

The eccentricity is determined by the auxiliary angle $\chi$, with values satisfying $-45° \leq \chi \leq +45°$:

Eccentricity: $$\tan\chi = \frac{A_2}{A_1}, \quad \text{where} \quad \chi = \frac{1}{2}\tan^{-1}\left[\frac{2 \cdot E_{ox} \cdot E_{ox}}{E_{ox}^2 - E_{oy}^2}(\sin\varphi)\right] \tag{2.20}$$

The **E** vector at each point is located on a plane that is perpendicular to the direction of propagation. On this plane, the vector's tip draws an ellipse that rotates with angular velocity $\omega$. This is **elliptically polarized light**. Depending on the angle $\varphi$, the **E** vector rotates either in a clockwise direction (when $0 < \varphi < \pi$ rad) and is right elliptically polarized (REP), or in a counterclockwise direction (when $\pi < \varphi < 2\pi$ rad) and is left elliptically polarized (LEP). The corresponding helix is counterclockwise for REP and clockwise for LEP.

If we rotate the reference system so that the $-x$ and $-y$ axes coincide with the major ellipse's axes [shown as $A_1$ and $A_2$ in Figure 2-25 (right)], then, although the corresponding amplitude magnitudess are different, the phase difference between the components is exactly $\pi/2$!

We thus realize that all of the states of polarization we have been discussing so far, linearly polarized, circularly polarized, etc., are specific cases of elliptically polarized light! Linearly polarized light is the case where the eccentricity is 0, while RCP and LCP have eccentricity values of +1.0 and −1.0, respectively.

**Elliptically polarized light**

Elliptically polarized light exists when the electric field circumscribes an ellipse on a plane perpendicular to its direction of propagation.

It may be a result of a combination of two orthogonal, linearly polarized waves with different amplitudes and a phase difference of $\pm\pi/2$.

While it is stated that elliptically polarized light results from a combination of two orthogonal, linearly polarized waves with different amplitude magnitudes and a phase difference of $\pm\pi/2$, the latter part of this definition can also be stated as '*equal amplitudes and a phase difference not equal to $\pi/2$.*'

Consider two orthogonal, linearly polarized waves with random amplitude magnitudes $E_{ox}$ and $E_{oy}$ and a random phase difference $\varphi$:

$$\mathbf{E}_x = E_{ox}\cdot\mathbf{x}\cdot\cos(\omega\cdot t - \mathbf{k}\cdot\mathbf{r}) \quad \text{and} \quad \mathbf{E}_y = E_{oy}\cdot\mathbf{y}\cdot\cos(\omega\cdot t - \mathbf{k}\cdot\mathbf{r} + \varphi) \tag{2.21}$$

The polarization state resulting from this composition depends on the phase difference and the ratio of amplitude magnitudes.

For $\varphi = 0°$, and generally, for an even multiple of 180° ($2\pi$, $4\pi$ rad, etc.), the result is linearly polarized light. The angle $\psi$ depends on the relative ratio of the amplitude magnitudes; for equal magnitudes, $\psi = 45°$. For a random magnitude ratio, the angle $\psi$ is expressed by Eq. (2.19).

For $\varphi = 180°$ and, generally, for an odd multiple of 180° ($\pi$, $3\pi$ rad, etc.), the result is also linearly polarized light. For equal amplitude magnitudes, $\psi = 135°$ (or −45°).

<u>Linearly Polarized</u>:

$$\mathbf{E} = E_{ox}\cdot\mathbf{x}\cdot\cos(\omega\cdot t - \mathbf{k}\cdot\mathbf{r}) + E_{oy}\cdot\mathbf{y}\cdot\cos(\omega\cdot t - \mathbf{k}\cdot\mathbf{r} \pm m\cdot\pi),\ m = 1, 2, 3, \ldots \tag{2.22}$$

When the amplitudes of the two components have equal magnitudes ($E_{ox} = E_{oy} = E_o$):

- if $\varphi = 90°$ ($+\pi/2$, $5\pi/2$, $9\pi/2$ ...) there is RCP, with eccentricity +1.0
- if $\varphi = 270°$ ($-\pi/2$, $3\pi/2$, $7\pi/2$ ...) there is LCP, with eccentricity −1.0.

Right Circularly Polarized:

$$\mathbf{E} = E_{ox}\cdot\mathbf{x}\cdot\cos(\omega\cdot t - \mathbf{k}\cdot\mathbf{r}) + E_{oy}\cdot\mathbf{y}\cdot\cos(\omega\cdot t - \mathbf{k}\cdot\mathbf{r} + \pi/2) \tag{2.23}$$

Left Circularly Polarized:

$$\mathbf{E} = E_{ox}\cdot\mathbf{x}\cdot\cos(\omega\cdot t - \mathbf{k}\cdot\mathbf{r}) + E_{oy}\cdot\mathbf{y}\cdot\cos(\omega\cdot t - \mathbf{k}\cdot\mathbf{r} + 3\pi/2) \tag{2.24}$$

Elliptically polarized light results from equal component amplitude magnitudes and/or a random phase difference ($\neq \pm\pi/2$, $\neq \pm\pi$, etc.). Positive eccentricity corresponds to REP, while negative eccentricity corresponds to LEP. If the amplitudes do not have equal magnitude, then regardless of the phase difference (as long as it is not $\pm m\cdot\pi$), the result is elliptically polarized light.

Elliptically Circularly Polarized:

$$\mathbf{E} = E_{ox}\cdot\mathbf{x}\cdot\cos(\omega\cdot t - \mathbf{k}\cdot\mathbf{r}) + E_{oy}\cdot\mathbf{y}\cdot\cos(\omega\cdot t - \mathbf{k}\cdot\mathbf{r} + \varphi) \tag{2.25}$$

where $E_{ox} \neq E_{oy}$, and/or $\varphi \neq m\cdot\pi$.

## 2.4.2 Generation of Circularly Polarized Light

The combination of two orthogonally polarized waves with equal amplitude magnitudes and a phase difference of $\varphi = \pi/2$ rad = 90° results in circularly polarized light. Therefore, to create circularly polarized light we need precisely these two 'building blocks.' Starting from unpolarized light, the first step is to convert to linear polarization using a linear polarizer. Next, we need to create two orthogonal components whose amplitudes have equal magnitude, and between them introduce a phase difference of $\pm\pi/2$, or a difference in optical path length equal to a quarter of a wavelength ($\lambda/4$).

This can be achieved with a specific **retardation plate**, a transparent material that introduces a phase difference between two waves. A **quarter-wave plate** ($\lambda/4$ plate) introduces a phase difference of $\pi/2$, or 90°. Often, this plate is a properly oriented birefringent crystal (§ 2.6.2). For waves traveling in a direction perpendicular to its optic axis, the crystal permits propagation of only two orthogonal polarization states along two axes—the **fast axis** ($F$) and the **slow axis** ($S$).

Regardless of the polarization of the incident light, the mathematical expression describing this light can be analyzed in two orthogonal states. For example, linearly polarized light of amplitude $\mathbf{E}_o$ with a polarization axis at angle $\vartheta$ relative to the crystal reference system (assuming that $F$ is parallel to the $-x$ axis) is resolved into two orthogonal components along the two axes $F$ and $S$ of the $\lambda/4$ plate, with component mangitudes $E_o\cos\vartheta$ and $E_o\sin\vartheta$, respectively. For the two magnitudes to be equal, angle $\vartheta$ must be 45°.

We also want one of the two linearly polarized components to advance by $\lambda/4$ over the other wave at the exit plane of the plate (hence, the name $\lambda/4$ plate). To achieve this, we need one linearly polarized wave—the one that oscillates parallel to the fast axis—to travel faster than the other one, which oscillates parallel to the slow axis. Indeed, the first wave encounters a lower refractive index $n_F$ and therefore travels at a greater speed, while the latter wave encounters a higher refractive index $n_S$ and therefore travels at a lower speed [Figure 2-26 (lower graph)]. In this way, the wave oscillating with a polarization parallel to axis $S$ is delayed compared to the wave oscillating parallel to axis $F$.

Inside a crystal (plate) of thickness $d$, the optical path difference between the two perpendicular waves, and the phase difference are given, respectively, by

Difference in Optical Path Length: $$(n_S - n_F)\cdot d \qquad (2.26)$$

Difference in Phase: $$(k_S - k_F)\cdot d = (2\pi/\lambda_o)\cdot(k_S - k_F)\cdot d \qquad (2.27)$$

Figure 2-26 illustrates the phase difference development between a slow wave and a fast wave propagating in a quarter-wave plate. As a reminder, the frequency of a wave does not change, so the reduced speed results in a wavelength change: $\lambda_S = \lambda_o/n_S$ and $\lambda_F = \lambda_o/n_F$, where $\lambda_o$ is the vacuum wavelength. Similar relationships exist for the wavevector magnitudes $k_S$ and $k_F$.

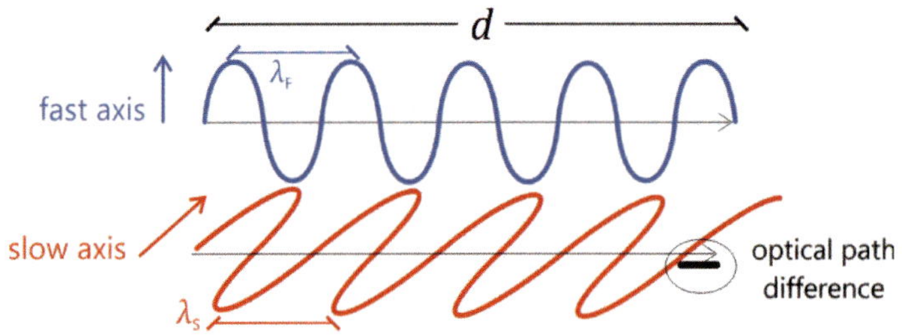


*Figure 2-26: Phase difference $\pi/2$ corresponding to an optical path difference of $\lambda/4$.*

In this simple example, the slow wave reaches the plate surface after three and one-quarter oscillation cycles, while the fast wave travels the same distance $d$ after only three of such cycles. This is because, for the fast wave with a longer wavelength $\lambda_F$, the same distance $d$ corresponds to fewer wavelengths. Thus, an optical path difference of $\frac{1}{4}\lambda_o$ develops. Equivalently, the developed phase difference is $\pi/2$ rad = 90°.

The refractive indices along the slow and fast axes depend on the material and the wavelength. We can manufacture plates with suitable thicknesses such that the optical path difference $(n_S - n_F) \cdot d$ equals $(m + ¼) \cdot \lambda_o$, where $m$ is an integer, for a $\lambda/4$ plate. If we want to make a half-wave ($\lambda/2$) plate, we simply set the optical path difference = $(m + ½) \cdot \lambda_o$.

Figure 2-27 illustrates the creation of circularly polarized light from linearly polarized light. Light polarized at 45° is incident on the $\lambda/4$ plate with respect to the fast (and the slow) axis. Thus, two components of equal amplitude ($\mathbf{E}_{oF} = \mathbf{E}_{oS} = \mathbf{E}_o \cos 45°$) and having an initial phase difference of $\varphi = \varphi_F - \varphi_S = 0$ are formed.

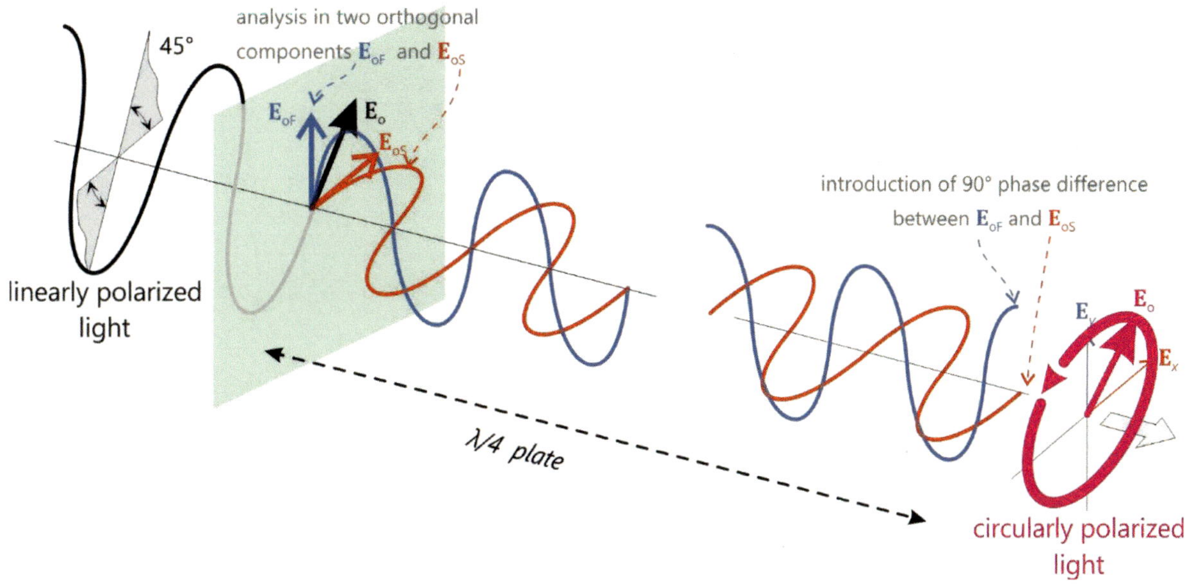


*Figure 2-27: A $\lambda/4$ plate in action: decomposition of the initial linearly polarized light into two components of equal amplitude, followed by introduction of a phase difference between the two components.*

Inside the plate, the component parallel to the slow axis moves progressively slower and a phase difference develops between the two components. (In Figure 2-27 and Figure 2-28, these two actions are illustrated separately.) Upon exiting, the amplitudes have again equal magnitude for minimum, or equal, absorption, but a phase difference of $\varphi = \varphi_F - \varphi_S = \pi/2$ has developed. Light leaving the $\lambda/4$ plate is circularly polarized.

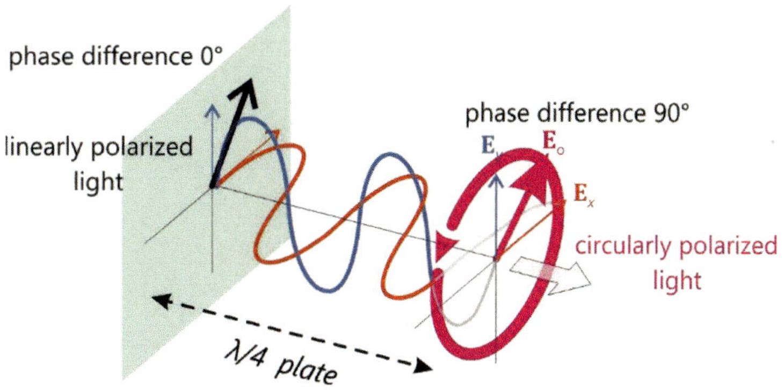


*Figure 2-28: Creation of circularly polarized light from linearly polarized light.*

If the relative angle $\vartheta$ between the initial linearly polarized light and the fast axis is not 45°, then the magnitudes of the two components are not equal ($E_{oF} = E_o\cos\vartheta \neq E_{oS} = E_o\sin\vartheta$). The phase difference is always $\varphi = \varphi_F - \varphi_S = \pi/2$, regardless of the relative component intensities. Thus, for an initial relative angle $\vartheta \neq 45°$, light emerging from the $\lambda/4$ plate is elliptically polarized.

**To create circularly polarized light** → First we create linearly polarized light, and place the axis of polarization at 45°. → Then we insert a $\lambda/4$ retardation plate, whose slow and fast axes are at 0° and 90°, respectively. → Light leaving the retardation plate is circularly polarized.

It is interesting to ask what state of polarization exists if we have a $\lambda/4$ plate with unpolarized light. The omission of the linear polarizer is important! Since 'random + $\pi/2$ = random,' the resulting beam is again unpolarized light!

The order in which we place the building blocks is critical; compare the results obtained if unpolarized light encounters the following:

- first a $\lambda/4$ plate, then a linear polarizer, and
- first a linear polarizer, then a $\lambda/4$ plate. The results are entirely different.

Note : Any given plate that is $\lambda/4$ for a specific wavelength $\lambda_o$ (such as the red) can also be $\lambda/2$ for another wavelength (such as the violet). Specifically, a $\lambda/4$ plate for the red ($\lambda_o \approx 700$ nm) introduces an optical path difference of 175 nm. This difference is nearly equal to half the wavelength for the violet ($\lambda_o' \approx 350$ nm); the exact plate can be $\lambda/2$ for the violet! The term '$\lambda/4$ plate' then applies to only a specific range of wavelengths.

## 2.4.3 Detection of Circularly Polarized Light

If natural light is incident on an analyzer, we note no changes in the detected light intensity over a complete rotation of the analyzer. This is because at any given time the electric field in natural light has a random orientation. What we are witnessing is the time-average of the square of its projection, a fixed (= ½) intensity.

The same happens with circularly polarized (CP) light, although for a different reason. In this case, at any given time, the electric field has a fixed amplitude magnitude and rotates with a constant angular velocity $\omega$ (several $10^{14}$ Hz). Imagine that we observe the blades of a propeller plane (Figure 2-29) spinning very quickly, at several hundreds of revolutions per second.

We no longer see the propeller blades but instead see only a whitish, hazy circle. In circularly polarized light, the rotation speed of the electric field vector is several orders of magnitude larger than that of the propeller. It is thus momentarily impossible to see the orientation of the electric field vector. Indeed, as the electric field circumscribes a circle, the time average of the detected light intensity over any given analyzer axis is fixed in time (= ½).

*Figure 2-29: The spinning propeller blades give the appearance of a hazy circle (photo by Ilias Diakoumakos used with permission).*

To distinguish circularly polarized light from unpolarized light, we use a key distinctive aspect: In circular polarization, the vector of the electric field is <u>not</u> random. At any given time, there can be two orthogonal, linearly polarized components with equal amplitude magnitudes (as in unpolarized light), but with a fixed phase difference of 90°. If *somehow* we can change this phase difference to 0° or 180°, their re-composition produces linearly polarized light!

We can do this by simply placing a $\lambda/4$ plate in the CP light path. For convenience (not necessity), we assume that the axes ($F_2$ and $S_2$) are parallel to those of the $\lambda/4$ plate. Thus, there is no further component analysis by this $\lambda/4$ plate, since the axes are already parallel to the initial axes. (It is a useful to prove that, even if there was further component analysis by the $\lambda/4$ plate, the two components would again have equal amplitude magnitudes and a 90° phase difference.)

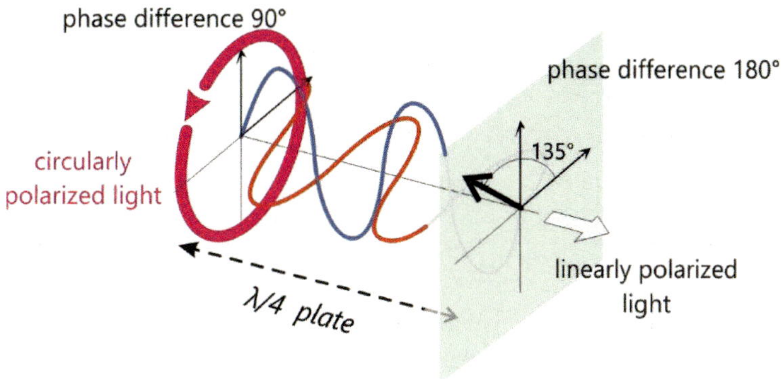


*Figure 2-30: Conversion from CP light into LP light with an additional $\lambda/4$ plate.*

This $\lambda/4$ plate adds (if $F_1 \parallel F_2$) or subtracts (if $F_1 \perp F_2$) 90° of the phase difference between the two orthogonally polarized components. Thus, the phase difference between the two components becomes 180° or 0°, and their re-composition upon exiting the second $\lambda/4$ plate results in linearly polarized light, which can be easily detected using an analyzer after the plate.

**To detect circular polarization, we place a $\lambda/4$ retardation plate along the beam, followed by an analyzer that is rotated.**

We observe the intensity of the light transmitted through the analyzer.

- If fluctuations are observed, the light is circularly or elliptically polarized.
  - Zero minimum
    - If it is on the same axis as the analyzer axis, the polarization is circular.
    - If it is on an axis other than the analyzer axis, the polarization is elliptical.
  - Minimum other than zero
    - If it is on the same axis as the analyzer axis, the polarization is partially circular.
    - If it is on an axis other than the analyzer axis, the polarization is partially elliptical.
- If no fluctuations are observed, the light is unpolarized.

We can now expand this exercise by adding not one, but two, retardation plates. The conversions will be: linearly polarized to circularly polarized (1st plate) and circularly polarized to linearly polarized (2nd plate). The light leaving the second plate is linearly polarized, but its axis is perpendicular to that of the light entering the first plate (45°); i.e., the $\pi$ phase difference causes a net phase difference of 135°.

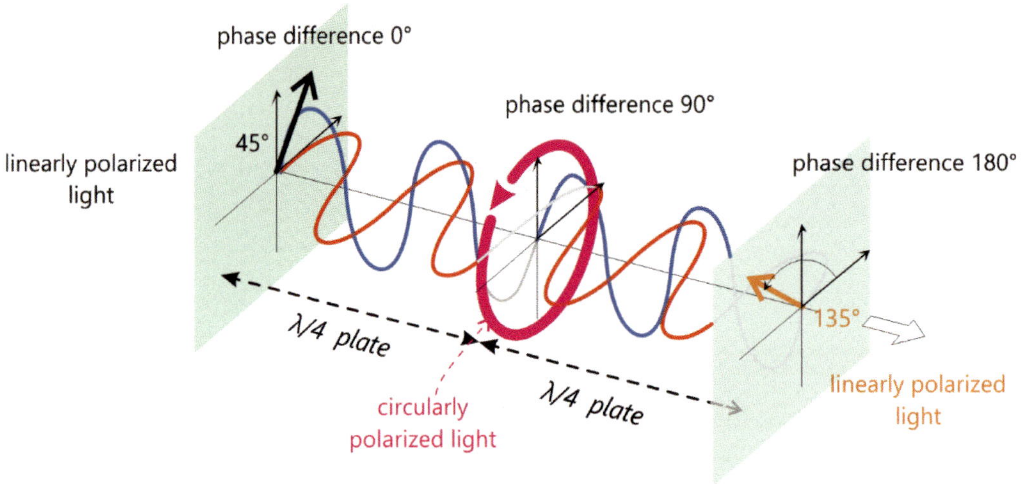


*Figure 2-31: Development of a phase difference of $\pi$ using two successive $\lambda/4$ plates.*

# 2.5 POLARIZATION AND NATURAL PHENOMENA

Polarization exists in nature! When wearing polarizing sunglasses, the fruit at the farmers' market on a sunny day seem less shiny, the sky looks more blue, and we can eliminate the annoying reflections from window glass. There are certain natural phenomena in which unpolarized sunlight is transformed to partially or completely polarized light. Such a phenomenon is the scattering of sunlight by the atmosphere's molecules, causing the deep blue light. Polarization can be also due to reflection; the reflected light is polarized and, in certain cases, is completely linearly polarized. Recall that reflection can be considered a form of scattering (as discussed in § 1.5.2).

## 2.5.1 Scattering in the Sky: The Color of Blue

A clear sky is mostly blue and is, to a large degree, linearly polarized. This is due to sunlight being scattered by atmospheric molecules. If there was no atmosphere, the sky viewed from Earth would look black, just like the sky that the Apollo 8 astronauts viewed from the moon.

*Figure 2-32: The famous 'Earthrise' photograph taken on Christmas Eve (24 December 1968) by the Apollo 8 astronaut, William A. Anders, while the spacecraft orbited the moon (image credit: NASA).*[6]

John Tyndall attempted to explain **scattering** (in 1859) by observing light propagation through a colorless liquid with small floating particles (a liquid such as soapy water), in which light is diffused to all directions. However, scattering is not just a random re-emission of light; the effect has certain distinctive properties. The shorter (blue) wavelengths scatter more than the red, primarily along the direction of light propagation; thus, the soapy water appears more blueish when viewed along the path of light propagation. Conversely, the longer (red)

[6] https://www.nasa.gov/multimedia/imagegallery/image_feature_1249.html.

wavelengths scatter less; thus, the soapy water appears more reddish when viewed from above, perpendicular to the path of light propagation.

**Elastic scattering** involves no frequency change (re-emission at the same wavelength) and no noticeable re-emission delay. **Elastic Rayleigh scattering**, named after William John Strutt, later known as Lord Rayleigh, who first described it in 1871,[7] is a type of scattering that involves scattering centers that are much smaller ($< \lambda/10$) than the average wavelength $\lambda$ of the radiation being scattered. For example, in the upper atmosphere, the air molecules $N_2$ and $O_2$ with diameters $d \approx 0.2$ nm = 0.0002 µm are much smaller than the average visible light wavelength of ≈ 0.5 µm. Therefore, the sunlight scattering that takes place in the upper atmosphere is an example of Rayleigh scattering.

A distinctive feature of Rayleigh scattering is that the scattering output is far stronger at short wavelengths, such as the blue,[8] than at longer wavelengths, such as the red. For an angle of observation $\vartheta$ with respect to the direction of the incident light, and for a wavelength $\lambda$, the Rayleigh-scattered luminant intensity observed at a distance $r$ is

Rayleigh-scattered luminant intensity:
$$I(\vartheta, r, \lambda) = \frac{1}{r^2}\frac{\sin^2\vartheta}{\lambda^4} \tag{2.28}$$

Note 🗎: The strong wavelength dependence favors short wavelengths; scatter intensity corresponding to the blue is much greater than that corresponding to the red. Because $\lambda_{blue} \approx \frac{1}{2}\lambda_{red}$, $I_{blue} \approx 16 \times I_{red}$!

Example ☞: What is the ratio of scatter intensity for the hydrogen blue line light ($\lambda_{blue}$ = 450 nm) relative to the hydrogen red line light ($\lambda_{red}$ = 635 nm)?

$I_{blue} / I_{red} = (\lambda_{red} / \lambda_{blue})^4 = (635/450)^4 = (1.35)^4 = 3.32$. The blue scatter is more intense by a factor of > 3.

Thus, Rayleigh scattering is responsible for the deep blue sky during the day and the reddish sky during sunrise/sunset: When looking at the sky at (mostly) right angles with respect to the sun, we receive the scattered light, which appears blue due to the dominance of the blue scattered component. The part of the sunlight that is not scattered propagates along the direction of the incident radiation; therefore, at sunset or sunrise, sunlight appears predominantly red.

---

[7] Strutt JW. On the light from the sky, its polarization and colour. Philos Mag Ser. 1871; 4(41):274-9.

[8] Strutt JW. On the transmission of light through an atmosphere containing small particles in suspension, and on the origin of the blue of the sky. Philos Mag Ser. 1899; 5(47):375-84.

Other interesting aspects of Rayleigh scattering are the following:

- Maximum scattering intensity occurs at angles $\vartheta = \pm 90°$ ($\sin\vartheta = 1.0$) in relation to the original light direction (for example, at right angles from the sun). The most intense blue light in the sky propagates in a direction almost perpendicular to the incident sunlight—this is the direction of the strongest scattering.
- Scattered light is linearly polarized. If we view this deep blue light through polarizing sunglasses, we will observe distinct variations of intensity, which is an indication of polarization. Along these directions of observation, the scattered light is linearly polarized. In any other direction, the light is partially polarized.

*Figure 2-33: The rich colors in the sky result from scattering (photo by Efstratios Prodromou).*

*Figure 2-34: The blue sky is caused by Rayleigh scattering off atmospheric molecules. The contrails, the cloudy streaks formed by the condensation of the engine exhaust, produce Mie scattering. (Left image by captain Ioannis Papadopoulos; right image by Lucien Schranz; both used with permission.)*

For larger-sized scattering particles, such as droplets in clouds, this analysis does not apply. Gustav Mie's analysis covers this effect, called **Mie scattering**. The water droplets that make up a cloud are much larger than the air molecules. This scattering is strongly forward (along the direction of the initial light propagation) and is almost independent of the wavelength in the visible range. This is why the clouds appear white. The light scatter from free electrons (which are much smaller than the wavelength of light) is described by Thomson scattering, a third type of elastic scattering.

Elastic scattering is summarized as follows: An electromagnetic wave incident on an atmospheric particle causes displacements in the electron distribution. These shifts follow the fluctuations of the electric field, so the electron distribution is brought into oscillation. This classical energy exchange corresponds to nonresonant absorption. The absorption creates an oscillating dipole, which is an elementary emission unit. The oscillating dipole re-emits the absorbed energy in all directions at the oscillating frequency of the original wave.

An atom or molecule is modeled as a system of distributed charged particles. In the center are the (assumed stationary) positive charges that form the nucleus, surrounded by the orbiting electrons. The electrons are considered to be bound under the influence of the attractive forces exerted on them by the nucleus.

If, in the absence of an external electric field, the average spatial distributions of the positive and negative charges coincide (Figure 2-35), the **electric polarization** in this charge system is zero. If there is a separation between the two centers, the system has electric polarization. Let us now distinguish between optical polarization, which is a property of light (and of electromagnetic waves, in general) relating to the stability of the orientation of the electric field, and electric polarization, which is a property of a charge system formed of a pair of positive and negative electric charges with displaced centers.

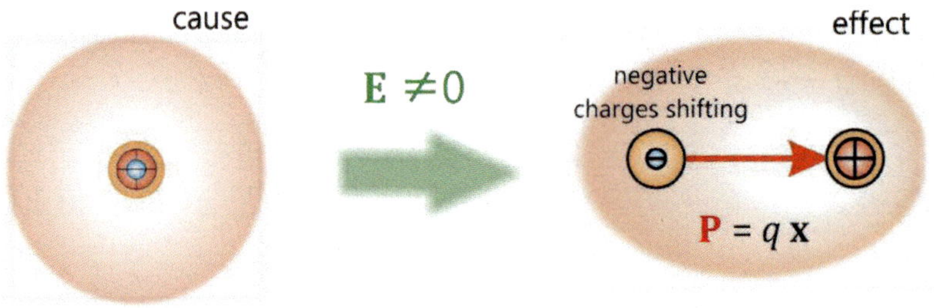


*Figure 2-35: Development of electric polarization in an atom under the influence of an electric field.*

In the simplest case, the system is nonpolarized. Initially, the center of the negative charge and the center of the positive charge coincide. Applied electric forces cause a separation in the distribution of positive and negative charges. Since the nucleus has considerably more mass than the electrons, there is a shift of the negative charges only. Then an induced electric polarization $\mathbf{P}$ develops, whose magnitude is the charge $q$ times the separation $\mathbf{x}$. The system becomes an **electric dipole**.

If the electric field oscillates with an angular frequency $\omega$ (for the visible, $\omega \approx 10^{14}$ Hz), it can induce an electric dipole oscillation (bipolar oscillation) with a frequency $\omega$, in the direction of the axis of the electric field. This electric dipole resembles an oscillating L-C circuit that can radiate electromagnetic radiation having exactly the same characteristics (oscillation frequency and electric field direction) as the generating field. The electric dipole is an element that emits light in all directions.

Consider now unpolarized sunlight propagating along the $-z$ direction. The axes of oscillation of the electric field are on a plane perpendicular to $-z$ (the direction of incident light from the Sun) such as $x$–$y$, which is perpendicular to $-z$ (Figure 2-36). The dipoles (oscillating atmospheric molecules) oscillate on an oscillation plane corresponding to the re-emitted (scattered) light: a plane perpendicular to $-z$. Thus, the orientation of the oscillation plane of the electric field of the re-emitted light is restricted to the $x$–$y$ plane.

The scattered light becomes optically polarized exactly due to this limitation: Light is a transverse wave. For observation along a direction parallel to that of incidence (observer A), the scattered light properties are such that the electric field retains all of the original directions of possible oscillation at $x$–$y$, which are always transverse to the direction of propagation $-z$. Thus, for unpolarized sunlight, its scatter along such a direction is also unpolarized. Direct scattered sunlight therefore is unpolarized.

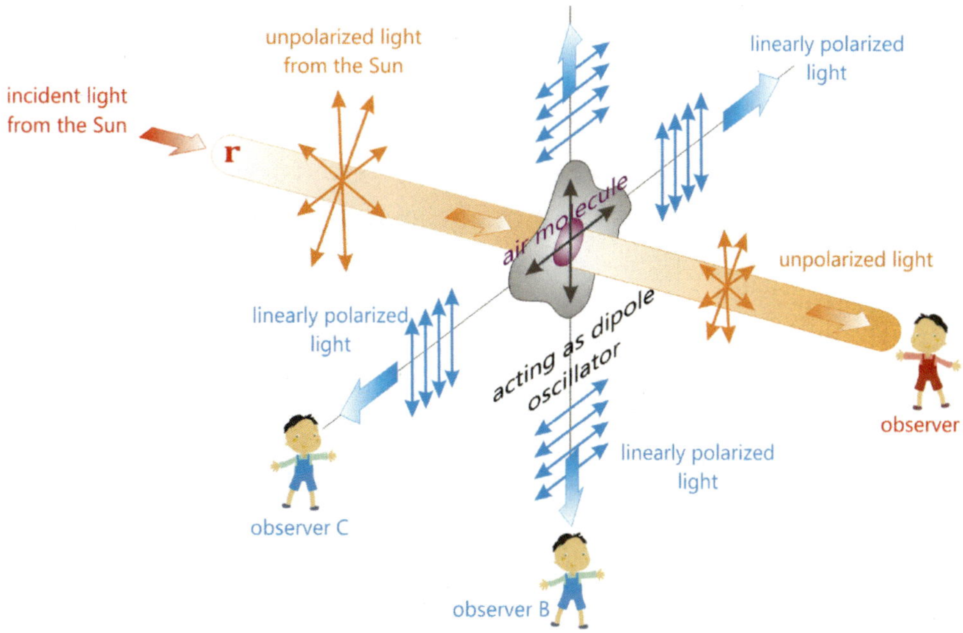


*Figure 2-36: Polarization of natural sunlight due to scattering by atmospheric molecules.*

For the light scattered in a direction perpendicular to that of incidence (observer B), the possible transverse orientations for the electric field oscillation are limited. Specifically, these directions are derived from the projection of the re-emission plane on the transverse plane to the direction of propagation $y$. For light observed by observer B, there exists only one of such directions, along the $-x$ axis. Therefore, for this scatter direction, light is linearly polarized ($-x$), and observer C receives linearly polarized light (restricted to propagating along the $-y$ axis). Thus, the scatter of atmospheric light displays linear polarization, and its propagation direction depends on the direction of observation.

Food for thought : What would be the state of polarization for light propagating in the directions of observers A and B if the original incident light was not unpolarized, but linearly polarized, along $-x$?

While observer A would detect vertically linearly polarized ($-x$) light, observer B would not detect any light!

### 2.5.2 Polarization by Reflection and Refraction

Reflection is another polarization mechanism. For specific angles (in the vicinity of 50°–60°), the reflected light can even prevent us from discerning what is behind a window glass. With an analyzer placed along the reflected light path, we can observe that there is an angle for which an appropriate rotation of the analyzer results in a complete attenuation of the reflected light. At this angle, the reflected light is linearly polarized; this is polarization created by refraction.

Geometrical optics is not concerned with the amount of light being reflected or refracted off a dividing surface (let alone the state of polarization). Here, we will derive analytical expressions for both qualitative and quantitative reflection and refraction, and will predict the state of polarization for the reflected and refracted beams. These expressions utilize an analytical consideration responsible for absorption and re-emission of light by electric dipoles along the media interface. This interface ($z$–$x$ plane, Figure 2-37) separates two dielectric media with refractive indices $n_1$ and $n_2$. Light is incident on this interface along a specific direction.

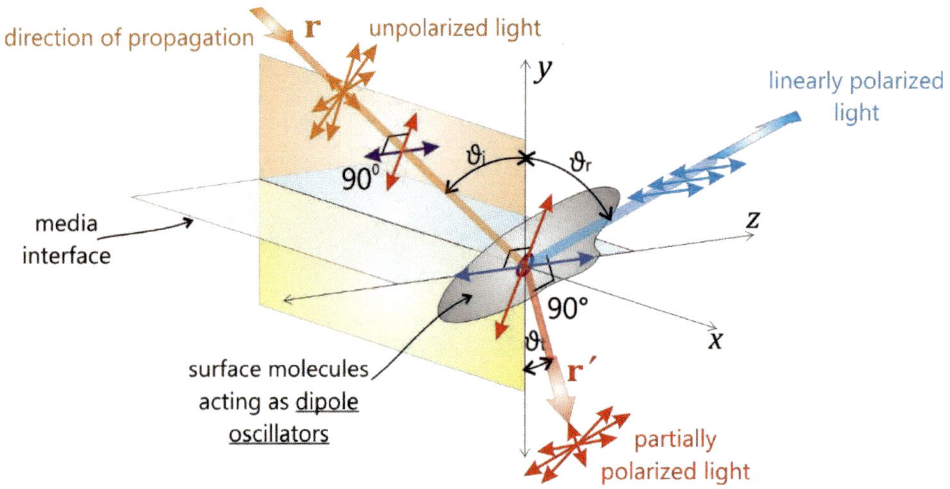


*Figure 2-37: Polarization by reflection ($\vartheta_i$, $\vartheta_r$, and $\vartheta_t$ denote angle of incidence, angle of reflection, and angle of refraction, respectively).*

Along the refracted ray inside the second medium (direction $\mathbf{r}'$), the electric field can have all possible oscillation directions; therefore, there is no linear polarization. However, along the direction of the reflected ray, the oscillating electric field is restricted, as it must be both perpendicular to the direction of propagation and parallel to the plane of dipole oscillation. The fields oscillate along a plane that is perpendicular to the reflected direction of propagation. The reflected light has a limited range of field oscillation planes and therefore is linearly polarized.

For a specific angle of incidence $\vartheta_i$ (later called Brewster's angle $\vartheta_B$) for which the reflected and refracted beams are orthogonal, the oscillation of the reflected electric field is limited to just one direction: along the axis of the intersection of the dividing surface plane with the dipole oscillation. Thus, the reflected light is linearly polarized, parallel to the reflecting surface. The angles of incidence $\vartheta_i$, reflection $\vartheta_r$, and refraction $\vartheta_t$ are related to each other according to the law of reflection ($\vartheta_i = \vartheta_r$) and the law of refraction [$n_1 \cdot \sin(\vartheta_i) = n_2 \cdot \sin(\vartheta_t)$][9]:

$$n_1 \cdot \sin(\vartheta_B) = n_2 \cdot \sin(90° - \vartheta_B) = n_2 \cdot \cos(\vartheta_B) \tag{2.29}$$

where $\vartheta_B$ is **Brewster's angle**, named after Sir David Brewster:

Brewster's Angle:

$$\vartheta_B = \tan^{-1}(n_2/n_1) \tag{2.30}$$

Under Brewster's angle of incidence, the reflected beam polarization is parallel to the surface (and simultaneously perpendicular to the plane of incidence).

#### 2.5.2.1 Fresnel Coefficients of Reflection

Apart from the above qualitative analysis, we can have a full quantitative study of reflection. We want to determine the exact portions of the incident radiation that are reflected or refracted for a specific angle of incidence and polarization state. We are interested in the analytical expressions for the **amplitude reflection coefficient** $\rho$, defined as the ratio of the reflected amplitude magnitude $E_r$ to the incident amplitude magnitude $E_i$, and the **amplitude transmission coefficient** $\tau$, defined as the ratio of the refracted (transmitted) amplitude magnitude $E_t$ to the incident amplitude magnitude $E_i$:

$$\rho = E_r/E_i \quad \text{and} \quad \tau = E_t/E_i \tag{2.31}$$

The detectable quantity is the intensity, not the electric field. Therefore, we define the ratio of the reflected intensity to the incident intensity, known as the **reflectance** or **reflectivity** $R$ ($R = \rho^2$)[10], and the ratio of the refracted (transmitted) intensity to the incident intensity, known

[9] *Introduction to Optics* § 3.2.2 The Law of Refraction.

[10] In a more accurate definition, reflectance is the fraction of electromagnetic power reflected from a specific sample, while reflectivity is a property of the material itself.

as the **transmittance** or **transmissivity** $T$ ($T = \tau^2$)[11]. These are the **Fresnel coefficients**, named after Augustin-Jean Fresnel. If there are no losses, then $R + T = \rho^2 + \tau^2 = 1.0$.

There are two distinct polarization eigenstates in relation to the plane of incidence:

- *Parallel Eigenstate*: The electric field vector is parallel to the plane of incidence ($x$–$y$ plane in Figure 2-38), while the magnetic field vector is perpendicular to the plane of incidence, so it is parallel to the surface [the $x$–$z$ plane in Figure 2-38 (left)]. This state is also called transverse magnetic (TM) polarization and parallel or $p$-polarization (‖).

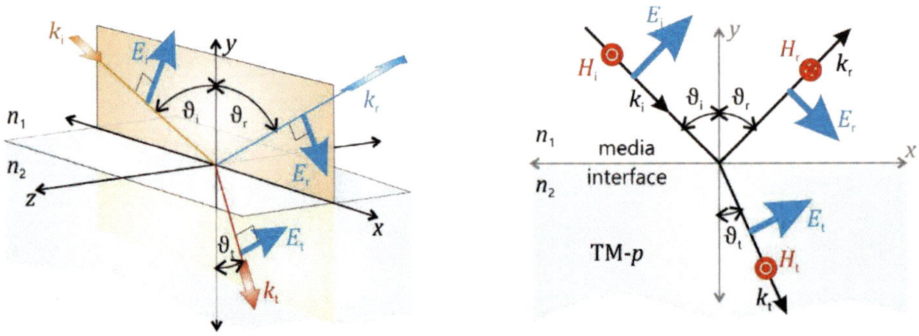


*Figure 2-38: Parallel polarization (p-polarization or TM-p): All electric field vectors are parallel to the plane of incidence x–y, and all magnetic field vectors are perpendicular to the plane of incidence x–y.*

- *Perpendicular Eigenstate*: The electric field vector is perpendicular to the plane of incidence ($x$–$y$ plane), and the magnetic field vector is parallel to the plane of incidence. Therefore, the magnetic field vector is perpendicular to the interface ($x$–$z$ plane) [Figure 2-39 (left)]. This state is also called transverse electric (TE) polarization and perpendicular or $s$-polarization (⊥), where $s$ stands for *Senkrecht*, the German word for perpendicular.

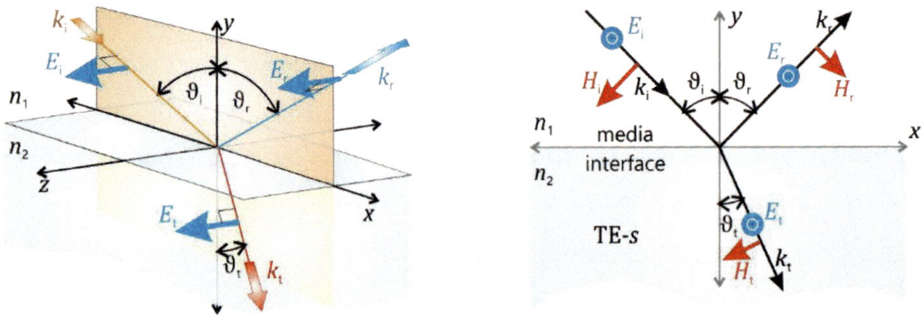


*Figure 2-39: Perpendicular polarization (s-polarization or TE-s): All electric field vectors are perpendicular to the plane of incidence x–y; all magnetic field vectors are parallel to the plane of incidence x–y.*

[11] Transmittance is the measured ratio of light at normal incidence, whereas transmissivity is the ratio of the total light that passes through the medium.

We apply the following simple physical principle, derived from fundamental electrodynamics:

> *If there are no static charges or currents on the dividing surface, then the tangential components of the electric and magnetic fields are equal before and after an interface.*

This principle is also known as the continuity condition for the parallel component of any field along the dividing surface *x–z*. (Equivalently, we can consider the continuity of the corresponding perpendicular projections of magnetic induction *B* and electric displacement *D*.)

The continuity condition provides a solvable set of four equations and four unknowns: the coefficients $\rho_\mathrm{p}$ and $\rho_\mathrm{s}$ for reflection, and $\tau_\mathrm{p}$ and $\tau_\mathrm{s}$ for transmission. These two orthogonal polarization states respond completely differently along the dividing surface because the field projections on the interface are different. For example, for the parallel-polarization state, the electric field vector continuity condition leads to the following relationship:

$$E_\mathrm{i}\cdot\cos\vartheta_\mathrm{i} + E_\mathrm{r}\cdot\cos\vartheta_\mathrm{i} = E_\mathrm{t}\cdot\cos\vartheta_\mathrm{t} \tag{2.32}$$

while for the perpendicular-polarization state, the relationship is

$$E_\mathrm{i} + E_\mathrm{r} = E_\mathrm{t} \tag{2.33}$$

Thus, the field continuity conditions provide different equations. The coefficients depend on: (1) the relative refractive index of the two media (their ratio), (2) the polarization state, and (3) the angle of incidence. In the general case, these coefficients are complex numbers.

For parallel polarization with a relative refractive index ratio $n_{21} = n_2/n_1$ and an angle of incidence $\vartheta_\mathrm{i}$, the coefficient $\rho_\mathrm{p}$ is

$$\rho_\mathrm{p}(\vartheta_\mathrm{i},n_{21}) = \left(\frac{E_\mathrm{r}}{E_\mathrm{i}}\right)_\mathrm{p} = \frac{-n_{21}^2\cos\vartheta_\mathrm{i} + \sqrt{n_{21}^2 - \sin^2\vartheta_\mathrm{i}}}{n_{21}^2\cos\vartheta_\mathrm{i} + \sqrt{n_{21}^2 - \sin^2\vartheta_\mathrm{i}}} \tag{2.34}$$

and the coefficient $\tau_\mathrm{p}$ is

$$\tau_\mathrm{p}(\vartheta_\mathrm{i},n_{21}) = \left(\frac{E_\mathrm{t}}{E_\mathrm{i}}\right)_\mathrm{p} = \frac{2\cdot n_{21}\cos\vartheta_\mathrm{i}}{n_{21}^2\cos\vartheta_\mathrm{i} + \sqrt{n_{21}^2 - \sin^2\vartheta_\mathrm{i}}} \tag{2.35}$$

The corresponding expressions for perpendicular polarization are

$$\rho_\mathrm{s}(\vartheta_\mathrm{i},n_{21}) = \left(\frac{E_\mathrm{r}}{E_\mathrm{i}}\right)_\mathrm{s} = \frac{\cos\vartheta_\mathrm{i} - \sqrt{n_{21}^2 - \sin^2\vartheta_\mathrm{i}}}{\cos\vartheta_\mathrm{i} + \sqrt{n_{21}^2 - \sin^2\vartheta_\mathrm{i}}} \qquad \tau_\mathrm{s}(\vartheta_\mathrm{i},n_{21}) = \left(\frac{E_\mathrm{t}}{E_\mathrm{i}}\right)_\mathrm{s} = \frac{2\cos\vartheta_\mathrm{i}}{\cos\vartheta_\mathrm{i} + \sqrt{n_{21}^2 - \sin^2\vartheta_\mathrm{i}}} \tag{2.36}$$

The above (rather complicated, indeed) expressions demonstrate the dependence of coefficients $\rho_p$ and $\rho_s$ on the angle of incidence $\vartheta_i$ and the relative refractive index $n_{21}$. It is helpful to visualize these relationships in their plotted formats as follows:

Figure 2-40 illustrates the dependence of the reflection coefficients $\rho_{p\text{-TM}}$ and $\rho_{s\text{-TE}}$ on the angle of incidence and polarization state for reflection from (left) an optically more dense medium ($n_{21} = n_2/n_1 > 1$), which is the case of external reflection, and for reflection from (right) an optically less dense medium ($n_{21} = n_2/n_1 < 1$), which is the case of internal reflection.

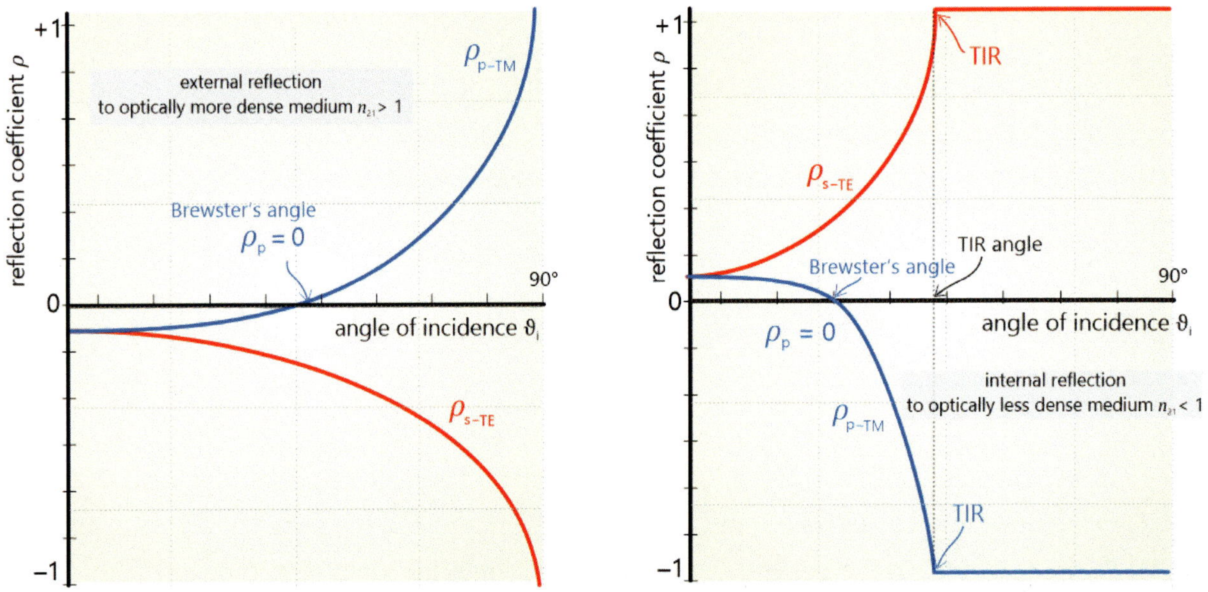


*Figure 2-40: Angle of incidence dependence of coefficients $\rho_{p\text{-}TM}$ and $\rho_{s\text{-}TE}$ for (left) external reflection and (right) internal reflection.*

Figure 2-41 illustrates the dependence of the reflectance ($R = \rho^2$) on the angle of incidence for the same parameters as in Figure 2-40.

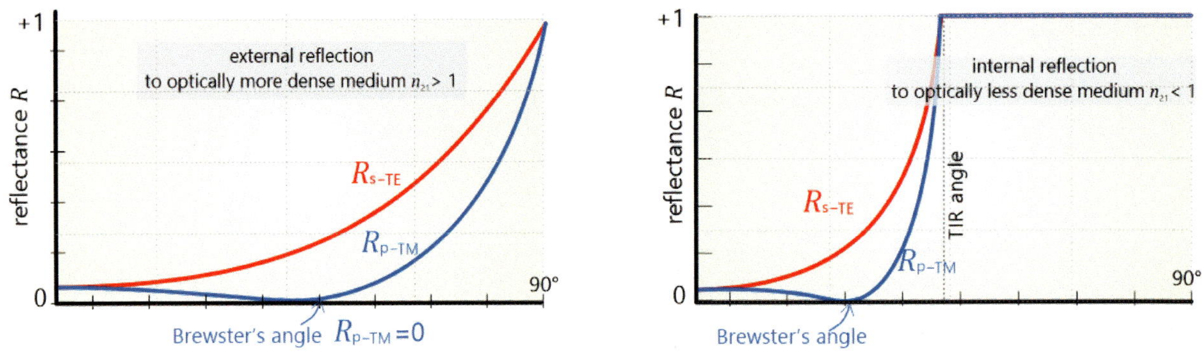


*Figure 2-41: Angle of incidence dependence of coefficients $R_{p\text{-}TM}$ and $R_{s\text{-}TE}$ for (left) external reflection and (right) internal reflection.*

We now summarize the properties of the coefficients $\rho_{\text{p-TM}}$, $\rho_{\text{s-TE}}$, $\tau_{\text{p-TM}}$, $\tau_{\text{s-TE}}$, $R_{\text{p-TM}}$, and $R_{\text{s-TE}}$:

- Coefficients $\rho_{\text{p-TM}}/\rho_{\text{s-TE}}$ or $\tau_{\text{p-TM}}/\tau_{\text{s-TE}}$ are, in general, complex numbers. In this case, reflectivity is computed by the products $R = \rho \cdot \rho^*$ and transmissivity by the product $T = \tau \cdot \tau^*$, where * indicates the conjugate of a complex number.

- Coefficients $\rho_{\text{p-TM}}$ or $\rho_{\text{s-TE}}$ can have negative values. In this case, there is a phase difference (shift) by $\pi$ rad with respect to the incident. Specifically, coefficient $\rho_{\text{s-TE}}$ is negative for all angles of incidence [the red line in Figure 2-40 (left)]. Coefficient $\rho_{\text{p-TM}}$ is also negative for small angles of incidence ranging from zero to Brewster's angle. This is modeled in mechanics as a reflection from a hard boundary. It occurs for $n_{21} = n_2/n_1 > 1$, when light is partially reflected upon entering a medium that is optically more dense, such as from air to glass.

- Coefficients $\rho_{\text{p-TM}}$ and $\rho_{\text{s-TE}}$ can have positive values, and no phase shift occurs for reflection from an optically less dense medium (internal reflection, $n_{21} < 1.0$) to an optically more dense medium, such as from glass to air. This applies to perpendicular polarization $\rho_{\text{s-TE}}$ for all angles of incidence [the red line in Figure 2-40 (right)], and to parallel polarization $\rho_{\text{p-TM}}$ up to Brewster's angle. This state of reflection is equivalent to reflection from a soft boundary.

- For normal incidence ($\vartheta_i = 0°$), coefficients $\rho_{\text{p-TM}}$ and $\rho_{\text{s-TE}}$ are equal:

$$\rho_{\text{s}\ \vartheta=0} = \rho_{\text{p}\ \vartheta=0} = \frac{1-n_{21}}{1+n_{21}} \quad \text{and} \quad R_{\text{s}\ \vartheta=0} = R_{\text{p}\ \vartheta=0} = \left(\frac{1-n_{21}}{1+n_{21}}\right)^2 \tag{2.37}$$

Example ☞: For normal incidence from air to glass ($n$ = 1.5), the reflection coefficients and the corresponding reflectivity coefficients are, respectively,

$$\rho_{\text{s}\ \vartheta=0} = \rho_{\text{p}\ \vartheta=0} = \frac{1-\frac{1.5}{1}}{1+\frac{1.5}{1}} = \frac{-0.5}{2.5} = -0.2 \quad \text{and} \quad R_{\text{s}\ \vartheta=0} = R_{\text{p}\ \vartheta=0} = (-0.2)^2 = 0.04\,.$$

This glass reflects 4% of the incident light. The reflected intensity from a shiny diamond with $n \approx 2.4$ is 17%.

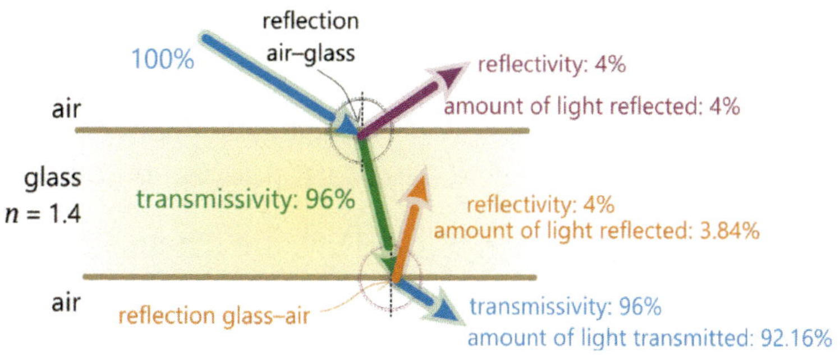


*Figure 2-42: Coefficients of reflection in a glass–air interface for nearly normal incidence (the incident ray is shown as oblique to differentiate it from the reflected and refracted rays).*

For nearly parallel incidence (angle of incidence $\vartheta_i$ = 90°) with respect to the dividing surface of the two media (grazing incidence), the coefficients $\rho_{p\text{-TM}}$ and $\rho_{s\text{-TE}}$, as well as the corresponding reflectivity coefficients, equal 1.0 for the two polarization eigenstates and are independent of the value of the refractive index:

$$\rho_{s\,\vartheta=90^\circ} = \rho_{p\,\vartheta=90^\circ} = 1.0 \quad \text{and} \quad R_{s\,\vartheta=90^\circ} = R_{p\,\vartheta=90^\circ} = 1.0$$

A smooth, flat glass (or water) surface acts as a 100% perfect mirror for incidence almost parallel to the surface, regardless of both internal/external reflection and the refractive index.

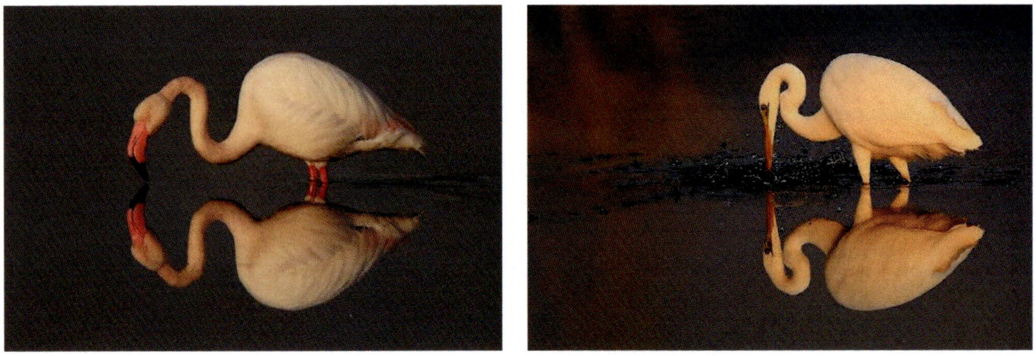

*Figure 2-43: Nature photographs showing strong reflection for very large angles of incidence (nearly parallel incidence). [Photos by Petros Tsakmakis of (left) a flamingo (Phoenicopterus roseus) and (right) a great white heron (Ardea alba), taken at the Kalloni Saltpans, Lesvos Island, Greece.]*

In internal reflection, where $n_2 < n_1$ or $n_{21} < 1.0$, there are values of $\vartheta_i$ such that $\sin(\vartheta_i) > n_{21}$ and $(n_{21}^2 \cdot \sin^2\vartheta_i) < 1.0$. This applies to angles of incidence $\vartheta_i > \vartheta_{CR}$. The mathematics for the critical angle of incidence $\vartheta_{CR}$ are such that the refracted ray is tangent (grazing emergence) to the dividing surface. Thus, the angle of incidence for which the angle of refraction is 90° is the critical angle $\vartheta_{CR}$. By Snell's law,

$$n_1 \cdot \sin(\vartheta_{CR}) = n_2 \cdot \sin(90^\circ) \;\Rightarrow\; \sin(\vartheta_{CR}) = \frac{n_2}{n_1} \;\Rightarrow\; \vartheta_{CR} = \sin^{-1}\left(\frac{n_2}{n_1}\right) = \sin^{-1}(n_{21})$$

This is the critical angle for total internal reflection.[12] In this case, the internal reflection (meaning the reflection toward the medium with the higher refractive index) is complete, and the reflectivity for both the parallel and the perpendicular polarization becomes 1.0.

For intermediate angles of incidence, 0° < $\vartheta_i$ < 90°, the component of perpendicular (s-TE) polarization has, compared to that of parallel polarization, higher reflectivity (Figure 2-41),

[12] *Introduction to Optics* § 3.2.4 Critical Angle of Incidence; Total Internal Reflection.

which gradually increases with the angle of incidence. For parallel polarization (p-TM), an angle of incidence with zero reflectivity occurs when

$$-n_2 \cos\vartheta_\mathrm{i} + n_1 \cos\vartheta_\mathrm{t} = 0 \;\Rightarrow\; \frac{n_2}{n_1} = \frac{\cos\vartheta_\mathrm{t}}{\cos\vartheta_\mathrm{i}}$$

We now apply Snell's law:

$$\frac{n_2}{n_1} = \frac{\sin\vartheta_\mathrm{i}}{\sin\vartheta_\mathrm{t}} \;\Rightarrow\; \frac{\cos\vartheta_\mathrm{t}}{\cos\vartheta_\mathrm{i}} = \frac{\sin\vartheta_\mathrm{i}}{\sin\vartheta_\mathrm{t}} \;\Rightarrow\; (\cos\vartheta_\mathrm{t})\cdot(\sin\vartheta_\mathrm{t}) = (\cos\vartheta_\mathrm{i})\cdot(\sin\vartheta_\mathrm{i})$$

In order for the above mathematical relationships to occur, the angles of incidence and refraction must be complementary: $\vartheta_\mathrm{i} + \vartheta_\mathrm{t} = 90°$, so $\cos(\vartheta_\mathrm{t}) = \sin(\vartheta_\mathrm{i})$. Now the angle of incidence is Brewster's angle, which was first introduced by Eq. (2.30):

**Brewster's Angle**: $$\vartheta_\mathrm{B} = \tan^{-1}(n_2/n_1) = \tan^{-1}(n_{21}) \quad (2.38)$$

For this particular angle, the parallel polarization reflection coefficient is zero; this polarization component is not reflected at all. We just use $\vartheta_\mathrm{i} = \vartheta_\mathrm{B}$ in the Fresnel relationships given by Eqs. (2.34) and (2.35):

$$\rho_{\mathrm{s}\,\vartheta=\vartheta\mathrm{B}} = \cos(2\vartheta_\mathrm{B}) \quad \text{and} \quad \rho_{\mathrm{s}\,\vartheta=\vartheta\mathrm{B}} = 0 \quad (2.39)$$

Regarding the perpendicular polarization, part of it is reflected and part is refracted, according to the coefficients described in the relationships given by Eq. (2.36).

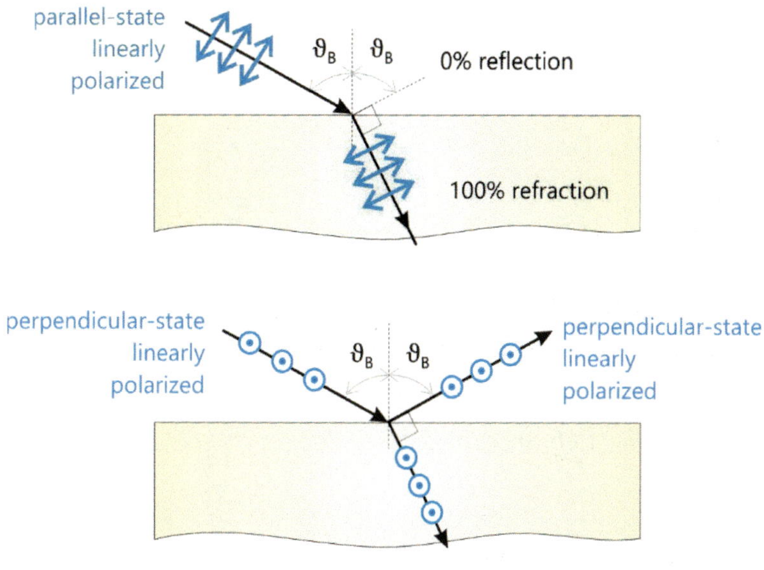


*Figure 2-44: Incidence under Brewster's angle: (top) parallel-state linearly polarized light and (bottom) perpendicular-state linearly polarized light.*

If the incident light is unpolarized, the reflected light contains only the perpendicular polarization state, linearly polarized perpendicular to the plane of incidence. For example, for reflection from a surface with $n_1 = 1.0$ and $n_2 = 1.4$ under an angle of incidence $\vartheta_i = \vartheta_B = \tan^{-1}(1.4) \approx 54°$, the reflected wave is linearly polarized (perpendicular state), and the refracted light is partially polarized, mostly in the parallel state (Figure 2-45).

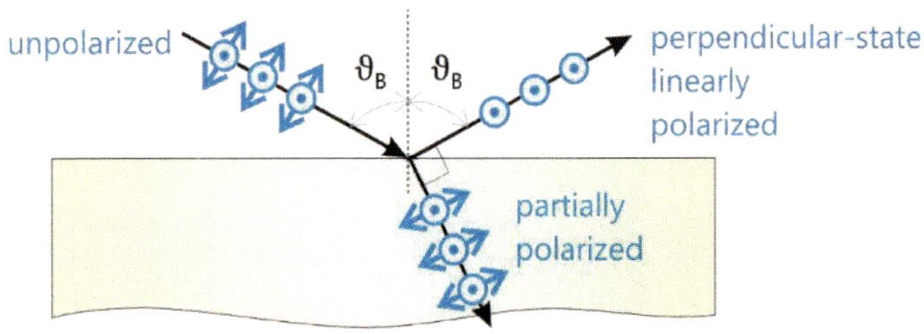


*Figure 2-45: Incidence under Brewster's angle when the incident light is unpolarized.*

Another application of this effect is polarizers operating by multiple refractions. An array of multiple parallel plates with successive reflections and refractions at Brewster's angle increases the degree of polarization in the refracted beam.

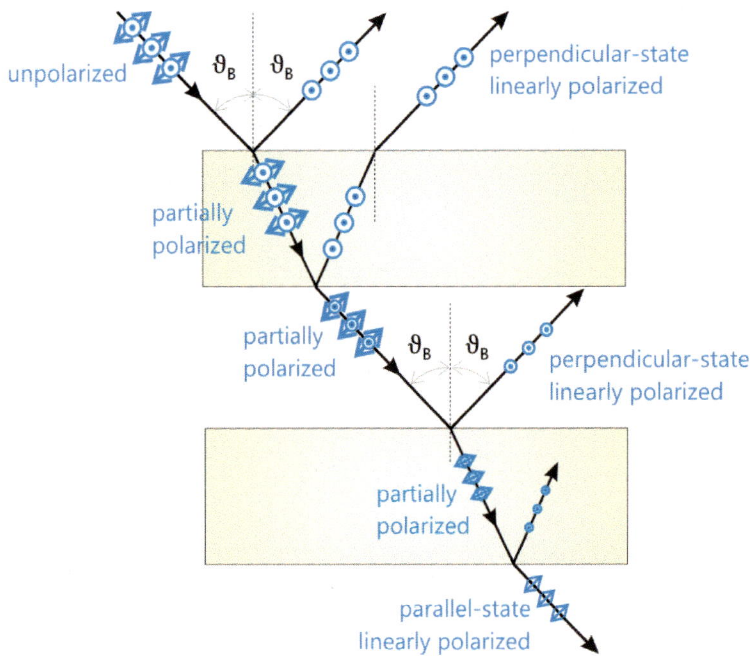


*Figure 2-46: Transmission through multiple plates results in linearly polarized transmitted light.*

Nature's most spectacular display, the rainbow, has a strong linear polarization state. This is because the rainbow ray emerges from the droplet at an angle of $\vartheta_t$ ($\approx$ 59°)[13], which is near Brewster's angle of 53.08° for water.

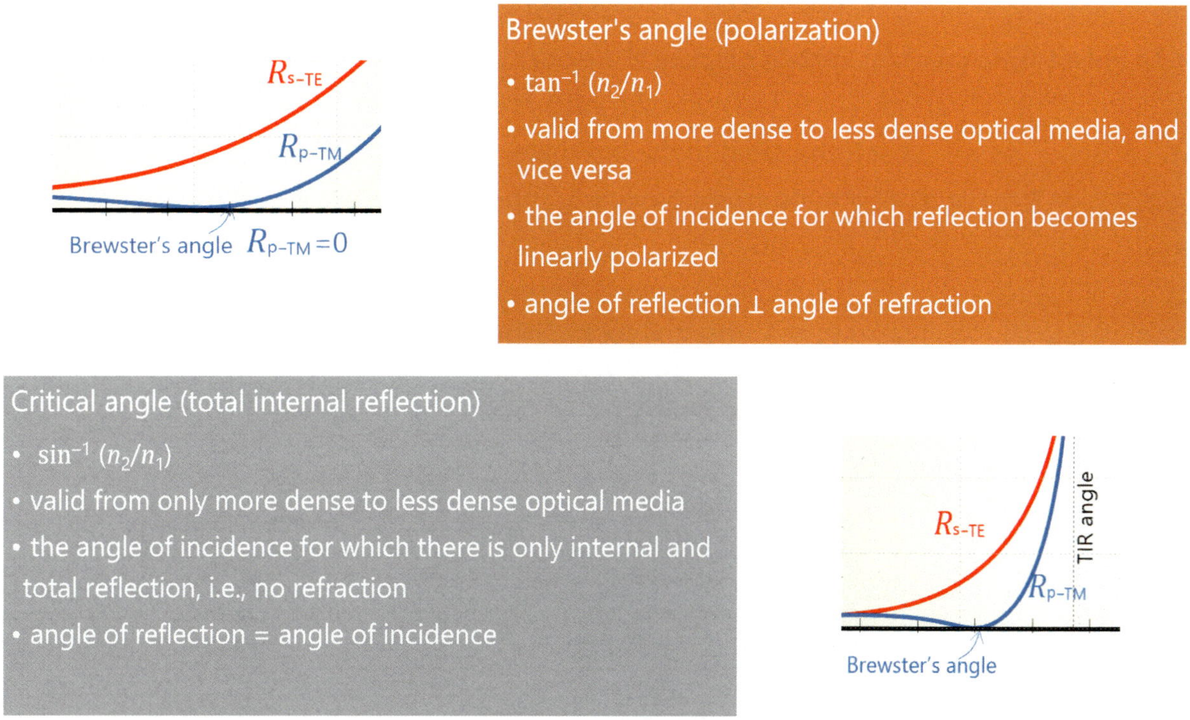


*Figure 2-47: Brewster's angle versus TIR critical angle.*

*Figure 2-48: Photographs (left) without and (right) with a linear polarizer. The strong surface reflection is eliminated once a linear polarizer is used, indicating that the surface reflection light (for example, from the windshield and the hood) is linearly polarized. (Photos by Manutsawee Buapet www.bmanut.com used with permission.)*

[13] *Introduction to Optics* § 3.5.2 Prismatic Atmospheric Phenomena.

# 2.6 POLARIZATION IN ANISOTROPIC MEDIA

## 2.6.1 Naturally Occurring Birefringence

In an **anisotropic medium**, certain properties are orientation dependent. Specific to light, a medium is optically anisotropic if the refractive index is dependent on the direction of propagation. As we will see, this owes to the fact that a property known as induced electric polarization $\mathbf{P}$ is not parallel to its causation, which is the applied electric field $\mathbf{E}$. The propagation of polarized light in anisotropic media has provided some early insights into the nature of the refractive index (which is discussed in its entirety in Chapter 3) in the form of **birefringence** or double refraction. The Danish scientist Erasmus Bartholin is credited with the first study of birefringence, published in *Experimenta Crystalli Islandici Disdiaclasici* (1669).

*Figure 2-49: Erasmus Bartholin (1625–1698) and his published work on birefringence in crystals.*

Today, birefringence is employed in medical diagnostics. In optical microscopes fitted with a pair of crossed polarizing filters, light from the source is polarized in the $-y$ direction after the first polarizer. An analyzer oriented in the $-y$ direction is placed over the specimen such that no light from the source is accepted by the analyzer, and the field appears dark. However, birefringent areas from the sample will couple some of the $-x$ polarized light into the $-y$ polarization; these areas will then appear bright against a dark background.

Binocular retinal birefringence screening of the Henle fibers (photoreceptor axons that grow radially outward from the fovea) provides a reliable detection of strabismus and possibly of anisometropic amblyopia.[14] In addition, scanning laser polarimetry utilizes the birefringence of the optic nerve fiber layer to indirectly quantify its thickness; this technique is used in the assessment and monitoring of glaucoma.

[14] Jost RM, Felius J, Birch EE. High sensitivity of binocular retinal birefringence screening for anisometropic amblyopia without strabismus. Journal of American Association for Pediatric Ophthalmology and Strabismus. JAAPOS. 2014; 18(4):e5–e6.

Unique manifestations of birefringence, as observed in a calcite crystal, are the following:

- There are two, instead of one, rays in the crystal. Thus, a double image appears (Figure 2-50).
- Rotation of the crystal results in a corresponding rotation of one of the two images.

When we direct a thin beam (such as a HeNe laser) through such a crystal,

- At the exit side of the crystal (parallel cut), the two rays are spatially separated.
- For propagation along a specific direction in the crystal (optic axis), there is only one ray.

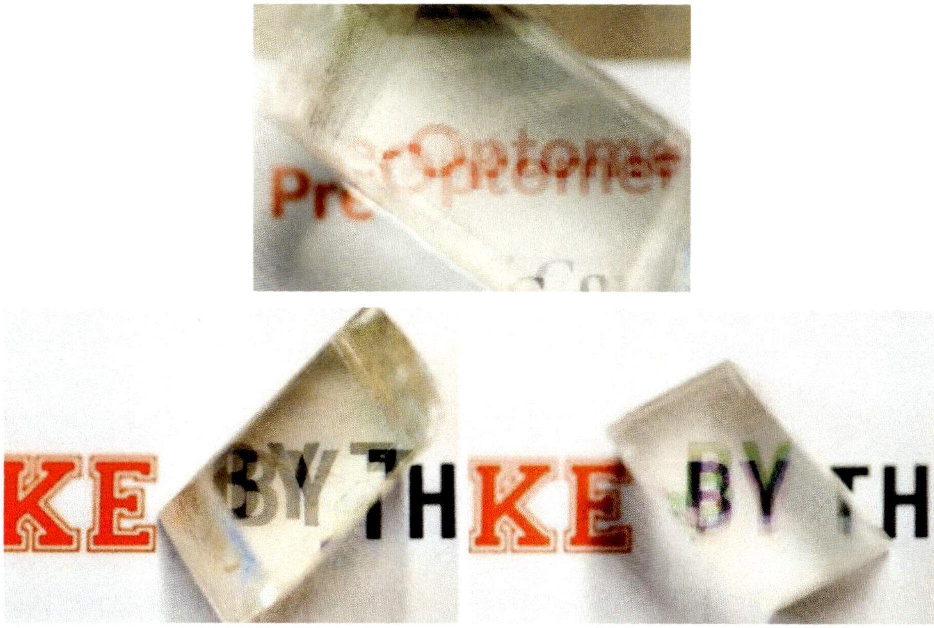


*Figure 2-50: Birefringence demonstration: (top) the word 'Optometry' appears as a double image. (bottom) Extraordinary image (left) adjacent to 'BY' and (right) above 'BY' as the crystal is rotated 90°.*

One of the two rays, the **ordinary ray** (angle of refraction $\vartheta_o$), appears to obey Snell's law. The other one, the **extraordinary ray** (angle of refraction $\vartheta_e$), appears not to obey Snell's law!

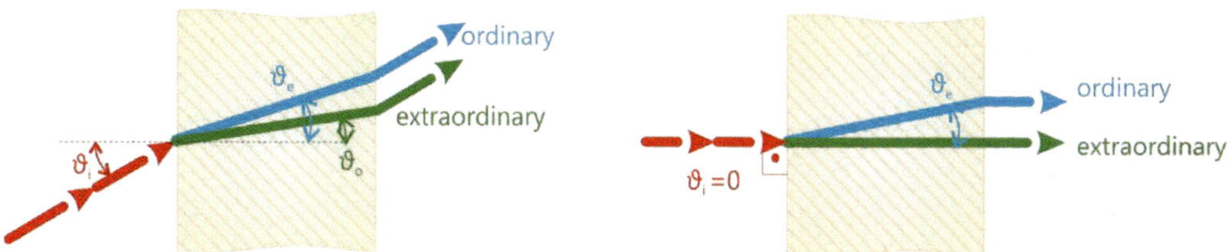


*Figure 2-51: Ordinary ray and extraordinary ray in a birefringent crystal with (left) oblique incidence and (right) normal incidence.*

Even in normal incidence [Figure 2-51 (right)], where the law of sines provides a zero refraction angle, we again see two rays, the ordinary ray, which obeys the law by not being deflected, and the extraordinary ray, which is deflected, in apparent 'defiance' of the law. (This is the ray that rotates with the crystal rotation.) Therefore, we cannot interpret the phenomenon by merely assuming that the two rays simply correspond to different refractive indices. Is the law of refraction applied selectively? What propagates inside the crystal, under what law, and in which directions?

The general principle obeyed by light is that of the least optical path—an equivalent expression of which is the continuation of the parallel component of the wavevector. This principle will be employed to interpret birefringence.

With Snell's law, we can find the direction of both the ray and the wavevector in a material. However, this only applies when light propagates in an **isotropic medium**—a medium that displays the same optical properties (e.g., same refractive index) for all directions and for all polarization eigenstates, and that retains the original polarization beam status, etc. Each point of an incident wavefront introduces a light-emitting point source inside the medium. Such a model was presented in § 2.5.2. The mechanism is nonresonant absorption, in which the wave's electric field oscillation sets in motion a dipole, which, in turn, re-emits in all directions an electromagnetic wave with the same frequency and oscillation field.

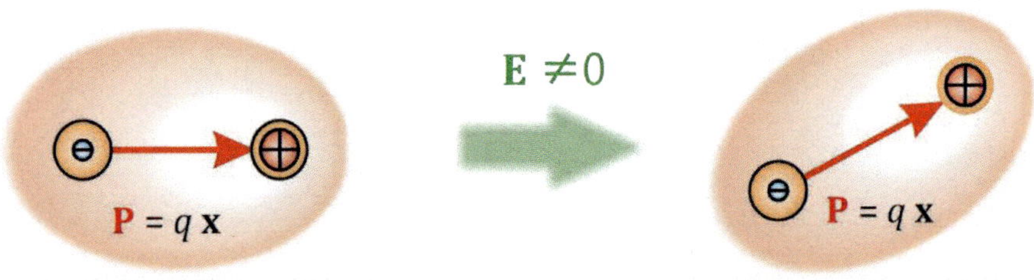


*Figure 2-52: Development of electric polarization in a dipole under an external field when the medium is (left) isotropic and (right) anisotropic.*

Whether a medium is optically isotropic or not is determined by its response to an external electric field. On a microscopic scale, the response is the induced electric polarization $\mathbf{P}$. If the induced electric polarization is parallel to the stimulus (the electric field) in any direction, the medium is isotropic. If it is not parallel, the medium is **anisotropic**. In such a system, the average response is expressed by the tensor matrix relationship:

$$\mathbf{P} = \varepsilon_o[\chi_{ij}]\cdot\mathbf{E} \tag{2.40}$$

while the electric displacement $\mathbf{D}$ is

$$\mathbf{D} = \varepsilon_o\mathbf{E} + \mathbf{P} = \varepsilon_o\mathbf{E} + \varepsilon_o[\chi_{ij}]\cdot\mathbf{E} = \varepsilon_o[\varepsilon_{ij}]\cdot\mathbf{E} \tag{2.41}$$

In an anisotropic medium, the vectors of the induced electric polarization $\mathbf{P}$ and of the electric field $\mathbf{E}$ are no longer parallel to each other. The elements of the electric susceptibility

3×3 tensor $[\chi_{ij}]$ are associated with the refractive index, and the refractive index is no longer a numerical constant. If the reference system is the system of the principal axes of the medium, i.e., the system where the tensor $[\chi_{ij}]$ can be diagonalized, then the refractive indices along the principal axes are expressed by the following nonzero elements:

$$n_x = n_{11} = \sqrt{1+\chi_{11}}, \quad n_y = n_{22} = \sqrt{1+\chi_{22}}, \quad \text{and} \quad n_z = n_{33} = \sqrt{1+\chi_{33}} \tag{2.42}$$

Any plane formed by the principal axes is a principal crystallographic plane. If that plane is also perpendicular to a crystal edge, it is a principal section.

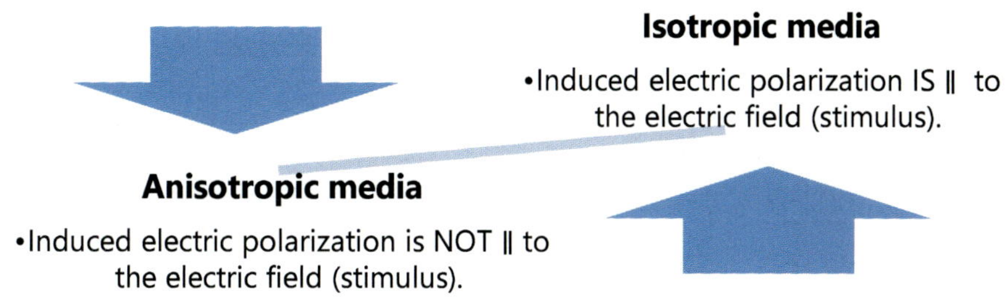


In an isotropic medium, $n_x = n_y = n_z$. Therefore, the propagation speed does not depend on the wavevector direction. Over time $t$, the equiphasic elementary surfaces are, for all directions, spherical with a radius $u \cdot t$, where $u = c/n$. The new wavefront results from the common tangent formed by the elementary spheres. If the incident wavefront is plane, the new wavefront is also plane and parallel to the incident wavefront [Figure 2-53 (left)].

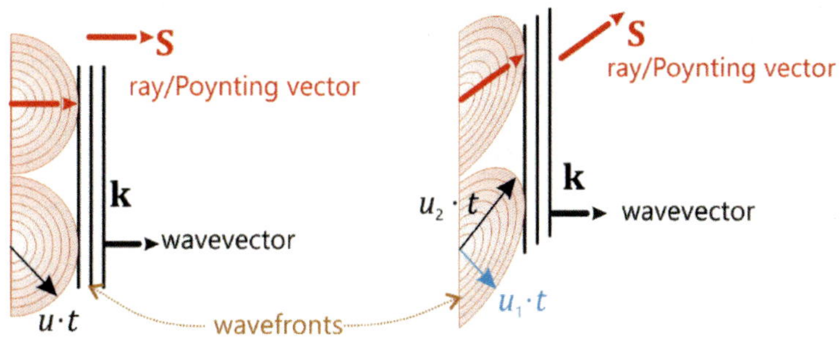


*Figure 2-53: Elementary equiphasic surfaces for (left) an isotropic medium and (right) an anisotropic medium.*

In an anisotropic medium [Figure 2-53 (right)], the optical properties depend on the direction of propagation, and the elementary equiphasic surfaces lose their symmetry. The simplest case of anisotropy under the system of principal axes $n_x = n_y \neq n_z$ involves a **uniaxial medium**. In general, $n_x \neq n_y \neq n_z$ describes a **biaxial medium**.

To study exactly how an electromagnetic wave propagates in a medium, we use the Lorentz oscillator model (see § 3.1.2), a mechanical analog of an electrical bipolar oscillator. The model employs a 3-D system, in which each axis corresponds to a spring whose constant impedance relates to the resonance frequency of the medium along this axis.

In an isotropic medium, the spring constant along all axes is the same, so $n_x = n_y = n_z$. In an anisotropic medium, each different value of $n$ is described by a different spring constant along the corresponding direction, the $-z$ direction in this case.

What materials exhibit such behavior? Certainly, not amorphous materials. A material that exhibits such behavior must be a crystal. Atoms in a crystal are arranged in strict sets of grid positions. The optical properties along the axes of the grid are not, by default, equivalent; the separation of atoms is often not the same for all directions. Suitable materials are crystals that have some degree of asymmetry in which the individual binding forces are not equal along the axes. Thus, different natural resonance vibration frequencies exist in certain directions; in other words, we have optical anisotropy.

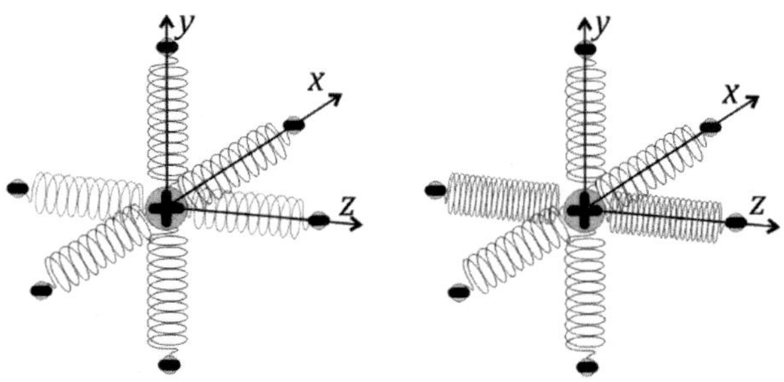


*Figure 2-54: (left) Isotropic and (right) anisotropic 3-D harmonic oscillator model. The spring constants are identical in the left image, while the constant along the –z axis is different in the right image.*

In cubic crystal systems, there is symmetry along the principal axes, so these systems do not exhibit anisotropy. In triangular, tetragonal, and hexagonal systems, one of the springs is different; these are uniaxial systems. Biaxial materials, in which all three pairs of springs along the principal axes have different elastic constants, belong to the triple, monoclinic, and orthorhombic systems. Monoclinic-class biaxial materials are mica minerals, which can be either muscovite or biotite. The best-known uniaxial crystal is the calcite called Iceland spar [chemical formula $CaCO_3$ (also known as Doppelspat), belonging to class $-3m$, or R-3$c$] with lattice constants[15] $a = b = 4.989$ Å and $c = 17.062$ Å, forming a rhomboid.

The peculiarity of this crystal is due to the geometry of the $CO_3$ group. The arrangement of oxygen atoms around the carbon atom is such that what is formed is an oblique plane

[15] Graf DL. Crystallographic tables for the rhombohedral carbonates. Am Mineral. 1961; 46(11-12):1283-316.

perpendicular to the axis that joins two opposing calcium atoms (Figure 2-55). Along this axis direction, the electrons are subjected to a strong spring. For this study, we consider a 3-D isotropic harmonic oscillator.

In such an anisotropic medium, the propagation speed of an electromagnetic wave for a particular direction of propagation depends on the direction of oscillation of the electric field. Consider the propagation of such a wave in the medium modeled in Figure 2-56. As in any propagation direction, in this direction, all oscillation field orientations are possible on the transverse plane, the plane that is perpendicular to the wavevector $\mathbf{k}$. From all possible oscillations, we distinguish two cases: (1) oscillations along the intersection of the main $x$–$y$ plane and the transverse plane (vector $\mathbf{E}_{\vartheta o}$) and (2) those perpendicular to the oscillation plane (vector $\mathbf{E}_{\vartheta e}$).

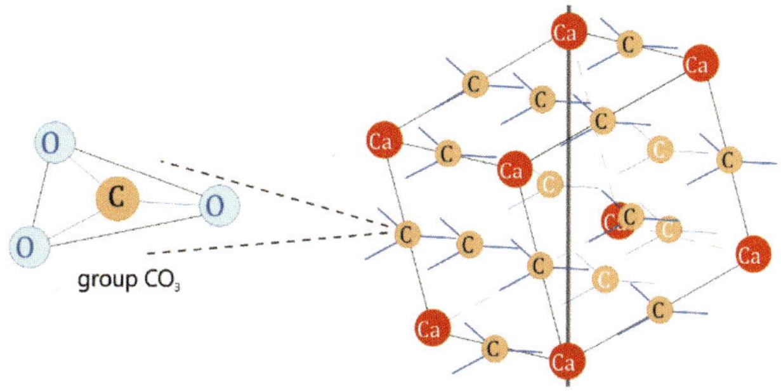


*Figure 2-55: Calcite unit cell.*

We can distinguish these two oscillations out of all of the other possible oscillations on the transverse plane because they represent completely different combinations of springs; oscillations parallel to $\mathbf{E}_{\vartheta o}$ (along the $x$–$y$ plane) correspond to the same spring constants along the –$x$ and –$y$ axes, and are not affected by a different spring constant along the –$z$ axis. In this case, –$z$ is the **optic axis**. Waves are only subject to identical spring constants along the –$x$ and –$y$ axes, so they are not affected by the crystal anisotropy.

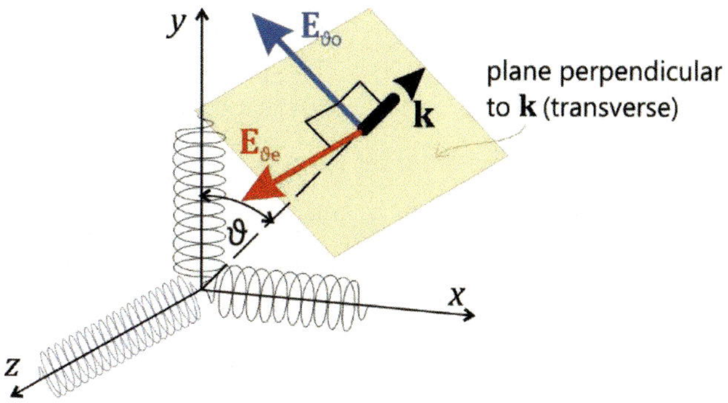


*Figure 2-56: Wave propagation along the direction of wavevector* $\mathbf{k}$ *causes two polarization eigenstates to develop.*

To the contrary, oscillations parallel to $\mathbf{E}_{\vartheta e}$ (along the $-z$ axis) are subject to this different spring constant. The different combinations of spring constants along the two polarization modes lead to the development of different propagation speeds. Thus, two polarization states (eigenstates) are developed for every direction of propagation within the medium.

It can be proven that along every direction of propagation there exist two orthogonal, linearly polarized eigenstates, each of which propagates under a different refractive index and, in general, follows different paths. In the general case in which a crystal displays both birefringence and optical activity in every direction, the initial propagation polarization eigenstates are also elliptically polarized.

The surfaces employed for visualization of wave propagation (in momentum space) are called **equiphasic surfaces**. The distance traveled in time is inversely proportional to the refractive index. For example, in an isotropic medium where there are spherical surfaces with radii $u \cdot t = \lambda = \lambda_o/n$, we note that the radii are inversely proportional to $n$. The intersection of each of such surfaces with a principal plane is a circle with a corresponding radius.

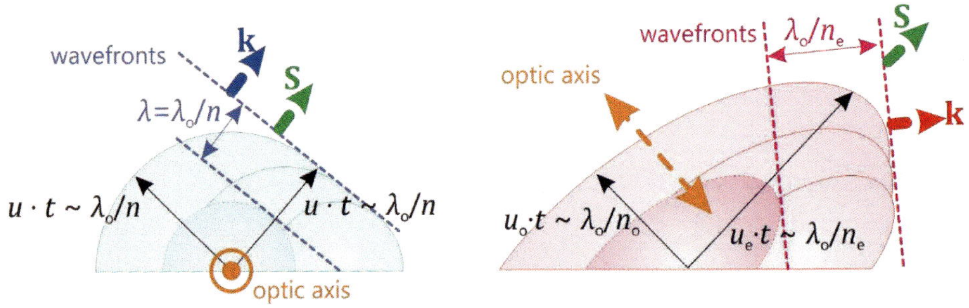


*Figure 2-57: Equiphasic surfaces for (left) an isotropic and (right) an anisotropic medium.* **S** *is the Poynting vector, and* **k** *is the wavevector.*

In an anisotropic medium, the equiphasic surfaces lose this symmetry. Suppose that $n_e < n_o$. In this case, an ellipse develops, elongated along a direction perpendicular to the optic axis with ellipticity $n_o/n_e$ (Figure 2-58). Along a particular direction (that of the major axis of the ellipse), the speed of light has the greatest value within that medium.

In accordance with the principle of the shortest optical path, rays turn toward this direction, and the new wavefronts are defined by the common tangent of all ellipses. If the incident wave is plane, the new wavefronts are again plane and parallel to the incident, but the direction of the ray is different. This direction is defined by the straight line joining an emitting point with the corresponding point adjacent to the ellipse having the new wavefront, and is roughly, but not exactly, the direction of the axis of the ellipse. This is the new extraordinary ray.

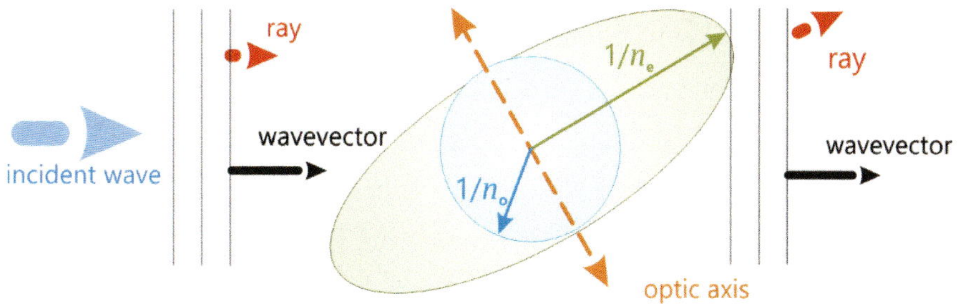


*Figure 2-58: Equiphasic surfaces for an anisotropic medium. Shown is a cross-section along the x–z plane for $n_e < n_o$. The optic axis is along the –z axis.*

On a plane of incidence that coincides with a major intersection, we consider the normal incidence of unpolarized light. We then analyze a light beam at this plane into two components, a perpendicular state [Figure 2-59 (left)] and a parallel state [Figure 2-59 (right)]. The perpendicular polarization corresponds to an oscillation along the main axis $-y$. For this state, the equiphasic surfaces are spherical; therefore, propagation follows the familiar laws of refraction—this is the **ordinary ray**.

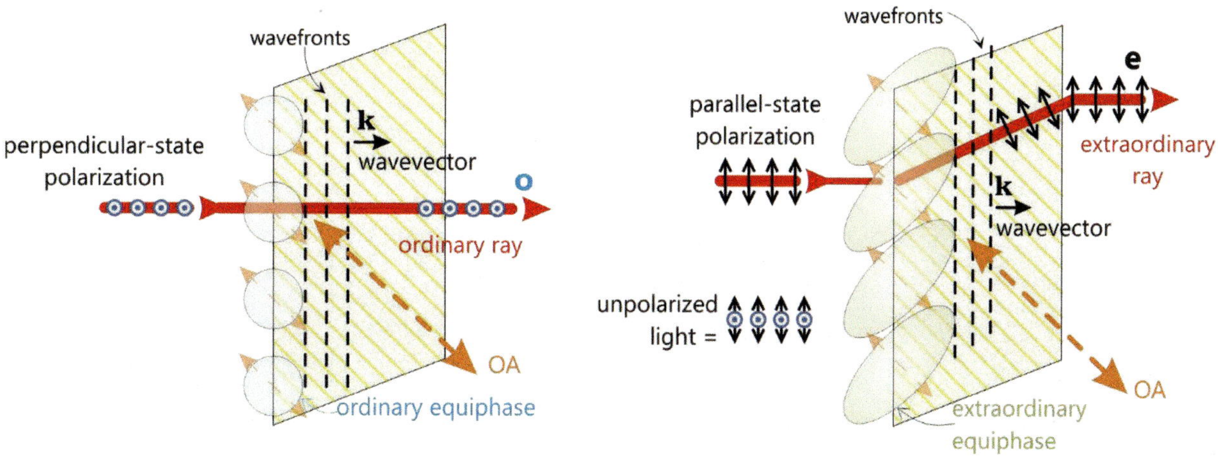


*Figure 2-59: Analysis of an unpolarized incident beam into two polarization eigenstates: (left) perpendicular state and (right) parallel state. OA denotes the optic axis.*

In contrast, the parallel polarization state is subject to a combination of $n_x$ and $n_z$, representing an anisotropic medium. The ray propagates as shown in [Figure 2-59 (right)] turning toward the direction that minimizes the propagation time—this is the **extraordinary ray**.

Careful observation shows that there is no violation of the law of refraction. The law refers to the directions not of rays, but of wavevectors, which, as shown in Figure 2-59, both remain perpendicular to the surface of the crystal. What differentiates the beams is that, in an isotropic medium, the wavevector direction coincides with the ray direction, while in an anisotropic medium (in which we may see an extraordinary beam), the directions of rays and wavevectors, in principle, do not coincide.

Thus, in a uniaxial crystal with parallel edges, if a beam is incident normally (incidence angle = 0), there exist two polarization eigenstates: one that is parallel to the plane that contains the optic axis, and one that is perpendicular to this plane. The polarization component that is perpendicular to the optic axis propagates normally, following the laws of refraction for an isotropic medium without a change in direction, and exits the crystal perpendicularly polarized with respect to the principal plane.

The horizontally polarized component—with a polarization plane that is parallel to the optic axis—propagates in another direction that is determined by the crystal orientation, and exits displaced and with a polarization parallel to the principal plane. The above applies for unpolarized or linearly polarized light with an axis ≠ 0° or 90° (as well as circularly polarized light). In general, incident light can be analyzed into two orthogonal eigenstates and treated according to the state of polarization in each eigenstate.

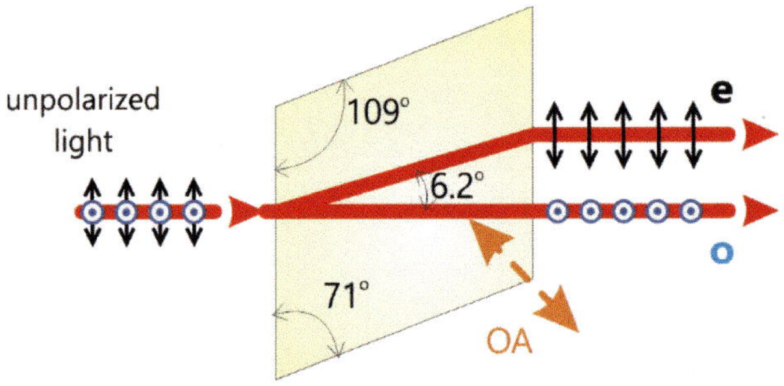


*Figure 2-60: Propagation of unpolarized light in a birefringent calcite crystal with a zero angle of incidence (normal incidence). OA denotes the optic axis.*

In the case of linearly polarized light with a polarization axis coincident with one of the two eigenstates (parallel or perpendicular to a crystal principal section),

- If the polarization state of the beam is perpendicular to the crystal principal plane, the beam propagates as shown in Figure 2-59 (left). The beam encounters an isotropic medium and propagates normally. There is, in other words, only one exiting beam, and it is an ordinary beam. Nothing is noted if the crystal rotates.

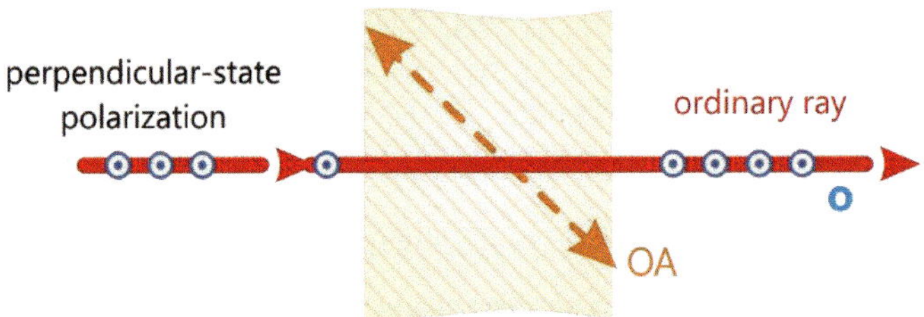


*Figure 2-61: Propagation of s-polarization light in a birefringent calcite crystal with a zero angle of incidence (normal incidence), perpendicular to a principal section.*

- If the beam polarization is parallel to the principal crystal plane, the beam propagates as shown in Figure 2-62, i.e., deflected by an angle, and upon exiting the crystal with a second refraction, it reverts to its original direction, with a parallel displacement. There is only one exiting beam, the extraordinary ray. If the crystal rotates, the optic axis, as well as the image seen through it, follows this rotation; the extraordinary beam displacement follows this rotation.

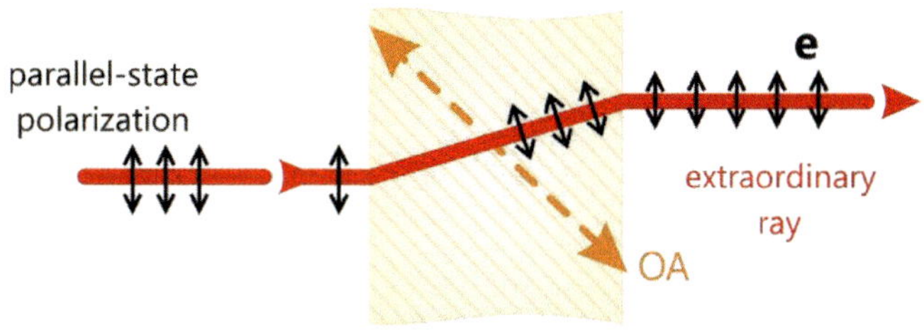


*Figure 2-62: Propagation of p-polarization light in birefringent calcite with a zero angle of incidence (normal incidence), perpendicular to a principal section.*

Uniaxial birefringent crystals with $n_e > n_o$ are positive, and those with $n_e < n_o$ are negative. Calcite at $\lambda = 589.3\,\text{nm}$ (yellow sodium line) has $n_o = 1658$ and $n_e = 1.486$, and therefore is negative, as is tourmaline with $n_o = 1.669$ and $n_e = 1.638$ (this material absorbs the extraordinary beam most strongly and, over a few millimeters thickness, also performs as a dichroic linear polarizer). Quartz is a positive uniaxial birefringent crystal with $n_o = 1.544$ and $n_o = 1.544$, as is zirconium with $n_o = 1.923$ and $n_e = 1.968$.

Refractive indices and therefore the extent of the birefringence depend on the wavelength. If we illuminate a birefringent crystal with white light and observe it with an analyzer or polarizing filter, it is likely that different extraordinary ray paths will correspond to different color components.

We distinguish two special cases of propagation for negative birefringent crystals such as calcite: parallel and perpendicular to the optic axis (Figure 2-63). For propagation parallel to the optic axis, the beam encounters exactly the same refractive index for both polarization states. This direction does not show any difference (deflection), as if we had an isotropic medium. This is an easy way to identify the optic axis!

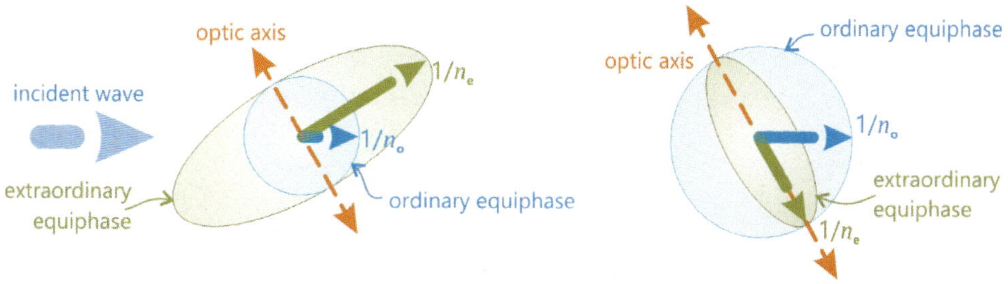


*Figure 2-63: Equiphasic surfaces for (left) negative and (right) positive uniaxial birefringent crystals.*

For propagation perpendicular to the optic axis, the ray encounters the largest difference in refractive indices between the two polarization states. The extraordinary ray encounters the lower refractive index $n_e$, while the ordinary ray encounters the higher refractive index $n_o$. If the beam is incident normally (perpendicularly) to the crystal, there is no apparent deflection for any ray because the direction of the minimum optical path is parallel to the original direction for both polarization states.

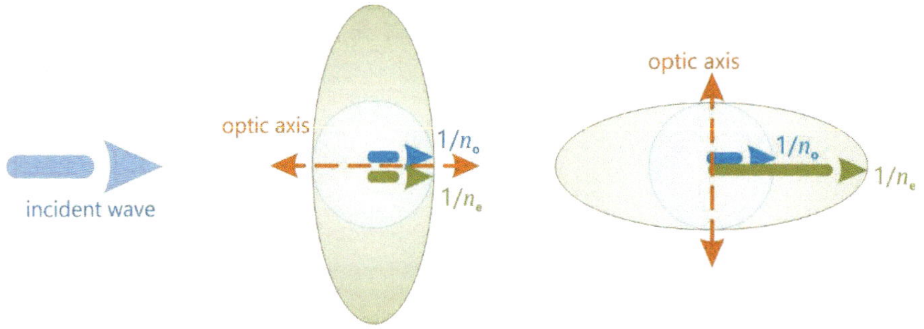


*Figure 2-64: Propagation (left) along the optic axis and (right) perpendicular to the optic axis—this is the principle of operation of retardation plates.*

However, the different refractive indices cause different propagation speeds in the medium: The ordinary beam propagates with $u_o = c/n_o$ and the extraordinary beam with $u_e = c/n_e$. In a negative crystal such as calcite, where $u_e > u_o$, the extraordinary ray is advanced. These different speeds are responsible for the development of a phase difference between the two polarization states. The extraordinary beam corresponds to a fast axis ($n_e = n_F$), while the ordinary beam corresponds to a slow axis ($n_o = n_S$). For thickness $d$ along the direction of propagation, the phase difference that develops between the slow axis and the fast axis is

$$\text{phase difference} = \frac{2\pi}{\lambda_o}\cdot(n_S - n_F)\cdot d = \frac{2\pi}{\lambda_o}\cdot(n_o - n_e)\cdot d \tag{2.43}$$

This is the principle of operation of retardation plates, such as a quarter-wave ($\lambda/4$) plate (as discussed in § 2.4.2). Thus, a $\lambda/4$ plate is constructed to develop a phase difference of 90° (or $\pi/2$). Depending on whether the crystal is negative or positive, the fast and slow axis orientations are determined in relation to the optic axis. The fast axis in negative crystals is parallel to the optic axis, while in positive crystals it is perpendicular to the optic axis.

## 2.6.2 Artificial Birefringence

Birefringence, as described so far, refers to naturally occurring, or intrinsic, birefringence. Birefringence can be caused by external factors that change (on a microscopic scale) the symmetry of the medium and modulate the dielectric tensor permeability.

Birefringence can be induced in certain materials by external or internal stress. This is mechanical birefringence, related to the photoelastic effect. Mechanical birefringence occurs in a wide range of materials such as synthetic polymers (e.g., a plastic ruler, the jelly in packaging products, or tempered glass), and in natural materials such as fibrous protein.

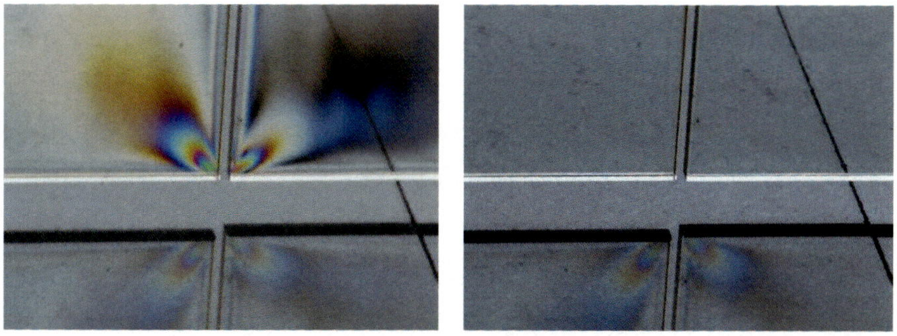

*Figure 2-65: (left) Artificial birefringence due to photoelastic stress, observed over the edge of a glass door through a polarizer. (right) The same photograph with the polarizer at crossed positions. The pattern is still seen in the reflection from the floor, indicating that the extraordinary photoelastic pattern's polarization axis forms at about 45° with respect to the horizontal.*

By placing a photoelastic material (a plastic ruler is fine) between two polarizers—or illuminating with linearly polarized light—we can observe internal stress variations, displayed as a spectrum of different colors. This is because the path of the extraordinary ray is wavelength dependent: The refractive index varies locally due to mechanical stress, thus causing different deviations between the ordinary and extraordinary rays for different wavelengths. We can observe this by viewing a tempered windshield through a set of polarizing filters.

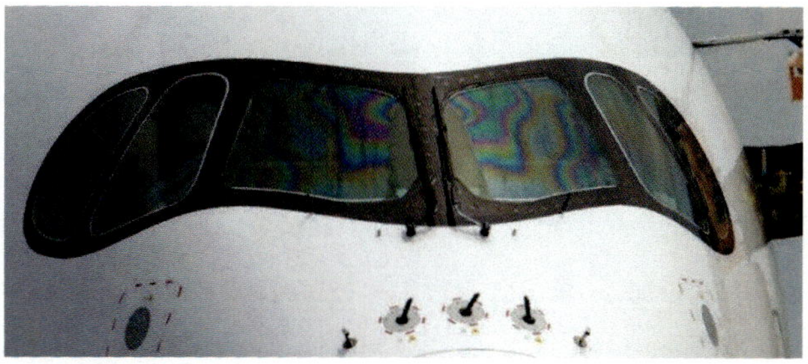

*Figure 2-66: The cockpit windows of the Airbus A350, one of the most-modern airliners, displaying artificial birefringence when viewed through a polarizer (photo by Weimeng from www.airliners.net used with permission).*

## 2.6.3 Liquid Crystal Display (LCD) Operation

A liquid crystal display (LCD) is a thin, flat panel employed for electronically displaying information such as text, images, and video. In today's world, LCDs are omnipresent. They are used as computer monitors, televisions, panels for smartphones, panels in aerospace, etc. A key element for their operation is polarization. Compared to the traditional cathode-ray tube (CRT) monitors, LCDs are far thinner and more lightweight, have significantly lower power consumption, emit no electromagnetic field, and have a longer service life.

**Liquid crystals** are (mainly organic) substances that do not melt directly to the liquid phase. There is an intermediate para-crystalline phase in which the molecules are partially oriented. In this stage, a liquid crystal is a translucent fluid, having some solid matter properties, as well. We can perceive liquid crystals as molecules 'swimming' in a liquid crystalline phase. An interesting property is that they can be aligned precisely when subjected to electric fields, just as metal shavings line up in a magnetic field. Another interesting characteristic is that they exhibit the property of **optical activity**, which is the ability of a chiral molecule to rotate the plane of polarization of linearly polarized light as it travels through the medium. Inherent anisotropy is related to the optical activity of liquid crystals.

Liquid crystals are encountered in nature, where they can be found in biological cell membranes. The slime secreted by certain slugs has liquid crystal properties; the rod-shaped molecules align in varying degrees to control their viscosity in order to adapt to different surface conditions. DNA and cell membranes have liquid crystal phases.

Types of liquid crystals are the smectic (*σάπων*, soap), whose molecules are parallel to one another, but with no periodic pattern. Nematic (*νήμα*, thread) liquid crystals are formed by rod-like molecules oriented parallel to each other, but they do not have a layered structure. Cholesteric liquid crystals are formed by parallel molecules whose layers are arranged in a spiral fashion. For liquid crystals to be applied in LCD technology, they have to be transparent and conductive.

The history of liquid crystals dates back to 1888, owing to the work of the Austrian botanist Friedrich Reinitzer. Reinitzer melted a cholesterol-like substance, which at first became a cloudy liquid and then cleared as its temperature rose. Upon cooling, the liquid turned blue, before crystallizing. The first experimental LCD was developed at the RCA Labs in 1968; the first LCD was produced a few years later, initially in quartz watches and early calculator displays.

Each LCD is composed of an assembly of many picture elements, or **pixels**, arranged in a matrix. The notion of a megapixel derives exactly from this arrangement and refers to the product of the number of pixels along the $-x$ axis times the number of pixels along the $-y$ axis. A computer screen of 2726 × 1824 pixels has nearly 5 million pixels. This is a 5 Mpixel screen.

*Figure 2-67: The cockpit instrument panels in the Airbus A 350. Six large (15-inch) LCDs display a multitude of information.*

Each pixel corresponds to a cell containing twisted nematic and cholesteric molecules. Inside such a pixel well, these nematic molecules are 'swimming' in a random orientation. With adequate optical thickness, the optical activity may result in a polarization plane rotation of exactly 90° [Figure 2-68 (left)]. If, however, the molecules are untwisted and properly oriented—this can be achieved by applying a voltage inside the well—they lose their optical activity [Figure 2-68 (right)]. Properly oriented molecules do not rotate the polarization axis.

The cell-cavity element is placed between a set of crossed linear polarizers and is illuminated from the backside with white unpolarized light (backlight illumination). (In addition to the backlight principle, the reflectivity principle is also involved). The first polarizer, oriented vertically, permits the transition of vertically polarized light inside the element cell. If no voltage is applied, the polarization plane rotates 90°. The second polarizer is aligned horizontally, so full light transmission takes place. The pixel is ON.

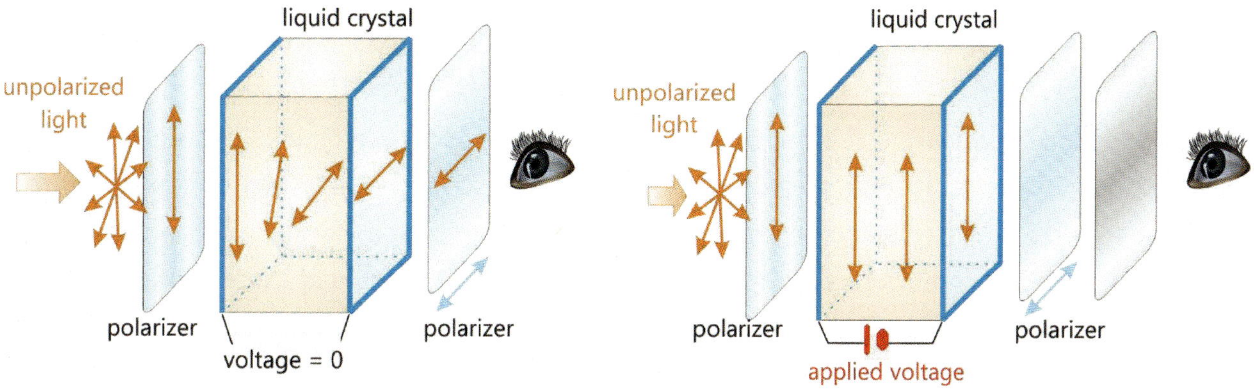


*Figure 2-68: Principle of operation of a liquid crystal display: (left) an ON pixel and (right) an OFF pixel.*

By applying a full voltage to the crystal, there is no polarization plane rotation, so the second—horizontal—polarizer does not let any light transition. The pixel is OFF. By application of a lower voltage, the rotation of the polarization plane inside the crystal is $< 90°$, and light passes through the second polarizer according to Malus' law (presented in § 2.3.2).

We summarize so far:

- Liquid crystal elements do not emit light of their own. [This contrasts with other displays such as plasma, CRTs, and organic light-emitting devices (OLEDs).] They modulate transmitted light by employing (1) polarization, (2) their inherent optical activity, and (3) an interaction with an applied electric field.
- In the liquid crystal element, application of an electric voltage inhibits the optical activity.
- The first polarizer creates a polarization state of light inside the crystal, while the second polarizer plays the role of an analyzer.
- Output light from an LCD is linearly polarized—we can see this by observing a computer screen through polarized sunglasses!
- The contrast ratio, the ratio of minimum brightness to maximum brightness, is affected by the polarizer's extinction coefficient (see § 2.3.2.1).

LCDs have evolved considerably in recent years. Initially, there were passive matrix displays, which were succeeded by active matrix displays. In the passive displays, pixels were addressed one at a time by a row-and-column matrix, resulting in a slow response time and poor contrast, with substantial cross talk (ghosting) as neighboring pixels affect each other, reducing grayscale. Active displays apply an individual switching transistor action that addresses each picture element separately and simultaneously.

Up to this point, we have described elements whose output is black and white. Light that transcends the crystal results in a white output, while light not passing results in a black output. Correspondingly, the screen output is grayscale, gray being an intermediate condition of partially transmitted light. In the digital world, grayscale typically means 256 shades of gray.

How do we achieve a color screen? Look carefully through a magnifying lens at such a screen (of a cell phone, for example; a drop of water on the screen will do—just do not sneeze on your screen!). Over a white pixel area, you will be surprised to see that the pixels are not white, but consist of three colors. In a white pixel area, there are groups of three adjacent subpixels that are usually red, green, and blue.

Each pixel, indeed, operates in a grayscale mode. By placing a colored filter (described in § 3.4.1) directly after the pixel exit, the grayscale variations become the brightness for the red, the green, or the blue (RGB). Thus, each red, green, and blue subpixel can display 256 different shades of its corresponding color.

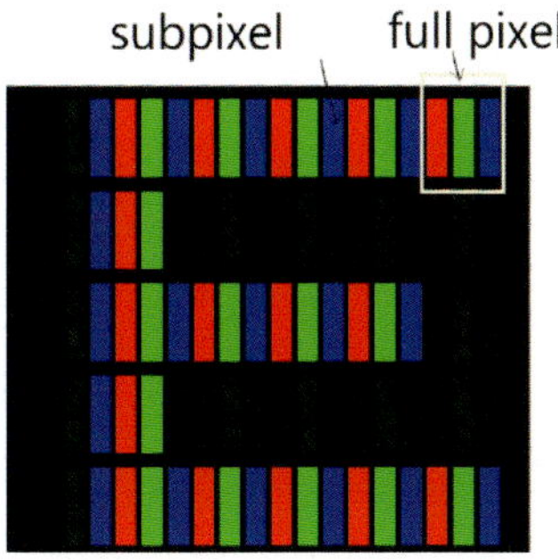


*Figure 2-69: A pixel is composed of three different-colored subpixels.*

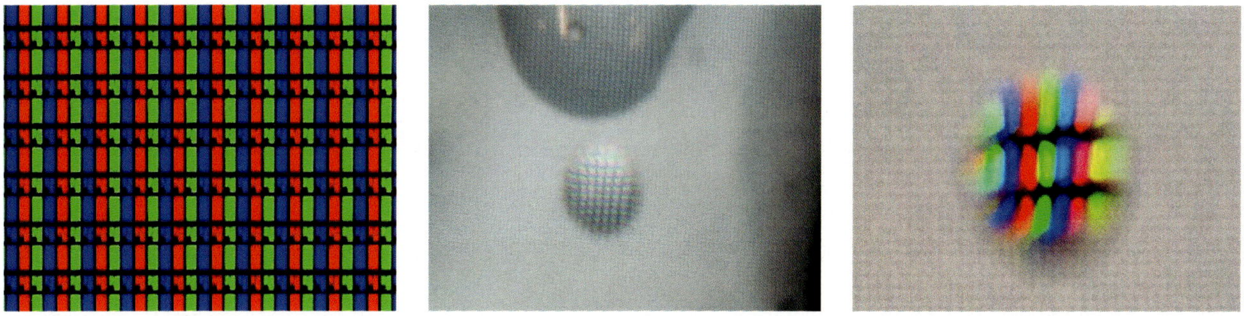

*Figure 2-70: (left) Microscopic image of pixels of a white area in a liquid crystal display. A magnifying lens or a drop of water on the screen (center and right) reveals the colors of the subpixels.*

To obtain white, the RGB values are all maxed out, or nearly so, and made equal to each other. If they are not even, the output can be tinted. To obtain black, the RGB values are all zero, or nearly so, and equal to each other. Again, if they are not even, the color will be tinted. By combining different brightnesses, through control and variation of the voltage applied on each of the three colored subelements, we obtain a palette of 16.8 million colors ($256^3$), explaining the impression of color on our screen. A simple, color SVGA (super video graphics array) screen has $800 \times 600 \times 3 = 1.44 \times 10^6$ of such subpixels.

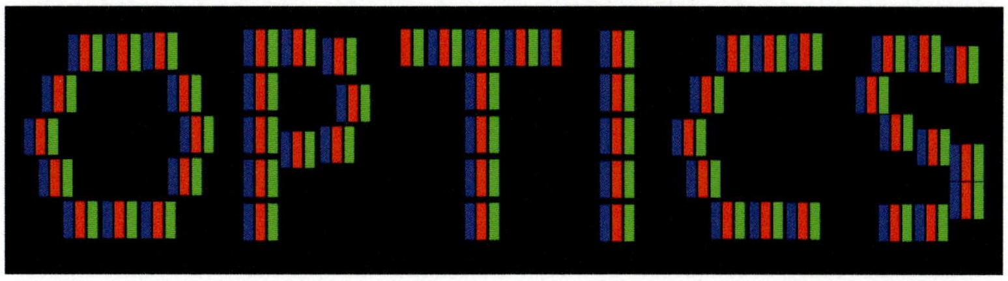

*Figure 2-71: From a distance, the word 'optics' appears white.*

# 2.7 Polarization Quiz

1) The concept of light being a transverse wave can be established by what relationships (two correct answers)?
   a) The electric field vector is perpendicular to the electric amplitude vector.
   b) The electric field vector is perpendicular to the magnetic field vector.
   c) The magnetic field vector is perpendicular to the magnetic amplitude vector.
   d) The electric field vector is perpendicular to the direction of propagation.
   e) The magnetic field vector is perpendicular to the direction of propagation.

2) Which of the following is an inherent property of an electromagnetic wave?
   a) The electric field vector is perpendicular to the electric amplitude vector.
   b) The electric field vector is perpendicular to the magnetic field vector.
   c) The magnetic field vector is perpendicular to the magnetic amplitude vector.
   d) The magnetic field amplitude magnitude equals the electric field amplitude magnitude.

3) Consider a natural (unpolarized) wave of light propagating along direction $-z$. The electric field vector oscillates in a direction that has components ...
   a) $-z$ but not $-x$ and/or $-y$
   b) $-x$ and/or $-z$ but not $-y$
   c) $-x$ but not $-y$ and/or $-z$
   d) $-x$ and/or $-y$ but not $-z$

4) In a linearly polarized wave of light propagating along direction $-z$, the electric field vector oscillates in a direction that has components ...
   a) $-z$ but not $-x$ and/or $-y$
   b) $-x$ and/or $-z$ but not $-y$
   c) $-x$ but not $-y$ and/or $-z$
   d) $-x$ and/or $-y$ but not $-z$

5) I am really confused. For both a natural, nonpolarized (NP) wave and a linearly polarized (LP) wave, when either propagates along direction $-z$, the electric field vectors oscillate in a direction in which the field has components $-x$ or $-y$ but not $-z$. What is their real difference (select two)?
   a) In the NP wave, there is always a $-x$ and/or $-y$ component; in the LP wave, there is either a $-x$ and/or $-y$ component, but not both components at the same time.
   b) In the NP wave, the relationship between the instantaneous $-x$ or $-y$ components is random and rapidly changing; in the LP wave, the relationship between the instantaneous $-x$ or $-y$ components is fixed.
   c) In the NP wave, on average, the proportion of the $-x$ to $-y$ components is about 50–50, while in the LP wave, there may be only a $-x$ component, or only a $-y$ component.
   d) In the NP wave, the relationship between the instantaneous $-x$ or $-y$ components is fixed; in the LP wave, the relationship between the instantaneous $-x$ or $-y$ components is random and rapidly changing.

6) In plane-polarized light propagating along direction $-z$, the electric field vector oscillates in a direction that has components only along the $-y$ axis. The polarization plane is ...
   a) plane $x$-$y$
   b) plane $y$-$z$
   c) plane $x$-$z$

7) The polarization axis of the wave described in Q 6 is ...
   a) axis $-x$
   b) axis $-y$
   c) axis $-z$

8) In plane-polarized light propagating along direction $-z$, the electric field vector oscillates in a direction that has equal-strength components along the $-x$ axis and the $-y$ axis. The polarization axis is ....
   a) on the $x$-$z$ plane, forming an angle of 45° with the $-x$ axis
   b) on the $x$-$y$ plane, forming an angle of 45° with the $-x$ axis
   c) on the $y$-$z$ plane, forming an angle of 45° with the $-z$ axis
   d) on the $x$-$z$ plane, forming an angle of 45° with the $-y$ axis

9) In partially polarized light, there is a mix of polarized intensity $I_P$ and an equal amount of nonpolarized intensity $I_{NP}$. What is the degree of polarization $p$?
   a) 0.00
   b) 0.33
   c) 0.50
   d) 0.90
   e) 1.00

10) Back to Q 9. When this light passes via an analyzer that completes a full 360° rotation, the maximum ($I_{MAX}$) light intensity detected is 1.0 (arbitrary units). What is the minimum ($I_{MIN}$) light intensity detected?
   a) 0.00
   b) 0.33
   c) 0.50
   d) 0.90
   e) 1.00

11) In partially polarized light, the degree of polarization is $p$ = 0.33. What is the relationship between the polarized intensity $I_P$ and the amount of nonpolarized intensity $I_{NP}$?
   a) $I_P = 0.3\ I_{NP}$
   b) $I_P = 0.5\ I_{NP}$
   c) $I_P = I_{NP}$
   d) $I_P = 2\ I_{NP}$

12) Linearly polarized light whose polarization axis is vertical (90° with the horizontal) passes via an analyzer whose axis forms 45° with the horizontal. If the polarized light intensity $I_P$ equals 100 units (arbitrary), what amount of light passes through the analyzer?
   a) 0 units
   b) 10 units
   c) 50 units
   d) 100 units

13) What is the polarization status of the light emerging from the analyzer in Q 12?
   a) unpolarized
   b) linearly polarized along the vertical (90° with the horizontal)
   c) linearly polarized along the 45° with the horizontal
   d) linearly polarized along the horizontal

14) Unpolarized light with an initial intensity of 100 units (arbitrary) passes via a linear polarizer A whose axis is along the vertical. What is the intensity of the light passing the polarizer, and what is the polarization status?
   a) 100 units, linearly polarized along the vertical
   b) 50 units, linearly polarized along the vertical
   c) 25 units, linearly polarized along the vertical
   d) 12.5 units, linearly polarized along the vertical
   e) 50 units, unpolarized
   f) 0 units (no light)

15) Unpolarized light with an initial intensity of 100 units (arbitrary) passes via two cascaded linear polarizers. Polarizer A has its axis along the vertical; polarizer B has its axis along the horizontal. What is the intensity of the light passing polarizer B, and what is the polarization status?
   a) 100 units, linearly polarized along the horizontal
   b) 50 units, linearly polarized along the vertical
   c) 25 units, linearly polarized along the horizontal
   d) 12.5 units, linearly polarized along the vertical
   e) 50 units, unpolarized
   f) 0 units (no light)

16) Back to Q 15. We now insert a third polarizer C between polarizers A & B. Polarizer C has its axis along the 45° direction with respect to the horizontal. What is the intensity of the light eventually passing polarizer B, and what is the polarization status?
   a) 50 units, linearly polarized along the 45°
   b) 25 units, linearly polarized along the 45°
   c) 12.5 units, linearly polarized along the horizontal
   d) 12.5 units, linearly polarized along the 45°
   e) 0 units (no light)

17) A twist (literally) to Q 16. Polarizer C is knocked out of its alignment along the 45° to an unknown angle with respect to polarizer A (and, subsequently, to polarizer B). The intensity of the light eventually passing polarizer B is now 5.16 units (down from 12.5). What are the probable angles of polarizer C with respect to the horizontal (two correct answers)?
   a) 20°
   b) 30°
   c) 40°
   d) 50°
   e) 60°
   f) 70°

18) Still on Q 17. As polarizer C is in a free-rotation mode, it is rotated from every possible angle with respect to the horizontal from 0° to 90°. What is the light intensity eventually passing polarizer B if that angle is 0°?

a) 0 units
b) 12.5 units
c) 25 units
d) 50 units

19) Still on Q 17. What is the light intensity eventually passing polarizer B if the angle formed by polarizer C is 90°?

a) 0 units
b) 12.5 units
c) 25 units
d) 50 units

20) Still on Q 17. What is the maximum light intensity eventually passing polarizer B if the angle formed by polarizer C can have any value between 0° and 90°?

a) 0 units
b) 12.5 units
c) 25 units
d) 50 units

21) Back to Q 20. This maximum light intensity occurs under what angle formed by polarizer C?

a) 25°
b) 35°
c) 45°
d) 55°
e) 65°
f) 75°

22) Unpolarized light with an initial intensity of 100 units (arbitrary) passes via two cascaded linear polarizers. Polarizer A has its axis along the vertical (90°); polarizer B has its axis along the 60° angle. What is the intensity of the light passing polarizer B, and what is the polarization status?

a) 50 units, linearly polarized along the 60°
b) 37.5 units, linearly polarized along the 60°
c) 25 units, linearly polarized along the 30°
d) 12.5 units, linearly polarized along the vertical
e) 12.5 units, linearly polarized along the 60°

23) Partially polarized light composed of a nonpolarized component $I_{NP}$ and a polarized component $I_P$ is passing via an analyzer that completes a full 360° rotation. We note the fluctuation between the maximum intensity $I_{MAX}$ and the minimum intensity $I_{MIN}$. The difference between the two ($I_{MAX} - I_{MIN}$) equals ...

a) the nonpolarized component $I_{NP}$
b) half of the nonpolarized component $\frac{1}{2}I_{NP}$
c) the polarized component $I_P$
d) half of the polarized component $\frac{1}{2}I_P$

24) Back to Q 23. Following a complete rotation of the analyzer, the maximum intensity $I_{MAX}$ is found to equal the minimum intensity $I_{MIN}$. This means that the light incident on the analyzer is ...

a) linearly polarized
b) 50% partially polarized ($I_{NP} = I_P$)
c) nonpolarized

25) Back to Q 23. The analyzer is rotated so that it reports the minimum intensity $I_{MIN}$. How many more degrees should the analyzer be turned in order to report the maximum intensity $I_{MAX}$?

a) 45°
b) 90°
c) 135°
d) 180°

26) Back to Q 23. Following a complete rotation of the analyzer, the maximum intensity $I_{MAX}$ is 1.00, while the minimum intensity $I_{MIN}$ is 0.60. The degree of partial polarization $p$ for the incident light is ...

a) 0.60
b) 0.50
c) 0.40
d) 0.25

27) When a polarizer is rotated while an observer is looking at the sky via the polarizer, the sky's blue color appears to darken at a specific angle. This is because the polarizer ...

a) attenuates significant amounts of blue at that specific angle, rendering it darker
b) absorbs almost all nonpolarized light
c) reflects back unwanted glare
d) saturates all colors

28) In linearly polarized light, the ____________ of the electric vector is fixed (three correct answers).

a) directional orientation
b) magnitude
c) plane of oscillation
d) $-x$ component
e) $-y$ component
f) ratio between the $-x$ and $-y$ components

29) In circularly polarized light, the ____________ of the electric vector is fixed (two correct answers).

a) directional orientation
b) magnitude
c) plane of oscillation
d) $-x$ component
e) $-y$ component
f) ratio between the $-x$ and $-y$ components

30) A linear polarizer acting as an analyzer completes a full 360° turn in front of a collimated beam of light stopping at 0° ($-x$ axis, horizontal). No intensity fluctuations are noted in the light intensity passing this polarizer. The light that is incident to the polarized light is (two correct) ...

a) unpolarized
b) linearly polarized along the horizontal
c) linearly polarized along the vertical
d) circularly polarized

31) Back to Q 30. Assuming that the light that is incident to the polarized light has 100 intensity units (arbitrary), what is the measured intensity of the light passing this polarizer?

a) 0
b) 25
c) 50
d) 100

32) The light leaving the polarizer described in Q 30 is ...

a) unpolarized
b) linearly polarized along the horizontal
c) linearly polarized along the vertical
d) circularly polarized

33) A twist to Q 32. The system is modified by inserting a quarter-wave plate between the initial incident beam and the polarizer. The light leaving the quarter-wave plate is (two correct) ...

a) linearly polarized if the initial beam is unpolarized
b) linearly polarized if the initial beam is circularly polarized
c) circularly polarized if the initial beam is unpolarized
d) circularly polarized if the initial beam is circularly polarized
e) unpolarized if the initial beam is unpolarized

34) Along the path of a clockwise circularly polarized beam, we insert a quarter-wave plate. The light emerging from that plate is ...

a) clockwise circularly polarized
b) counterclockwise circularly polarized
c) nonpolarized
d) linearly polarized

35) Twist to Q 34. Now we insert a half-wave plate (two cascaded quarter-wave plates instead of one). The light emerging from that plate is ...

a) clockwise circularly polarized
b) counterclockwise circularly polarized
c) nonpolarized
d) linearly polarized

36) Which wavelength of light scatters more than deep-red 680 nm by a factor of 5?

a) 1521 nm
b) 1017 nm
c) 455 nm
d) 304 nm

37) This yellow wavelength of 572 nm scatters how much more than this red wavelength of 680 nm?

a) 0.5× (half as much)
b) 1.4× (about one & one-half more)
c) 2.0× (twice as much)
d) 2.8× (almost three times as much)

38) A light component with wavelength $\lambda_A$ scatters 3× more than light with wavelength $\lambda_B$. This means that wavelength $\lambda_A$ is ...

a) one-third of $\lambda_B$
b) three times $\lambda_B$
c) about 0.76 of $\lambda_B$
d) about 1.31 of $\lambda_B$

39) An underwater scuba diver is looking up at the sea–air surface and sees a shark reflected on that surface. Scared to death, he dives deeper, when suddenly the reflection is gone. This happens at what angle of incidence ($n_{water}$ = 1.33, $n_{air}$ = 1.0)?

a) critical angle, $\vartheta_{CR}$ = 36.8°
b) critical angle, $\vartheta_{CR}$ = 48.6°
c) critical angle, $\vartheta_{CR}$ = 53.1°
d) Brewster's angle, $\vartheta_B$ = 53.1°

40) When light is reflected at Brewster's angle, reflection is ...

a) 100% reflected, i.e., all is reflected
b) lightly reflected but 100% polarized

c) 0%, i.e., nothing is reflected
d) 50% reflected, 50% polarized

41) To observe 100% linearly polarized light from an air–glass interface ($n_{\text{air}}$ = 1.0, $n_{\text{glass}}$ = 1.7), one has to be observing light reflected at what angle?
a) This is not possible.
b) Brewster's angle $\vartheta_B$ = 30.5°
c) critical angle $\vartheta_{CR}$ = 36.0°
d) Brewster's angle $\vartheta_B$ = 59.5°

42) Back to Q 41. We now reverse the order: Light travels from glass into air ($n_{\text{glass}}$ = 1.7, $n_{\text{air}}$ = 1.0). What should be the observation angle of the reflected angle for 100% linearly polarized light?
a) This is not possible.
b) Brewster's angle $\vartheta_B$ = 30.5°
c) critical angle $\vartheta_{CR}$ = 36.0°
d) Brewster's angle $\vartheta_B$ = 59.5°

43) Which interface has the strongest reflectivity when light is striking at a normal angle to that interface?
a) medium 1 $n_1$ = 1.00; medium 2 $n_2$ = 1.333
b) medium 1 $n_1$ = 1.00; medium 2 $n_2$ = 1.40
c) medium 1 $n_1$ = 1.333; medium 2 $n_2$ = 1.50
d) medium 1 $n_1$ = 1.333; medium 2 $n_2$ = 1.9

44) An air–glass interface ($n_{\text{air}}$ = 1.0, $n_{\text{glass}}$ = 1.5) is illuminated normally. Out of 100 (arbitrary) units of incident light reaching the interface, how many are reflected?
a) none
b) 0.4
c) 4
d) 40

45) Flipping Q 44. We now flip the interface: Light travels from glass into air. Out of 100 (arbitrary) units of incident light reaching the interface, how many are reflected?
a) none
b) 0.4
c) 4
d) 40

46) A twist to Q 44. Now this glass block is submerged in water, creating a water–glass interface ($n_{\text{water}}$ = 1.333, $n_{\text{glass}}$ = 1.5). Out of 100 (arbitrary) units of incident light reaching the interface, how many are reflected?
a) none
b) 0.4
c) 4
d) 40

47) In an anisotropic crystal exhibiting birefringence, a phase difference without a physical separation may develop between two rays ...
a) that are both incident on the crystal, parallel to the optic axis
b) that are both incident on the crystal, perpendicular to the optic axis
c) that are both incident on the crystal at an oblique angle with respect to the optic axis
d) in which one ray is incident parallel, and one ray is incident perpendicular, to the optic axis

48) In an anisotropic birefringent crystal, a physical separation may develop between two rays ...
a) that are both incident on the crystal, parallel to the optic axis
b) that are both incident on the crystal, perpendicular to the optic axis
c) that are both incident on the crystal at an oblique angle with respect to the optic axis
d) in which one ray is incident parallel, and one ray is incident perpendicular, to the optic axis

49) You are observing a birefringent quartz crystal forming a double image. As the crystal rotates, one of the two images turns along the crystal. The part of the image that turns is formed by ...
a) the extraordinary ray, in defiance of the law of refraction
b) the extraordinary ray, in compliance with the law of refraction
c) the ordinary ray, in defiance of the law of refraction
d) the ordinary ray, in compliance with the law of refraction

50) Did you just say 'in compliance'? Why does the ray turn when it is not supposed to turn?
a) The medium is not isotropic; the law of refraction should be restated with respect to the wavevectors, not the rays.
b) The medium is isotropic; the law of refraction allows the rotation of this double image.
c) The medium is not isotropic; the law of refraction should be restated with respect to the rays, not the wavevectors.
d) The medium is isotropic; the law of refraction should be restated with respect to the wavevectors, not the rays.

# 2.8 POLARIZATION SUMMARY

Light is a transverse electromagnetic wave in which the electric and magnetic fields are oscillating perpendicularly (at right angles) to the direction of propagation. Polarization effects relate to light intensity variations due to electric field vector associations.

## Classification of Polarization

- In unpolarized (natural) light, there is no restriction on where the electric (or magnetic) field oscillates, other than the requirement to be perpendicular.
- In polarized light, the direction of oscillation of the electric field is restricted.
- In linearly polarized light, the tip of the electric field vector oscillates along a horizontal, vertical, diagonal (or anything between) line, called the axis.
- In circularly polarized light, the tip of the electric field vector runs the circumference of a circle. The rotation can be clockwise or counterclockwise.
- In elliptically polarized light, the tip of the electric field vector runs the circumference of an ellipse.
- Partially polarized light is a mix of unpolarized and polarized light.

## Detection

To detect linearly polarized light, we place an additional polarizer in the path of the beam. We observe possible fluctuations in the detected light intensity passing through the analyzer. In linearly polarized light, the maxima and minima (darkness) alternate every 90°. In partially polarized light, the minima are not dark.

Malus' law describes the amount of light passing through two cascaded linear polarizers (or between a linearly polarized beam and an analyzer): The fraction of light passing through is proportional to the squared cosine of the relative angle between the two axes (or the axis of the polarized light and the analyzer).

To detect circularly polarized light, we place an additional quarter-wave plate in the path of the beam. If the light is circularly polarized, it is converted to linearly polarized light and is detected with an analyzer.

## Mechanisms of Polarization

Linearly polarized light can be produced by:

- Transition through a linear polarizer (dichroic crystals). Light is linearly polarized along the axis of the polarizer.
- Reflection off a dielectric surface. Reflected light is linearly polarized when the reflected ray is perpendicular to the refracted ray (in this case, the tangent of the angle of incidence, known as Brewster's angle, equals the relative index of refraction).
    - The amount of light being reflected (and, consequently, refracted) by an optical interface is dependent on the state of polarization of the incident beam, the angle of incidence, and the relative ratio of the refractive indices between the two media. The Fresnel coefficients provide the ratios of the reflected (and also the refracted) amplitude magnitude and intensity to the incident amplitude magnitude and intensity.
- Scattering. Scattered light is linearly polarized in directions at right angles to the incident beam when the scattering is governed by the Rayleigh criterion. Scatter is dependent on the wavelength: Shorter wavelengths (the blue) exhibit much more scatter than longer wavelengths (the red).
- Transition through a naturally occurring birefringent medium (such as calcite). Under specific orientations of a birefringent crystal and the incident beam, one or two linearly polarized beams may be formed upon exiting the crystal.
- Transition through stress-induced birefringent media and media with optical activity.

Circularly polarized light can be produced using the following methods:

- If natural light is used, use a linear polarizer and a quarter-wave plate. Place the linear polarizer axis at 45° with respect to the principal axes of the quarter-wave plate.
- If linearly polarized light is used, orient the polarization axis at 45° with respect to the principal axes of a quarter-wave plate.

In both cases, the light leaving the quarter-wave plate is circularly polarized. The light is elliptically polarized if the angle of the polarizer with respect to the axes of the quarter-wave plate is other than 45°.

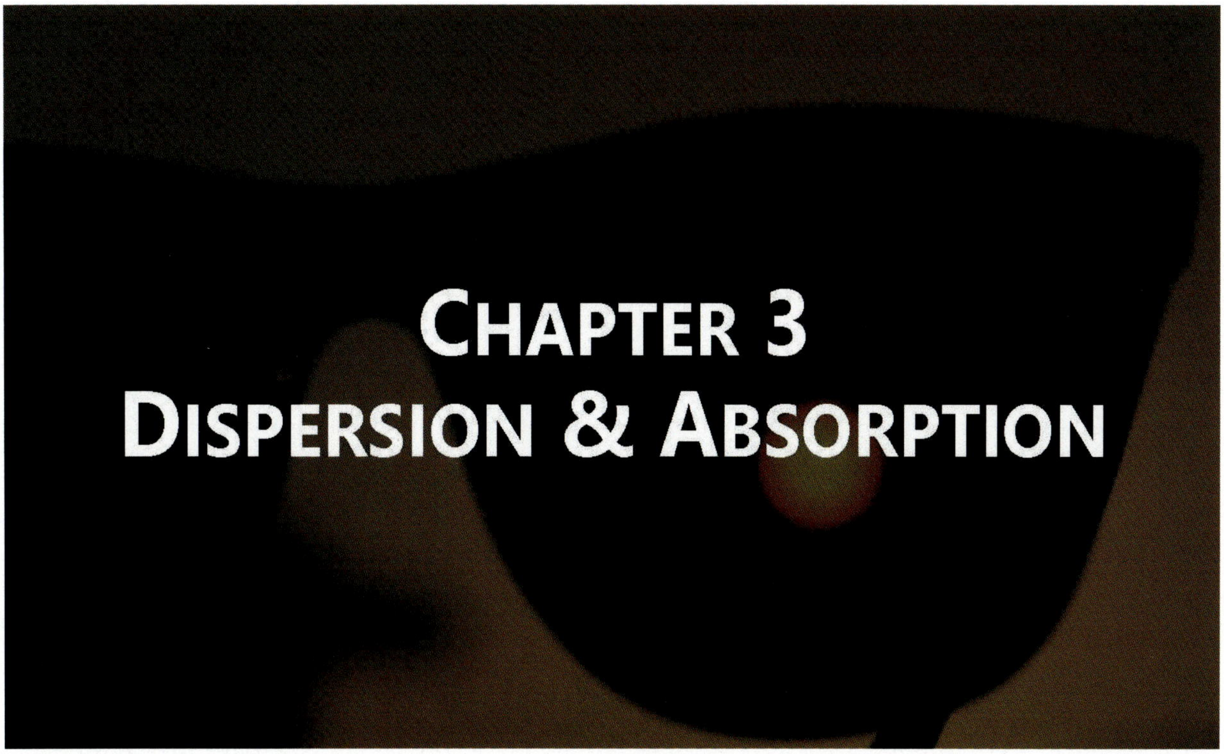

# CHAPTER 3
# DISPERSION & ABSORPTION

## 3.1 REFRACTIVE INDEX: A COMPLEX NUMBER

Geometrical optics accepts certain convenient simplifications. One of these approaches is to treat the refractive index as a fixed, real number. This simplification is adopted despite strong indications that the refractive index is wavelength dependent. What more would we need than the analysis of colors through a prism or the chromatic aberrations in lenses? In addition, birefringence indicates that the refractive index in optical anisotropic materials is not just a number, but a 3×3 tensor; in other words, it is not one, but nine numbers, which, in principle, means that the numbers are complex! However, this simplification is quite a useful approach.

The simplification that the refractive index is a just a number is justified by macroscopic considerations. In the microcosm of the atomic scale, however, this simplification is not valid. This is evident when we examine in detail what happens on a small scale, when light, a high-frequency electromagnetic wave, propagates in a medium. Can we just assume that the light is simply delayed, and that the propagation phenomena can be described by a fixed, real number called the refractive index? What is the mechanism of this slowdown of the light?

Two effects, **dispersion** (the dependence of the phase velocity of a wave or, equivalently, the refractive index of the material, on the frequency or wavelength) and **absorption** (represented as a frequency-dependent decrease in amplitude magnitude) provide the basis to understanding the mechanism of light propagation through a medium. Ultimately, it turns out that the refractive index is a complex number!

General Expression for the Complex Refractive Index: $$\underbrace{\tilde{n}}_{} = \underbrace{n}_{\text{real part}} - i \underbrace{\kappa}_{\text{imaginary part}} \qquad (3.1)$$

The real part of this complex number relates to dispersion, and the imaginary part relates to absorption. Dispersion and absorption occur whenever electromagnetic radiation interacts with matter. Only in vacuum, where there is no matter to interact with light, is there no dispersion and no absorption. Instead, in every material, there is a different speed of light and a different degree of absorption, depending on the wave's frequency. This dependence of absorbance explains, among other, the colors in many filters and objects.

### 3.1.1 The Origin of the Refractive Index

The statement that 'the refractive index is higher than 1.0' is equivalent to saying that 'light slows down when propagating within a medium other than vacuum.'

Why is that? We know that light, an electric field oscillating at a very high frequency, can cause bound electron-clouds to oscillate in an atom or a crystal lattice—this is scattering (introduced in § 2.5.1). An atom thus becomes an electrical dipole, oscillating at the same frequency as the stimulus. There are some caveats, however. First, there is a time-delay. The oscillating dipole re-emits a light wave along the same direction as the incident light, and this light wave, in turn, stimulates a neighboring atom. This relay mechanism is a simple 1-D representation of the mechanism that governs light propagation in a dielectric medium.

Within the medium, in the empty space between individual atoms, the propagation speed of the disturbance is $c$ because there is no reason for any delay. The delay is only caused by the response time of the electron-cloud to the stimulus. This is the microscopic governance of light propagation within a medium. The speed of light within a dielectric is affected by the 'how' and the 'when' of the elementary dipole responses. This explains why light, an electromagnetic disturbance, slows down when propagating inside a dielectric medium.

The delay in the electron-cloud response is not the same for all stimulation frequencies. This difference might be due to inertia and the likely presence of friction forces, which can cause a stimulus–response phase difference. This phase difference might facilitate the process or not, and is also likely to attenuate the propagating wave. For example, if the phase difference is 0 or $2\pi$, the response is immediate and the field does not encounter noticeable resistance when propagating through the medium. Completely different results arise when the frequency of the field coincides with the resonance oscillation frequency of the electron-cloud. We know that if we try to stimulate a spring with a force oscillating at the same frequency as any of the system's natural resonance frequencies, the maximum energy is absorbed! This corresponds to a phase difference of $\pi$.

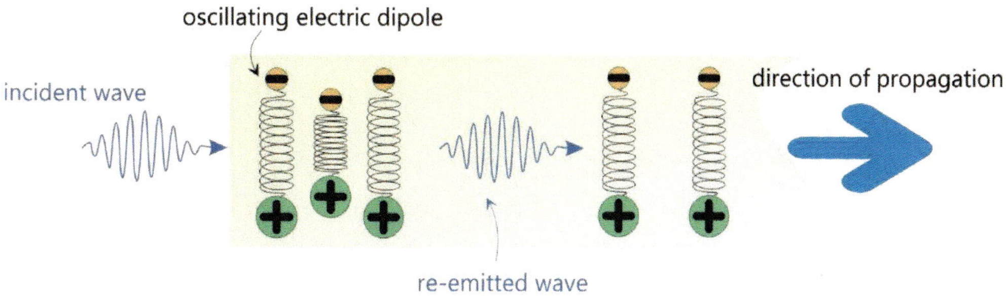


*Figure 3-1: Light propagation mechanism within a material: the mechanical analog model.*

The (first) Tacoma Narrows suspension bridge collapse in Washington State is presented as an example of elementary forced resonance, with the wind providing an external periodic frequency (oscillating stimulus) that matched the bridge's natural resonant frequency—although later theories support a wind-driven amplification of the torsional oscillation. Regardless of the argument, this is an example of a significant energy transfer from a wave to a structure.

*Figure 3-2: (left) The (first) Tacoma Narrows Bridge twisting and vibrating violently under 40 mph (64 km/h) winds on the day of the collapse. (right) Still frame from a 16 mm Kodachrome motion picture film taken by Barney Elliott on 7 November 1940.*

For a detailed expression of the light propagation speed in different media, we study the reaction of a bound electron-cloud to an oscillating electric field. Light corresponds to a propagating electric field disturbance with frequencies on the order of $10^{14}$ Hz. Optics can often be expressed with standard electromagnetism (EM) relationships, as long as we take into account this frequency range. The propagation of the oscillation of an electric field in vacuum is described by the wave equation [Eq. (1.3)]:

Wave Equation (in vacuum):
$$\nabla^2 \mathbf{E} - \varepsilon_o \mu_o \frac{\partial^2}{\partial t^2} \mathbf{E} = 0 \tag{3.2}$$

A similar relationship exists for the magnetic field $\mathbf{H}$. Comparing Eq. (3.2) to the simple wave equation [Eq. (1.1)], we conclude that the propagation speed of the electromagnetic wave in vacuum is

EM Wave Speed (in vacuum):
$$u = c = \frac{1}{\sqrt{\varepsilon_o \mu_o}} \tag{3.3}$$

where the electric permeability $\varepsilon_o$ and magnetic $\mu_o$ permeability refer to vacuum. In a dielectric material, the corresponding permeabilities, electric $\varepsilon$ and magnetic $\mu$, have different values. Thus, the wave equation for the electric field [Eq. (3.2)] in a dielectric medium is

Wave Equation (in medium other than vacuum):
$$\nabla^2 \mathbf{E} - \varepsilon \mu \frac{\partial^2}{\partial t^2} \mathbf{E} = 0 \tag{3.4}$$

In such a medium, the propagation speed of an electric field disturbance is

EM Wave Speed (in medium):
$$u = \frac{1}{\sqrt{\varepsilon \mu}} \neq c \tag{3.5}$$

If the incident wave corresponds to a harmonic electric field, as is described by Eq. (1.7), then $\nabla = -i\,\mathbf{k}$ and $\partial/\partial t = i\omega$. It is easy to conclude that

EM Harmonic Wave Speed (in medium):
$$u = u_{\text{phase}} = \omega/k \tag{3.6}$$

This ratio of the angular frequency $\omega$ to the wavevector magnitude $k$ is the **phase velocity**. The refractive index $n$ [Eq. (1.18)] is defined as the ratio of the speed of a harmonic wave in vacuum $c$ to the magnitude of the phase velocity $u$ in the medium:

Refractive Index:
$$n = c/u_{\text{phase}} \tag{3.7}$$

Combining the above with the dielectric constant $\kappa_e = \varepsilon/\varepsilon_o$ and the relative magnetic permeability $\kappa_m = \mu/\mu_o$, we express the refractive index as

Refractive Index:
$$n = \frac{c}{u_{\text{phase}}} = \frac{\dfrac{1}{\sqrt{\varepsilon_o \mu_o}}}{\dfrac{1}{\sqrt{\varepsilon \mu}}} = \sqrt{\frac{\varepsilon \mu}{\varepsilon_o \mu_o}} = \sqrt{\kappa_e \cdot \kappa_m} \approx \sqrt{\kappa_e} \tag{3.8}$$

This approximation is in very good agreement with experimental data for thin materials, transparent gasses, and nonpolar dielectrics such as glass. These are usually diamagnetic, so $\kappa_m = 1.0$. Here we will study these materials, which are indeed isotropic dielectrics; certainly, this study can be extended to a nonisotropic material (such as birefringent materials, discussed in § 2.6). We now write the refractive index in the following form:

$$n = \sqrt{\kappa_e} \Rightarrow n^2 = \kappa_e = 1 + \chi_e \tag{3.9}$$

where $\chi_e$ is the electric susceptibility of the material, which is always > 0, except in vacuum, where $\chi_e = 0$. The refractive index, a macroscopically observed quantity, directly relates to a fundamental material property, the electric susceptibility $\chi_e$, or equivalently, the electric permittivity $\varepsilon = \varepsilon_o \cdot \chi_e$.

Almost all materials, including glass, have positive values for both the electric permittivity $\varepsilon$ and the magnetic permeability $\mu$ at optical frequencies. When both values are positive, the medium has a positive refractive index and, as will see later, is transparent to the respective electromagnetic radiation frequency.

In metals, the values of the electric permeability for the visible are negative. For a material where one of the two permeability values (but not both) is negative, the refractive index becomes purely imaginary, per Eq. (3.8), causing the material to be opaque. The electromagnetic radiation of the corresponding frequency can propagate only superficially.

Let's see what happens on a microscopic scale in a dielectric material under the influence of an oscillating electric field. The field exerts forces on the positive and negative charges. The electron cloud, being much more agile than the nucleus, is shifted from its equilibrium position. This shift represents a separation of charges and creates an electric dipole. The electric dipole moment is a vector pointing from the negative charge to the positive charge. If the offset is expressed as a displacement $\mathbf{x}$, then the electric dipole moment is

$$\mathbf{p} = q \cdot \mathbf{x} \tag{3.10}$$

Many of such excited dipoles contribute to a macroscopically observed **induced** electric polarization $\mathbf{P}$. If $\bar{N}$ is the number of such dipoles per unit volume (charge density), the induced electric polarization is

Induced Electric Polarization: $$\mathbf{P} = \bar{N} \cdot \mathbf{p} = \bar{N} \cdot q \cdot \mathbf{x} \tag{3.11}$$

For isotropic materials, the induced electric polarization depends on the stimulus according to a simple linear relationship, where the coefficient is the electric susceptibility:

Induced Polarization (in isotropic media): $\mathbf{P} = \varepsilon_0 \chi_e \cdot \mathbf{E}$ (3.12)

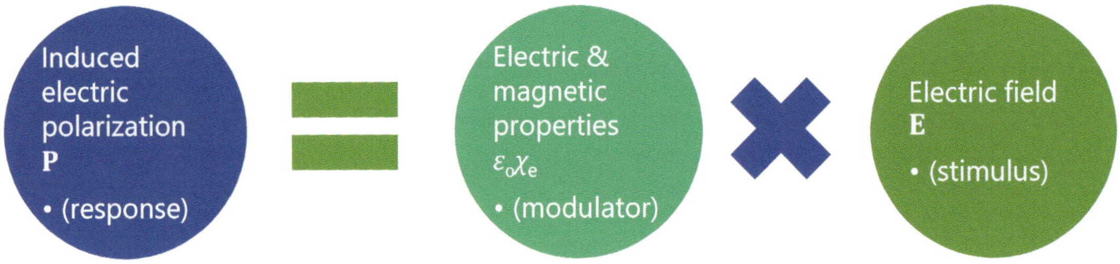


The above relationship is the key to understanding the optical properties of a medium. The exact form of $\chi_e$ provides a good predictor of this optical behavior.

- If $\chi_e$ is constant for any electric field orientation, the material is isotropic. If this is not the case, then $\varepsilon_0\chi_e$ becomes a tensor, and the induced polarization is, generally, not parallel to the electric field. This material is **anisotropic**.
- If there are other terms beyond the first order (such as a $\chi^2_e$ term), the material displays nonlinearity: The response—the magnitude of the induced polarization—is not in a simple linear relationship (proportionality) with the stimulus (electric field) (more on this in § 6.3.3).
- The most interesting relationship of all is the possible dependence of $\chi_e$ on the frequency $\omega$ of the stimulus (harmonically oscillating electric field). This is the focus of the analysis herein.

For simplicity, we consider a linear and isotropic medium. The induced electric polarization is proportional and collinear to the externally applied electric field:

$$\mathbf{P}(\omega) = \varepsilon_0 \chi_e(\omega) \cdot \mathbf{E}(\omega) \tag{3.13}$$

The index of refraction is calculated using the following simple relationship:

$$n^2(\omega) = \kappa_e(\omega) = 1 + \chi_e(\omega) \tag{3.14}$$

Food for thought: Shortly before entering a detailed mathematical development, a good point to ponder is: Since the propagating wave, refracted inside the medium, results from successive, dipole-relay re-emissions between adjacent atoms inside the medium, where does the reflected wave derive from? In geometrical optics, we accept that an abrupt event occurs precisely at the dividing interface. A detailed analysis of Fresnel's coefficients (presented in § 2.5.2.1) demonstrates that the reflected and refracted beams arise due to a re-emission mechanism from the surface dipoles. In fact, the reflected wave, just like the refracted wave, results from successive dipole re-emissions within the entire volume of the medium, from opposite directions and not solely from the dipole re-emissions from the surface.

The oscillating dipoles in a medium such as glass emit elementary waves not only along the direction of propagation of the incident wave but in all directions. Many of the 'all direction' waves mutually cancel out. An extant wave corresponds to directions along which the elementary waves cannot be canceled on a large scale; such is the reflected wave, which is a contribution not only from the surface, but also from dipoles from the entire material volume. Macroscopically, the result is exactly the same as if we had considered only the dividing surface.

The same is true for the refracted wave, which is also a result of comprehensive re-emissions by all dipole oscillators (elementary waves) from dipoles within the medium.

### 3.1.2 The Lorentz Mechanical Analog Model

To express the induced electrical polarization due to an external oscillating electric field, we employ the Lorentz mechanical analog model. This model, which connects electromagnetic theory and atomic physics, was proposed in 1880 by Hendrik Antoon Lorentz,[16] a Nobel Prize winner in Physics (1902). This is a 3-D oscillator atomic model: The positively charged particles, concentrated in the nucleus, which is also the center of mass and charge, are connected by pairs of springs to the evenly distributed, negatively charged particles.

Such mechanical springs do not exist, nor is there a way to pinpoint the electron, let alone to 'tie' it with a spring. This mechanical analog is phenomenological and provides a very useful approach. The elastic forces are mutual Coulomb forces between the charges.

If the elastic constants of each spring are identical for all three pairs along the $x$–$y$–$z$ axes, we have an isotropic, 3-D harmonic oscillator. The response of the electron-cloud to an external electric field is dependent on the field frequency, just as the forced oscillation of a spring depends on the oscillating frequency of the force.

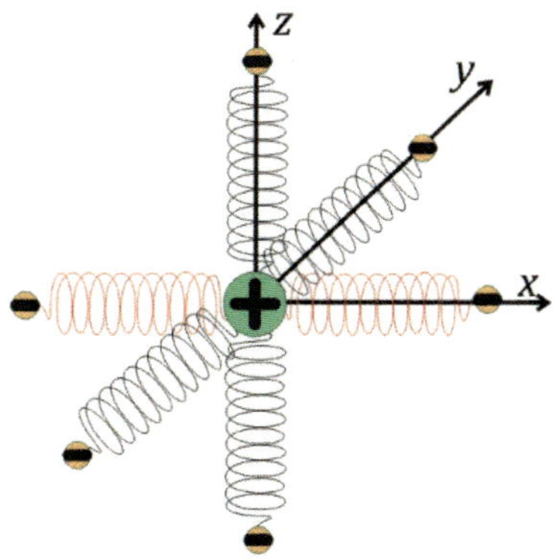


*Figure 3-3: Isotropic 3-D harmonic oscillator model.*

[16] Lorentz HA. Über die Beziehungzwischen der Fortpflanzungsgeschwindigkeit des Lichtes der Körperdichte. Ann Phys. 1880; 9:641-65.

At certain frequencies, it is easy for the spring to follow the oscillating force in phase; for other frequencies that coincide with the medium's resonant frequencies, the spring's response is quite different. The natural spring oscillation frequency $\omega_o$ relates to the elastic constant $\kappa_s$ and the effective mass $m$ according to

$$\omega_o = \sqrt{\frac{\kappa_s}{m}} \qquad \text{or} \qquad \kappa_s = m \cdot \omega_o^2 \tag{3.15}$$

We assume an isotropic oscillator and therefore can restrict the analysis to one dimension. An oscillating electric field $\mathbf{E}$ exerts a Coulomb force on a charge $q$:

Coulomb Force:
$$\mathbf{F}_e = q \cdot \mathbf{E} = q \cdot \mathbf{E}_o \exp(i\omega \cdot t) \tag{3.16}$$

This force displaces the average electron charge distribution by $\mathbf{x}$. The spring reacts with a spring-back elastic force:

Elastic Force:
$$\mathbf{F}_s = -\kappa_s \cdot \mathbf{x} = -m \cdot \omega_o^2 \cdot \mathbf{x} \tag{3.17}$$

We also consider a damping force that is proportional to the speed. The proportionality constant is the friction coefficient $\gamma$:

Damping Force:
$$\mathbf{F}_T = -\gamma \cdot m \cdot \mathbf{u} = -\gamma \cdot m \cdot \frac{d\mathbf{x}}{dt} \tag{3.18}$$

The net force acting on the charge is

Net Force:
$$\mathbf{F} = \mathbf{F}_e + \mathbf{F}_s + \mathbf{F}_T = q \cdot \mathbf{E} - m \cdot \omega_o^2 \cdot \mathbf{x} - \gamma \cdot m \cdot \frac{d\mathbf{x}}{dt} \tag{3.19}$$

Thus, the equation of motion becomes

$$m \cdot \frac{d^2\mathbf{x}}{dt^2} = q \cdot \mathbf{E}_o \exp(i\omega t) - m \cdot \omega_o^2 \cdot \mathbf{x} - \gamma \cdot m \cdot \frac{d\mathbf{x}}{dt} \tag{3.20}$$

or, it can be re-arranged to become

$$\frac{d^2\mathbf{x}}{dt^2} + \gamma \cdot \frac{d\mathbf{x}}{dt} + \omega_o^2 \cdot \mathbf{x} = \frac{q \cdot \mathbf{E}_o}{m} e^{i\omega t} \tag{3.21}$$

The solutions of Eq. (3.21) take the following form:

$$\mathbf{x}(\omega) = \mathbf{x}_o e^{i\omega t} = \frac{q \cdot \mathbf{E}_o e^{i\omega t}}{m \cdot \left(\omega_o^2 - \omega^2 + i\gamma\omega\right)} = \frac{q}{m \cdot \left(\omega_o^2 - \omega^2 + i\gamma\omega\right)} \mathbf{E}(\omega) \tag{3.22}$$

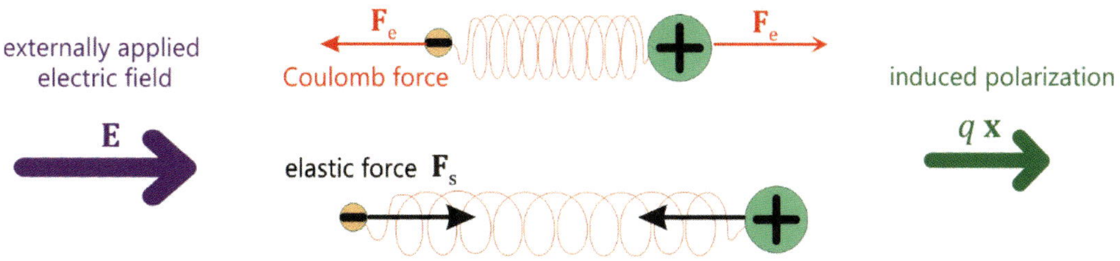


*Figure 3-4: Forces on oscillating charges: the mechanical analog model.*

Thus, we describe, in terms of the classical mechanical model, the effects of an electromagnetic wave with a frequency $\omega$, on the electron-cloud. The cloud response depends on the frequency $\omega$ and on the natural/resonant spring vibration frequency $\omega_o$. The induced electric polarization is

$$\mathbf{P} = \bar{N}\cdot\mathbf{p}(\omega) = \bar{N}\cdot q\cdot\mathbf{x}(\omega) = \frac{\bar{N}q^2}{m\cdot\left(\omega_o^2-\omega^2+i\gamma\omega\right)}\mathbf{E}(\omega) \tag{3.23}$$

which can be rearranged as:

$$\mathbf{P} = \varepsilon_o \underbrace{\frac{\bar{N}q^2}{m\varepsilon_o}\cdot\frac{1}{\left(\omega_o^2-\omega^2+i\gamma\omega\right)}}_{\text{permeability } \chi_e(\omega)}\mathbf{E}(\omega) \tag{3.24}$$

This is an analytical expression describing the dependence of the induced electric polarization on an electric field stimulus. The friction coefficient $\gamma$ is none other than the electric susceptibility, and it is clear that the stimulus frequency $\omega$ is the key parameter. We then combine the above result and the analytical expression for the refractive index:

$$\tilde{n}^2(\omega) = 1+\chi_e(\omega) = 1+\frac{\bar{N}q^2}{m\varepsilon_o}\cdot\frac{1}{\left(\omega_o^2-\omega^2+i\gamma\omega\right)} = 1+\frac{\omega_p^2}{\omega_o^2-\omega^2+i\gamma\omega} \tag{3.25}$$

where $\omega_p$ is the plasma frequency of the medium:

$$\omega_p = \sqrt{\frac{\bar{N}q^2}{m\varepsilon_o}} \tag{3.26}$$

The plasma frequency of the medium is the natural resonant frequency of a free-electron gas inside the material and is a characteristic physical property of that medium. It depends on the charge density and the effective mass.

The above relationships [Equations (3.23) through (3.25)] express the dependence of the refractive index on the frequency—this is dispersion. It is obvious that the refractive index is a complex number, as its square contains an '$i$' part. In the form of $\tilde{n} = n - i\kappa$, we can express the real ($n$) and the imaginary ($\kappa$) parts of the refractive index. The analytical expression of $n$ and $\kappa$

for thin media, where the approximation $\tilde{n} \approx 1$ is facilitated by the **thin** (or rare-field) **media approximation** is

$$\tilde{n}^2 - 1 = (\tilde{n} - 1)\cdot(\tilde{n} + 1) \approx 2\cdot(\tilde{n} - 1) \tag{3.27}$$

We now equate the real and imaginary parts of the refractive index $\tilde{n}$:

Real Part:
$$n(\omega) = 1 + \frac{\omega_p^2}{2}\cdot\frac{\omega_o^2 - \omega^2}{\left(\omega_o^2 - \omega^2\right) + (\gamma\omega)^2} \tag{3.28}$$

Imaginary Part:
$$\kappa(\omega) = \frac{\omega_p^2}{2}\cdot\frac{\gamma\omega}{\left(\omega_o^2 - \omega^2\right) + (\gamma\omega)^2} \tag{3.29}$$

In this model, we assume that all of the electrons are associated with the same type of elastic spring forces and that there is a single resonance frequency. In a more realistic case, there are different ways of associating electrons with the nucleus or the ionic lattice; in other words, there are different natural resonance frequencies $\omega_{o1}$, $\omega_{o2}$, etc. This applies even to the simplest atom, the hydrogen atom.

Using statistical weights ($f_1$% with resonant frequency $\omega_{o1}$, $f_2$% with $\omega_{o2}$, ... $f_j$% with $\omega_{oj}$),

$$\tilde{n}^2(\omega) = 1 + \omega_p^2\cdot\sum_j \frac{f_j}{\omega_{oj}^2 - \omega^2 + i\gamma_j\omega} \qquad \sum_j f_j = 1 \tag{3.30}$$

We therefore conclude that the dependence of the refractive index on the frequency (and wavelength) relates to the atomic and molecular structure of the medium. The complex form of the refractive index is due to the combination of inertia and frictional forces introduced by a phase difference between the electric forces exerted on the system and the system's response. Friction forces cause a nonreversible energy transfer from oscillating charges to the medium, which manifests macroscopically as radiation absorption.

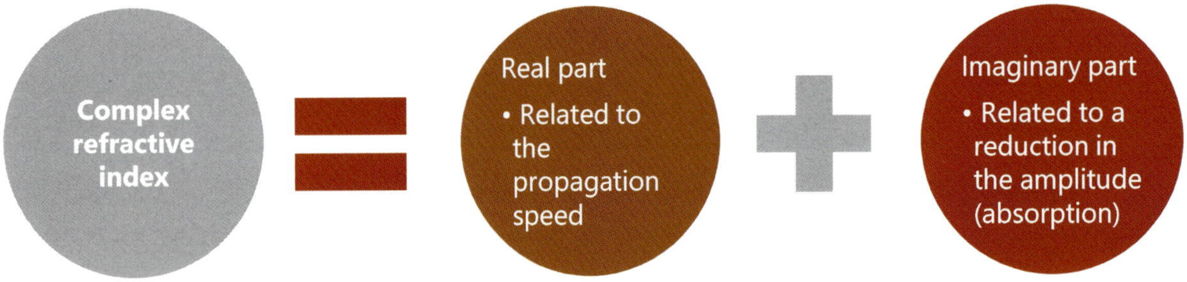

## 3.2 THE IMAGINARY PART OF THE REFRACTIVE INDEX

If the refractive index is complex, the wavevector is also complex:

$$\tilde{k} = \frac{\omega}{c}\cdot\tilde{n} = \frac{\omega}{c}(n - i\kappa) = k - i\left(\frac{\omega}{c}\kappa\right) \tag{3.31}$$

We insert this new wavevector into the expression for a harmonic wave that propagates in a dielectric along the –$z$ direction:

$$\mathbf{E} = \mathbf{E}_0\cdot\exp\left[i(\omega\cdot t - \mathbf{k}\cdot\mathbf{r} + \varphi)\right] = \underbrace{\mathbf{E}_0\cdot\exp\left[-\left(\frac{\omega\cdot\kappa(\omega)\cdot z}{c}\right)\right]}_{\text{new amplitude}}\cdot\underbrace{\exp\left[i(\omega\cdot t - k\cdot z + \varphi)\right]}_{\text{propagating wave}} \tag{3.32}$$

This is a traveling wave with an amplitude that decreases exponentially along the direction of propagation $z$ (note the $k{\cdot}z$ term in the propagating wave phasor):

New Amplitude:
$$\mathbf{E}_0\cdot\exp\left[-\left(\frac{\omega\cdot\kappa(\omega)\cdot z}{c}\right)\right] \tag{3.33}$$

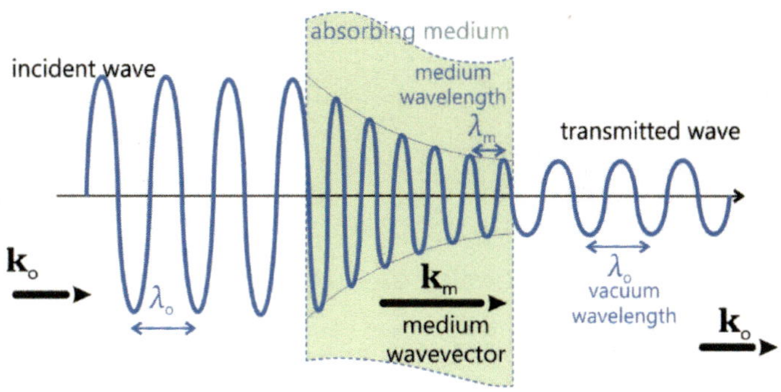


*Figure 3-5: Exponential amplitude magnitude reduction of an electromagnetic wave due to absorption.*

The intensity, which is proportional to the square of the electric field in such harmonic waves, also decreases exponentially, according to the Beer–Lambert law of absorption (using the exponent):

$$I_z = \langle\mathbf{E}\cdot\mathbf{E}^*\rangle = E_0^2\cdot\exp\left[-2\left(\frac{\omega\cdot\kappa(\omega)\cdot z}{c}\right)\right] = I_0\cdot\exp\{[-\alpha(\omega)]\cdot z\} \tag{3.34}$$

The coefficient
$$\alpha(\omega) = 2\frac{\omega\cdot\kappa(\omega)}{c} \tag{3.35}$$

is the material **absorption** or **extinction coefficient**, which describes the intensity logarithmic loss per unit length for a specific frequency or wavelength and has units of inverse length.

The quantity
$$\frac{1}{\alpha(\omega)} = \frac{c}{2\omega \cdot \kappa(\omega)} \tag{3.36}$$

is called the **absorption depth**, with units of length, and is the reciprocal of the absorption coefficient. The absorption depth expresses the length at which (as light propagates) the intensity drops by a factor of $1/e = 1/\ 2.71828 \approx 36.6\%$. An expression that is equivalent to Eq. (3.34) and uses the natural logarithm (ln) is

Beer–Lambert Law of Absorption (using the logarithm):
$$\ln\left(\frac{I_z}{I_o}\right) = -\alpha(\omega)\cdot z \tag{3.37}$$

The relationships in Eqs. (3.34) and (3.37) constitute the **Beer–Lambert absorption law**, named after the German physicist August Beer and the Swiss polymath Johann Heinrich Lambert:

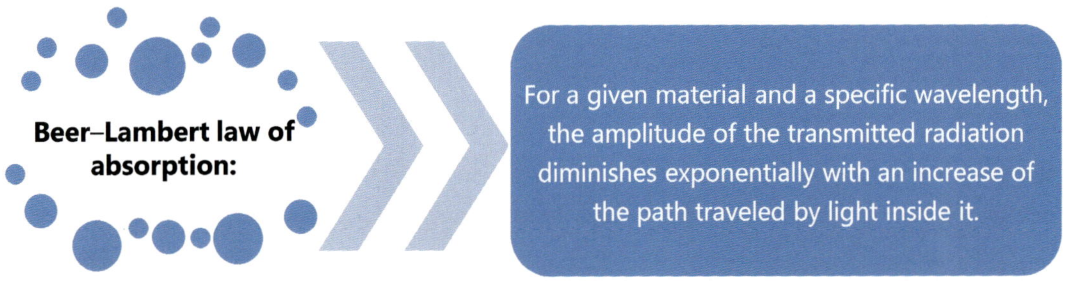


The imaginary part of the refractive index $\kappa(\omega)$, also termed the **absorption index** $\kappa(\lambda)$, as well as the absorption or extinction coefficient, is strongly dependent on the frequency, or equivalently, on the wavelength. This is expected because the friction term in the differential equation [Eq. (3.19)] is responsible for the imaginary component, representing a nonreversible energy transfer from the oscillating charges to the medium—a transfer that is demonstrated macroscopically as absorption.

The specific absorption coefficient for any given frequency $\omega$ (or wavelength $\lambda$) can be measured using a simple experimental setup. A beam of known intensity $I_o$ and wavelength is incident on a plate, such as a gelatin of thickness $d_1$ [Figure 3-6 (left)].

Part of the beam is reflected, the amount depending on the refractive index, the angle of incidence, and the polarization state of the beam. Quantitative relationships are provided by the Fresnel coefficients. Recall that for nearly perpendicular incidence, an air–glass interface reflects approximately 4% of the incident light (as discussed in § 2.5.2.1). It is certain, however, that the reflected part does not depend on the material thickness, under the condition that the plate thickness is not on the order of magnitude of the incident radiation wavelength. Antireflection coatings are discussed as interference effects in § 4.2.3.1.

The specific material thickness absorbs part of the beam, with the amount absorbed being dependent on the color. The material thus acts as a filter (see § 3.4.1). A red filter absorbs relatively little between 680 and 590 nm, which is the part of the visible spectrum that we perceive as red. The remaining part is the transmitted beam.

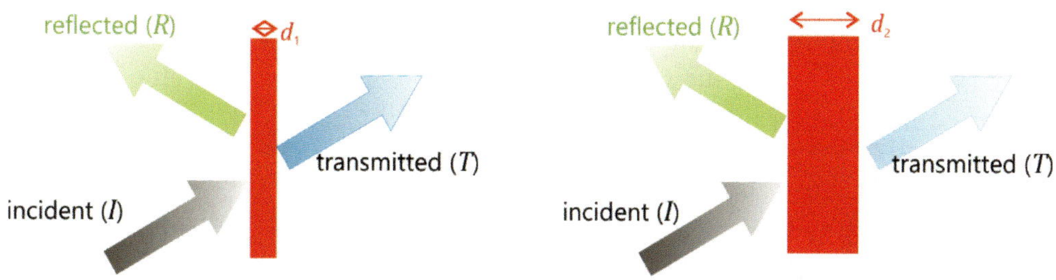


*Figure 3-6: Setup for measurement of the absorption coefficient.*

We record the transmitted beam intensity $I_{d1}$, which is in an exponential relationship with the incident beam intensity, for thickness $d_1$ along the direction of axis $-z$. We then change the gelatin tile thickness to $d_2$ [Figure 3-6 (right)]. The transmitted intensity $I_{d2}$ is again in an exponential relationship [Eq. (3.34)] with the incident intensity, where $z$ is now the new thickness $d_2$. The curve of the ratio $I_d/I_o$ (the relative transmitted intensity) versus the material thickness $d$ is an exponentially decreasing curve, in accordance with the Beer–Lambert law.

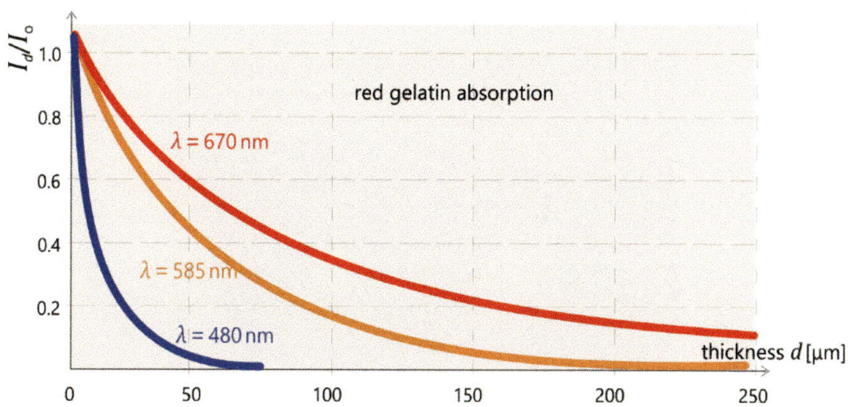


*Figure 3-7: Exponential decrease of the transmitted beam intensity due to an increased material thickness.*

The three curves (experimental data) in the red gelatin absorption graph (Figure 3-7) correspond to wavelengths of 670 nm (the red), 585 nm (the orange), and 480 nm (the blue). The rate of decline per curve depends only on the material absorption coefficient per each wavelength, not on its thickness. This is better visualized after converting to a logarithmic form in which the data curves are now straight lines with a negative slope.

The slopes of these straight lines in the logarithmic representation shown in Figure 3-8 provide the absorption coefficient via the relationship in Eq. (3.37). Thus, $\alpha_{670nm} = 9.6\ mm^{-1}$, $\alpha_{585nm} = 18.8\ mm^{-1}$, and $\alpha_{480nm} = 159.2\ mm^{-1}$. We note an increase in the rate of absorption from the red to the orange, and even more so in the blue. That is why this is a red filter! It absorbs relatively little in the red and absorbs strongly in other wavelengths.

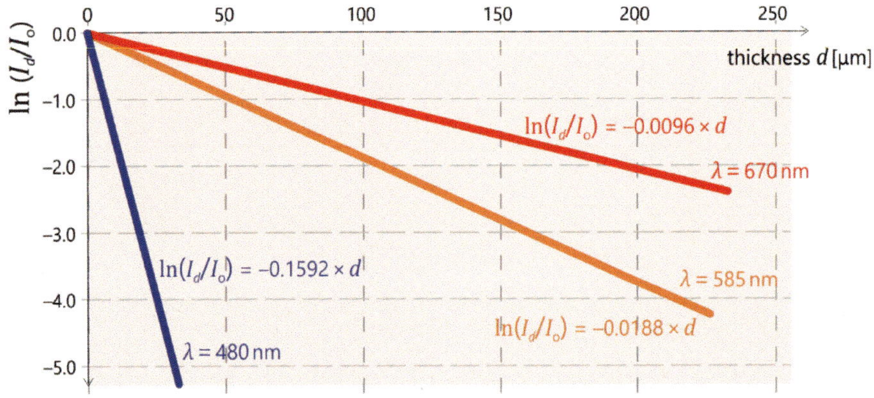


*Figure 3-8: Logarithmic form of the Beer–Lambert relationship employed for calculating the absorption coefficient.*

**A red filter**

- Absorbs relatively little in the red
- Absorbs significantly in other wavelengths, such as blue and green

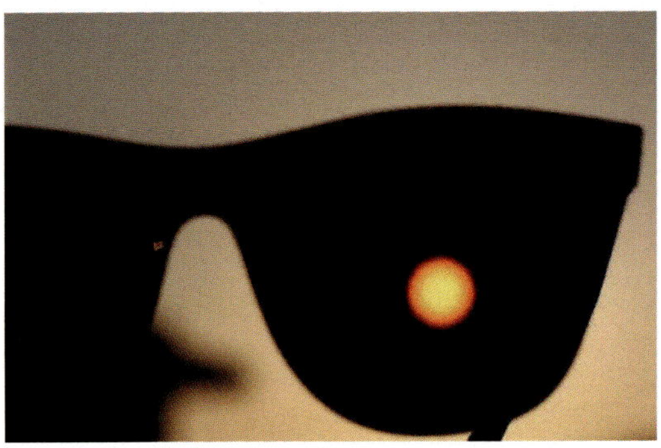

*Figure 3-9: Sunglasses are strongly absorbing, broadband filters (photo by Efstratios I. Kapetanas used with permission).*

# 3.3 THE REAL PART OF THE REFRACTIVE INDEX

The dependence of the material's refractive index on the frequency (wavelength), termed **dispersion**, is expressed mathematically as

Dispersion:
$$\frac{dn(\lambda)}{d\lambda} \neq 0 \quad \text{or} \quad \frac{dn(\omega)}{d\omega} \neq 0 \tag{3.38}$$

The angular frequency $\omega$ is associated with the frequency $\nu$ and wavelength $\lambda$ via the relationships $\omega = 2\pi\nu$ and $\omega = 2\pi c/\lambda$. For this reason, we consider the expressions $n(\omega)$, $n(\nu)$, and $n(\lambda)$ as equivalent. Also, from now on we will call the real part $n(\omega)$ simply the refractive index and the imaginary part $\kappa(\omega)$ the absorption index.

## 3.3.1 Dispersion in Thin Media

Thin (or rare-field) media are those for which $n \approx 1.0$. Approximations are then possible.

Real Part:
$$n(\omega) = 1 + \frac{\omega_p^2}{2} \cdot \frac{\omega_o^2 - \omega^2}{\left(\omega_o^2 - \omega^2\right) + (\gamma\omega)^2} \tag{3.39}$$

Imaginary Part:
$$\kappa(\omega) = \frac{\omega_p^2}{2} \cdot \frac{\gamma\omega}{\left(\omega_o^2 - \omega^2\right) + (\gamma\omega)^2} \tag{3.40}$$

The above quantities are not independent. This is a consequence of the fact that they result from the real and imaginary parts of the same physical quantity, the complex index of refraction. The relationships that link them are the **Kramers–Krönig relations**, named after the Dutch physicist Hendrik Anthony Kramers and the German physicist Ralph de Laer Krönig.

To ignore all friction forces, we set $\gamma = 0$. The refractive index becomes a real number and its imaginary part is zero. To determine whether the absorption is eliminated, the real part [Eq. (3.39)] is written as

$$n_{\gamma=0}(\omega) = 1 + \frac{\omega_p^2}{2} \cdot \frac{\omega_o^2 - \omega^2}{\left(\omega_o^2 - \omega^2\right)} = 1 + \frac{\omega_p^2}{2} \cdot \frac{1}{\omega_o^2 - \omega^2} \tag{3.41}$$

We note the refractive index dependency on frequency (dispersion). We can explore some aspects of this dependency: The refractive index for a frequency equal to a material eigenfrequency $\omega_o$ is undefined. Specifically, for $\omega < \omega_o$ and $\omega > \omega_o$,

$$n_{\gamma=0}(\omega=\omega_o) = \text{undefined} \qquad \lim_{\gamma\to+0} n(\omega)_{\omega<\omega_o} = +\infty \qquad \lim_{\gamma\to+0} n(\omega)_{\omega>\omega_o} = -\infty \tag{3.42}$$

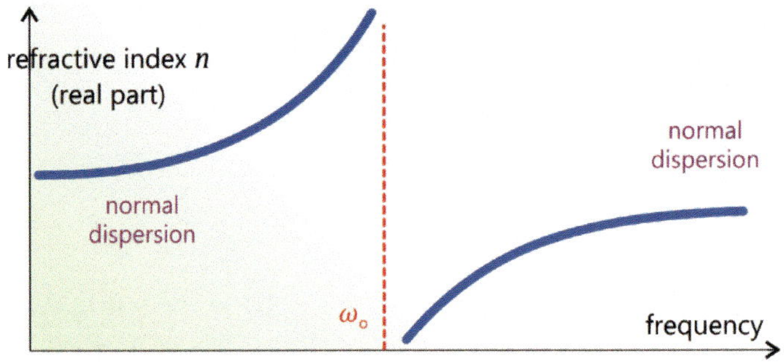


*Figure 3-10: Real part of the refractive index under the assumption of zero friction forces.*

For the resonant frequencies $\omega_o$, the refractive index $n_{(\omega=\omega o)}$ curve introduces an unspecificity (discontinuity). With the exception of these uncertainty shifts, the value of the refractive index increases with the frequency, or equivalently, decreases with the wavelength. This corresponds to **normal dispersion**. Glass, for example, which has resonance in the ultraviolet ($\lambda_o \approx 100$ nm), displays normal dispersion in the visible frequencies.

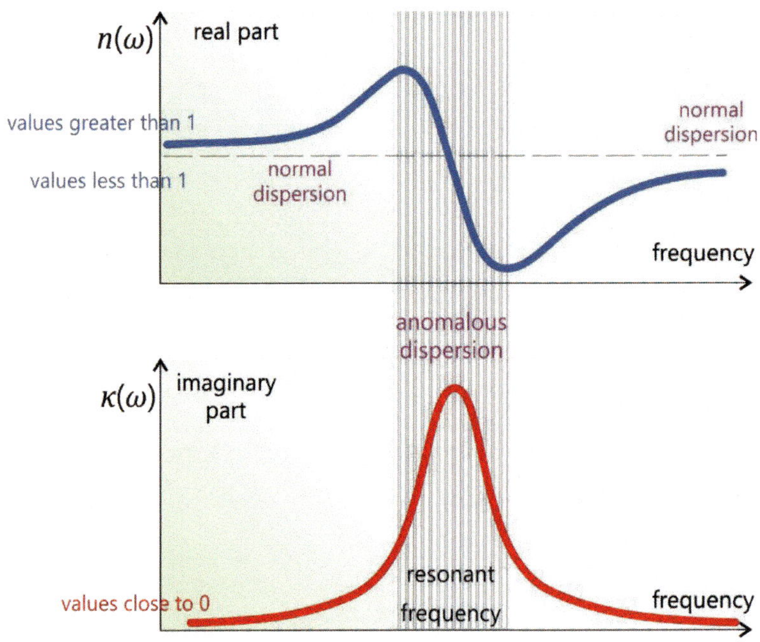


*Figure 3-11: Refractive index in the vicinity of a resonant frequency's (top graph) real part and (lower graph) imaginary part.*

Figure 3-11 shows the dependence of $n(\omega)$ and $\kappa(\omega)$ on the frequency, considering the friction forces in the relationships of Eqs. (3.39) and (3.40). For frequencies/wavelengths near the resonant frequency/wavelength ($\omega_o/\lambda_o$), a completely different behavior is observed, even if friction forces are completely absent. Just as in the resonance of classical mechanics, where the greatest possible energy transfer occurs, so in optics, there is intense absorption around such frequencies—those that coincide with the material's resonant frequencies. This frequency corresponds, for example, to the photon energy absorbed by a permissible bipolar transition from a low energy level to another, higher one, such as in quantum absorption (§ 6.2.1.3). There is absorption, and indeed intense absorption, precisely around the frequencies $\omega = \omega_o$.

The refractive index $n(\omega)$ is a continuous function, but the inflection in the vicinity of a resonant frequency is **anomalous dispersion**. Concurrently, the value of the imaginary part presents the maximum, which means that the absorption is strong.

We proceed with another simplification. Specifically, at the vicinity of the resonance, where $\omega \approx \omega_o$, so $|\omega - \omega_o| \ll \omega_o$ and $\omega_o^2 - \omega^2 \approx 2\omega_o \cdot (\omega_o - \omega)$, we express Eqs. (3.39) and (3.40) as

$$n(\omega) = 1 + \frac{\omega_p^2}{4\omega_o} \cdot \frac{\omega_o - \omega}{(\omega_o - \omega) + \left(\gamma/2\right)^2} \quad \text{and} \quad \kappa(\omega) = \frac{\omega_p^2}{8\omega_o} \cdot \frac{\gamma}{(\omega_o - \omega) + \left(\gamma/2\right)^2} \tag{3.43}$$

The distribution of $\kappa(\omega)$ near the absorption band is symmetrical and follows a Lorenz statistical distribution. The maximum of the absorption index $\kappa(\omega)$ occurs when $\omega = \omega_o$, as expected. The extrema, either maximum or minimum, of the refractive index $n(\omega)$ are at the values of $\omega$ where the derivative of Eq. (3.43) (the first equation) with respect to $\omega$ becomes zero:

$$n(\omega)_{\text{MAX}} = 1 + \frac{\omega_p^2}{4\omega_o\gamma} \quad \text{and} \quad n(\omega)_{\text{MIN}} = 1 - \frac{\omega_p^2}{4\omega_o\gamma} \tag{3.44}$$

which correspond to $\omega = \omega_o \mp \gamma/2$, respectively. These extremes, which differ by $\gamma$, define an absorption zone (band). We note that the bandwidth is proportional to the absorption coefficient of friction $\gamma$. From the second relationship in Eq. (3.43), we can also see that the absorption index maximum is inversely proportional to the coefficient $\gamma$. When $\gamma$ is very small, the distribution peak depicted in Figure 3-11 (lower graph) becomes very sharp:

$$\kappa(\omega)_{\omega=\omega_o} = \frac{\omega_p^2}{2\omega_o\gamma} \tag{3.45}$$

At the limit $\gamma \to +0$ ($\gamma$ tends to zero from the positive values), there is unspecificity, as suggested in the relationships of Eq. (3.41). If the frequency equals $\omega_o$, $n_{(\omega=\omega o)} = 1.0$, exactly. For $\omega < \omega_o$, $n_{(\omega<\omega o)} > 1.0$, while for $\omega > \omega_o$, $n_{(\omega>\omega o)} < 1.0$.

Let's pause for a moment. We just read that the refractive index can be less than unity. This is a rather unexpected surprise; the refractive index should be higher than 1.0. Here, the refractive index (real part) is less than unity because, for $\omega_o < \omega \rightarrow (\omega_o^2 - \omega^2) < 0$, the bipolar oscillator shift is in a phase difference of $\pi$ with the inducing field.

However, this anomaly applies only to a small range on the right side of the band, where the anomalous dispersion phase velocity is greater than $c$. It should not be a big concern, in the sense that it is violating the laws of physics. The phase velocity corresponds to a propagation of an ideal harmonic wave, which by itself does not exist in nature. Instead, there are impulses traveling at the group velocity, which thus has a physical content and is always less than $c$. In this special band, the magnitude of the complex refractive index is still greater than 1.0.

Another simplification is in place for frequencies much lower than the resonant frequency(-ies). In this region, $\omega^2 \ll \omega_o^2$, and we can re-write Eqs. (3.39) and (3.40) as

$$n(\omega) = 1 + \frac{\omega_p^2}{2\omega_o^2} \quad \text{and} \quad \kappa(\omega) = 0 \tag{3.46}$$

Thus, for frequencies moderately lower than the resonant frequency, the refractive index is a real number that is slightly greater than 1.0 and is independent of the frequency.

We further simplify for much higher frequencies than the resonant frequency(-ies). At this limit, the approximations $\omega^2 \gg \omega_o^2$ are valid, and Eq. (3.39) becomes

$$n(\omega) = 1 - \frac{\omega_p^2}{2} \cdot \frac{1}{\omega^2 + \gamma^2} \tag{3.47}$$

For these frequencies, which are much higher than the resonant frequency, the refractive index is a real number slightly less than 1.0, approaching 1.0 at very high frequencies. Figure 3-12 depicts a general case of the dependence of $n$ on the frequency $\omega$ in a dielectric with three resonant frequencies $\omega_1$, $\omega_2$, and $\omega_3$. The resonant bands coincide with the absorption (and the anomalous dispersion) bands. We note the limits of low and high frequencies, in which $n$ is slightly greater than 1 (far left) and slightly less than 1 (far right), respectively.

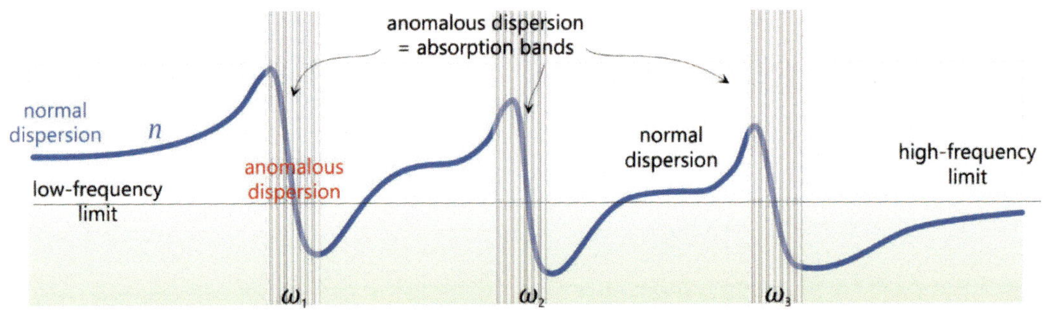


*Figure 3-12: Variation of n with respect to the resonant frequencies in a thin dielectric medium.*

## 3.3.2 Dispersion in Optical Glass

The dispersion curve depicts $n(\lambda)$ versus $\lambda$, i.e., the dependence of the refractive index on the wavelength. As shown in Figure 3-13, a specific material (flint glass) presents a refractive index distribution that ranges from 1.685 for the violet to 1.645 for the red. Note that with increasing wavelength, the value of the refractive index is decreasing—this is normal dispersion.

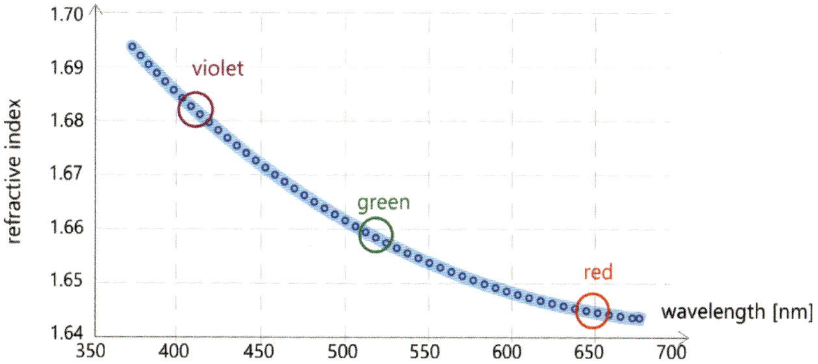


*Figure 3-13: Normal dispersion curve in a transparent material (flint glass) for the visible spectrum.*

The dimensionless **Abbe number** (named after the German physicist Ernst Karl Abbe), also known as constringence, is an expression of the medium's dispersion and is defined as

Abbe Number:
$$V = \frac{n_{Y/d} - 1}{n_{B/f} - n_{R/C}} \tag{3.48}$$

where $n_{R/C}$ represents the red hydrogen spectral line ($\lambda_R = 656.3\,\text{nm}$), $n_{Y/d}$ represents the yellow sodium line ($\lambda_Y = 587.6\,\text{nm}$), and $n_{B/f}$ represents the blue hydrogen line ($\lambda_B = 486.1\,\text{nm}$).

If the refractive indices $n_B$ and $n_R$ are quite different, the Abbe number $V$ has a relatively small value (such as $V < 55$), indicating that the material is strongly dispersive (such as flint glass). If the indices $n_B$ and $n_R$ differ slightly, then $V$ has a relatively large value (such as $V > 55$), indicating that the material has a low dispersion.

*Table 3-1: Refractive index and Abbe number values for certain types of optical materials.*

| Lens Material | Refractive Index | Abbe Number |
|---|---|---|
| crown glass | 1.523 | 58 to 60 |
| CR-39 (plastic spectacle lens material) | 1.498 | 58 |
| Trivex® (aka Phoenix, NXT®, Trilogy®) | 1.523 | 43 to 45 |
| polycarbonate (plastic spectacle lens material) | 1.586 | 30 |
| dense flint glass | 1.61 | 36.8 |

### 3.3.2.1 Measurement of the Refractive Index in Glass

The refractive index of optical materials such as glass can be experimentally measured via a variety of techniques, using wavelengths in the visible and near-visible regions. The most common measurement technique uses refraction: We know that, in a prism of apical angle $A$, the minimum angle of deviation $\vartheta_{\text{E MIN}}$ can be used to calculate the refractive index $n(\lambda)$ for a spectral window centered around a narrow range between $\lambda - \delta\lambda$ and $\lambda + \delta\lambda$ via a relationship presented in *Introduction to Optics* § 3.3.2, the minimum angle of deviation:

Minimum Angle of Deviation: $$n \cdot \sin\frac{A}{2} = \sin\frac{A+\vartheta_{\text{E MIN}}}{2} \Rightarrow n(\lambda) = \frac{\sin\dfrac{A+\vartheta_{\text{E MIN}}(\lambda)}{2}}{\sin\dfrac{A}{2}} \tag{3.49}$$

Most often, an isosceles triangular prism (meaning that its apical angle $A$ equals 60°) made of the glass is used, placed in air, while scanning the deviated light in order to progressively identify the different spectral parts of the visible spectrum. The white light from a source has a known discrete spectrum (for example, a hydrogen lamp), so the select lines observed at a given time correspond to well-known values of the wavelength. The prism is rotated until the minimum value of the deviation angle for the select wavelength is measured; the process is repeated for different available wavelengths.

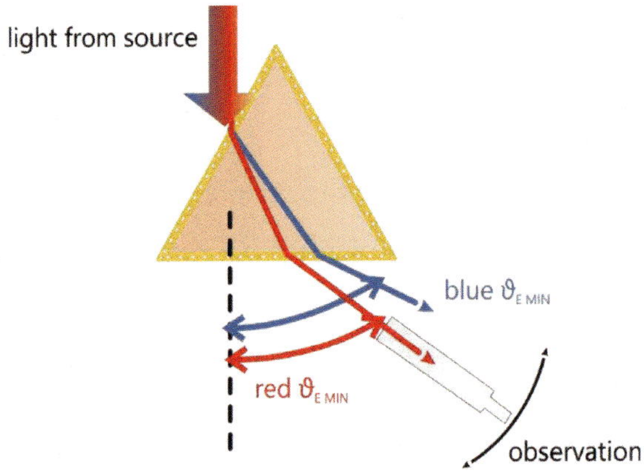


*Figure 3-14: Measurement of the minimum angle of deviation $\vartheta_{\text{E MIN}}$ for different wavelengths for calculating the refractive index for that wavelength.*

Example ☞: Calculate the refractive index for the glass material in an isosceles triangular prism if, for a certain wavelength, the minimum angle of deviation is $\vartheta_{\text{E MIN}} = 46.25°$.

$$n \cdot \sin\left(\frac{60^\circ}{2}\right) = \sin\left(\frac{60^\circ + 46.25^\circ}{2}\right) \Rightarrow n \cdot 0.5 = \sin(53.125^\circ) \Rightarrow n = 2 \cdot 0.80 = 1.6$$

The minimum deviation angle technique is suitable for the visible part of the spectrum, as it relies on the observation of light ray deviation. Other techniques that can be used to measure the refractive index of glass involve interferometric applications called spectral-line observing.[17] The principle of operation in spectral-line observing is that multiple light reflections between the faces of a flat piece of glass create interference fringes whose maxima occur for wavelengths for which the optical thickness of the glass plate is an integral number of a half wavelength [see § 4.2.5 and Eq. (4.55)]. Thus, from wavelength and thickness measurements, the index of refraction of the glass can be determined. Other techniques for measuring the refractive index of glass—those that extend to other parts of the electromagnetic spectrum neighboring the visible—include infrared spectroscopy[18] and focal displacement.[19]

## 3.4 EMISSION AND ABSORPTION SPECTRA

We distinguish two specific cases of light absorption by a dielectric (more cases are presented in § 1.5.2):

- If the frequency of the incident wave coincides with any of the medium's resonant frequencies ($h \cdot \nu = E_2 - E_1$), there is quantum resonant absorption (an area of anomalous dispersion). The absorbed energy can be attributed to some form of non-radiative process, such as thermal, as is the case primarily in solids, so the material is **opaque** to the particular frequency. In thin media and gasses, after photon absorption, the atom can emit a photon with the exact same frequency via a spontaneous emission mechanism.

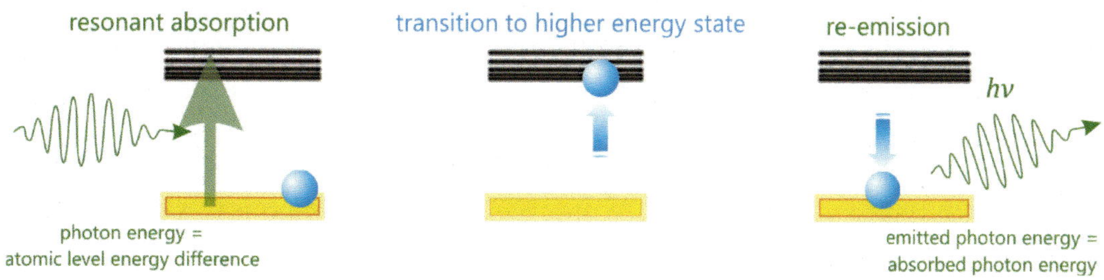


*Figure 3-15: Mechanism of resonant (quantum) absorption and photon re-emission.*

---

[17] Randall CM, Rawcliffe RD. Refractive indices of germanium, silicon, and fused quartz in the far infrared. Appl Opt. 1967; 6(11):1889-95.

[18] Tan C. Determination of refractive index of silica glass for infrared wavelengths by IR spectroscopy. J Non-Cryst Solids. 1998; 223: 158-63.

[19] Pandey N, Singh MP, Pant LM, Ghosh A. A simple method to measure refractive index of optical glasses using focal displacement method. Proc. SPIE 9654, 96540L (2015) [doi: 10.1117/12.2181509].

• If the frequency of the incident wave does not coincide with any of the resonant frequencies of the medium (nonresonant wave, $h \cdot \nu \neq E_2 - E_1$), we do not have quantum absorption. For small photon energies, the absorption mechanism, which is a classical mechanistic absorption, may lead to simulated bipolar oscillation. The oscillation may, in turn, lead to same-frequency light re-emission via a scattering mechanism. This is an area of normal dispersion. The material appears **transparent** in that specific frequency.

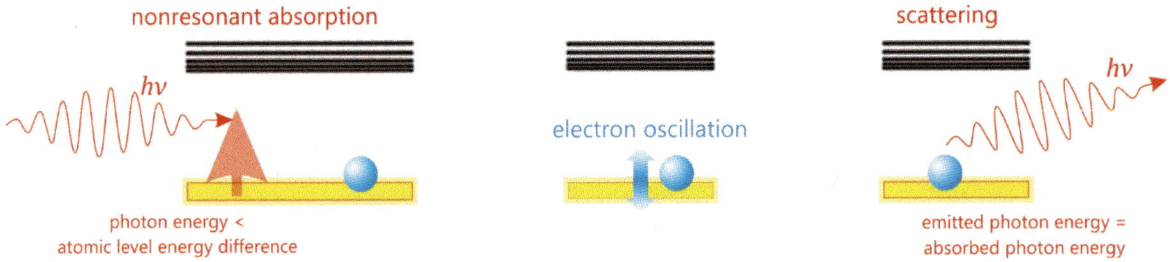


*Figure 3-16: Mechanism of nonresonant absorption and photon re-emission.*

## 3.4.1 Spectra and Filters

All wavelengths emitted from a source form the source's **emission spectrum**. A spectrum can be continuous [Figure 3-17 (top)] if it consists of uninterrupted, successive frequencies/wavelengths. Such is the case in white light from an incandescent lamp. By applying a voltage to the filament, the accelerated free electrons collide with a metal lattice that absorbs part of their energy, which, when re-emitted, results in continuous-spectrum lattice oscillations.

An object whose spectral content has several color components, which when combined form white or nearly white light, is a **white light source**. We emphasize here that by saying radiation-specific color we are referring to radiation that correlates to a specific wavelength.

On the other hand, a spectrum is **linear** or discrete if there are lines, actually narrow bands, that correspond to specific colors. In practice, what is recorded is the image of the instrument entrance pupil, convoluted for the color components. In spectral analyzers, such as a monochromator (presented in § 5.7.1), the entrance is a slit, which is why we see the spectral lines. A light source that has a linear emission spectrum, such as light from a gas-discharge fluorescent lamp, can also appear white, or nearly white.

An emission spectrum [Figure 3-17 (top and bottom)] results from analysis of light emitted by a source. Conversely, an absorption spectrum is produced if we illuminate the material with a continuous-light spectrum and examine the resulting transmission spectrum. The medium absorbs certain spectral lines [Figure 3-17 (middle)]. Thus, a linear absorption spectrum arises from a continuous spectrum, from which the spectral regions (colors) in the medium having significant absorption are removed. These areas are the material absorption bands. Both absorption and emission bands depend exclusively on the material and, specifically, on the structure of the permissible electronic transitions.

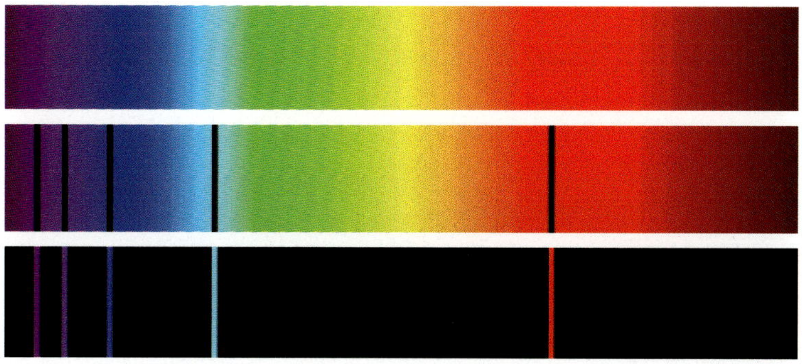

*Figure 3-17: (top) Continuous spectrum of a white emission source, (middle) hydrogen absorption spectrum, and (bottom) hydrogen linear emission spectrum.*

The absorption and emission bands of low-pressure gasses correspond to thin, well-defined lines that are characteristic of the element. In 1814 Joseph von Fraunhofer observed in the sun's outer atmosphere thin black absorption bands from a—then unknown—element he named helium, from the Greek *ἥλιος*, for Sun. A photon is absorbed only if its energy corresponds to an exact transition, which, described simply, is an available and permissible electron transition between orbitals (see § 6.1.1) that permits electron migration to an unoccupied level. The difference in energy is emitted as light. In any particular gas, the absorption transitions match the emission transitions.

The absorption bands of solids, on the other hand, are expansive due to degeneration, which results in line broadening. It is possible to correlate the percentage of absorbance or permeability with the material color. If, for example, an object illuminated by white light appears red, it is quite possible that the absorbance is strong for almost all visible frequencies up to the red, around 600 nm. Concurrently, permeability is virtually zero for all wavelengths, up to the vicinity of the red. Such are the attributes of a **red filter**. Respectively, for a blue filter, there is little absorption at wavelengths up to 500 nm, while absorption increases sharply for wavelengths greater than 500 nm. For a green filter, there is absorption around 550 nm.

Reflecting objects can also be classified in terms of permeability and reflectivity. A black object absorbs all wavelengths, while a white object reflects all wavelengths. For classifying colored reflecting objects, we consider the spectral distribution of their reflectivity. An object can strongly absorb red and blue; what is not absorbed is reflected, so when the object is illuminated by white light, it appears green.

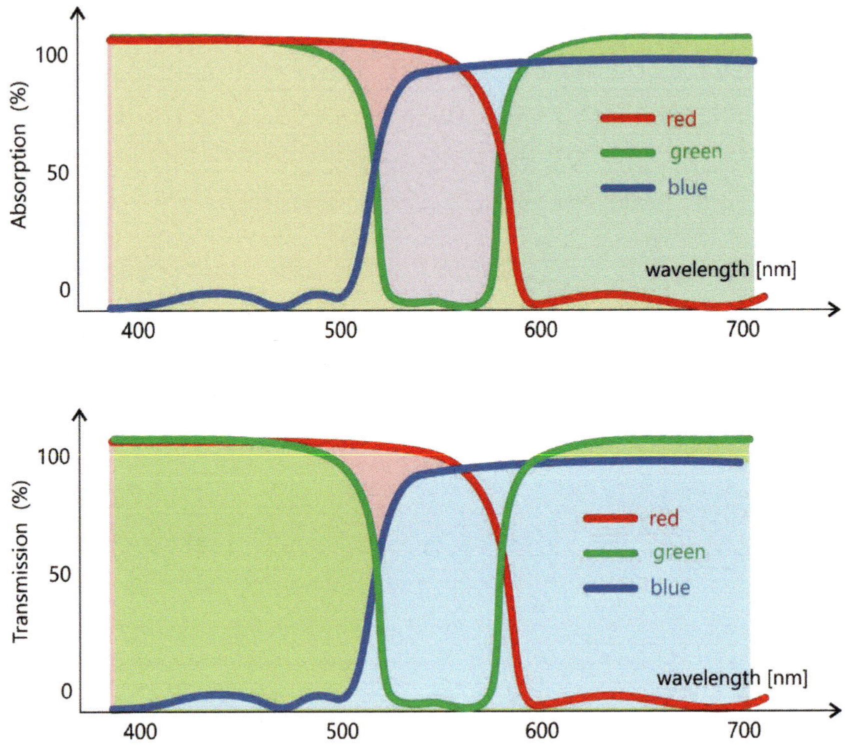


*Figure 3-18: Spectral distribution of (top graph) absorption and (lower graph) transmission for various filters.*

*By the diſcovered Properties of Light to explain the permanent Colours of natural Bodies.*

Theſe Colours ariſe from hence, that ſome natural Bodies reflect ſome ſorts of rays, others other ſorts more copiouſly than the reſt. Minium reflects the leaſt refrangible or red-making rays moſt copiouſly, and thence appears red. Violets reflect the moſt refrangible, moſt copiouſly, and thence have their Colour, and ſo of other Bodies. Every Body reflects the rays of its own Colour more copiouſly than the reſt, and from their exceſs and predominance in the reflected Light has its Colour.

*Figure 3-19: Newton's perspective on the colors of objects [reprinted from* Opticks, *pg. 135 (1704)].*

Tree leaves are considered reflecting objects whose strong green color is associated with chlorophyll. The chlorophyll absorption spectrum peaks at both 450 nm and 650 nm, meaning that chlorophyll absorbs strongly in the blue and red; therefore, the leaves appear green, as they strongly reflect the green, which is not absorbed. This, of course, assumes daytime white sunlight; if we use, for example, red or blue light, these color components will be absorbed, so the leaves will appear black!

Tree leaves contain other pigments whose action is largely suppressed during the spring and summer. With the gradual drop in fall temperatures, chlorophyll gradually degrades, so the action resulting from the other pigments becomes apparent. For example, the bright yellow foliage is due to carotenoids pigments, which strongly reflect orange and red, while absorbing blue and green. The carotenoid spectrum is therefore complementary to the chlorophyll spectrum.

*Figure 3-20: Colorful autumn leaves in nature.*

Thus, the answer to 'Why do leaves change color in the fall?' has to do with their mission: They are the food-making factory of the tree, and they change color in autumn as a final effort to collect as much of the increasingly less available solar energy as possible (the direct sunlight spectrum presents a maximum near the green). Red leaves are actually in a special state where sugars are still being produced but do not exit the leaves, as the leaves are sealed off in preparation for winter. The sugars in a red leaf end up as anthocyanins (*ἀνθό-* for flower & *-κυανό* for blue), which have a strong red reflection.

Food for thought: We usually play cards under white light. Figure 3-21 (left) is a familiar view of a card. Under which light should we play in order to obtain the (center) and (right) views?

*Figure 3-21: These French playing cards may appear quite ... different, depending on the type of illumination used: (left) illuminated with white light, (center) illuminated with red light, and (right) illuminated with blue light.*

## 3.4.2 Absorption Properties of the Optical Glass

Optical glass is perhaps among the few solids that transmit visible light. This means that absorption across the visible spectrum is considerably low (but not zero). A glass' absorbance, which is dependent on the wavelength, is used to describe the decrease in intensity of light as it travels through the glass volume.

In addition, absorbance varies by the glass type and composition. Even in clear glass, which contains no additives other than those involved in the manufacturing process, there is absorption. Often, iron oxides produce significant absorption in the infrared. Absorption in glass also depends on the tint, which may be gray, suggesting a uniform drop in transmittance (an increase in absorbance), or it may have a hue, suggesting absorption properties similar to those of a color filter. Absorption depends on the thickness and composition of the glass, as well as the wavelength of the light.

The dependence of absorption on the glass thickness is a property deriving from Eqs. (3.34) and (3.37), which state that the greater the glass thickness $z$, the greater the absorption, in an exponential dependence. By reducing the thickness of the material, you increase the amount of light that can pass through. To facilitate calculations, the transmittance factor $q$ is used, which is the fraction of light transmitted per unit length of optical material. In other words, a large transmittance factor is associated with a low absorption per unit thickness length. Due to the exponential dependence of absorption on glass thickness, the transmissivity is calculated by the transmittance factor raised to the power of the fractional length for which the transmittance factor is quoted (glass thickness / unit length of transmittance factor).

Example ☞: A given glass material has a transmittance factor $q$ = 0.8 per 1 mm of material. What is the transmissivity for 1 mm of lens thickness? For 2 mm of thickness? For 4 mm of thickness?

For the 1 mm lens thickness, we simply use $q$: $T = q^1 = 0.8$.
For the 2 mm lens thickness, we raise $q$ to the power of 2: $T = q^2 = 0.64$.
For the 4 mm lens thickness, we raise $q$ to the power of 4: $T = q^4 = 0.41$.

The dependence of absorbance on the glass composition relates not to the bulk material but rather to the impurities incorporated during its manufacture and its chemical formulation. This dependence is intricately related to the dependence on the wavelength: For most of the visible spectrum, glass has near-zero absorbance, which, however, peaks at the UV part of the spectrum. Despite being low, absorption metrics increase significantly for shorter wavelengths. Absorption in the blue is far stronger than absorption in the red; certain elements present in the glass composition also affect absorption in other parts of the spectrum, such as in the infrared.

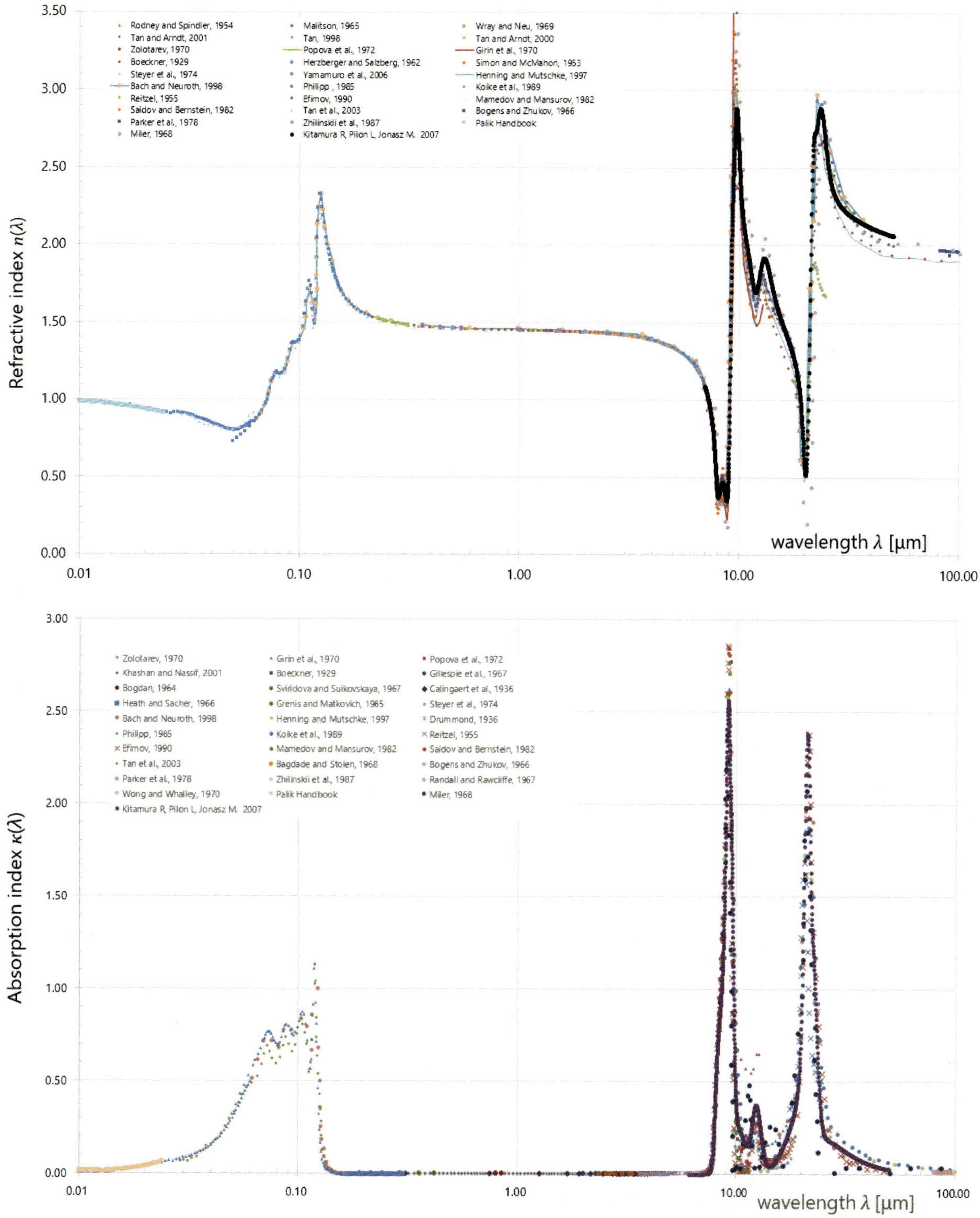


*Figure 3-22: (top graph) Refractive index n(λ) and (lower graph) absorption index κ(λ) in silica glass. Note the near-zero absorption coefficients for a significant part of the spectrum, from the UV (peaking at around 0.15 μm) to the near-IR (peaking at around 9 μm). Between the UV and the near-IR there is negligible absorption and normal dispersion. [Data for these plots, courtesy of Dr. Laurent Pilon, are reported in the literature and published in R. Kitamura, L. Pilon, M. Jonasz, Optical constants of silica glass from extreme ultraviolet to far-infrared at near room temperature. Applied Optics. 2007; 46(33):8118–33.]*

# 3.5 DISPERSION AND ABSORPTION QUIZ

1) As a wave propagates, its phase and amplitude change. The change in phase is associated with the ...
   a) propagation speed in that medium
   b) source frequency
   c) field strength
   d) energy attenuation

2) The refractive properties of a wave propagating in a medium are dependent on the ___________ of the refractive index.
   a) real part
   b) imaginary part
   c) sum of real and imaginary parts
   d) difference between real and imaginary parts

3) The reduction in amplitude magnitude of a wave as it propagates in a medium is dependent on the ___________ of the refractive index.
   a) real part
   b) imaginary part
   c) sum of real and imaginary parts
   d) difference between real and imaginary parts

4) A certain medium has an absorption (extinction) coefficient $a$ = 10 mm$^{-1}$. If 100 (arbitrary) units of light intensity are incident on a filter made of this medium with a thickness (along the path of light propagation) of 0.1 mm, how much intensity (same arbitrary units) exits this filter (disregard reflection losses, which are in the vicinity of less than 3% per incident radiation on each interface)?
   a) 90
   b) 60
   c) 40
   d) 14
   e) 8
   f) 1

5) A twist to Q 4. Now we double the path along the light propagation, which becomes 0.2 mm. How much intensity exits this filter?
   a) 90
   b) 60
   c) 40
   d) 14
   e) 8
   f) 1

6) Damn the torpedoes! Damn the exponentials! They are just so hard. Let's talk about logarithms, which are so much nicer; I know it, you know it, everybody knows it. Back to Q 4. What is the natural logarithm of the ratio of the intensity leaving the 0.1 mm thick filter to the intensity incident on that filter?
   a) +1.0
   b) +0.5
   c) 0.0
   d) –0.5
   e) –1.0
   f) –2.0

7) Back to Q 5. What is the natural logarithm of the ratio of the intensity leaving the 0.2 mm thick filter to the intensity incident on that filter?
   a) +1.0
   b) +0.5
   c) 0.0
   d) –0.5
   e) –1.0
   f) –2.0

The following six questions (Q 8 to Q 13) discuss a specific filter with the following absorption coefficients: frequency $\omega_A$, $a(\omega_A)$ = 20 mm$^{-1}$, frequency $\omega_B$, $a(\omega_B)$ = 10 mm$^{-1}$, and frequency $\omega_C$, $a(\omega_C)$ = 5 mm$^{-1}$. The term 'natural logarithm' specifically pertains to the ratio of the intensity leaving the filter to the intensity incident on that filter.

8) At what frequency does this filter absorb more strongly (for the same length of material)?
   a) frequency $\omega_A$
   b) frequency $\omega_B$
   c) frequency $\omega_C$
   d) not dependent on frequency

9) The filter we are discussing is a blue filter. Which of these frequencies is closest to the blue?
   a) frequency $\omega_A$
   b) frequency $\omega_B$
   c) frequency $\omega_C$
   d) not dependent on frequency

10) What length of material (thickness $d$) is required such that the intensity drops to about 36.6% if the incident radiation corresponds to $\omega_A$?
   a) 0.05 mm
   b) 0.10 mm
   c) 0.20 mm
   d) 0.25 mm
   e) 0.40 mm

11) What length of material (thickness $d$) is required such that the natural logarithm drops to –2.0 when illuminating the filter with radiation frequency $\omega_C$?
   a) 0.05 mm
   b) 0.10 mm
   c) 0.20 mm
   d) 0.25 mm
   e) 0.40 mm

12) If, over a length of 0.1 mm, the ratio of transmitted intensity to incident intensity is 0.36, then over a length of 0.2 mm, the ratio of transmitted intensity to incident intensity is (radiation frequency $\omega_B$) ...
   a) 0.61
   b) 0.22
   c) 0.18
   d) 0.14
   e) 0.08

13) What is the required filter thickness in order to absorb equal amounts of radiation (95% in all three) for the three different frequencies?
   a) frequency $\omega_A$: 0.15 mm; frequency $\omega_B$: 0.30 mm; frequency $\omega_C$: 0.60 mm
   b) frequency $\omega_A$: 0.20 mm; frequency $\omega_B$: 0.40 mm; frequency $\omega_C$: 0.60 mm
   c) frequency $\omega_A$: 0.30 mm; frequency $\omega_B$: 0.60 mm; frequency $\omega_C$: 0.90 mm
   d) frequency $\omega_A$: 0.60 mm; frequency $\omega_B$: 0.30 mm; frequency $\omega_C$: 0.15 mm

14) In normal dispersion, the phase velocity is subject to a refractive index that is typically (two correct answers) ...
   a) $< 1.0$
   b) $> 1.0$
   c) increasing for increasing frequency
   d) increasing for increasing wavelength

15) Glass, in the majority of the visible spectrum, has remarkable transparency. This suggests that (two correct answers) ...
   a) absorption is significant
   b) absorption is minimal
   c) dispersion is normal
   d) dispersion is anomalous
   e) dispersion is minimal
   f) dispersion is significant

16) As the dispersive properties of glass for the visible part of EM radiation decrease, the Abbe number...
   a) increases
   b) is not affected
   c) decreases

17) A distinct black band is noted in the hydrogen absorption spectrum at 656 nm. This means that in atomic hydrogen, the incident radiation in the vicinity of 656 nm exhibits (two correct) ...
   a) significant absorption
   b) minimal absorption
   c) normal dispersion
   d) anomalous dispersion
   e) minimal dispersion
   f) significant dispersion

18) Negligible absorption is usually accompanied by ...
   a) normal dispersion
   b) anomalous dispersion
   c) minimal dispersion
   d) significant dispersion

19) In a spectral band associated with normal dispersion, as the wavelength increases, the refractive index typically ...
   a) decreases
   b) remains stable
   c) increases

20) The spectacle lens material CR-39 has Abbe number 58, while polycarbonate, another plastic lens material, has Abbe number 30. Which of the two materials exhibits the greater refractive index difference between the violet ($\lambda_V \approx 400$ nm) and the red ($\lambda_B \approx 700$ nm)?
   a) CR-39
   b) polycarbonate
   c) They are not different.
   d) The data are not sufficient.

# 3.6 DISPERSION AND ABSORPTION SUMMARY

The refractive index is not just a simple number. It is a complex number that is dependent on the frequency (wavelength) of the light (in general, radiation). Dispersion is the dependence of the real part of the refractive index on the frequency (wavelength). Absorption is the dependence of the imaginary part of the refractive index on the frequency.

The dependence of the refractive index on the frequency relates to the atomic and molecular structure of the medium. The complex form of the refractive index is due to the combination of inertial and frictional forces introduced by a phase difference between the electric forces exerted on the system and the system's response. Frictional forces cause a nonreversible energy transfer from oscillating charges to the medium; this energy transfer manifests macroscopically as radiation absorption.

According to the Beer–Lambert law, for a given material and a specific wavelength, the amplitude magnitude of the transmitted radiation diminishes exponentially with an increase in the path traveled by light within it. The absorption coefficient is strongly dependent on the material properties and the wavelength. Certain filters strongly absorb select parts of the visible spectrum. Strong absorption occurs around resonant frequencies.

Typically, the (real part of the) refractive index increases with the frequency—or, alternatively, it decreases with the wavelength. Thus, the refractive index is higher for the shorter-wavelength parts of the visible spectrum, such as the violet, and lower for the longer-wavelength parts of the visible spectrum, such as the red. This is normal dispersion. The Abbe number is an expression of a material's dispersive properties. The greater the number, the lower the dispersion.

In the vicinity of resonant frequencies, dispersion becomes anomalous. The material is opaque to the particular frequencies. In the spectral areas where frequencies are not resonant, the material typically displays normal dispersion and is transparent to the particular frequencies.

An emission spectrum results from analysis of light emitted by a source. An absorption spectrum is produced if we illuminate the material with a continuous light spectrum and examine the resulting transmission spectrum.

# CHAPTER 4 INTERFERENCE

Interference phenomena appear when at least two 'synchronized' waves meet and the intensity distribution is (locally) different from the sum of the individual wave intensities. The phase, which is the internal clock of the wave disturbance, plays a key role in this occurrence. Two waves are synchronized if there is a fixed relationship between their internal clocks, in which case they are coherent; otherwise, they are incoherent.

Interference phenomena are encountered in all types of waves, under the proper conditions. If we throw two pebbles onto a water surface (such as in Figure 4-1), we may note in the space between the two ripples a zero, or a periodically reinforced, disturbance. Many effects in nature display light interference; for example, the opalescent colors in abalone shells and the iridescence in oil slicks result from thin-film interference (§ 4.2.3). The colorful feathers in many birds are also due to interference.

The study of interference phenomena sparked a revolution in the wave theory of light. Until 1800, the classical corpuscular theory provided a complete interpretation of all known optical effects. This was the case, despite previously observed interference effects, such as the rings (§ 4.2.4) studied by Isaac Newton. Newton's belief in the corpuscular theory was so strong that it led him to propose—without proof, however—some 'fits of easy transmission' to explain interference.

A fundamental contribution to the understanding of interference was Thomas Young's experiment (§ 4.2.1), in which two coherent light sources interfere. The key aspect in that experiment was the phase parameter. The physical entity that plays the key role in interference is not the intensity, which is proportional to the energy, but the vectorial, phase-dependent electric field. Summation of the field vectors at every point in space gives different results, depending on each vector's phase difference. The photon theory of light (introduced in § 1.4.6) also provides a good interpretation of interference effects. Properly added wavevectors (§ 4.1.4) explain the alternating constructive and destructive interference fringes.

## 4.1 ADDITIONS OF LIGHT PRODUCES DARKNESS

For interference to occur, at least two waves must be simultaneously present at the same point. However necessary, this wave confluence condition is not by itself sufficient for interference effects to develop. Some other conditions must be in place.

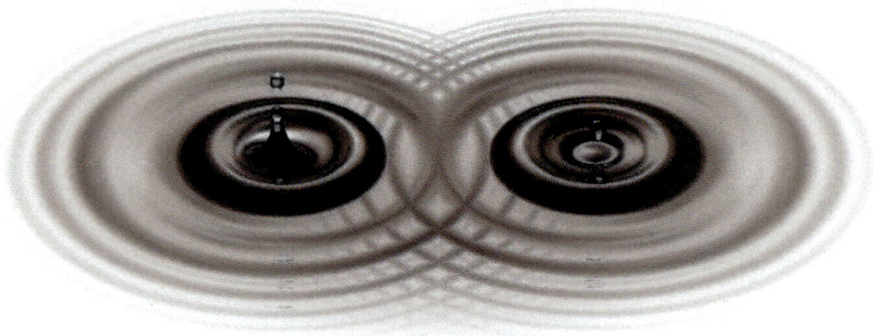

*Figure 4-1: Interference effects in water ripples.*

In many interference experiments, we note areas with zero-intensity areas, dark areas, and very bright areas. It is possible, in other words, to have darkness if we add light to light. This results in interference fringes. The questions are, why and how does the meeting of two of such waves produce zero disturbance? The answers are provided by the wave theory. To calculate the resultant disturbance, we apply the **principle of linear superposition**, developed by Augustin-Jean Fresnel.

The electric field $\mathbf{E}$ produced jointly by two or more waves with fields $\mathbf{E}_1$ and $\mathbf{E}_2$ is their vector sum:

Principle of Linear Superposition : $$\mathbf{E} = \mathbf{E}_1 + \mathbf{E}_2 + \mathbf{E}_3 + \ldots \quad (4.1)$$

For the principle of linear superposition, we note the following:

- The disturbance at every point in space is the vector sum of the fields that correspond to the interfering waves.

• The contribution of each wave to the resultant disturbance is independent of the contribution of another wave.

• The superposition principle does not apply to intensities: The resultant intensity is not the sum of the individual intensities (we will soon see why).

• The superposition principle applies to both the electric field and the magnetic field, and is a direct consequence of the inherent linearity of Maxwell's equations (§ 1.1.4.1). We use the electric field for convenience but can reach the same conclusions using the magnetic field instead.

Fresnel: The resultant field is the vector sum of the corresponding electric fields, considering their relative phase difference.

Note ✐: The principle of superposition is violated when the waves are so strong that they can alter the material properties of the medium in which they propagate. This is nonlinear optics (see § 6.3.3).

There are two interesting cases of interference: (1) The added waves of equal amplitude magnitude [Figure 4-2 (top)] result in a net disturbance with twice the amplitude magnitude—this is **constructive interference**. (2) The added waves of equal amplitude magnitude [Figure 4-2 (bottom)] result in a net disturbance with zero amplitude—this is **destructive interference**.

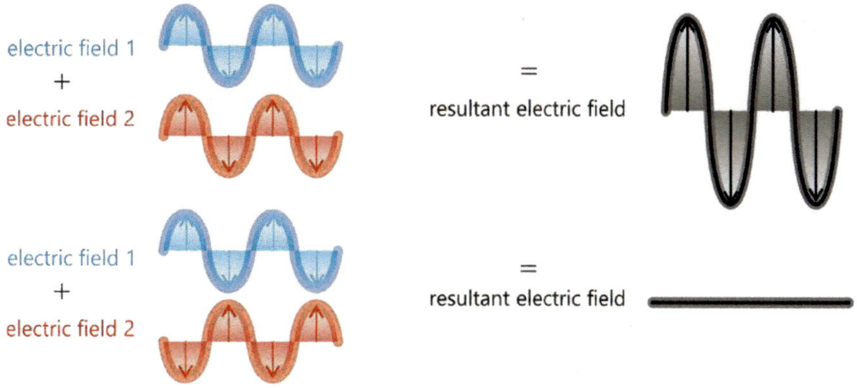


*Figure 4-2: Wave interference (top) in phase and (bottom) with opposing phases.*

It is noted that the interference of two light waves results in a maximum or a minimum (zero); in the latter case, we add light to light and get darkness. This is due to the phase—the internal clock of the disturbance that determines its instantaneous magnitude—which is repeated circularly in time (frequency) and in space (wavevector). We use plane harmonic waves not because being flat or harmonic is an interference condition (not at all!), but because the calculations are vastly simplified. Nevertheless, every wave can be analyzed into such harmonic components, which, following the notation introduced in Eq. (1.7), become

$$\mathbf{E}_1 = \mathrm{Re}\,\{\mathbf{E}_{1o}\cdot \exp[i\,(\omega\cdot t - \mathbf{k}\cdot\mathbf{r} + \varphi_{1o})]\} \quad \text{and} \quad \mathbf{E}_2 = \mathrm{Re}\,\{\mathbf{E}_{2o}\cdot \exp[i\,(\omega\cdot t - \mathbf{k}\cdot\mathbf{r} + \varphi_{2o})]\} \tag{4.2}$$

Here, the phase is the argument in the trigonometric function that describes the disturbance $\omega\cdot t - \mathbf{k}\cdot\mathbf{r} + \varphi_{1o}$. These two electric vector waves have the following phase difference:

$$\delta\varphi_{12} = (\omega_1 - \omega_2)\cdot t - (\mathbf{k}_1 - \mathbf{k}_2)\cdot\mathbf{r} + (\varphi_{o1} - \varphi_{o2}) \tag{4.3}$$

The addition of these two waves depends on this phase difference. The waves described in Figure 4-2 (left) have the same frequency, amplitude, and the exact same phase at every point; in other words, they have a zero and fixed phase difference. In Figure 4-2 (right) the waves are similar to those on the left, but they are shifted by half a wavelength. Their phase difference is $\pi$ rad: This is the opposite phase. In the first case, the sum of the two fields results in a wave with twice their initial amplitude, while the second case results in a wave with zero amplitude.

The physical quantity that is directly detectable in optics is the intensity $I$ (introduced in § 1.1.4). Intensity is the time average of the square of the corresponding electric field amplitude [Eq. (1.11)]:

$$I = c\varepsilon_o\langle\mathbf{E}\cdot\mathbf{E}\rangle = \tfrac{1}{2}\, c\varepsilon_o\, E_o^{\,2} \tag{4.4}$$

where $\langle\,\rangle$ indicates the mean temporal value of the enclosed quantity.

Example ☞: Two waves have amplitude magnitudes of 5 and 10 units. Because the intensity is proportional to the amplitude magnitude squared, the second wave has 4× the intensity of the first wave.

To calculate the intensity resulting from wave interference, we follow these steps:

1. The electric field is determined by adding the respective fields.
2. The intensity is calculated as the square of the resultant field.

Specifically, for the waves described in Eq. (4.2), the intensity distribution pattern that results from their interference can be described by

Interference Equation:

$$I_{TOT} = I_1 + I_2 + 2c\varepsilon_o\cdot\langle\mathbf{E}_1\cdot\mathbf{E}_2\rangle\cdot\langle\cos(\delta\varphi)\rangle \tag{4.5}$$

Intensity therefore does not equal the simple sum of the two intensities ($I_1 + I_2$); an additional term is included, the **interference term**. This term determines the interference:

Interference Term:

$$2c\varepsilon_o\cdot\langle\mathbf{E}_1\cdot\mathbf{E}_2\rangle\cdot\langle\cos(\delta\varphi)\rangle \tag{4.6}$$

We can now define interference:

| Interference: | The interaction of two or more waves that results in an intensity distribution that differs from the simple sum of the waves' individual intensities. |
|---|---|

We emphasize that in interference the resultant intensity is calculated by first adding the component electric fields, thus finding the resultant field, and then squaring the resultant field. This is a direct consequence of the wave nature of light. If the reverse path were to be applied—which is the case in the realm of classical particle theory—we would first find the intensities for every wave ($I_1$ and $I_2$) and then add them. In this case, the total intensity would have no interference factor and would simply equal $I_1 + I_2$. Had light been a classical particle, we would never have obtained the result 'light + light = darkness.'

Take-home message for interference ✉: The simple confluence of two waves is not an adequate condition for interference. What is absolutely necessary is a nonzero interference factor.

## 4.1.1 Temporal and Spatial Coherence

In order for the interference factor to be nonzero, there must co-exist at least two nonzero intensity beams: $I_1 \neq 0$ and $I_2 \neq 0$. It takes two to do the interference tango. With a single beam, it is impossible to have interference.

There is no interference when $\mathbf{E}_1 \cdot \mathbf{E}_2 = 0$, which states that the internal product of the electric fields is zero. This occurs when the waves are perpendicularly polarized (for example, linearly polarized with orthogonal polarization planes). Thus, a second condition is the existence of non-mutually perpendicularly polarized waves. This can be ensured with parallel, linearly polarized (or circularly polarized) waves, or simpler yet, with unpolarized light. From now on, for simplicity, we will assume waves with parallel polarizations.

The interference factor is also zero when the term $\langle\cos(\delta\varphi)\rangle$ is zero. If the difference $\delta\varphi$ varies randomly from 0 to $\pi$, the term $\cos(\delta\varphi)$ has random values from +1.0 to −1.0, and its mean temporal value $\langle\cos(\delta\varphi)\rangle$ is zero. Therefore, the phase difference must not vary randomly. Ideally, we would prefer this phase difference to be fixed (and why not?) and equal to zero always, everywhere. Indeed, we just stated the requirement for ideal coherence.

Waves whose phase difference varies randomly over time or space are **incoherent**. A basic prerequisite for interference effects is that the waves must be **coherent**, maintaining a fixed phase difference between them.

**Incoherent waves**: The phase difference varies randomly in time and space.

**Coherent waves**: The phase difference is constant in time and space.

**Coherence** is a fundamental condition for interference.

Let us examine the factors affecting the phase difference:

$$\delta\varphi_{12} = (\omega_1 - \omega_2)\cdot t - (\mathbf{k}_1 - \mathbf{k}_2)\cdot\mathbf{r} + (\varphi_{o1} - \varphi_{o2}) \tag{4.3}$$

The magnitude, as well as the time and space dependence of the phase difference, depends on the difference of three parameters:

- the initial phase difference $\varphi_{o1} - \varphi_{o2}$
- the frequency difference $\omega_1 - \omega_2$, or equivalently, the wavelength difference $\lambda_1 - \lambda_2$
- the wavevector direction difference $\mathbf{k}_1 - \mathbf{k}_2$

If the waves originate from two independent sources, the initial phase values $\varphi_{1o}$ and $\varphi_{2o}$ are random, and vary randomly and independently. Thus, a fixed phase difference cannot be achieved. The phase difference is randomly changing: The conditions for constructive or destructive interference may change every $10^{-8}$ s, which is the mean pulse (photon) duration.

Thus, it is impossible to have coherent waves and therefore interference from two independent sources. To obtain two coherent light waves, it is necessary to have one initial source and (by some means) to obtain two waves from this initial source by splitting. Any random change in the phase of the initial source propagates in exactly the same manner in both resulting beams. Thus, the phase difference term $(\varphi_{o1} - \varphi_{o2})$ can be zero or, in general, can be fixed.

To maintain a zero frequency difference ($\omega_1 - \omega_2 = 0$) so that this initially fixed phase difference is maintained, ideally, the two frequencies must be absolutely equal. This is called ideal **temporal coherence**. If we know the field in one wave at a given time $t$, then the other wave has exactly the same value at any time $t+\tau$, regardless of the length of the time interval $\tau$. Temporal coherence therefore relates to the wave differences as time develops.

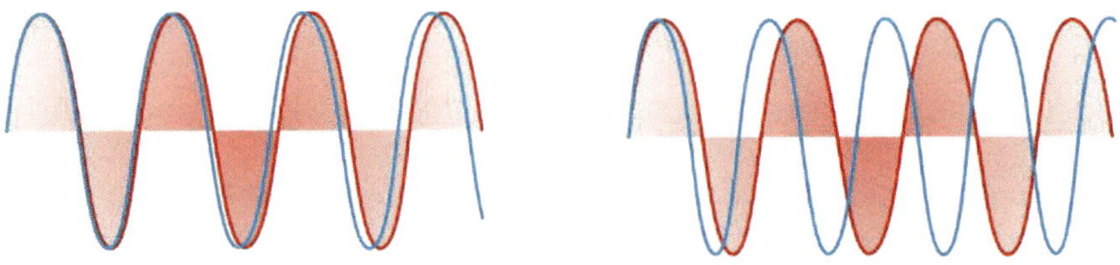

*Figure 4-3: Waves with (left) high and (right) low temporal coherence.*

Another type of coherence relates to the wave properties that concern the direction of propagation. The phase difference depends on the vectorial difference $\mathbf{k}_1 - \mathbf{k}_2$. For this term to be zero, the differences between the two wavevectors should be zero, or at least as small as possible.

Even if the wavevector magnitude is fixed (constant $\lambda$), its direction changes because it is always perpendicular to the wavefront. The flatter the wavefront, the more fixed the wavevector direction. In an ideal plane wavefront, the phase is fixed and the wavevector has a specific direction. Thus, any two wavefront points contribute equally to the phase difference. This is **spatial coherence.**

The concept of coherence is relative. There is coherence between two waves and coherence within a wave emitted by a source. In the latter case, a wave originating from the source is compared to another wave that originated from the same source but slightly later.

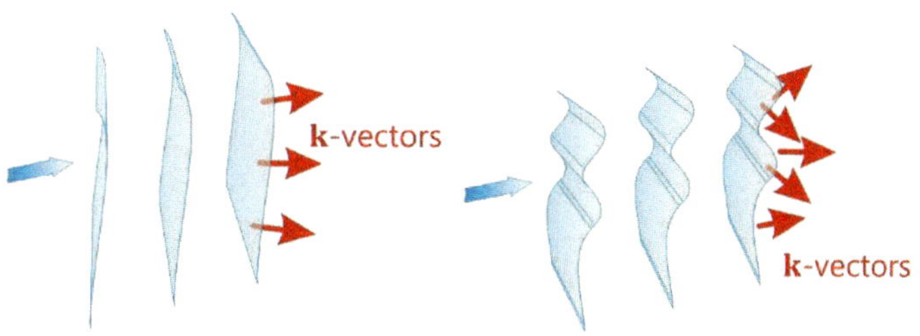


*Figure 4-4: Waves with (left) high and (right) low spatial coherence.*

In an ideal monochromatic source, we expect significant temporal coherence. This source has a small spectral linewidth of $\pm\delta\lambda$ around an average wavelength $\lambda_\mu$ with $\delta\lambda/\lambda_\mu \ll 1$. Absolute monochromatic sources do not exist; nature does not allow it. Even in lasers, which are known for their monochromaticity (§ 6.2.4), there is a (small perhaps) range of $\delta\lambda$. A source is satisfactorily quasi-monochromatic when the spectral range is satisfactory small.

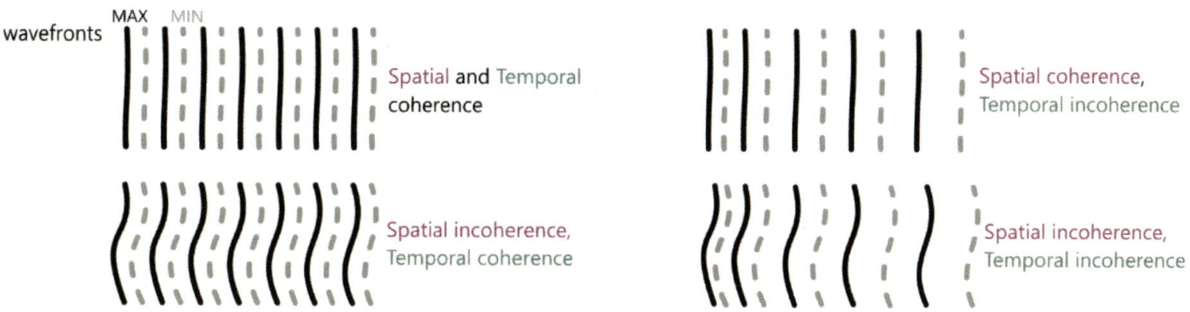


*Figure 4-5: (top left) Perfect spatial and temporal coherence. (lower left) Spatial incoherence and temporal coherence. (top right) Spatial coherence and temporal incoherence. (lower right) No coherence, or spatial and temporal incoherence.*

The degree of temporal coherence expresses the field difference at a moment in time $t'$ compared to a moment later in time, $t'+\tau$. It is calculated by the time autocorrelation integral:

$$\gamma(\tau) = \langle \mathbf{E}(t') \cdot \mathbf{E}(t'+\tau) \rangle \tag{4.7}$$

The degree of coherence is, in general, a complex number. It can be expressed as

$$\gamma_{12}(\tau) = |\gamma_{12}(\tau)| \cdot \exp[i\,\varphi_{12}(\tau)] \tag{4.8}$$

The modulus $|\gamma_{12}(\tau)|$ has values from 0.0 to 1.0. For $|\gamma_{12}(\tau)| = 1.0$ there is ideal coherence, and for $|\gamma_{12}(\tau)| = 0.0$ there is incoherence, while for $0.0 < |\gamma_{12}(\tau)| < 1.0$ there is partial coherence. If $\varphi_{12}(\tau) = \pi$, then the imaginary part causes contrast visibility reversal (see § 4.1.3).

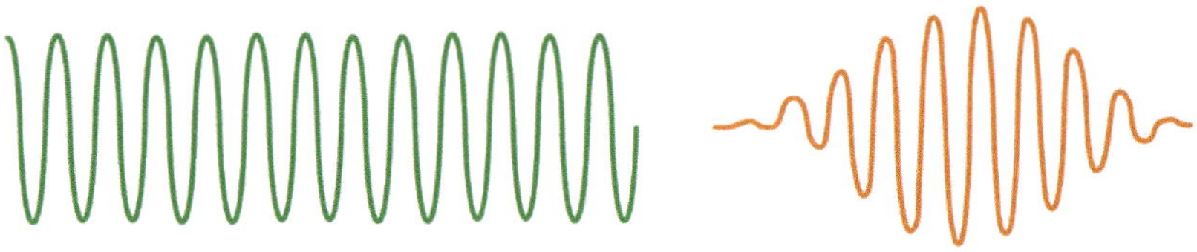

*Figure 4-6: Representation in time of (left) an ideal harmonic wave of only one frequency and (right) a light pulse resulting from a summation of many harmonic waves with different frequencies.*

The higher the degree of the source's temporal coherence, the smaller its spectral bandwidth. These quantities are related via a Fourier transform described by the Weiner–Khinchin theorem (named after Norbert Wiener and Aleksandr Yakovlevich Khinchin). Recall the principle of uncertainty: A wave confined in the frequency domain is extended in time, and vice versa.

A harmonic wave [with a magnitude expressed as $E_o\cos(\omega \cdot t - k \cdot x)$] is an idealization. It has no origin and no end (it has values for every $t$ and $x$), and has only one frequency $\omega_o$. This is expressed in the frequency domain (§ 5.2.2) by the delta function $\delta(\omega - \omega_o)$. On the other hand, a pulse with a limited time duration is composed of many harmonic waves with slightly different frequencies. Consider a source that emits two ideal, continuous harmonic waves with frequencies $\omega_o$ and $\omega_o + \delta\omega$. The composition of these waves is a light pulse [Figure 4-6 (right)].

The time span of this pulse is $\tau_c = 2\pi/\delta\omega$, a time interval that is inversely proportional to the spectral range $\delta\omega$. This **coherence time** $\tau_c$ expresses the time duration of the emission during which the pulse is continuous. The **coherence length** $l_c$ expresses how far can we 'walk' within the wave while the wave still maintains its coherence. Thus, the coherence length corresponds to the coherence time and is expressed as: $l_c = u \cdot \tau_c \approx u/\delta\nu$. From the analogy $\delta\nu/\nu_m = \delta\lambda/\lambda_m$, we have, for an average refractive index $n$,

Coherence Length:

$$l_c = \frac{\lambda_m^2}{n \cdot \delta\lambda} \tag{4.9}$$

For white light with $\lambda_m \approx 0.5$ µm and linewidth $\delta\lambda \approx \pm 0.3$ µm, the coherence length is $\approx 0.8$ µm, while for a common HeNe laser, also with $\lambda_\mu \approx 0.5$ µm but with $\delta\lambda \approx \pm 10^{-4}$ nm, the coherence length is $\approx 2.5$ m. <u>Coherence time and length both describe temporal coherence</u>.

Ideal spatial coherence exists when the wavefronts are flat or perfectly spherical. Such are the waves from a point source, where the wavefronts are concentric spheres, or from a source so far away (a perfectly collimated beam) that all of the wavefronts are ideally flat. Of course, an ideal plane wavefront or point source are both conceptual idealizations.

Just as in the case of temporal coherence, we can express an uncertainty principle: The more confined a wave is in space, the more the wave extends spatially as it propagates. These quantities are given, again, by a Fourier transform described by the Van Cittert–Zernike theorem, which relates the degree of spatial coherence of a source to its spatial extent. The degree of spatial coherence states the field difference between a point $z'$ of a wavefront and another point $z'+\zeta$, and may be expressed by the integral of the mutual coherence function:

Mutual Coherence Function: $$\gamma(\zeta) = \langle \mathbf{E}(z') \cdot \mathbf{E}(z'+\zeta) \rangle \qquad (4.10)$$

The two coherence types are expressions of the same physical property, coherence in space and time. Their similarities are not a coincidence. Over space there is spatial coherence, and over time there is temporal coherence.

Ideally, a monochromatic point source has high spatial and temporal coherence. An extended monochromatic source has high temporal coherence but low spatial coherence, and a multicolor point source has high spatial coherence but low temporal coherence.

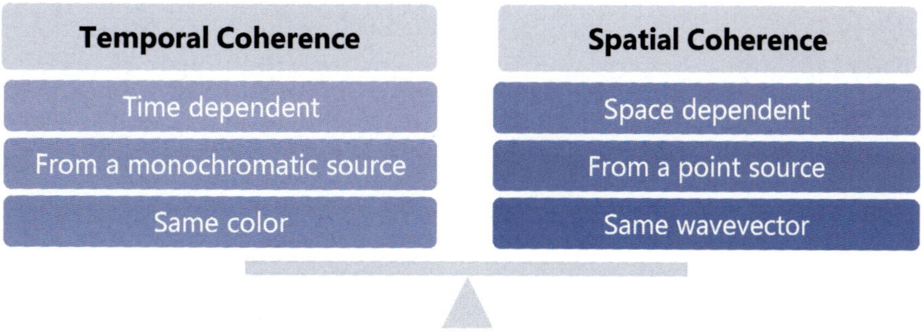


## 4.1.2 Phase Difference and Optical Path Difference

Assuming that all interference conditions are satisfied, the intensity is derived from the interference equation [Eq. (4.5)], and the interference term [Eq. (4.6)] can be re-written as

Interference Equation: $$I_{TOT} = I_1 + I_2 + 2\sqrt{(I_1 \cdot I_2)} \cdot \langle \cos(\delta\varphi) \rangle \qquad (4.11)$$

The loci of the maximum or minimum intensity $I$ are determined by a very simple condition: $\cos(\delta\varphi) = +1$ for maxima, and $\cos(\delta\varphi) = -1$ for minima. Specifically,

- **Minimum intensity $I_{MIN}$ (dark fringes)** occurs when $\cos(\delta\varphi) = -1$, or $\delta\varphi = (2m+1)\cdot\pi$:

**condition for minimum phase**: phase difference = $(m+½)\cdot 2\pi$ (4.12)

where $m$ is an integer 0, ±1, ±2, … . The minimum intensity is

$$I_{MIN} = I_1 + I_2 - 2\sqrt{(I_1 \cdot I_2)} = (\sqrt{I_1} - \sqrt{I_2})^2 \quad (4.13)$$

- **Maximum intensity $I_{MAX}$ (bright fringes)** occurs when $\cos(\delta\varphi) = +1$, or $\delta\varphi = (2m)\cdot\pi$:

**condition for maximum phase**: phase difference = $(m)\cdot 2\pi$ (4.14)

where $m$ is, again, an integer 0, ±1, ±2, … . The maximum intensity is

$$I_{MAX} = I_1 + I_2 + 2\sqrt{(I_1 \cdot I_2)} = (\sqrt{I_1} + \sqrt{I_2})^2 \quad (4.15)$$

Discussion: Consider two equal-amplitude interfering waves, each having 50 photons. At areas of destructive interference, the intensity is zero, which corresponds to 0 photons. At areas of constructive interference, the intensity corresponds to 200 photons, which is 4×, not 2×, the initial number of photons. At an intermediate point, the sum of 50+50 may yield any value from 0 to 200. Do not rush to accuse optics of … rigged math. The fallacy is due to the fact that, in reality, the intensities (the number 50) should not be summed. What is added are the vectors that correspond not to the intensity, but to the electric field, which has a magnitude and a phase. This aspect is presented in more detail in § 4.1.4.

We emphasize that the total intensity (over a large area) neither increases nor decreases. It is simply spatially redistributed: Intensity that 'disappears' from points of destructive interference 'appears' at the respective points of constructive interference. Moreover, this redistribution is fixed in time. The spatial succession of maxima and minima ensures that the total energy (intensity) is preserved. The opposite would be a serious violation of a number of principles in physics.

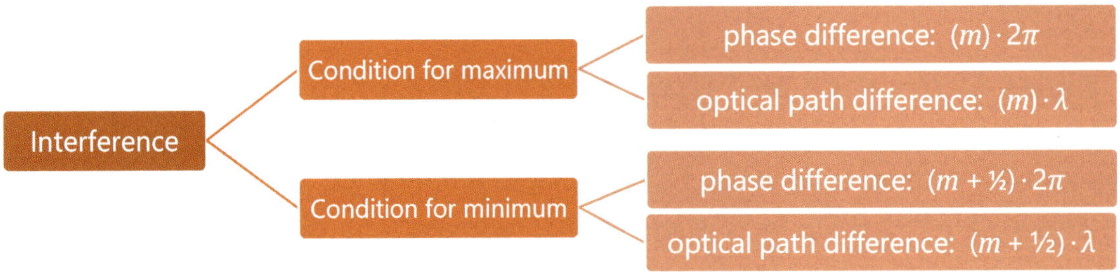

If the two interfering waves do not have equal amplitude magnitudes, the interference conditions are not affected; however, at the locations for which destructive interference is taking place [which corresponds to Eq. (4.13)], a minimum, nonzero intensity is observed.

We consider now two coherent, monochromatic waves that originate from two different points (soon we will see how). There is a fixed initial phase difference $\varphi_{1o} - \varphi_{2o}$ and a zero frequency difference because $\omega_1 = \omega_2$. Since the two waves originate from different points, they have different wavevector orientations, even if their magnitudes are equal. The loci for the intensity minimum and maximum conditions are the solutions of the vectorial equation,

$$(\mathbf{k}_1 - \mathbf{k}_2) \cdot \mathbf{r} = \text{constant} \tag{4.16}$$

The maximum-intensity (bright fringes) surfaces and minimum-intensity (dark fringes) surfaces (Figure 4-7) are perpendicular to the vector of the wavevector difference.

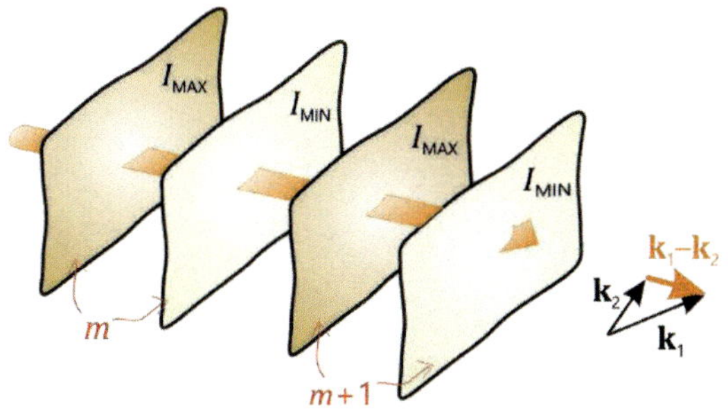


*Figure 4-7: Alternating planes of maximum and minimum intensity.*

If the vector $(\mathbf{k}_1 - \mathbf{k}_2)$ has a fixed, constant direction, the maximum/minimum surfaces (shown as planes $I_{MAX}$ and $I_{MIN}$ for two successive orders in Figure 4-7) are flat. If the vector is constant in magnitude, the surfaces alternate along this vector with equal separation steps. The latter condition describes **fringes of equal thickness**.

We can express Eqs. (4.12) and (4.14) using either optical path difference expressions or phase difference expressions. The generally applicable relationship that associates phase difference with optical path difference is

$$\frac{\text{phase difference}}{2\pi} = \frac{\text{optical path difference}}{\lambda} \tag{4.17}$$

This is because if we 'walk over' a wavelength, for an optical path difference of $\lambda$, the phase changes by 360° or $2\pi$ rad. In general, the optical path difference expressions and the phase difference expressions are equivalent. Therefore, the condition for the minimum optical path difference [Eq. (4.12)] is expressed as follows:

**Condition for Minimum Optical Path Difference**: optical path difference = $(m + ½)\cdot\lambda$ (4.18)

where $m$ is an integer 0, ±1, ±2,.... The condition for maximum optical path difference [Eq. (4.14)] is

**Condition for Maximum Optical Path Difference**: optical path difference = $m\cdot\lambda$ (4.19)

Regardless of the specific interference setup (we will present a few) in which the geometrical expression of the optical path difference or phase difference varies, the maximum and minimum conditions will always be expressed by the simple form of Eqs. (4.12) and (4.14) for the phase difference, or Eqs. (4.19) and (4.18) for the optical path difference.

### 4.1.3 Fringe Visibility and Contrast

Fringe visibility (contrast) expresses the difference between the maximum and minimum intensities in interference fringes. Quantitatively, it is measured by a modulation index $\mathcal{V}$, a unitless quantity that is defined, according to Michelson, as[20]

Fringe Visibility: $$\mathcal{V} = \frac{\text{luminance difference}}{\text{average luminance}} = \frac{I_{\text{MAX}} - I_{\text{MIN}}}{I_{\text{MAX}} + I_{\text{MIN}}} = \frac{2\sqrt{I_1 \cdot I_2}\langle\cos(\delta\varphi)\rangle}{I_1 + I_2} \quad (4.20)$$

The modulation index is directly dependent on the interference factor. The maximum visibility for ideal interference conditions ($I_1 = I_2$ and $\delta\varphi = 0$) is 1.0, and the minimum ($I_1 = I_2$ and $\delta\varphi = \pi/2$) is zero. In the case of partial coherence, we express the interference factor as

$$2c\varepsilon_0\cdot\langle \mathbf{E}_1\cdot \mathbf{E}^*_2\rangle\cdot\langle\cos(\delta\varphi)\rangle = 2\sqrt{(I_1\cdot I_2)}\cdot \text{Re}\{\,|\gamma_{12}(\tau)|\cdot\exp[i\,\varphi_{12}(\tau)]\,\}$$

$$= 2\sqrt{(I_1\cdot I_2)}\cdot \text{Re}\{|\gamma_{12}(\tau)|\}\cdot\cos[\,\varphi_{12}(\tau)] \quad (4.21)$$

Based on the above, the modulation index is: $$\mathcal{V} = \frac{2\sqrt{I_1\cdot I_2}}{I_1 + I_2}\cdot|\gamma_{12}(\tau)| = |\gamma_{12}(\tau)|_{I_1 = I_2} \quad (4.22)$$

If the intensities are equal ($I_1 = I_2$), the degree of source coherence equals the fringe visibility.

### 4.1.4 Interference, the Vector Synthesis Aspect

The complex wave representation (using an exponential phase factor) is a versatile tool for managing linear superposition math, and by extension, interference. The expressions of Eq. (4.2) have amplitude and phase, represented by a vector that rotates in space. The angular speed is

[20] The Weber contrast, an alternative definition of modulation, is the ratio of the luminance difference to the background luminance.

the angular frequency $\omega$ (the positive direction is counterclockwise). This is also applicable for photon probability summation (Feynman's vectors, discussed in § 1.4.7). The magnitude $\rho$ of this vector corresponds to the electric field amplitude. We re-write the wave as

$$\mathbf{E} = \mathbf{E}_o \cdot \exp[i(\omega \cdot t - \mathbf{k} \cdot \mathbf{r} + \varphi_o)] = \boldsymbol{\rho} \cdot \exp[i(\alpha)] \tag{4.23}$$

The argument in the exponential is the phase, the angle $\alpha$ formed with the horizontal axis. The real part (Re {}) is a vector that projects on the horizontal axis (–$x$) a magnitude $E_{ox}$; the imaginary part (Im {}) projects on the vertical axis (–$y$) a magnitude $E_{oy}$.

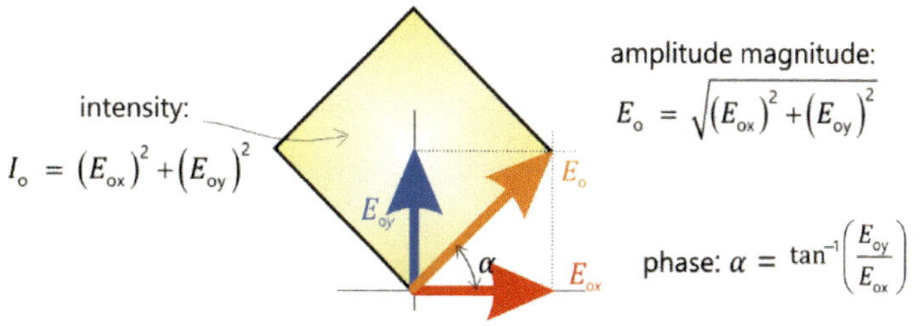


*Figure 4-8: Vector representation of a wave disturbance.*

The measurable quantity in light is the intensity, which is proportional to the square of the magnitude ($E_o$) of the field amplitude. Intensity is therefore represented by the area of the regular quadrilateral whose sides are formed by $E_o$ (such as the yellow square in Figure 4-8).

In addition to assisting in visualizing intensity, this representation is very helpful in visualizing the principle of linear superposition. Recall that this principle states that the resultant disturbance is the vector sum of all respective disturbances, i.e., the electric fields. The sum of two or more disturbances (linear superposition) can be expressed as

$$E_x = E_{1x} + E_{2x} + E_{3x} + \cdots + E_{Nx} \quad \text{and} \quad E_y = E_{1y} + E_{2y} + E_{3y} + \cdots + E_{Ny} \tag{4.24}$$

where $$E_{Nx} = E_N \cdot \cos(\alpha_N), \quad E_{Ny} = E_N \cdot \sin(\alpha_N), \quad \text{and} \quad E_o = \sqrt{E_{ox}^{\ 2} + E_{oy}^{\ 2}} \tag{4.25}$$

The simple algebraic sum of the two vector projections equals the projection of their vector sum. Thus, the vector sum is easily calculated if the vectors are drawn using the head-to-tail method, preserving the initial phase arguments of each (angle $\alpha$ with respect to the horizontal axis). This is a **phasor diagram**. The procedural steps are as follow:

**1.** Construct a series of consecutive vectors, placing the tail of the second at the head of first.

**2.** Calculate the vector sum by first adding their projections, using Eqs. (4.24) and (4.25). The magnitude of this vector is the amplitude magnitude. The phase of this vector is determined by the angle formed with the horizontal axis; the vector's projections on the horizontal and vertical axes are the real and imaginary parts, respectively.

**3.** The intensity of this field interaction is the area of the square whose sides are formed with the amplitude magnitude.

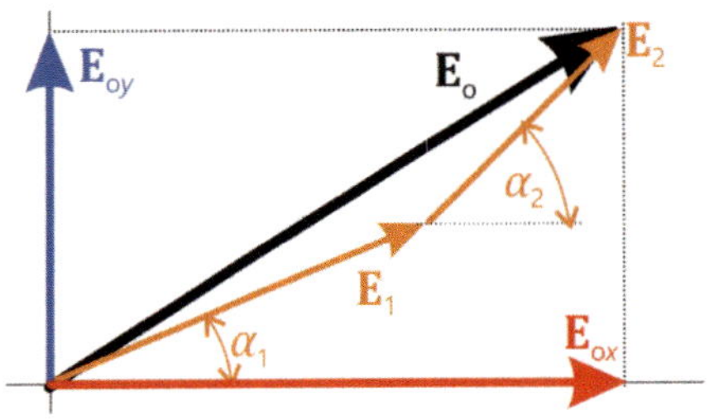


*Figure 4-9: Vector representation of the summation of two wave disturbances.*

We now schematically represent some very simple interference cases. First, we investigate vector combinations, in which we conveniently set $E_{o1} = E_{o2}$, for the following cases:

First case: Phase difference $\varphi_{o1} - \varphi_{o2} = 0°$ (= 0 rad) or an even multiple of $\pi$ rad: e.g. $2\pi$, $4\pi$ rad.

We select zero as the initial phase for both waves. The resultant wave has 2× the amplitude magnitude and 4× the intensity, which amounts to ... 200 photons. No, there is no energy gain out of nowhere, as we will soon see.

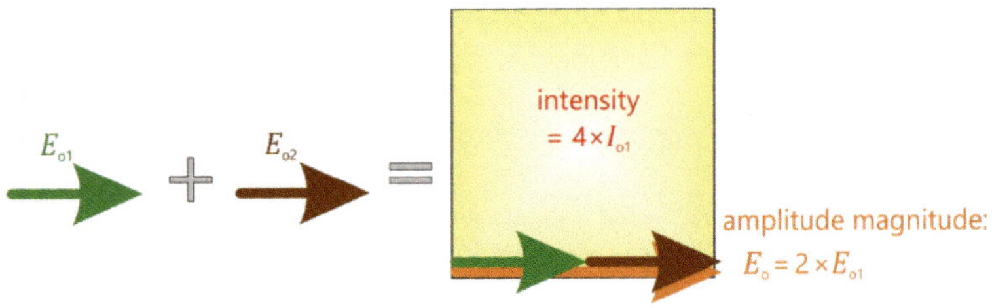


*Figure 4-10: Vector summation of two wave disturbances in phase.*

Second case: Phase difference $\varphi_{o1} - \varphi_{o2} = 180°$ (= $\pi$ rad) or an odd multiple of $\pi$ rad: e.g., $3\pi$, $5\pi$ rad.

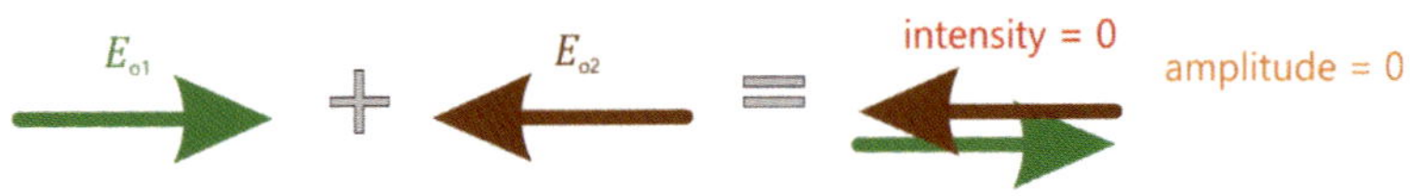


*Figure 4-11: Vector summation of two wave disturbances with opposite phases.*

The resultant wave has zero amplitude, so it has zero intensity—the zero-photon case. What happened to the energy of these photons? Why do 50 + 50 photons = 0 photons? Also, look at Figure 4-10, in which 50 + 50 photons produce 200 photons! This is exactly the redistribution that takes place in interference: There is a localized rearrangement of intensity (energy) from the destructive to the constructive interference areas. Energy does not disappear but is rearranged; the average spatial intensity within at least two fringes does not change.

Third case: Phase difference $\varphi_{o1} - \varphi_{o2} = 90°$ ($\pi/2$ rad) or an odd multiple of $\pi/2$ rad: e.g., $3\pi/2$, $5\pi/2$ rad.

The resultant vector is shown in Figure 4-12. The amplitude has magnitude $E_o = \sqrt{2} \cdot E_{o1}$, so the intensity is $I_o = 2 \cdot I_{o1}$, which is simply the sum of the respective intensities. There are exactly 100 photons. This simply indicates that there is no interference! Indeed, for a phase difference of 90°, and generally, in waves with orthogonal polarizations, the interference factor is zero.

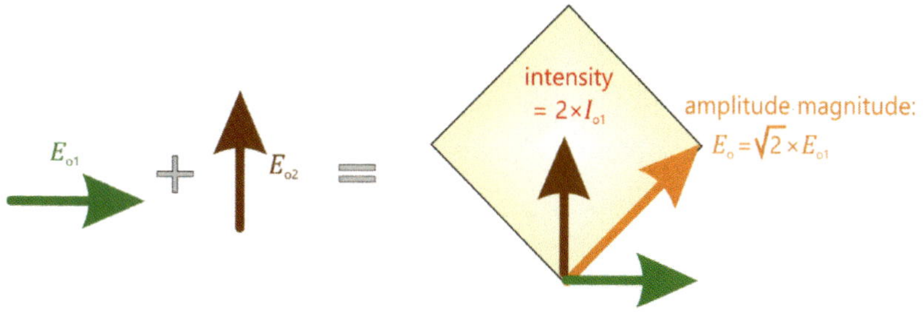


*Figure 4-12: Vector summation of two wave disturbances with a phase difference of 90°.*

By reversing this technique, we can determine what phase difference the two waves should have in order for the interference between 50 + 50 photons to result in any number between 0 and 200, e.g., 120 photons. If the area of the parallelogram should be 120, then we can use geometry to identify the angle between the two vectors, which is their relative phase.

This technique is not limited to interference effects from only two components. Recall that interference is defined as the interaction of at least two nonzero intensity waves. It is possible to have a very large number of waves interfering at the same point. The vector sum rules apply for any number of interfering waves; this fact will be particularly useful in multiple-beam interference (§ 4.2.5). For example, let's add the following vectors:

$$\mathbf{E}_1 = \mathbf{E}_{o1} \cdot \mathrm{Re}\{\exp(i\varphi_1)\}, \quad \mathbf{E}_2 = \mathbf{E}_{o2} \cdot \mathrm{Re}\{\exp(i\varphi_1 - \delta)\},$$

$$\mathbf{E}_3 = \mathbf{E}_{o3} \cdot \mathrm{Re}\{\exp(i\varphi_1 - 2\delta)\}, \quad \mathbf{E}_4 = \mathbf{E}_{o4} \cdot \mathrm{Re}\{\exp(i\varphi_1 - 3\delta)\}... \tag{4.26}$$

Their vector sum, shown in Figure 4-13, has an amplitude magnitude $E_o$ equal to the vector magnitude and forms an angle $\alpha$ with the horizontal axis that determines its phase.

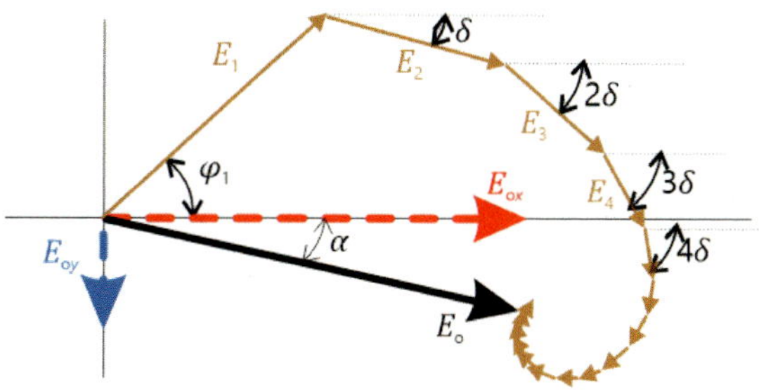


*Figure 4-13: Vector sum of multiple wave disturbances.*

# 4.2 Interference Setups

## 4.2.1 Young's Experiment

We need at least two coherent light sources for interference. However, it is impossible to have coherence from two independent sources. Therefore, we start from an initial source and (somehow) obtain two waves. There are two ways to do this. The first involves a wavefront split, for example, through two small apertures. This is wavefront division. The second involves a ray split, for example, with partial reflection. This is amplitude division. Using either method, we obtain two coherent waves. The degree of coherence of the initial source determines to a large extent the degree of coherence of the two resulting waves.

The Thomas Young interference experiment (in 1801) was the first to provide crucial evidence in support of the wave nature of light—and the first to measure the wavelength of light as well! The arrangement uses wavefront division via a pair of small apertures $S_1$ and $S_2$ separated by $d$ on an opaque screen. The two sources are part of the same wavefront from a monochromatic point source $S$ (in Young's experiment it was solar light filtered through a small aperture). Interference is observed on an observation screen at distance $z$.

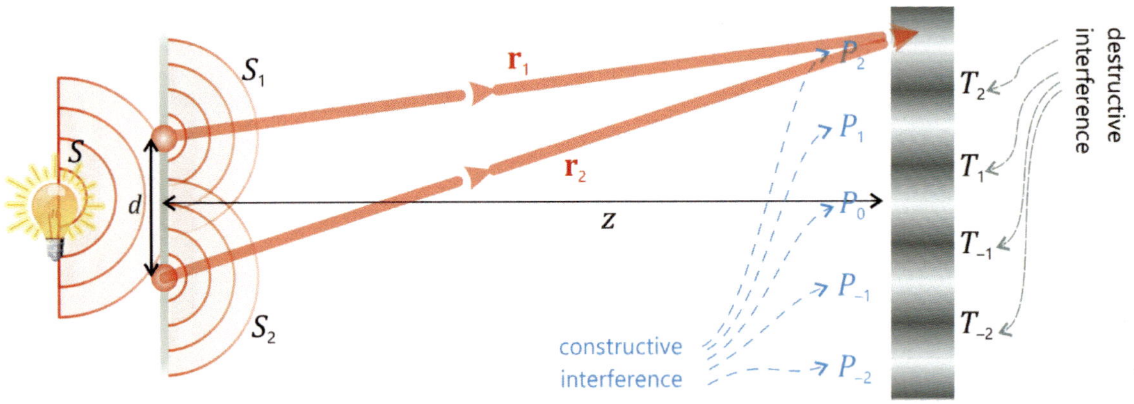


*Figure 4-14: Setup of Young's interference experiment.*

Source $S$ emits spherical waves that pass through apertures $S_1$ and $S_2$. These waves, according to the Huygens–Fresnel principle, become sources of secondary spherical waves propagating in all forward directions. We are definitely not in the realm of geometrical optics, which states that rays passing by an obstacle or through a small aperture do not diverge. Indeed, light diffracts past the two apertures. Because the characteristic size of the small apertures is now comparable with the wavelength, there is an elementary diffraction effect through sources $S_1$ and $S_2$.

*Figure 4-15: Thomas Young (1773–1829).*

Because $S_1$ and $S_2$ share the same source $S$ and are spaced equally from it, they have the same initial phase, so $\varphi_{o1} - \varphi_{o2} = 0$. The two sources are therefore coherent; specifically, they have the same degree of coherence as the initial source. Of course, there are many ways to manipulate this phase difference, such as by adding a glass plate in front of any of the two apertures.

Interference fringes on the observation screen form a stationary intensity distribution with periodically alternating maxima and minima. Maximum bright fringes are always bright, and minimum dark fringes always remain dark. Bright fringes correspond to constructive interference (waves in phase), while dark fringes correspond to destructive interference (waves in opposite phase). The key question is: Given that the two sources $S_1$ and $S_2$ are coherent, what causes a phase difference between the two waves arriving at different points on the observation screen?

On the screen, at point $P_0$ situated across the bisector from the two sources, the two waves travel exactly equal distances: $r_1 = r_2$. The optical path lengths are equal, so the waves arrive at point $P_0$ after traveling the same number of wavelengths—regardless of how large this number may be. Consider a wave disturbance with crests and troughs. Regardless of the wave snapshot that arrives from $S_1$, the exact same wave snapshot arrives from $S_2$. These can be either crests or troughs, or any intermediate phase; in any case, this quickly changes with time. The key aspect is that these waves are always in phase, and constructive interference occurs at point $P_0$, which is the center of symmetry of the interference fringe pattern.

Constructive interference also exists at other points, for example, at points $P_1$ and $P_{-1}$. However, the waves that arrive at these two points have traveled different distances, so they have different optical paths. In order to produce constructive interference, the difference between their optical paths must be an integer multiple of the wavelength [Eq. (4.19)]. In the simplest case, the difference is $r_1 - r_2 = \lambda$. The same applies to all other points for which the distance difference between the two sources is a multiple integer ($m$) of the wavelength: $r_1 - r_2 = m \cdot \lambda$. At these points, the waves arrive with a phase difference that is a multiple integer ($m$) of $2\pi$. The integer $m$ is the **interference order** and has values of 0, ±1, ±2, ... .

We now investigate some other points, for example, point $T_1$, whose distance $r_1$ from source $S_1$ is slightly smaller than the distance $r_2$ from source $S_2$. Specifically, we consider the case where the distances $r_1$ and $r_2$ differ by $\lambda/2$. Thus, when a 'crest' arrives from $S_1$, a 'trough' arrives at the same point from $S_2$. The resultant field is zero, as is the intensity. At this point, the waves mutually cancel out to darkness through destructive interference (minimum intensity). This also occurs at all other points for which the optical path difference equals an odd multiple of $\lambda/2$ such that $r_1 - r_2 = (2m+1)\cdot\lambda/2$, so the waves arrive with a phase difference of $(2m+1)\cdot\pi$. On the screen, the maxima alternate with the minima in a characteristic periodic fashion.

#### 4.2.1.1 Considerations Regarding the Wave and the Corpuscular Nature of Light in view of Young's Experiment

The appearance of interference fringes in Young's experiment proved beyond any doubt the wave nature of light. Now that we know better, how is this compatible with the photonic aspect? Particles do not interfere, this is clear. What happens here? Does light lose (perhaps temporarily) its photon nature?

Let's take if from the start. Assume that light is, indeed, composed of classical particles. We set up a 'light gun' over the first screen with the two small apertures, $S_1$ and $S_2$ [Figure 4-16 (left)]. On an observation screen, a detector records the number of bullets arriving at various points on the screen over time. Recording a large number of bullets, practically we find the arrival probability of a bullet at every screen point. Thus, we obtain a distribution curve $N_{12}$ that indicates that most bullets arrive at points exactly across from each aperture, forming small piles. This practically means that the bullets passing through the openings do not spread in other directions.

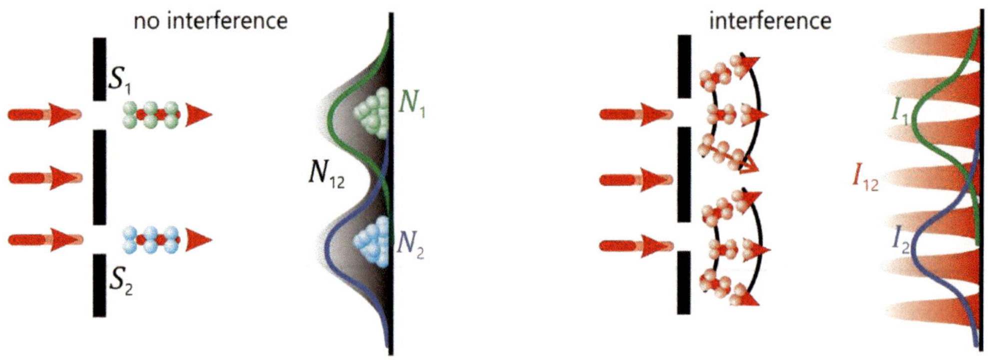


*Figure 4-16: Application of (left) the corpuscular theory and (right) the wave theory of light in interference.*

The distribution curve $N_{12}$ could be the simple sum of two independent distributions $N_1$ and $N_2$ that result as follows: Curve $N_1$ expresses the number of bullets arriving from $S_1$ when we close down $S_2$ (say with a cover); we obtain curve $N_2$, accordingly.

We conclude that the number of bullets passing through both apertures is the sum of the number of bullets that pass from $S_1$ and from $S_2$. Let's call this 'distribution with no interference:'

$$N_{12} = N_1 + N_2 \text{ (no interference)} \tag{4.27}$$

Now we consider coherent light waves. The source sends a wave across the two apertures, and we observe the intensity distribution $I$ across the screen. If we cover $S_2$, light passes only from $S_1$ and forms a 'sugarloaf' of light $I_1$. Another 'sugarloaf' $I_2$ is obtained if we cover $S_1$. These distributions appear quite similar to $N_1$ and $N_2$.

The case, however, is much different when we leave both apertures open. We obtain a curve $I_{12}$ [such as the one shown in Figure 4-16 (right)] that is quite different from curve $N_{12}$. The key question is: What causes this difference? The fundamental differences between particles and waves are the following:

- In particles, the recorded quantity is their population. Therefore, we add particles one at a time, with no chance that one particle cancels out another. When the particles pass through the two apertures, we just sum their population.
- To the contrary, in coherent waves, the recorded quantity is the square of the sum of the individual disturbances. The total wave disturbance (electric field) is the vector sum of the fields. Because the waves are coherent, there is a specific (at a point) and stationary (instantaneous) phase difference.
- The result of the vector sum depends on this phase difference. Thus, $I_1 + I_2$ does not equal $I_{12}$ and changes from point to point on the observation screen. This is interference. Therefore, the formation $I_{12}$ is entirely different from $N_{12}$.

The resultant electric field at every point on the screen results in the vector sum of the two fields—the field arriving from source $S_1$ and the field arriving from $S_2$:

$$\mathbf{E}_{12} = \mathbf{E}_1 + \mathbf{E}_2 \text{ (interference)} \tag{4.28}$$

The intensity is proportional to the square of the field:

$$I_{12} = \left(\mathbf{E}_1 + \mathbf{E}_2\right)^2 = I_1 + I_2 + 2\sqrt{I_1 \cdot I_2} \cdot \left\langle \cos\left(\delta\varphi\right)\right\rangle \tag{4.29}$$

We note that $I_{TOT}$ differs from the simple sum of the independent $I_1$ and $I_2$:

$$I_{12} \neq I_1 + I_2 \text{ (interference)} \tag{4.30}$$

It is therefore possible that the summation of the respective fields becomes a zero disturbance (darkness), or a disturbance that is twice as large (intensity = $4 \times I_O$). This distribution has a specific spatial periodicity.

If, for any reason, the waves from the sources $S_1$ and $S_2$ do not fulfill the interference conditions, the waves do not interfere. While we still add the electric field, the resultant intensity is simply the sum of the respective intensities. In Young's experiment setup, using noncoherent sources, we would see a distribution just like the one in Figure 4-17 (left), while using coherent sources, we would obtain interference fringes, as in Figure 4-17 (right).

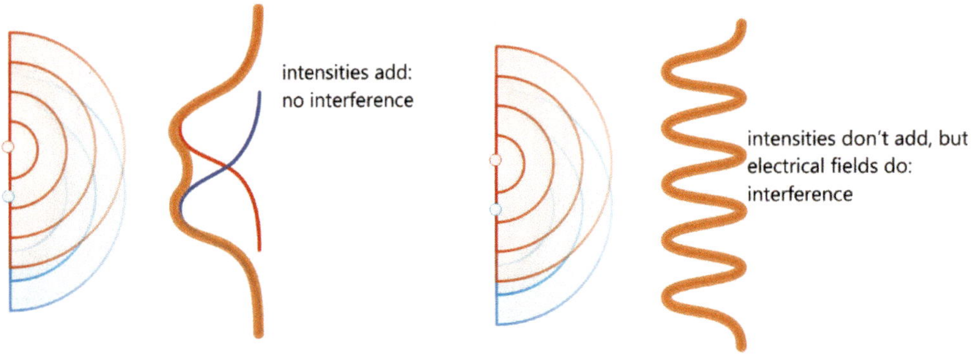


*Figure 4-17: (left) Summation of light intensities of noncoherent sources. (right) Summation of electric fields of coherent sources.*

Food for thought: What would the intensity distribution be if we:

☞ place two orthogonally crossed polarizers in front of $S_1$ and $S_2$, for example, a polarizer at 0° in front of $S_1$ and a polarizer at 180° in front of $S_2$?

☞ place a phase retardation plate of a half-wavelength ($\lambda/2$) in front of $S_1$?

## 4.2.2 Measurements in Young's Experiment

To develop a mathematical model for Young's experiment, we investigate the geometrical parameters and seek to identify:

- the interfering beams
- the geometrical factor that causes the optical path difference and therefore the phase difference.

The interfering beams originate from the coherent sources $S_1$ and $S_2$. They run distances $r_1$ and $r_2$, respectively, arriving at a random observation point $P(x)$ on the observation screen. The geometrical factor is the length difference between the two rays.

The magnitudes of the electric fields may be expressed as

$$E_1 = \frac{E_o}{r_1} \cdot \cos\left( \omega \cdot t - \frac{2\pi}{\lambda} r_1 + \varphi_1 \right) \quad \text{and} \quad E_2 = \frac{E_o}{r_2} \cdot \cos\left( \omega \cdot t - \frac{2\pi}{\lambda} r_2 + \varphi_2 \right) \tag{4.31}$$

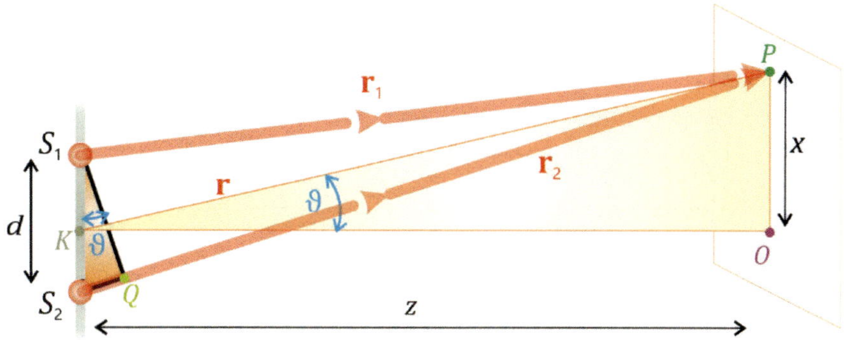


*Figure 4-18: Young's interference experiment model.*

We approximate the amplitude magnitudes as $A \approx E/r$, where $r$ is the average of $r_1 \approx r_2$. In addition, since the two electric fields share the same original wavefront, we set $\varphi_1 = \varphi_2 = 0$:

$$E_1 = A \cdot \cos\left( \omega \cdot t - \frac{2\pi}{\lambda} r_1 \right) \quad \text{and} \quad E_2 = A \cdot \cos\left( \omega \cdot t - \frac{2\pi}{\lambda} r_2 \right) \tag{4.32}$$

The geometrical factor responsible for the optical path difference is the difference in lengths between paths $r_1$ and $r_2$ from $S_1$ and $S_2$, respectively, to the screen:

Optical Path Difference: $$n \cdot (S_2P - S_1P) = n \cdot S_2Q \tag{4.33}$$

where $n$ is the refractive index of the optical medium between the source and the screen. We draw the right-angle triangle $S_1S_2Q$ to approximate the length difference $S_2P - S_1P$ with $S_2Q$.

We can use either the optical path difference or the phase difference, which are related by Eq. (4.17):

Phase difference / $2\pi$ — Conversion between phase difference and optical path length: — Optical path difference / $\lambda$

Phase Difference: $$\frac{2\pi}{\lambda} n \cdot S_2Q \tag{4.34}$$

where $\lambda$ is the wavelength in vacuum. In a more general case, we may consider that the two sources initially have a fixed but nonzero phase difference $\varphi_{o1} - \varphi_{o2} \neq 0$. Then we simply add this non-zero phase difference in Eq. (4.34).

The geometrical expression of the optical path difference (Figure 4-19) is $S_2Q = d\cdot\sin\vartheta$, so the optical path difference [Eq. (4.33)] is expressed as

Optical Path Difference: $n\cdot d\cdot\sin\vartheta$

Optical Path Difference (in air): $d\cdot\sin\vartheta$ (4.35)

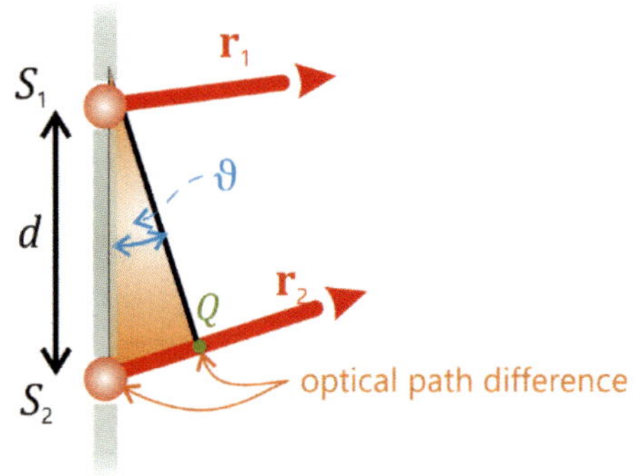


*Figure 4-19: Optical path difference in Young's interference experiment.*

From triangle $KPO$ (see Figure 4-18), $\sin\vartheta = OP/KP$. When distance $z$ is large compared to $d$, angle $\vartheta$ is relatively small, and we can set $KP \approx KO = z$:

$$\sin\vartheta = \frac{OP}{KP} \approx \frac{OP}{KO} = \frac{x}{z} \tag{4.36}$$

Then, the optical path difference and phase difference <u>in air</u> become

Optical Path Difference (in air): $$d\cdot\sin\vartheta = d\cdot\frac{x}{z} \tag{4.37}$$

Phase Difference (in air): $$\frac{2\pi}{\lambda}\cdot\frac{x\cdot d}{z} \tag{4.38}$$

To find the maxima $x_{MAX}$, we set the optical path difference equal to a multiple integer of the wavelength, or equivalently, we set the phase difference equal to a multiple integer of $2\pi$.

**Condition for maxima** (bright fringes):

optical path difference = $m\cdot\lambda$ or phase difference = $m\cdot 2\pi$ (4.39)

Their loci are: $$\frac{x_{MAX}\cdot d}{z} = m\cdot\lambda \quad \text{or} \quad x_{MAX} = m\cdot\frac{\lambda z}{d} \tag{4.40}$$

Minima $x_{MIN}$ result if the optical path difference is a multiple integer of $\lambda + ½\lambda$, or equivalently, if the phase difference is a multiple integer of $2\pi$ plus $\pi$:

**Condition for minima** (dark fringes):

$$\text{optical path difference} = (m+½)\cdot\lambda \quad \text{or} \quad \text{phase difference} = (m+½)\cdot 2\pi \qquad (4.41)$$

Their loci are:

$$\frac{x_{\text{MIN}}\cdot d}{z} = \left(m+\frac{1}{2}\right)\cdot\lambda \quad \text{or} \quad x_{\text{MIN}} = (2m+1)\cdot\frac{\lambda z}{2d} \qquad (4.42)$$

The integer $m$ (0, ±1, ±2, ... .) is the **interference order**. The fringe spacing, the separation between any two successive maxima or any two successive minima, is

Fringe Spacing:

$$\Delta x = x_{m+1} - x_m = \frac{\lambda z}{d} \qquad (4.43)$$

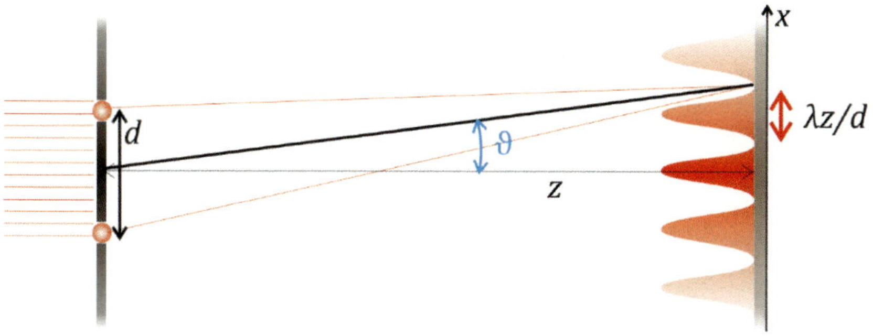


*Figure 4-20: Interference fringes from two point sources separated by d. The fringe spacing is λz/d.*

The separation between two successive interference fringes is independent of the order, so the fringes are equally spaced. This applies to both bright and dark fringes. The thickness of a bright fringe (also called a lobe), defined as the distance between its full-width-at-half-maximum (FWHM) points, is approximately one-half of the separation between two successive fringes. Thus, the lobes have also equal thickness.

Expressions using the observation angle $\vartheta$ with respect to the perpendicular bisector are often more useful than expressions using the distance $x$ on the screen. The characteristic quantity is the angle $\lambda/d$ (expressed in radians), which is independent of the observation distance $z$. Maxima appear for $\vartheta = 0$, $\pm\lambda/d$ and, in general, for $(m)\lambda/d$. Minima appear for $\pm½\lambda\cdot z/d$, $\pm 3/2\lambda\cdot z/d$, and, in general, for $[(2m+1)/2]\cdot\lambda/d$, $m = 0, \pm 1, 2, \ldots$ .

The fringe pattern displays a characteristic periodicity: $\Delta x = \lambda z/d$. It is therefore proportional to the wavelength $\lambda$, proportional to the screen distance $z$, and inversely proportional to the source separation $d$. This periodic pattern can be made broader by placing the screen farther away (increasing $z$), reducing the source separation $d$, or increasing the wavelength. If we measure the fringe separation, the distance to the screen $z$, and the source separation $d$, we can determine the wavelength $\lambda$. For distances $z$ that are large enough in relation to $d$, we practically measure the wavelength magnified by $z/d$.

*Table 4-1: Interference fringes formed by two slits separated by d on a screen placed z meters away.*

| | Distance $x$ on screen | Angle $\vartheta$ ($\sin\vartheta \approx \vartheta$, in radians) |
|---|---|---|
| fringe spacing $\Delta x$ | $\lambda \cdot z/d$ | $\lambda/d$ |
| minima | $\pm{}^1/_2\lambda \cdot z/d,\ \pm{}^3/_2\lambda \cdot z/d, \ldots\ (2m+1)/2\,\lambda \cdot z/d$ | $\pm{}^1/_2\lambda/d,\ \pm{}^3/_2\lambda/d, \ldots\ (2m+1)/2\,\lambda/d$ |
| maxima | $0,\ \pm\lambda \cdot z/d,\ \pm 2\lambda \cdot z/d, \ldots\ (m)\lambda \cdot z/d$ | $0,\ \pm\lambda/d,\ \pm 2\lambda/d, \ldots\ (m)\lambda/d$ |

Example ☞: Two slits 1 mm apart produce interference on an observation screen placed $z = 1$ m away. If the fringe separation is 0.5 mm, what is the light wavelength? Assume air between the slits and the screen.

Fringe separation is proportional to wavelength by a factor of $z/d$ = 1 m / 1 mm = 1 m / $10^{-3}$ m = $10^3$. Thus, the wavelength is proportional to the fringe separation by $10^{-3}$, so $\lambda = 10^{-3} \times 0.5$ mm = 0.5 µm = 500 nm.

Example ☞: Two narrow slits separated by $d$ = 3 mm = 3 ×$10^{-3}$ m produce interference fringes on a screen placed $z$ = 3 m away. What is the fringe spacing (separation) if blue light ($\lambda$ = 400 nm) is used? What is the fringe spacing if red light ($\lambda$ = 800 nm) is used? Assume air between the slits and the screen.

Blue: Fringe spacing is $\Delta x = \lambda z/d = 0.4\times10^{-6}$ m·3 m / 3×$10^{-3}$ m = 0.4×$10^{-3}$ m = 0.4 mm.
Red: Fringe spacing is $\Delta x = \lambda z/d = 0.8\times10^{-6}$ m·3 m/ 3×$10^{-3}$ m = 0.8×$10^{-3}$ m = 0.8 mm.

Example ☞: The sodium D line (at $\lambda$ = 589.3 nm) is used to form interference fringes on a screen placed $z$ = 2 m away. What is the fringe spacing if the slits are separated by $d$ = 0.1 mm? What is the fringe spacing if the slits are separated by $d$ = 0.4 mm? Assume air between the slits and the screen.

Separation $d$ = 0.1 mm: $\Delta x = \lambda z/d = 0.5893\times10^{-6}$ m·2 m / 0.1×$10^{-3}$ m = 11.79×$10^{-3}$ m = 11.79 mm.
Separation $d$ = 0.4 mm: $\Delta x = \lambda z/d = 0.5893\times10^{-6}$ m·2 m / 0.4×$10^{-3}$ m = 2.95×$10^{-3}$ m = 2.95 mm.

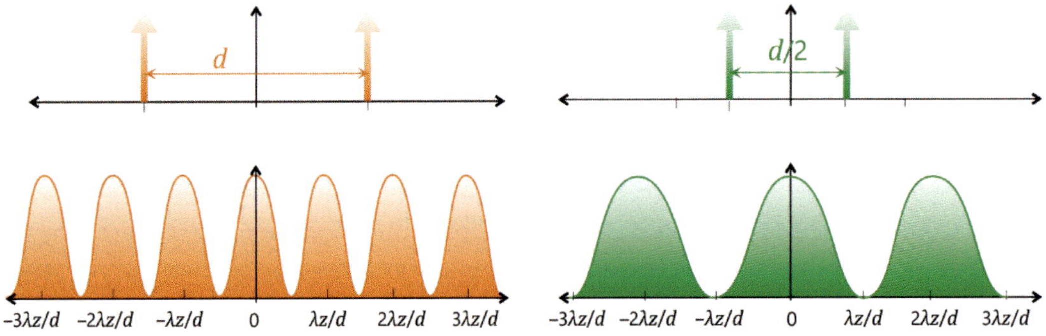


*Figure 4-21: (left) Fringe intensity periodic distribution versus source separation d. (right) The source separation is reduced by a factor of 2, and the periodic fringe distribution broadens by a factor of 2.*

The fringe spacing is usually measured experimentally by counting not just one, but many, consecutive fringes and dividing the total spacing (span) by their count. For example, we can measure the spacing of 20 fringes with much less error than if we measure the spacing of just one or two of them.

Example ☞: Interference fringes are formed by two narrow slits separated by $d = 1$ mm on a screen placed $z = 2$ m away using green ($\lambda = 550$ nm) light. What is the span of 20 consecutive fringes?

A single fringe spacing is $\Delta x = \lambda z/d = 0.55\times10^{-6}\,\text{m} \cdot 2\,\text{m} / 1\times10^{-3}\,\text{m} = 1.1$ mm. The 20 consecutive fringes span a distance of $20\cdot\Delta x = 22$ mm.

So far, we have assumed that the medium between the slits and the screen is air. If this is not the case, we substitute $\lambda$ with $\lambda/n$, where $n$ is the refractive index of this medium.

Example ☞: Blue light ($\lambda = 400$ nm) is used to form interference fringes from two narrow slits separated by $d = 1$ mm on a screen placed 2 m away. What is the fringe separation if the medium between them is air? What is the fringe separation if the medium between them is water with $n = 1.33$?

Air $n = 1$: $\Delta x = \lambda z/d = 0.400\times10^{-6}\,\text{m} \cdot 2\,\text{m} / 1\times10^{-3}\,\text{m} = 0.8$ mm.

Water $n = 1.33$: $\Delta x = \lambda z/nd = 0.400\times10^{-6}\,\text{m} \cdot 2\,\text{m} / (1.33 \cdot 1\times10^{-3}\,\text{m}) = 0.6$ mm.

### 4.2.2.1 Alternative Development of Young's Experiment Model

We apply superposition to the waves described by Eq. (4.31). The net field is

$$\mathbf{E}_{\text{TOT}} = \mathbf{E}_1 + \mathbf{E}_2 = E_o \cdot \cos\left(\omega \cdot t - \frac{2\pi}{\lambda} r_1\right) + E_o \cdot \cos\left(\omega \cdot t - \frac{2\pi}{\lambda} r_2\right) \tag{4.44}$$

Distances $r_1$ and $r_2$, where $r$ is the bisector distance of $r_1$ and $r_2$ (see Figure 4-18), are

$$r_1 = r - \frac{d}{2}\sin\vartheta = r - \frac{d}{2}\frac{x}{z} \quad \text{and} \quad r_2 = r + \frac{d}{2}\sin\vartheta = r + \frac{d}{2}\frac{x}{z} \tag{4.45}$$

The field is

$$\begin{aligned}\mathbf{E}_{\text{TOT}} &= \mathbf{E}_o \cdot \text{Re}\left\{\exp\left[i\left(\omega \cdot t - \frac{2\pi}{\lambda} r\right)\right] \cdot \left[\exp\left(i\pi\frac{xd}{\lambda z}\right) + \exp\left(-i\pi\frac{xd}{\lambda z}\right)\right]\right\} = \\ &= \underbrace{\mathbf{E}_o \cdot \text{Re}\left\{\exp\left[i\left(\omega \cdot t - \frac{2\pi}{\lambda} r\right)\right]\right\}}_{\text{propagating wave}} \cdot \underbrace{2\cos\left(\pi\frac{xd}{\lambda z}\right)}_{\text{amplitude modulation in space}}\end{aligned} \tag{4.46}$$

This relationship expresses a wave with the same characteristics as the initially interfering waves: It propagates along $\mathbf{r}$, has an angular frequency $\omega$ and a wavelength $\lambda$, but does not have a constant amplitude magnitude, which varies spatially along the transverse axis $x$ in a sinusoidal fashion. The temporally fixed but locally varying intensity corresponds to the resultant field squared:

$$I = \langle \mathbf{E}_{\mathrm{TOT}} \cdot \mathbf{E}_{\mathrm{TOT}} \rangle = (E_{\mathrm{TOT}})^2 = 4I_o \cdot \cos^2\left(\pi \frac{xd}{\lambda z}\right) \tag{4.47}$$

This distribution is periodic with a step (periodicity) $\Delta x = \lambda z/d$, which is exactly the periodic distribution we can obtain analytically from the vectorial wavevector sum! We consider wavevector $\mathbf{k}_1$ from source $S_1$ and wavevector $\mathbf{k}_2$ from $S_2$. The two vectors have an equal magnitude because $\lambda_1 = \lambda_2$. Using the unit vectors along each direction $\boldsymbol{\upsilon}$, their difference is

$$\mathbf{k}_1 = \boldsymbol{\upsilon}_1 \frac{2\pi}{\lambda}, \quad \mathbf{k}_2 = \boldsymbol{\upsilon}_2 \frac{2\pi}{\lambda} \quad \text{and} \quad |\mathbf{k}_1 - \mathbf{k}_2| = \frac{2\pi}{\lambda z} \tag{4.48}$$

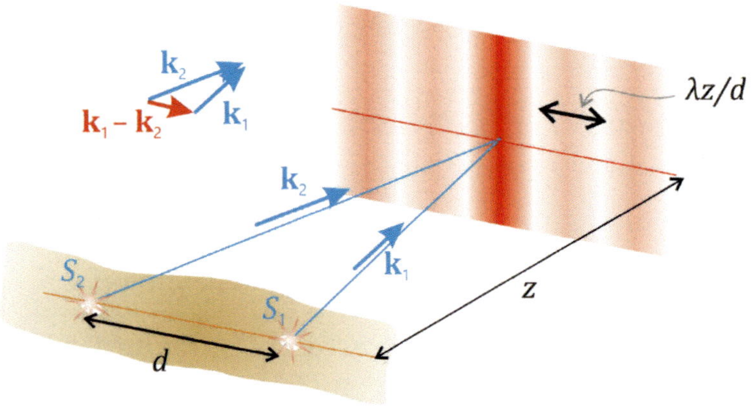


*Figure 4-22: Interference pattern resulting from the wavevector sum.*

### 4.2.2.2 Interference by a Non-monochromatic Source

Strictly monochromatic sources do not exist. The spectral range of any source is never infinitesimal. We assume that a source is monochromatic if $\delta\lambda/\lambda \ll 1$. How does this affect interference? For sure, the temporal coherence of the source is affected. This is reflected in the fringe sharpness (discussed in § 4.1.3). In addition, the higher the interference order and, equivalently, the longer the optical path difference between the two interfering waves, the more likely it is that these values will exceed the source coherence length and therefore that the fringe extent will restricted.

If the optical path difference [even if it satisfies the conditions for maxima Eq. (4.39)] exceeds the coherence length, the two waves might not interfere. If the coherence length is short, only a few low-order fringes appear. The fringe sharpness is limited, as the sharpness property relates to the source coherence.

In a simple approximation, a source emits two wavelengths $\lambda$ and $\lambda' = \lambda + \delta\lambda$, creating two independent but overlapping fringe formations (Figure 4-23), one with $\lambda$ and one with $\lambda'$. The separations between the zeroth-order and first-order fringes are $\Delta x = \lambda z/d$ and $\Delta x' = \lambda' z/d$, respectively. **Fringe broadening** is caused by the overlap of these fringes. Its extent can be described as

$$m \cdot \frac{\lambda' z}{d} - m \cdot \frac{\lambda z}{d} = m \cdot \frac{z}{d} \cdot \delta\lambda \tag{4.49}$$

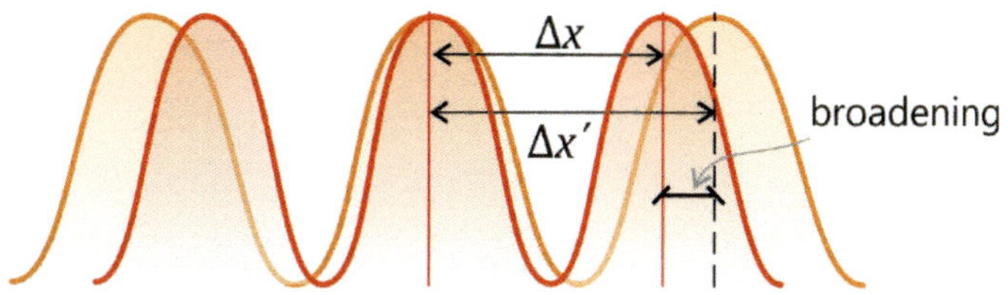


*Figure 4-23: Changes in fringe periodic distribution with wavelength.*

Fringe broadening increases linearly with the interference order $m$ and with the source spectral width $\delta\lambda$. There is an interference order $m$ such that a maximum of one distribution ($\lambda'$) overlaps a minimum of the other distribution ($\lambda$) such that fringes are no longer visible. This occurs when

$$x_{\text{MIN}}(\lambda) = x_{\text{MAX}}(\lambda') \;\Rightarrow\; (2m+1) \cdot \frac{\lambda z}{d} - (2m) \cdot \frac{\lambda' z}{d} \tag{4.50}$$

The interference order past which fringes are not visible can be expressed by the relationship

$$m = \lambda/(2\,\delta\lambda) \tag{4.51}$$

With a monochromatic or pseudo-monochromatic source, we can distinguish fringes up to orders (integer) $m$ that comply with Eq. (4.51), i.e., fringes on the order of $m < \lambda/(2\,\delta\lambda)$.

Example ☞: For red monochromatic light with $\lambda = 0.6$ μm and $\delta\lambda = 10$ nm, there can be approximately $m = \lambda/(2\delta\lambda) = 30$ visible fringes.

**Interference problem-solving strategy:**

1. Identify the interfering beams.
2. Identify the geometrical factor that causes the optical path difference. Convert to phase difference, if necessary.
3. Implement the constructive and destructive interference conditions:
   - Constructive interference (maxima) phase difference: $(m) \cdot 2\pi$, optical path difference: $(m) \cdot \lambda$
   - Destructive interference (minima) phase difference: $(m + ½) \cdot 2\pi$, optical path difference: $(m + ½) \cdot \lambda$

## 4.2.3 Transparent Plate: Thin-Film Interference

In many credit cards, there is an iridescent band that changes color depending on the observation angle. This is similar to the iridescent colors on oil slicks. These are examples of interference effects in thin, transparent slabs.

Figure 4-24 illustrates a glass plate with parallel surfaces, a fixed thickness $d$, and a refractive index $n$, surrounded by a medium with a refractive index $n_o$ (such as air). The plate is considered a **thin film** if the optical path inside it is smaller than the coherence length of the beam. It is considered transparent if the absorption inside the plate is minimal such that, after a pass through the glass, the exiting beam has an intensity comparable with that of the incident beam. Such a parallel-surface glass plate is illuminated with a thin monochromatic beam. We study the effect using the same method we applied in Young's experiment. We seek:

- the interfering beams
- the geometrical factor that causes the optical path difference and therefore the phase difference.

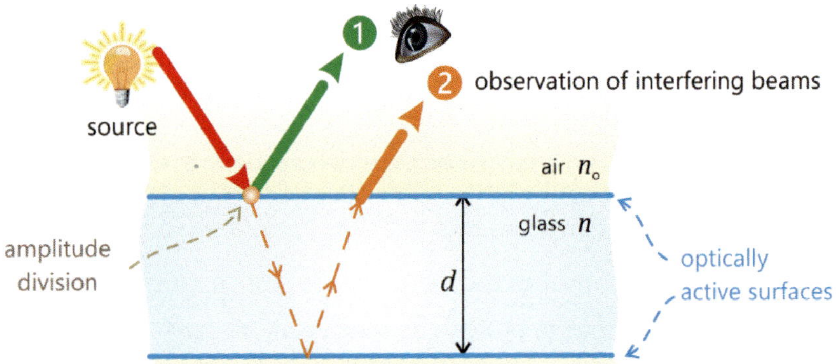


*Figure 4-24: Interfering beams by reflection off a thin transparent plate.*

The interfering beams derive from the incident beam that is partially reflected off the upper surface (beam ❶) and partially refracted off this surface. The refracted beam is subsequently (partially) reflected off the inner surface (which is parallel to the upper surface) and exits the upper surface by refraction. This is beam ❷. The two surfaces, upper and lower, are termed **optically active surfaces**. Since it is the same beam that splits into two parts, this is interference with amplitude division.

Geometry indicates that the optical path difference between beams ❷ & ❶ (see Figure 4-25) is

Optical Path Difference (2 – 1): $$n \cdot (AB + BC) - n_o(AD) \tag{4.52}$$

From triangle $ABE$: $$AB = BC = \frac{d}{\cos\delta} \quad \Rightarrow \quad n\cdot(AB+BC) = \frac{2nd}{\cos\delta} \tag{4.53}$$

From triangles $ACD$ and $ABE$, and by applying the law of refraction ($n_o \cdot \sin\vartheta = n \cdot \sin\delta$),

$$AD = AC\cdot\sin\vartheta = 2AE\cdot\sin\vartheta = 2\left(d\frac{\sin\delta}{\cos\delta}\right)\cdot\sin\vartheta = 2\left(d\frac{\sin\delta}{\cos\delta}\right)\cdot\frac{n}{n_o}\cdot\sin\delta \tag{4.54}$$

Combining Eq. (4.54) with Eq. (4.52), the optical path difference for the reflected beams takes the following simple form:

Optical Path Difference between reflected beams (2 – 1): $2nd\cdot\cos\delta$ (4.55)

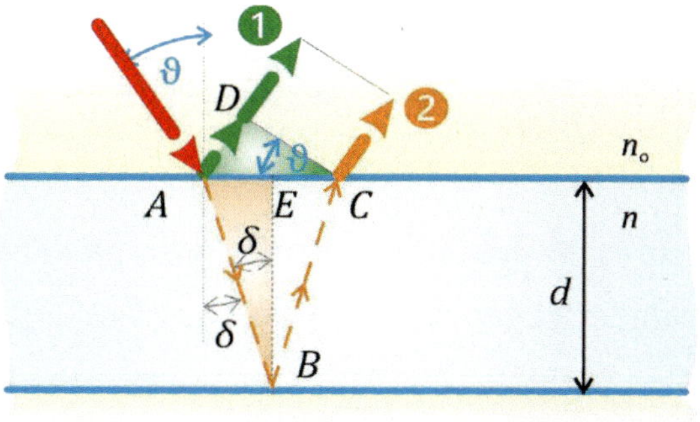


*Figure 4-25: Calculation of the optical path difference for the reflected beams.*

Now we convert the optical path difference to a phase difference. There is a peculiarity, however. Reflected beam ❶ is subject to an <u>additional phase shift</u> $\pi$. This phase shift occurs because beam ❶ is reflected off the surface that separates the optically thinner external medium (air, $n_o$) from an optically denser medium (glass, $n > n_o$). If the slab medium was optically thinner than the external medium (a thin layer of air inside glass), we would have $n < n_o$, but the phase shift would appear in reflected beam ❷.

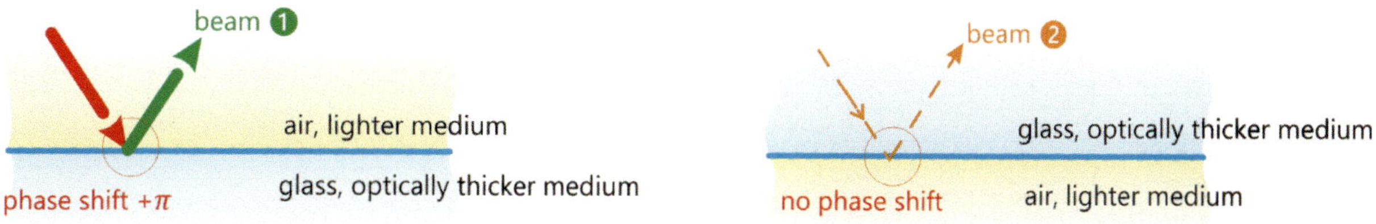


*Figure 4-26: Phase difference $+\pi$ is added to the beam reflected from an optically denser medium.*

Since we seek the phase difference ❷ – ❶, the phase of ❶ subtracted from the phase of ❷, a term ($-\pi$) must be added. Thus, the phase difference for the <u>reflected</u> beams is

$$\text{Phase Difference (2 – 1):} \quad \frac{2\pi}{\lambda}\cdot(\text{optical path difference})-\pi = \frac{4\pi\cdot nd\cdot\cos\delta}{\lambda}-\pi \tag{4.56}$$

Now we apply the conditions for constructive interference (maxima) and destructive interference (minima). Maxima appear for the phase difference $(m)\cdot 2\pi$, while minima appear for the phase difference $(m+½)\cdot 2\pi$. Again, an integer $m$ expresses the interference order and has values of 0, ±1, ±2, ... .

**Condition for constructive interference, reflected:**

$$\text{Phase Difference (2 – 1)} = (m)\cdot 2\pi \;\Rightarrow\; \frac{4nd\cdot\cos\delta}{\lambda} = (2m+1) \tag{4.57}$$

**Condition for destructive interference, reflected:**

$$\text{Phase Difference (2 –1)} = (m+1/2)\cdot 2\pi \;\Rightarrow\; \frac{4nd\cdot\cos\delta}{\lambda} = (2m) \tag{4.58}$$

Interference also exists in the space beneath the glass plate because of the interaction between the refracted (transmitted) beams. These are beams ❶′ and ❷′, which separate at point $B$ (Figure 4-27).

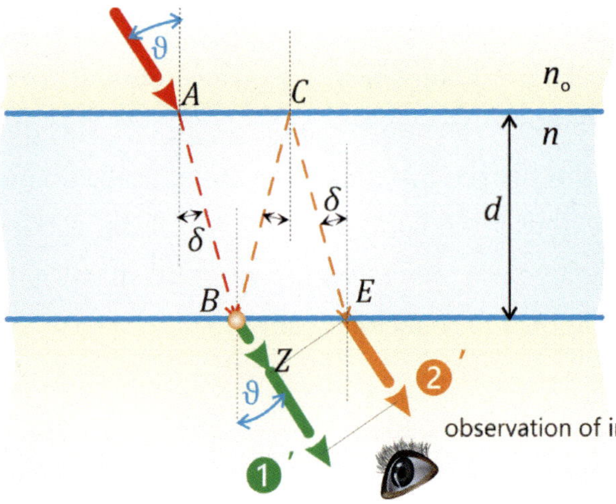


*Figure 4-27: Calculation of the optical path difference for the transmitted beams.*

At point $B$, beam ❶′ is subject to a refraction, while beam ❷′ is subject to two additional internal reflections from a denser medium to a thinner medium. Thus, there is no phase shift in any beam. (If the slab was of an optically thinner medium, the two reflections would result in an additional phase shift of $2\pi$ each, which, in the end, would still not affect the conditions.) The optical path difference and phase difference for the <u>transmitted</u> beams are the following:

Optical Path Difference (2′ – 1′): $n\cdot(BC+BE) - n_o(BZ) = 2nd\cdot\cos\delta$ (4.59)

Phase Difference (2′ – 1′): $\frac{2\pi}{\lambda}\cdot(\text{optical path difference}) = \frac{4\pi\cdot nd\cdot\cos\delta}{\lambda}$ (4.60)

**Condition for constructive interference, transmitted**:

Phase Difference (2′ – 1′) $= (m)\cdot 2\pi \Rightarrow \frac{4nd\cdot\cos\delta}{\lambda} = (2m)$ (4.61)

**Condition for destructive interference, transmitted**:

Phase Difference (2′ – 1′) $= (m+1/2)\cdot 2\pi \Rightarrow \frac{4nd\cdot\cos\delta}{\lambda} = (2m+1)$ (4.62)

We note that the interference conditions for the transmitted beams are complementary to those for the reflected beams; therefore, when there is minimum interference in the reflected side, we have, at the same time, maximum interference in the transmitted side, and vice versa.

The bright (and the dark) fringes form concentric circles and are termed **fringes of equal inclination** because they originate from a plate with parallel (equal slope), optically active surfaces. The shape of the fringes, which is determined by the loci of maxima or minima, depends on the interference condition. As always, the interference order $m$ must be an integer number. Specifically, for reflection interference, minima (zeros) appear when

Dark fringes, reflected: $m = \frac{2nd}{\lambda}\cos\delta = \text{integer}$ (4.63)

The parameters that determine the fringe shape are the plate refractive index $n$, the plate thickness $d$, the wavelength $\lambda$, and the incidence angle $\vartheta$ that relates to the refraction angle $\delta$ via the relationship $n_o\cdot\sin\vartheta = n\cdot\sin\delta$.

The shape of the refraction fringes is symmetrical to that of the reflection fringes since the interference conditions are complementary. The condition that applies to dark reflection fringes [Eq. (4.63)] also applies to bright transmission fringes. Thus, while the reflection fringes present at the center have a dark minimum, the refraction fringes have a bright maximum.

| Phase change upon reflection: | ☞ There is a $\pi$ (180°) phase shift if the second medium has a higher refractive index (hard end). | ☞ There is no phase shift if the second medium has a lower refractive index (soft end). |
|---|---|---|

Note: The quantity $n \cdot d$ is omnipresent in the relationships describing the optical path length. Recall that the property known as **optical path length**, or **optical thickness**, is the distance traveled by light multiplied by the 'degree of difficulty' of travel in that medium, which is the refractive index.[21]

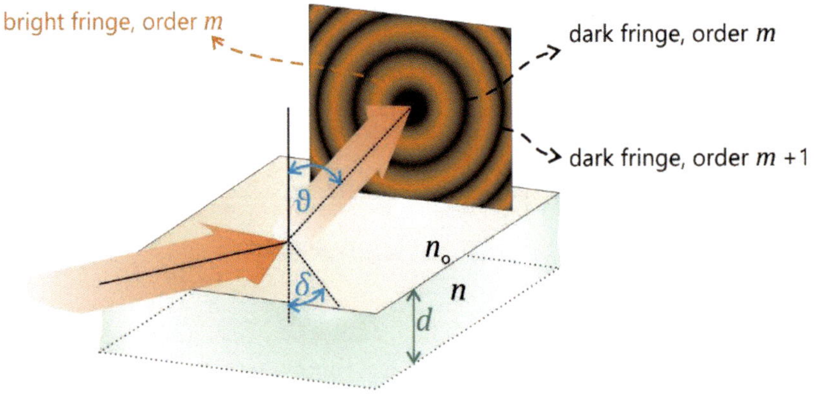


*Figure 4-28: Interference fringes by reflection off a thin transparent plate. The dark center indicates destructive interference, meeting the condition for phase difference (2 – 1) = (m + 1/2)·2π.*

If we illuminate the glass plate with a thin coherent beam, minima (zeros) appear for values of the observation angle $\vartheta$ for which $\cos\delta$ renders an integer having the quantity expressed in Eq. (4.63). For the dark-fringe order $m$, $\cos\delta_m = m \cdot (\lambda/2nd)$, for the next order, $\cos\delta_{m+1} = (m^{+1}) \cdot (\lambda/2nd)$, and so on.

Despite the fact that these fringes are termed 'equal inclination,' their radii do not have equal steps; rather they have steps that produce an integer increase of the quantity $\cos\delta = \lambda/2nd$. Inversely solving the problem, we find angle $\delta$, and from the law of refraction, we find angle $\vartheta$. It can be proven that, initially, the fringes gradually compress with decreasing steps, but then expand.

For a fixed angle, the simplest case is normal incidence ($\vartheta = 0$, $\delta = 0 \Rightarrow \cos\delta = 1$). The interference Eqs. (4.57) and (4.60) now take the form

Normal Incidence, reflection interference:

$$\text{maxima: } \frac{4nd}{\lambda} = (2m+1) \quad \text{and} \quad \text{minima: } \frac{4nd}{\lambda} = (2m) \tag{4.64}$$

Normal Incidence, refraction (transmission) interference:

$$\text{maxima: } \frac{4nd}{\lambda} = (2m) \quad \text{and} \quad \text{minima: } \frac{4nd}{\lambda} = (2m+1) \tag{4.65}$$

[21] *Introduction to Optics* § 1.4.1 Optical Density and Optical Path Length.

The condition for constructive interference with a transmitted wave—or, equivalently, destructive interference with a reflected wave—is a very a simple relationship: $d = m \cdot (\lambda/2n)$. Layer thickness $d$ is a multiple integer $m$ of one-half of the wavelength inside the medium $(\lambda/2n)$. Thus, for the double path, the reflected wave from the upper active surface interferes with a wave from the lower active surface; the latter wave carries an additional optical path length equal to a wavelength.

A simple and common interference application with variable thickness occurs in a soap bubble. Because of the gravitational force, the layer thickness increases toward the bottom.

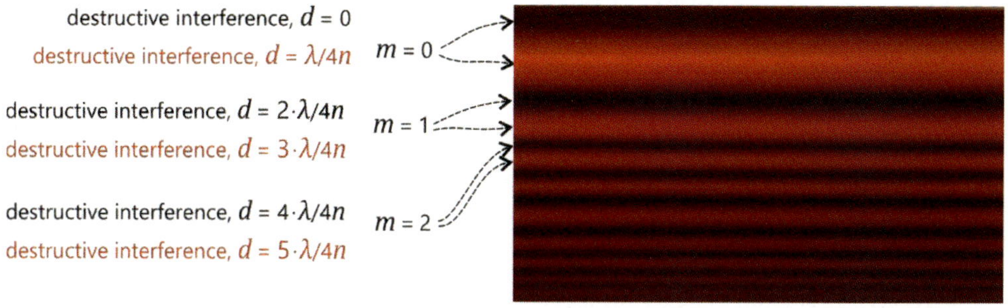


*Figure 4-29: Interference effects in a soap bubble with increasing slab thickness.*

We illuminate this layer with an extended, coherent HeNe laser beam (Figure 4-29). Bright fringes (maxima) on the reflection side appear under the condition of constructive interference [Eq. (4.64) (left)], and dark fringes (minima) appear under the condition of destructive interference [Eq. (4.64) (right)].

Because the layer thickness $d$ increases toward the lower part of the slab, the fringes condense. At the topmost part, where the thickness may be even smaller than half a micron, practically, $d = 0$, and the zeroth order ($m = 0$) of destructive interference occurs. This is why the film appears black.

### *4.2.3.1 Soap Bubbles and Antireflection (AR) Coatings*

The iridescent colors in oil slicks result from interference in thin, transparent layers! The 'plate' is the floating thin oil layer. The same occurs in colorful, iridescent soap bubbles, where the 'plate' is a thin layer of soapy water.

These examples are well represented in the setup of Figure 4-24. The only difference is that we now have white light, which means many, not just one, wavelength. Equations (4.64) and (4.65) are still applicable with the use of a select wavelength. Due to the many possible wavelengths, there is a multitude of outcomes. It is possible to have constructive interference for certain wavelengths such as the red and the blue, while simultaneously—under exactly the same conditions—having destructive interference for the other wavelengths such as the green.

Figure 4-30: Thin-film interference occurs in our daily-life experience (right photo by Coralie, SweetMellowChill used with permission).

Exactly the complementary, negative effect appears for the light that passes through this thin plate. If we apply constructive interference conditions to the reflected waves for some wavelengths, the opposite conditions apply simultaneously to the transmitted waves (see Figure 4-31). If, on one side, there is destructive interference for the green and constructive interference for the red and blue, then, on the other side, there is constructive interference for the green and destructive interference for the red and blue.

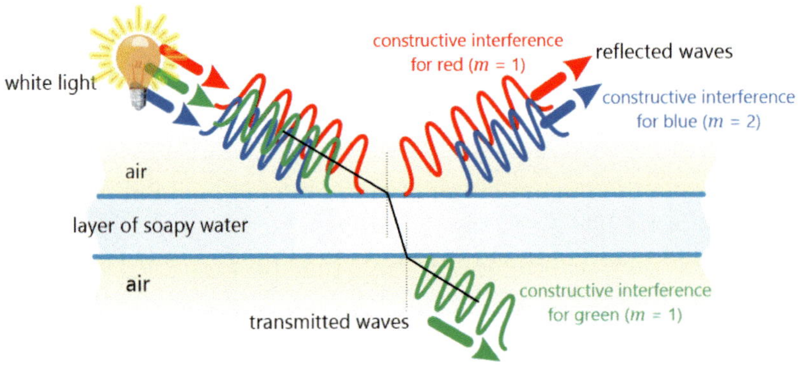


Figure 4-31: White light interference in a thin film of a layer of soapy water.

We can simultaneously have maxima for two different wavelengths (blue and red) when Eq. (4.64) (left) is satisfied for two different integers. A two-equation system with one unknown (thickness $d$) and the condition that both $m$ and $m'$ must be integers renders the system solvable.

Constructive Interference, reflected waves: $$\lambda_1 : d = (2m+1)\frac{\lambda_1}{4n} \qquad \lambda_2 : d = (2m'+1)\frac{\lambda_2}{4n} \qquad (4.66)$$

For just a slightly different layer thickness or a slightly different observation angle, the interference conditions produce a different wavelength and therefore a different color! This is the beauty of iridescence, which can present the illusion of so many colors, depending on the observation angle or the locality of the observation (the layer thickness).

Example ☞: A water–soap bubble ($n$ = 1.33) in air has thickness $d$ = 320 nm. If it is illuminated normally ($\vartheta = 0$, $\delta = 0$) with white light, constructive interference (maxima) occurs in the reflected path for $\lambda = 4n \cdot d/(2m+1)$, where $m$ is an integer. Destructive interference (minima) occurs for $\lambda = 4n \cdot d/(2m)$.

Using the layer thickness, maxima are observed for $\lambda = 1700/(2m+1)$ and minima for $\lambda = 1700/(2m)$, where $\lambda$ is expressed in nanometers. For small integers, we find that this thin layer has maxima at 570 nm ($m$ = 1) and 340 nm ($m$ = 2), while minima occur for 850 nm ($m$ = 1) and 425 nm ($m$ = 2), wavelengths for which reflection is entirely zero. For this reason, the layer appears yellow-green ($\lambda$ = 570 nm), which shows a strong constructive reflection interference in the visible.

If we illuminate the film in Figure 4-29 with white light, again, for near-zero layer thickness (top), destructive interference appears for all wavelengths ($m = 0$). As the thickness increases, the successive interference orders display multicolor interference fringes. In each interference order (same $m$), there is a successive appearance of the source spectral components. Assume only the red and blue: Red is seen where there is destructive interference for the blue within the same order (same $m$). As the layer thickness increases, the interference condition becomes destructive for the red and the film appears blue. With white light, the fringe sharpness—the intensity ratio of bright to dark—decreases quickly, since a broadband source has a much lesser degree of coherence than a monochromatic source.

*Figure 4-32: Interference in soap film with (left) white light and (right) monochromatic light.*

The **antireflection** (AR) **coating** is an interesting thin-film interference application that aims to reduce reflectivity via broad-spectrum destructive interference. As light passes through an interface, a small part of it, approximately 4%, is reflected off of each surface for normal incidence, and a larger part is reflected for oblique incidence. A lens system with many refracting surfaces (a zoom lens with five lenses may have more than ten refracting surfaces), or a laser cavity involves multiple interactions between the optical elements. In such systems, the total intensity loss due to each reflection may add up substantially. A multiple-lens system like those found in microscopes, telescopes, binoculars, or photographic lenses should always come with a high-performance antireflective coating.

An antireflection coating is a thin layer of a special material deposited over glass. This layer has, ideally, an intermediate refractive index $n_f = \sqrt{(n_1 \cdot n_L)}$, where $n_1$ is the refractive index of the initial medium (typically, it is air), and $n_L$ is the refractive index of lens material. A typical example of a material used in antireflection coatings is magnesium fluoride ($MgF_2$, commercially known as Sellaite and Afluon) with $n = 1.38$ at 552 nm. This coating is ideal for broadband use, although it is best when used between air ($n_1 = 1$) and very high-refractive-index glass ($n_L = 1.9$).

Note: Why is $MgF_2$ an ideal single-layer AR coating material when used for high-index glass?

Because the square root of ($n_1 \cdot n_T$), where $n_1$ is the refractive index of air, and $n_T$ is the refractive index of the substrate's high-index glass: $\sqrt{(1 \times 1.9)} = 1.378$, is very close to the refractive index of $MgF_2$: $n = 1.38$.

Example ☞: What would be the ideal refractive index for a material if we wish to construct a single-layer AR coating over a glass lens with a refractive index of 1.66?

The coating will be situated between air, $n_1 = 1.0$ and glass, $n_T = 1.66$. The coating's ideal refractive index would be $\sqrt{(n_1 \cdot n_T)} = \sqrt{(1 \times 1.66)} = 1.288$.

Other AR materials are VIS 0° and VIS 45°. VIS 0° (0° referring to the angle of incidence) and VIS 45° (45° angle of incidence) provide optimized transmission over most of the visible spectrum, eliminating reflection down to 0.4% (425 nm) and 0.75% (675 nm).

*Figure 4-33: (left) A photographic lens with many elements that have anti-reflection coatings. (right) The effect of antireflection coating over a screen (top) without an AR coating and (bottom) with an AR coating.*

The AR coating principle of operation (Figure 4-34) uses the same concept as is depicted in Figure 4-24, which shows thin transparent layers. However, at the second active surface, which is the coating–substrate interface, there is reflection from an optically less dense (rare) material to an optically more dense material. Here a phase shift occurs in both interfering beams, so Eqs. (4.64) and (4.65) are interchanged. We now use these relationships as interference conditions for the reflected beams.

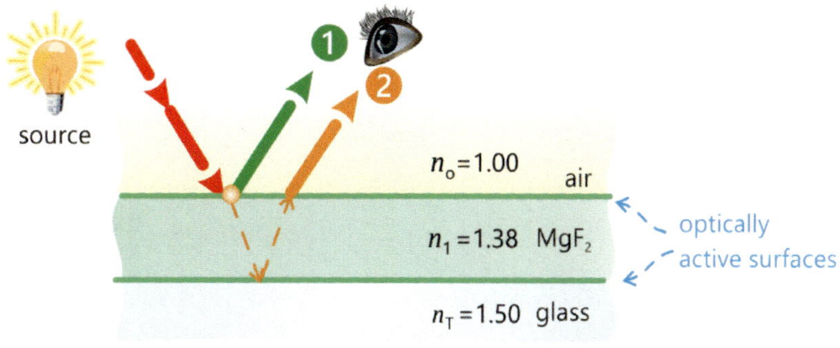


*Figure 4-34: Principle of operation of antireflection coatings.*

To completely eliminate reflection, we apply destructive interference between beam ❷ and beam ❶. For normal incidence ($\vartheta = 0$, $\delta = 0 \longrightarrow \cos\delta = 1$), the relationship that now provides the condition for destructive interference for the reflected beams ❷ and ❶ is $4n_1 \cdot d = (2m+1) \cdot \lambda$.

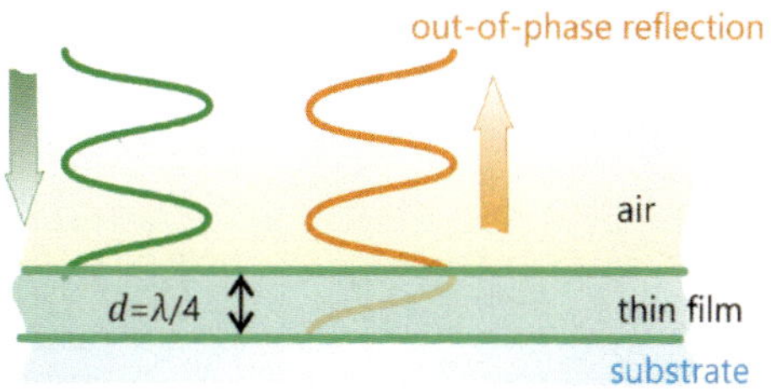


*Figure 4-35: Development of a phase difference in antireflection coatings.*

For minimum-order destructive interference ($m = 0$), we obtain a minimum coating thickness $d = \lambda/4n_1$. The visible ($\lambda = 0.55\,\mu\text{m}$) requires a coating with a thickness of 100 nm whose optical density is approximately one-quarter of the wavelength in that medium. Thus, the wave reflected from the coating–substrate interface (double path) has a phase opposite to that of the incident wave, and the interference of the two waves is destructive.

Example ☞: What is the minimum $MgF_2$ layer thickness that can be used as an antireflection coating over a glass lens?

We use $d = \lambda/4n_1 = 552\text{ nm}/(4 \cdot 1.38) = 100\text{ nm}$.

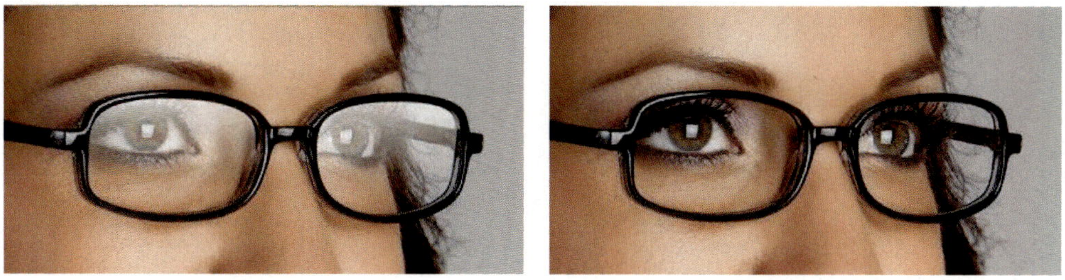

*Figure 4-36: Spectacle glasses (left) without and (right) with antireflection coatings.*

Since one coating layer is designed to eliminate one wavelength, other wavelengths (or colors) might not be eliminated by that layer. Additional layers of coatings with alternating refractive index values called **multilayer antireflection coatings** help to eliminate more wavelengths. These coatings consist of several dielectric layers; the more layers, the more wavelengths over which the AR coating is optimized; therefore, they are known as **broadband**.

The material type and layer thickness depend on the kind of substrate glass, the wavelength range, the properties of the incident light, and the tolerance requirements. The choices of material and layer thickness are subject to complex engineering formulation. Such coatings are common in optical elements used in photography lenses, AR shades for computer screens, etc. When applied in spectacle glasses, they appear to have minimal reflection over a wide range of angles of incidence, offering visual and cosmetic improvements to the wearer. Particularly when higher-refractive-index glass is used (it can be as high as 1.9), reflection is a problem: It can be as high as 9.6%. Antireflection coatings are part of a larger family of thin-film coatings that can be deposited on a lens or another transparent media by vacuum deposition.

### *4.2.3.2 Wedge-Shaped Plate*

In a vertically oriented soap film (Figure 4-32), the thin film does not have a fixed thickness, as its thickness increases progressively. Here we will examine this effect in detail, using a wedge-shaped plate, where the surfaces are no longer parallel but form a specific inclination angle $\alpha$. Thus, the thickness increases with distance $x$ from one end as $d = d_o + \alpha x$. The active surfaces are the upper (inclined) surface and the lower (horizontal) surface. For normal incidence ($\vartheta = 0$, $\delta = 0$ $\rightarrow \cos\delta = 1$), the relationships for the reflected wave are

Wedge-Shaped Plate Interference:

$$\text{maxima: } \frac{4n(d_o + \alpha x)}{\lambda} = (2m+1) \quad \text{and} \quad \text{minima: } \frac{4n(d_o + \alpha x)}{\lambda} = (2m) \tag{4.67}$$

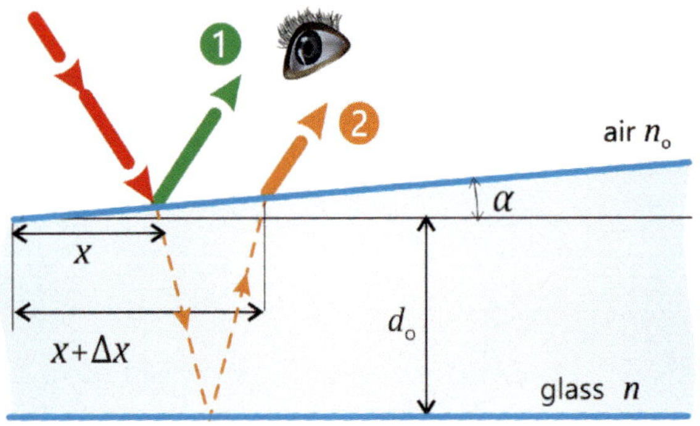


*Figure 4-37: Interference geometry in a wedge-shaped plate.*

To find the distance $\Delta x$ that separates two successive orders of constructive interference, we apply Eq. (4.67) for two consecutive integers: interference order $m$ at distance $x$ and interference order $m+1$ at distance $x+\Delta x$:

Interference Order $m$: $$4n[d_o+\alpha x] = [2m+1]\cdot\lambda \tag{4.68}$$

Interference Order $m+1$: $$4n[d_o+\alpha(x+\Delta x)] = [(2m+1)+1]\cdot\lambda \tag{4.69}$$

By subtraction, we find that the separation of two successive bright fringes is

$$\alpha\cdot\Delta x = \lambda/2n \tag{4.70}$$

The added optical path length $(a\cdot\Delta x)$ equals one-half of a wavelength in the medium ($\frac{1}{2}\lambda/n$) such that a complete path corresponds to one wavelength ($\lambda/n$). This is not surprising at all—this is the case in constructive interference.

We note in Eq. (4.70) that the separation $\Delta x$ between consecutive-order fringes does not depend on the order (they are equispaced) and that this distance is inversely proportional to the (fixed) slope $\alpha$. If the slope is not fixed, the spacing between fringes is affected accordingly.

Because the periodic distribution of monochromatic fringes is inversely proportional to the slope, this effect can be employed to investigate surface flatness. Condensed fringes correspond to increased inclination (angle $\alpha$), and expanded fringes correspond to reduced inclination. These fringes are called **Fizeau fringes**, named after the French physicist Armand Hippolyte Louis Fizeau, who applied interference to measure crystal dilation.

One common method used to test a polished surface's flatness is interference pattern analysis, in which the surface is placed against another polished surface of known flatness. When the two surfaces are in full contact, a pattern of concentric dark and bright circles appears. However, when the surfaces are separated by a very thin, wedge-shaped layer of air, equispaced, equal-thickness, parallel fringes are produced—these are Fizeau interference fringes.

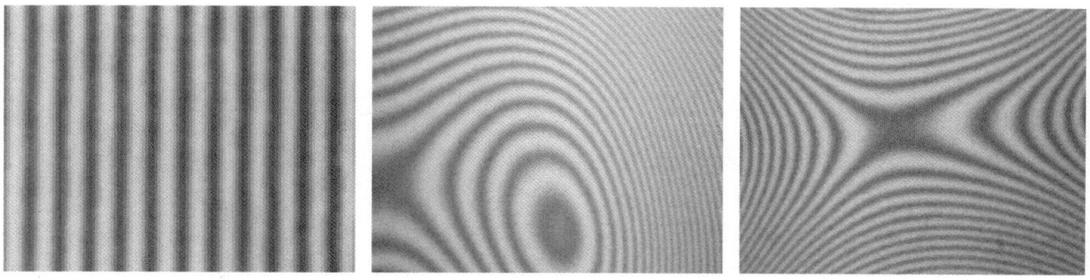

*Figure 4-38: Fizeau interference fringes of (left) equal thickness and (center and right) variable spacing. A large tilt between the plates corresponds to denser fringes.*

## 4.2.4 Newton's Rings

An interesting interference configuration is the rings that bear the name of Isaac Newton, although these rings were initially reported by Robert Hooke in his book *Micrographia* (1664). The setup is a curved surface (of a planoconvex lens) in contact with a flat surface. Iluminating the lens with an extended source results in alternating concentric bright and dark fringes, or **Newton's rings**.

This effect is a variation of the case with a wedge, with two differences. The first is that the wedge here is composed of a layer of air surrounded by two surfaces of optically denser material, such as glass. The second difference is that the slope is not constant. In Eq. (4.70), it is noted that the fringe separation is determined by the wedge slope: A fixed slope $\alpha$ results in fringes of equal thickness. Here the slope of the wedge is not constant, and the fringes are not equispaced. Specifically, because the slope increases progressively, the fringes condense.

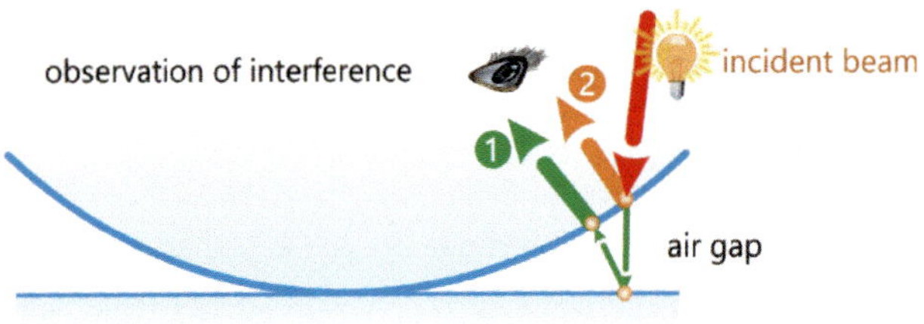


*Figure 4-39: Newton's rings interference setup.*

We now implement the interference problem-solving technique. We seek:

- the interfering beams
- the geometrical factor that causes the optical path difference and therefore the phase difference.

The interfering beams are the reflected beams ❶ and ❷. To model the developing phase difference, we consider the optically active surfaces, which are the convex surface and the flat, upper glass surface (Figure 4-40).

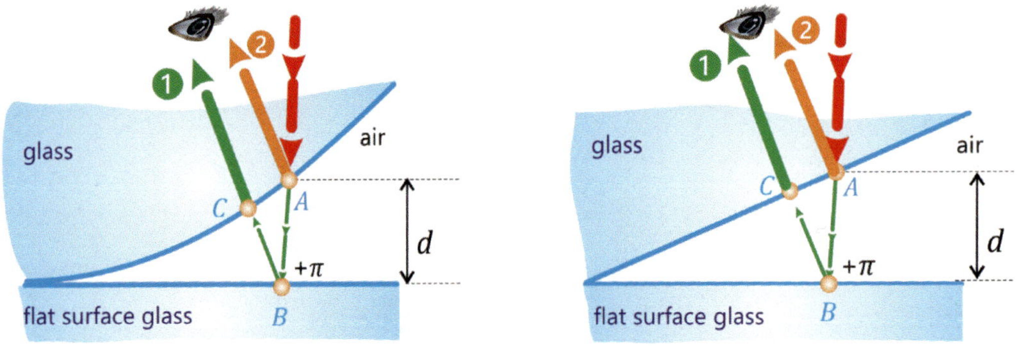


*Figure 4-40: Optical path difference in (left) Newton's rings and (right) an air-gap wedge.*

There is amplitude division at point $A$. Beam ❷ is reflected, while beam ❶ travels an additional path $AB+BC$ in the air gap, which serves as a thin wedge. External reflection occurs at point $B$, so there is a phase shift. Geometry indicates that the optical path difference is

Optical Path Difference (1 – 2): $$n_o \cdot (AB+BC) \approx 2d \cdot n_o \tag{4.71}$$

where $d$ is the air layer thickness along the path of ray ❶. This approximation applies to normal incidence ($\vartheta = 0$, $\delta = 0 \longrightarrow \cos\delta = 1$). Here $n_o$ is the refractive index of the wedge medium, in this case, $n_o = 1$, although it may be, in general, the refractive index of any other medium. Finally, because there is a phase shift due to an external reflection from the optically denser glass surface at point $B$, there is an additional phase shift $\pi$ in beam ❶. Therefore,

Phase Difference (1 – 2): $$= \frac{2\pi}{\lambda}(\text{optical path difference}) + \pi \approx \frac{2\pi}{\lambda}(2d \cdot n_o) + \pi \tag{4.72}$$

Here we are expressing an approximate thickness $d$. Since $r^2 = R^2 - (R-d)^2 \approx 2R \cdot d$ in triangle $ABC$ (Figure 4-41), if $R \gg d$, then $d \approx r^2/2R$, and for refractive index $n_o = 1$ (air), the phase difference is

Phase Difference (1 – 2): $$\frac{2\pi}{\lambda}(2d) + \pi = \frac{2\pi}{\lambda} \cdot \frac{r^2}{R} + \pi \tag{4.73}$$

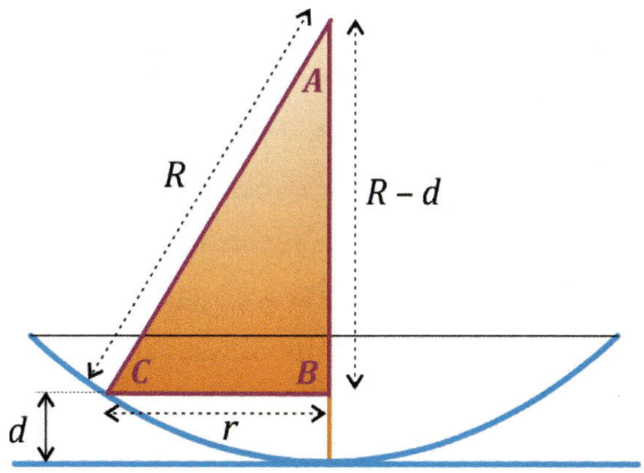


*Figure 4-41: Thickness d and radius of curvature.*

We now apply the interference conditions. For maxima (bright fringes), the phase difference is $2m \cdot \pi$, and for minima (dark fringes), the phase difference is $(2m+1) \cdot \pi$, where the integer $m$ is the interference order:

Maxima: phase difference (1 – 2) = $(2m) \cdot \pi \Rightarrow$ $$\frac{2\pi}{\lambda} \cdot \frac{r^2}{R} + \pi = (2m) \cdot \pi \tag{4.74}$$

Minima: phase difference (1 – 2) = (2$m$+1) · $\pi$ ⇒ $$\frac{2\pi}{\lambda}\cdot\frac{r^2}{R}+\pi = (2m+1)\cdot\pi \quad (4.75)$$

where $r$ corresponds to the radii (semi-diameters) of bright (or dark) fringes for constructive or destructive interference. The fringes form concentric circles with gradually denser radii, in proportion to the square root of the order. This is because the slope of the wedge increases progressively. For example, in destructive interference, the radii of dark fringes are

$$r^2_{\text{dark},\,m} = \lambda\cdot R\cdot m, \quad \text{where } m = 0, \pm1, \pm2, \ldots \quad (4.76)$$

For surfaces in contact, the zeroth-order ($m$ = 0) dark fringe appears at the center ($d$ = 0).

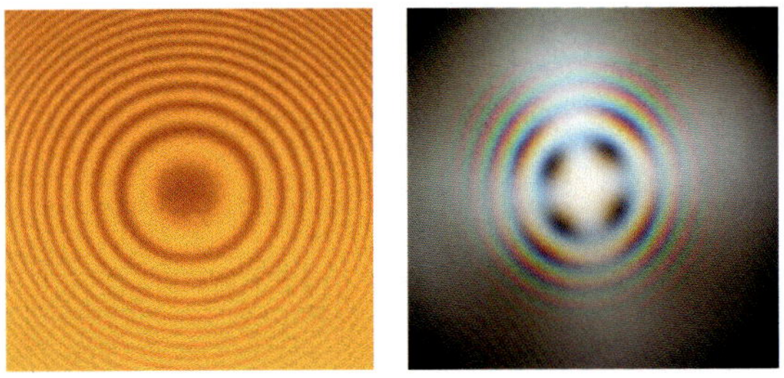

*Figure 4-42: Interference fringes in Newton's rings using (left) a sodium lamp and (right) a mercury lamp.*

### 4.2.5 Multiple-Beam Interference

In thin-film interference (discussed in § 4.2.3), the two interfering beams result from reflection (or refraction) by the two optically active surfaces. In reality, interference results from multiple reflections by the active surfaces (Figure 4-43). For interference at the upper part of the film, we consider beams 1, 2, 3, 4, ... , and for interference at the lower part, we consider beams 1′, 2′, 3′, 4′, ... . We then investigate the phase difference, as well as the relative amplitude magnitudes between the interfering beams, which are no longer equal.

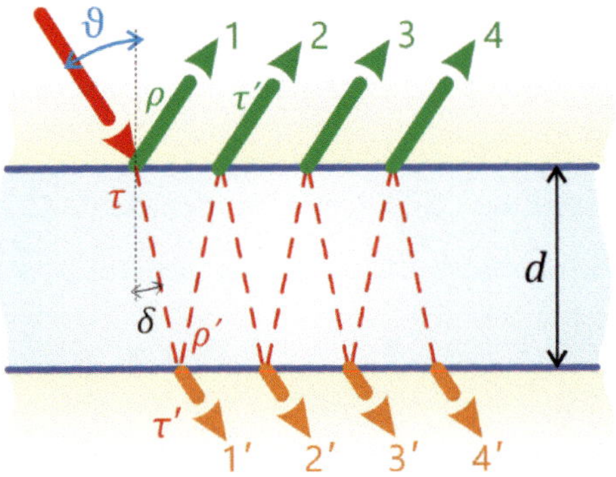


*Figure 4-43: Multiple-beam interference in a thin film.*

The phase difference results from the phase retardation $\varphi_\delta$, which is attributed to both the double path inside the medium and the possible phase shifts due to reflection from an optically denser medium. The amplitude magnitudes of the interfering beams are no longer equal because they are subject to a large number of reflections and refractions by the active surfaces. The primary reason for the beam attenuation is that the surface reflectivity is no longer considered equal to 1.0. A second parameter is the absorption inside the medium; however, in the first approximation, this is (still) ignored.

We consider an initial incident beam with amplitude magnitudes $E_o$ and phase $\varphi_o$, $E_o \cdot \cos(\varphi_o)$. Beam 1 results from reflection at the upper surface. The amplitude magnitude of beam 1 is $E_o \cdot \rho$, where $\rho$ is the reflection coefficient for this configuration, which includes the angle of incidence $\vartheta$, the relative refractive index $n$ between the surrounding medium and the material, and the polarization state of the incident beam. The analytical expressions for the reflection coefficients $\rho$ are provided by the Fresnel coefficients, Eqs. (2.34) to (2.36). In addition, beam 1 is subject to a phase shift ($+\pi$) due to reflection from the optically denser glass–air interface.

Incident Beam: $$E_o \cdot \cos(\varphi_o) \tag{4.77}$$

Reflected Beam 1: $$E_1 = E_o \cdot \rho \cdot \cos(\varphi_o - \pi) = -E_o \cdot \rho \cdot \cos(\varphi_o) \tag{4.78}$$

Reflected beam 2 has amplitude magnitude $E_o \cdot \tau \cdot \rho' \cdot \tau'$ since it derives from refraction at the upper surface of the air–glass interface ($\tau$), reflection from the lower surface ($\rho'$), and refraction at the glass–air interface ($\tau'$). The transmission ($\tau$ & $\tau'$) and internal reflection $\rho'$ coefficients are expressed analytically by their respective Fresnel coefficients (§ 2.5.2.1).

In addition, beam 2 is subject to a phase retardation $\varphi_\delta$, which is attributed to the double path inside the glass. Therefore, we express the beam 2 amplitude magnitude as $E_o \cdot \tau \cdot \rho' \cdot \tau' \cdot \cos(\varphi_o - \varphi_\delta)$. The phase retardation $\varphi_\delta$ for the double path inside the glass is exactly the same as in two-beam interference and is expressed by the relationship

Phase Difference (2′– 1′): $$= \frac{2\pi}{\lambda} \cdot (\text{optical path difference}) = \frac{4\pi \cdot nd \cdot \cos\delta}{\lambda} \tag{4.79}$$

where $(n \cdot d)$ is the optical path length of the glass plate (optical thickness), $\delta$ is the angle of refraction inside the medium, and $\lambda$ is the wavelength.

Beam 3 has amplitude magnitude $E_o \cdot \tau \cdot \rho' \cdot \rho' \cdot \rho' \cdot \tau'$ since its path includes two additional internal reflections at the upper and the lower surfaces ($\rho'$). This beam is subject to an additional phase retardation $\varphi_\delta$. Therefore, we express the beam 3 amplitude magnitude as $E_o \cdot \tau \cdot \rho' \cdot \rho' \cdot \rho' \cdot \tau' \cdot \cos(\varphi_o - 2\varphi_\delta)$. We conclude the following:

Reflected Beam 1: $E_1 = E_o \cdot \rho \cdot \cos(\varphi_o - \pi) = -E_o \cdot \rho \cdot \cos(\varphi_o)$

Reflected Beam 2: $E_2 = E_o \cdot \tau \cdot \rho' \cdot \tau' \cdot \cos(\varphi_o - \varphi_\delta)$

Reflected Beam 3:

$$E_3 = E_o \cdot \tau \cdot \rho' \cdot \rho' \cdot \rho' \cdot \tau' \cdot \cos(\varphi_o - 2\varphi_\delta) = \underbrace{E_o \cdot \tau \cdot \rho' \cdot \tau' \cdot \cos(\varphi_o - \varphi_\delta)}_{\text{reflected beam 2}} \cdot \left[\rho'^2 \cdot \cos(-\varphi_\delta)\right] \tag{4.80}$$

We can generalize Eq. (4.80) as

Reflected Beam $N$:

$$E_N = \left[\text{reflected beam } (N-1)\right] \cdot \left[\rho'^2 \cdot \cos(-\varphi_\delta)\right] \tag{4.81}$$

Now we simply need to add all of these interfering beams using a vector sum as described in § 4.1.4. [Indeed, Figure 4-13 refers to this exact case.] Thus, the resultant amplitude magnitude for multiple-beam reflection interference is the following sum:

$$\begin{aligned} E_{\text{Ref TOT}} &= -E_o \cdot \rho \cdot \cos(\varphi_o) + E_o \cdot \tau \cdot \rho' \cdot \tau' \cdot \cos(\varphi_o - \varphi_\delta) \\ &\quad \cdot \left[1 + \left[\rho'^2 \cdot \cos(-\varphi_\delta)\right] + \left[\rho'^2 \cdot \cos(-\varphi_\delta)\right]^2 + \left[\rho'^2 \cdot \cos(-\varphi_\delta)\right]^3 + \cdots\right] \\ &= -E_o \cdot \rho \cdot \cos(\varphi_o) \cdot \left[1 - \frac{\tau \cdot \tau' \cdot \cos(-\varphi_\delta)}{1 - \rho'^2 \cdot \cos(-\varphi_\delta)}\right] = E_o \cdot \rho \cdot \cos(\varphi_o) \cdot \left[\frac{\rho'\left[1 - \cos(-\varphi_\delta)\right]}{1 - \rho'^2 \cdot \cos(-\varphi_\delta)}\right] \end{aligned} \tag{4.82}$$

If there are no losses, then $\rho^2 + \tau \cdot \tau' = 1$. We also use $\rho = -\rho'$ and the identity

$$1 + x + x^2 + \cdots + x^N = \frac{1 - x^N}{1 - x} \approx \frac{1}{1 - x} \tag{4.83}$$

under the condition that $x < 1$, which also leads to $x^N \ll 1$.

Note: We assumed that the surface reflection coefficients are equal ($\rho = -\rho'$). This is true if the surrounding medium is the same. The coefficients may be different if there is a medium (below the second active surface) with a refractive index other than that of the medium above the first active surface. In this case, instead of $R = \rho^2$, we use the more general relationship $R = \rho \cdot \rho'$.

Using a similar line of thought, we can find the resultant amplitude magnitude for multiple-beam interference due to refraction (the transmitted wave), which is expressed as

$$E_{\text{Trans TOT}} = E_o \cdot \rho \cdot \cos(\varphi_o) \cdot \left[\frac{1 - \rho'^2}{1 - \rho'^2 \cdot \cos(-\varphi_\delta)}\right] \tag{4.84}$$

The corresponding intensities are

$$I_{\text{Ref TOT}} = E_o^{\,2} \cdot \frac{2\rho^2\left[1-\cos(\varphi_\delta)\right]}{\left(1+\rho^4\right)-2\rho^2\left[1-\cos(\varphi_\delta)\right]} \quad \text{and} \quad I_{\text{Trans TOT}} = E_o^{\,2} \cdot \frac{\left(1-\rho^2\right)^2}{\left(1+\rho^4\right)-2\rho^2\left[1-\cos(\varphi_\delta)\right]} \tag{4.85}$$

Employing trigonometric processing, the above relationships can be re-arranged as

$$I_{\text{Ref TOT}} = I_o \cdot \frac{\left(\frac{2\rho}{1-\rho^2}\right)^2 \cdot \sin^2\left(\frac{\varphi_\delta}{2}\right)}{1+\left(\frac{2\rho}{1-\rho^2}\right)^2 \cdot \sin^2\left(\frac{\varphi_\delta}{2}\right)} \quad \text{and} \quad I_{\text{Trans TOT}} = I_o \cdot \frac{1}{1+\left(\frac{2\rho}{1-\rho^2}\right)^2 \cdot \sin^2\left(\frac{\varphi_\delta}{2}\right)} \tag{4.86}$$

The unitless quantity

$$F \equiv \left(\frac{2\rho}{1-\rho^2}\right)^2 = \frac{4R}{(1-R)^2} \tag{4.87}$$

is the **coefficient of finesse**, which directly relates to surface reflectivity, $R = \rho^2$. If the reflectivity is nearly 1.0, the coefficient of finesse reaches very large values. For example, for $R = 0.87$, the coefficient of finesse has a value of 200. $F$ can be considered as an expression of the active number of waves that participate in multiple interference. The higher the reflectivity, the lower the attenuation in each double pass, so the greater the population of participating waves.

We also use the **coefficient of reflecting finesse**:

$$\mathfrak{F} \equiv \frac{\pi\sqrt{R}}{1-R} \tag{4.88}$$

These two coefficients are related by:

$$\mathfrak{F} \equiv \frac{\pi^2}{4} F \tag{4.89}$$

We will use these coefficients in laser optical resonator multiple-beam interference (§ 6.2.2.1).

Applying the coefficient of finesse [Eq. (4.87)], the expressions for the intensity resulting from multiple-beam interference are quite simplified:

$$I_{\text{Ref TOT}} = I_o \cdot \frac{F \cdot \sin^2\left(\frac{\varphi_\delta}{2}\right)}{1 + F \cdot \sin^2\left(\frac{\varphi_\delta}{2}\right)} \quad \text{and} \quad I_{\text{Trans TOT}} = I_o \cdot \frac{1}{1 + F \times \sin^2\left(\frac{\varphi_\delta}{2}\right)} \tag{4.90}$$

Just as in the case of two-beam interference, interference expressions for the transmitted and the reflected sides are complementary. The condition for the appearance of maxima is exactly the same because the condition for maxima/minima depends on the (same) phase difference $\varphi_\delta$. Specifically, when the phase difference $\varphi_\delta$ is a multiple integer of $2\pi$, there is a maximum for the transmitted wave and, simultaneously, a minimum for the reflected wave. Indeed, the relationship $\varphi_\delta = 2m \cdot \pi$, where $m$ is the interference order, gives $\sin(\varphi_\delta/2) = 0$; therefore,

$$I_{\text{Ref MIN}} = 0 \quad \text{and} \quad I_{\text{Trans MAX}} = I_o \tag{4.91}$$

Accordingly, the relationship $\varphi_\delta = (2m+1)\pi$ leads to $\sin(\varphi_\delta/2) = 1$, so

$$I_{\text{Ref MAX}} = I_o \cdot \frac{F}{1+F} \quad \text{and} \quad I_{\text{Trans MIN}} = I_o \cdot \frac{1}{1+F} \tag{4.92}$$

Thus, in multiple-beam interference, what is different in comparison to simple two-beam interference is the sharpness of the fringes. The sharpness depends on the coefficient of finesse $F$, a unitless quantity represented by the Airy function, which is named after Sir George Biddell Airy and expresses the relative variation of fringe intensity due to multiple-beam interference for the transmitted beam:

$$\text{Airy Function} = \frac{1}{1+F \cdot \sin^2\left(\dfrac{\varphi_\delta}{2}\right)} \tag{4.93}$$

Figure 4-44 illustrates three different Airy functions in relation to the phase difference $\varphi_\delta$ for three different values of reflectivity $R$ and therefore three different coefficients of finesse $F$.

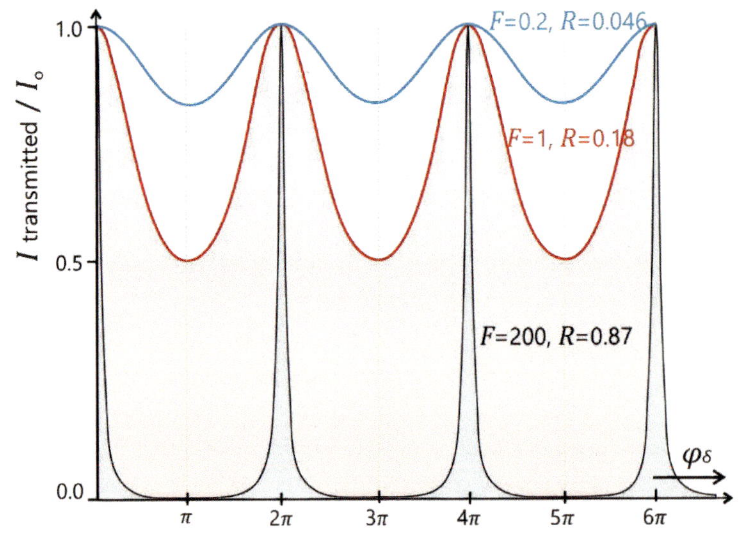


*Figure 4-44: Graph of the Airy function for three different values of the coefficient of finesse F and their corresponding reflectivity R values.*

As in two-beam interference, the periodic distribution of the Airy function depends exclusively on the phase difference due to the double path: Maxima appear when $\sin\varphi_\delta/2 = 0$ ($\varphi_\delta = 2m\pi$), exactly as in the case of two-beam interference.

The new aspect is the sharpness of the interference maxima, described by the **fringe sharpness** $\mathcal{V}$. Fringe sharpness can be defined as the ratio between the maximum and the minimum intensities. The sharpness depends on the value of the coefficient of finesse $F$, which depends on the surface reflectivity $R$. The minimum Airy function value (for phase difference $\varphi_\delta = \pm\pi, \pm3\pi, \ldots$) becomes zero for a large $F$. Thus, even for ideal coherent sources, the coefficient of finesse affects the fringe sharpness $\mathcal{V}$:

Fringe Sharpness $\mathcal{V}$:

$$\frac{I_{\text{Trans MAX}}}{I_{\text{Trans MIN}}} = \frac{I_o}{\dfrac{I_o}{1+F}} = 1+F \tag{4.94}$$

In thin maxima, there are many thinner bright lines for the transmitted wave and many thinner dark lines for the reflected wave. For comparison, the distribution we obtain from only two interfering beams has a sinusoidal form.

A second parameter is the **fringe thickness**, expressed by the FWHM. Fringe sharpness can then be expressed as $I_{\text{MAX}}/2\cdot(\text{FWHM})$. Fringe thickness may be conventionally reported in units of angle (phase difference), and the FWHM is the **angular fringe width** $\varepsilon$. To the left of the FWHM, for point $A$ (see Figure 4-45),

$$\frac{I_A}{I_{\text{MAX}}} = \frac{1}{1+F\cdot\sin^2\left(\dfrac{\varphi_A}{2}\right)} = \frac{1}{2} \tag{4.95}$$

where the value of the phase difference $\varphi_A$ is $\varphi_A = 2m\cdot\pi-\varepsilon/2$. We can also state for point $B$, to the right of the FWHM, that $\varphi_B = 2m\cdot\pi+\varepsilon/2$. In general,

$$\frac{1}{1+F\cdot\sin^2\left(m\cdot\pi\pm\dfrac{\varepsilon}{4}\right)} = \frac{1}{2} \Rightarrow \sin^2\left(m\cdot\pi\pm\frac{\varepsilon}{4}\right) = \frac{1}{F} \tag{4.96}$$

For large values of the coefficient of finesse (and accordingly small values of $\varepsilon$),

$$\sin^2\left(m\cdot\pi\pm\frac{\varepsilon}{4}\right) = \sin^2\left(\frac{\varepsilon}{4}\right) \approx \left(\frac{\varepsilon}{4}\right)^2 = \frac{1}{F} \Rightarrow \varepsilon \approx \frac{4}{\sqrt{F}} = \frac{2\pi}{\mathfrak{F}} \tag{4.97}$$

It is obvious that the larger the coefficient of finesse, the lower the value of the maximum $\varepsilon$, or equivalently, the higher the fringe sharpness, as illustrated in Figure 4-45.

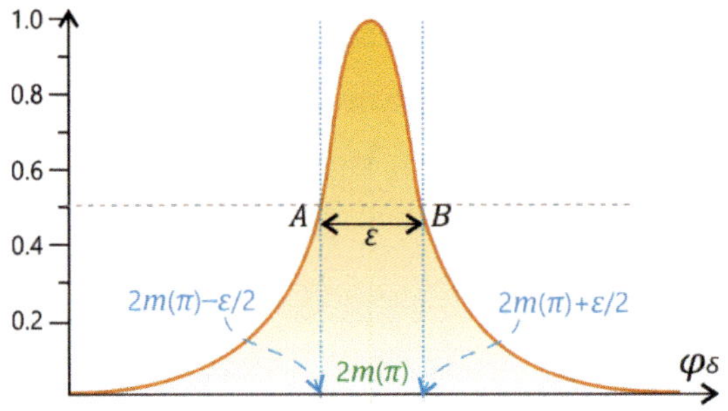


*Figure 4-45: Calculation of the angular extent for the Airy function principal maximum.*

In **multiple-beam reflection interference**, high reflectivity results in:

- an increase in the coefficient of finesse,
- thinning of the bright fringes, and
- an increase in the contrast between the maximum value of the bright fringe and the minimum value of the dark fringe.
- Conversely, multiple-beam reflection interference does not affect the maximum appearence condition. The maxima appear exactly at the same locations, even for much lower reflectivity, or, equivalently, if only two beams 'participate' in the effect.

Glossy, iridescent feathers on birds is nature's most impressive manifestation of interference. A striking example is the purple starling, whose shimmering colors are not a result of pigments but of interference owing to the physical structure and arrangement of melanin in the feather barbules. In sunbird feathers the arrangement is like a stack, in peacock feathers there is a lattice structure, and in hummingbird feathers the arrangement has air gaps.

*Figure 4-46: (left) The purple starling (Lamprotornis purpureus), perhaps the shiniest bird on the planet (photo by Steve Bidmead used with permission). (right) Iridescence is often seen in insect shells (photo by Yegor Kamelev used with permission).*

### 4.2.6 Interference and the Principle of Least Time

We can now address a fundamental principle employed in geometrical optics. The principle of least optical path is the constitution of all laws pertaining to light propagation and image formation, such as rectilinear propagation in free space, the law of reflection, and the law of refraction. The principle of least time is accepted on a phenomenological basis by virtue of its obvious agreement with experimental data. Geometrical optics, however, does not offer any proof for this.

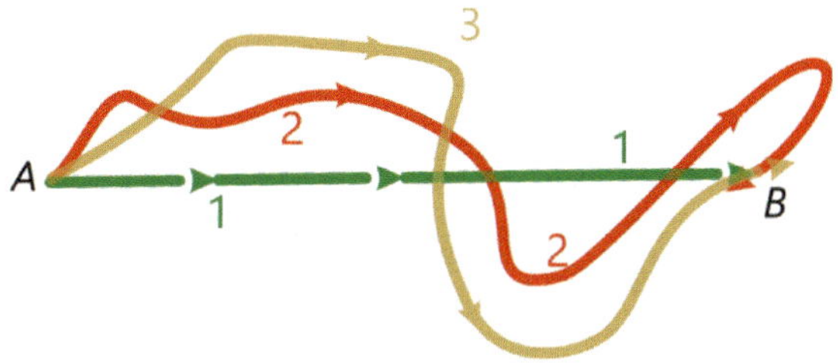


*Figure 4-47: Probable random paths from point A to point B. Path 1 is the shortest.*

The principle of the least optical path may be proven in terms of photon interference. This is true despite the fact that photons obey the principle of uncertainty, which does not allow us to follow a photon closely as it propagates in space. The prediction of quantum electrodynamics is that a photon that follows from point *A* to point *B* can follow any random path, each path having the very same probability.

There is a seeming contradiction between the statements '*a photon may follow any path with the same probability*' and '*light will follow only path 1 from A to B (Figure 4-47) because it is the shortest.*' In fact, the two statements are in agreement. In addition, the principle that 'a photon may follow any path with the same probability' has as a consequence the fact that 'light will follow only the shortest path from *A* to *B*.'

We accept that there are many probable paths for a random photon. In each probability, there is a corresponding wavevector, such as those presented in § 4.1.4. These vectors are described by a complex number, $\rho \times \exp(i\vartheta)$, where the angle $\vartheta$ is proportional to the time required to travel this path, and $\rho$ is the probability amplitude. The observed probability for this path is the square of its amplitude, just as the intensity in a wave is the square of the field.

To find the path that light (a large photon population) follows from *A* to *B*, we place at point *A* a light source that (of course) emits a very large photon population in all directions. For each of these photons, we construct wavevectors for all probable paths. Since each photon is likely to follow a random path, its vector $\rho \times \exp(i\vartheta)$ has a random phase $\vartheta$. The statement 'a photon may follow any path with the <u>same</u> probability' dictates that the vector magnitudes $\rho$ are equal: $\rho_1 = \rho_2$, but $\vartheta_1 \neq \vartheta_2$.

Still referring to Figure 4-47, path 1, which is the closest to a straight line, appears to require the least time compared to the other paths. Let us assign to it a vector $\rho_1 \times \exp(i\vartheta_1)$. Paths that are very close to path 1 (those that are directly neighboring it) correspond to nearly the same time intervals, so all of these paths have nearly the same phase $\vartheta_1$. We can then expect many vectors with nearly the same phase $\vartheta_1$.

Paths 2 and 3 obviously require a much longer—and therefore a very different—time interval. Accordingly, their probability vectors have very different, and therefore random, phases between 0 and $2\pi$. In the phase space, their distribution is random.

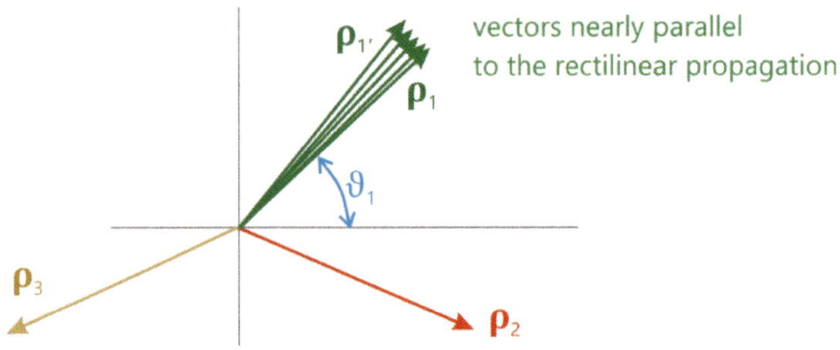


*Figure 4-48: Probability vectors for random paths from A to B.*

With the methodology employed in § 4.1.4, we add all of the vectors of all of the probable paths. The sum describes the probability vector with magnitude $\rho_\tau$ and phase $\vartheta_\tau$ that correspond to the actual light path! The total summation might be as shown in Figure 4-49.

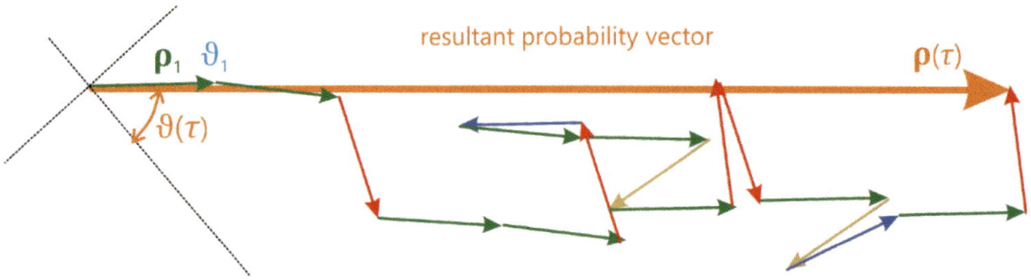


*Figure 4-49: Summation of the probability vectors for paths from A to B.*

Figure 4-49 illustrates the probability vector that indicates the path to be followed. We note, not surprisingly, that the vector sum of all of these probability vectors has nearly the same phase as the one corresponding to the rectilinear propagation between points *A* and *B*. This is because the probability that a photon propagates along path 2 annuls the probability that a photon propagates along path 3, while the probability that a photon propagates along the path of minimum time is intensified by the probabilities of many other photons propagating along almost identical paths.

This resembles another quantum electrodynamics colloquium, which states that the probability of a photon in a specific state increases with the presence of other photons already in this state. In other words, the path of a light beam is determined by the interference of all of the paths potentially followed by each photon, and light travels along the directions in which these probabilities interfere positively, or constructively. Light interferes with itself as it propagates! The principle of the least optical path can therefore be interpreted as a result of multiple-beam interference.

## 4.3 MICHELSON INTERFEROMETRY

With Young's experimental setup or the Michelson interferometer developed by Albert Abraham Michelson in 1880, we can measure very small displacements such as the wavelength of light.

*Figure 4-50: Armand Louis Fizeau (1819–1896) and Albert Abraham Michelson (1852–1931).*

The basic principle of Michelson interferometry is illustrated in Figure 4-51. A beam is incident on a beamsplitter, which is a surface that reflects half of the incident beam and allows transmission of the other half.

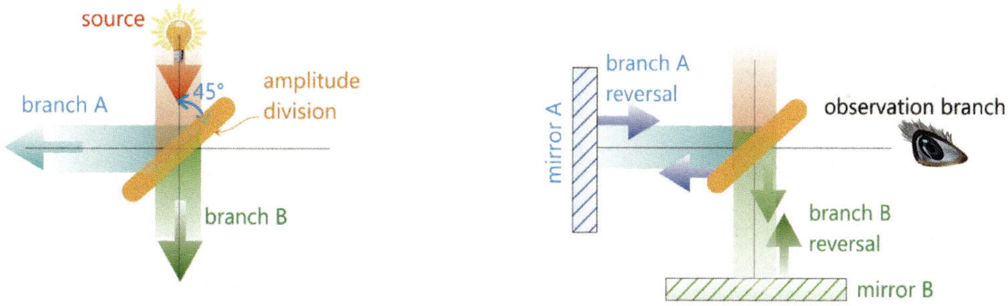


*Figure 4-51: (left) Amplitude division and interfering beams. (right) Light propagation along the two branches in a Michelson interferometer.*

This partially reflective glass surface produces two coherent beams of equal amplitude magnitudes: half that of the initial beam. This is an amplitude-division interference configuration.

If the relative orientation of the beamsplitter is 45°, the two beams propagate in two perpendicular paths, branches A and B. We use two mirrors, A and B, to reverse the beams' paths. The mirrors are exactly normal (perpendicular) to the beams, so after the beams pass through the beamsplitter for a second time (again reflecting and transmitting half the beam), they will be exactly parallel at the observation branch. These parallel beams are the two interfering beams, if, of course, they satisfy the interference conditions (i.e., they are not orthogonally polarized, they have not exceeded their coherence length, etc.).

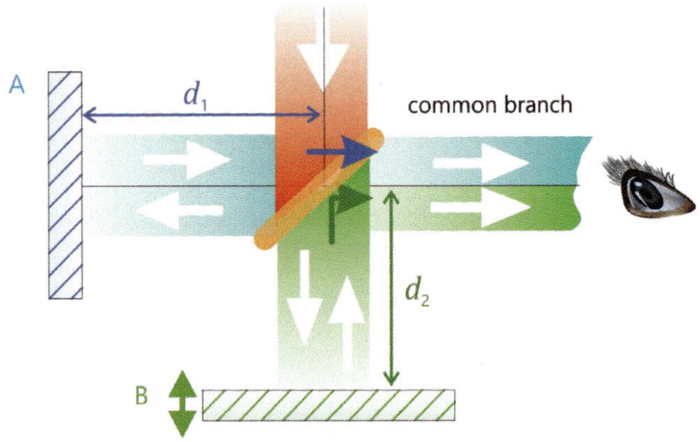


*Figure 4-52: Interfering beams propagate along a common branch. Mirror A is fixed, while mirror B can move in a mechanically controlled fashion.*

Now that we have identified the interfering beams, we seek a geometric factor that results in the development of an optical path difference between them. Referring to Figure 4-52, we see that the interfering beams propagate along a common branch. Mirror A is fixed, while mirror B can move in a mechanically controlled fashion.

From the point where the beams separate, beam A follows path $d_1$ twice, while beam B follows path $d_2$ twice. These distances are, in general, different. Thus, if we consider that the beams propagate in an optical medium with the same refractive index $n$ (this may not be always the case), the optical path difference between them is

Optical Path Difference (1 – 2): $$n \cdot 2d_1 - n \cdot 2d_2 = 2n \cdot (d_1 - d_2) \tag{4.98}$$

Now we apply the interference conditions. For constructive interference, the optical path difference is $m \cdot \lambda$, and for destructive interference, the optical path difference is $m \cdot \lambda + \frac{1}{2}\lambda$, where the integer $m = 0, \pm 1, \pm 2, \ldots$ is the interference order:

Constructive Interference: optical path $(1-2) = 2n\cdot(d_1 - d_2) = m\cdot\lambda$ (4.99)

Destructive Interference: optical path $(1-2) = 2n\cdot(d_1 - d_2) = (m + \tfrac{1}{2})\cdot\lambda$ (4.100)

If we displace mirror B from $d_2$ to $d'_2$, the optical path length changes by $2n\cdot(d'_2 - d_2)$. The interference condition changes from constructive to destructive, and the central bright lobe appears dark and then again bright. Let us assume that the conditions satisfy a bright order $m$. If we move mirror B $\tfrac{1}{4}\lambda$, the optical path length increases by $\tfrac{1}{2}\lambda$ (for $n = 1$) and a minimum appears instead. If we move the mirror an additional $\tfrac{1}{4}\lambda$, for a total of $\tfrac{1}{2}\lambda$, the optical path length increases by $\lambda$ and a new maximum order $(m + 1)$ appears. Accordingly, for displacement by $-\tfrac{1}{2}\lambda$, the new maximum is order $(m-1)$.

By simply counting the number of these fringe successions, we can measure the wavelength. If for branch length $d_2$ there is a condition of constructive interference (bright center fringe) of order $m$, and for length branch $d'_2$ (again) a bright fringe of order $m'$, then

$$2n\cdot(d_2 - d'_2) = (m - m')\cdot\lambda \tag{4.101}$$

Example ☞: In a Michelson interferometer, if the adjustable mirror 2 is displaced 0.05 mm, we observe 150 alternating fringes. What is the wavelength (assuming air $n = 1$)?
The known quantities are $(m - m') = 150$ and $(d - d') = 0.05$ mm. Then, $\lambda = 2\cdot(d - d')/(m - m') = 0.666$ μm.

Example ☞: How many alternating fringes can we observe in a Michelson interferometer using red light with $\lambda = 0.633$ μm if we move mirror 2 $(d - d') = 0.096$ mm?
The answer is $m - m'$. In air with $n = 1$, $(m - m') = 2\cdot(d - d')/\lambda = 2\cdot(0.096\text{ mm})/0.633\times10^{-3}\text{ mm} = 303$ fringes.

The integer $m$ (we never know $m$ precisely; what we measure is the difference $m - m'$) might not take any arbitrarily large value; we can alter the optical path length as much as permitted by altering the source coherence length. With a high-coherence laser source, it is possible to alter the optical path length over many meters. With a low-coherence source (even white light), interference fringes appear only for a much smaller difference in optical path length.

In Figure 4-53 (left), the central fringe corresponds to constructive interference, in which a bright maximum is observed; the right part of the figure shows destructive interference, in which a dark central minimum is observed. The central fringe is along the optical axis if we assume that the beam is thin, running along the common branch. The fringes are called longitudinal because we can see them by moving mirror B, thus increasing or decreasing the optical path length along the optical axis. Simultaneously, however, we have many concentric circular (transverse) fringes. We note a near similarity to thin-film fringes (Figure 4-28). How is this explained?

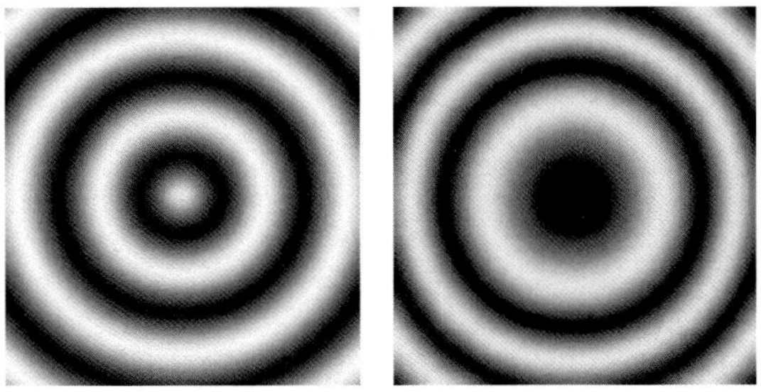

*Figure 4-53: Interference fringes in a Michelson interferometer: (left) center bright maximum and (right) center dark minimum.*

If the optical path length satisfies the condition of constructive interference, there should be only a bright maximum, and for destructive interference, only a dark minimum (a minimum-intensity fringe). But these conditions apply only along the optical axis. For a relative angle $\vartheta$ with respect to the optical axis, the optical path difference is expressed as

Optical Path Difference (1 – 2): angle $\vartheta = 2n\cdot(d_1 - d_2)\cdot\cos\vartheta$ (4.102)

In this case, the condition for constructive interference is

$$2n\cdot(d_1 - d_2)\cdot\cos\vartheta = m\cdot\lambda \;\Rightarrow\; m = 2(n/\lambda)\cdot(d_1 - d_2)\cdot\cos\vartheta = \text{integer} \qquad (4.103)$$

Here, $m$ corresponds to the fringe order that is used for a specific wavelength and a fixed mirror distance when the observation plane is perpendicular to the optical axis. There are maxima for angle $\vartheta$ with respect to the optical axis, for which the value of $\cos\vartheta$ renders an integer for the parameter $m$ in Eq. (4.103). This explains the presence of concentric circular maxima: Similar to the fringes from plates with parallel surfaces, these, too, are fringes of equal inclination. The slopes here are equal because the mirrors are perpendicular.

The fringes depend on the following parameters: the refractive index $n$ of the medium, the difference in length between the branches ($d_1 - d_2$), the wavelength $\lambda$, and the angle $\vartheta$ with respect to the optical axis. We set the transverse fringe order $m$ at the center ($\vartheta = 0$). The next bright transverse fringe (order $m$+1) forms at angle $\vartheta$ (with respect to the optical axis) such that $\cos\vartheta_{m+1} = \lambda/[2n(d_1 - d_2)]$; the next bright transverse fringe (of order $m$+2) forms at angle $\vartheta$ (again, with respect to the optical axis) such that $\cos\vartheta_{m+2} = 2\cdot\lambda/[2n(d_1 - d_2)]$, and so on.

Similar to thin-plate fringes, in this case, an equal increase in $\cos\vartheta$ between consecutive transverse fringes also occurs. However, the radii of the fringes do not have equal increments. Solving with respect to the angle $\vartheta$ that corresponds to the fringe and therefore also corresponds to its radius, we may find that the fringes gradually compress (their spacing becomes smaller) by increments that reduce nearly exponentially.

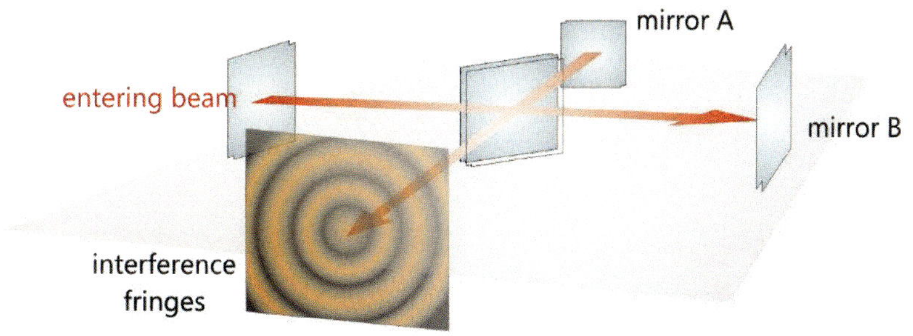


*Figure 4-54: Geometry of equal-inclination fringes in a Michelson interferometer.*

These radii depend on the refractive index $n$, the difference in length between the branches ($d_1 - d_2$), and the wavelength $\lambda$. If instead of red light [Figure 4-55 (left)] we use blue light [Figure 4-55 (right)], we note that the fringes compress; in other words, their spacing becomes smaller, and we can measure this difference using the shorter radii of the corresponding dark circles.

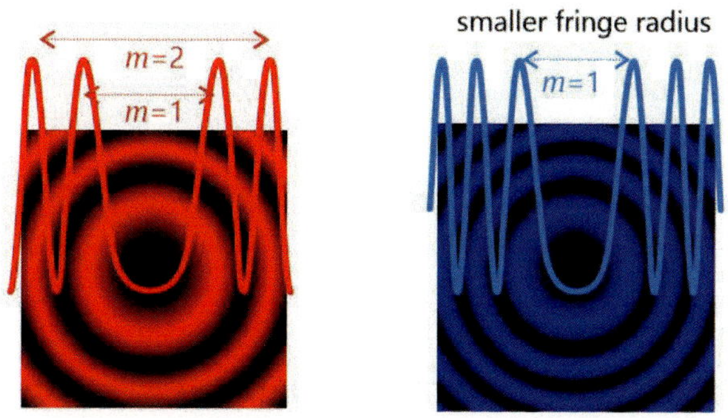


*Figure 4-55: Fringe thickness dependence on wavelength: (left) red wavelength and (right) blue wavelength with a smaller spacing.*

If mirrors A and B are exactly perpendicular, the corresponding beams along their common branch are parallel. If mirrors A and B are not exactly perpendicular, the beams diverge.

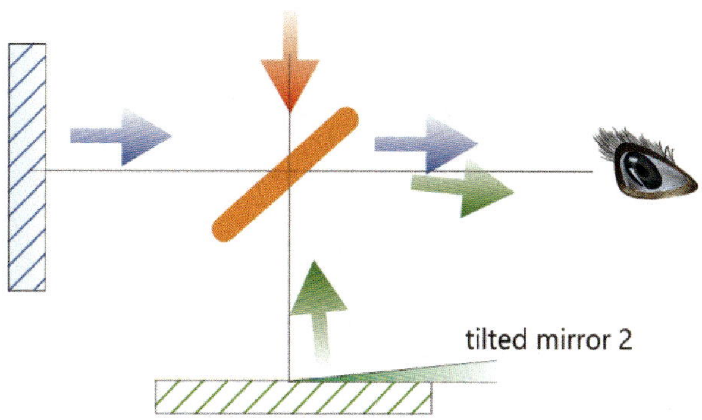


*Figure 4-56: Mirror tilt in a Michelson interferometer.*

This effect is equivalent to an effect observed with a wedge-shaped plate. The fringes are not symmetric with respect to the center, but display behavior similar to that observed with a wedge, like parallel Fizeau fringes whose spacing is determined by the inverse of the relative inclination of the mirrors.

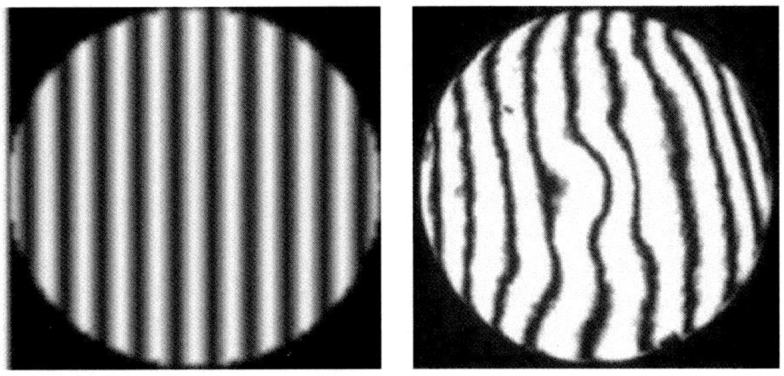

*Figure 4-57: Interference fringes in a Michelson interferometer with mirrors at a tilt: (left) mirrors with absolute flatness and (right) mirrors with local imperfections.*

With his interferometer, Michelson measured the length of the then-prototype measure of length, the platinum meter bar, at the International Bureau of Weights and Measures (Bureau International des Poids et Mesures) in Sèvres, outside of Paris, France. He compared the length of an interference branch, in wavelengths of cadmium red (0.6438 μm), with the length of the prototype meter bar.

This resulted in a fundamental change to the prototype: The magnitude of the model meter (the unit of length measurement) was redefined. The model meter was thereafter defined as a multiple of a wavelength of a properly selected monochromatic radiation. Indeed, until 1983, the unit of length was defined as a multiple (1,650,763.73) of the wavelength of the orange-yellow krypton-86 line (0.60578 μm). This emission line, which corresponds to the transition $2p_{10} \rightarrow 5d_5$ (electronic transitions are described in § 6.1.1), offers one of the most stable wavelengths ($\delta\lambda \ll 1$). After 1983, the model meter was defined as the distance light travels in vacuum during a time equal to 1/299,792,458 of a second.

Albert Abraham Michelson was the recipient of the 1907 Nobel Prize in Physics.

# 4.4 INTERFERENCE QUIZ

1) Interference is purely a wave phenomenon because the added quantities are ...
   a) vectors that carry both magnitude and phase
   b) scalar quantities such as energy
   c) field strength magnitudes
   d) photon counts

2) The principle of linear superposition involves the addition of the (select two) ...
   a) electric field vector
   b) magnetic field vector
   c) electric field amplitude
   d) magnetic field amplitude

3) Interference of light provides evidence that ...
   a) the speed of light is too fast to be measured
   b) light is composed of transverse electric fields
   c) light has a wave nature
   d) conservation of energy is not applicable in light

4) In order for the superposition of two field vectors to result in complete attenuation, the two fields must have (select two) ...
   a) identical amplitudes
   b) opposite amplitudes
   c) the same phase
   d) opposite phases

5) Wave A has amplitude magnitude of $E_{oA} = 2 \cdot E_{oB}$, where $E_{oB}$ is the amplitude magnitude of wave B. Wave B has a phase that is opposite to the phase of wave A. The linear superposition of these two waves results in a wave whose amplitude magnitude equals ...
   a) $E_{oA}$
   b) $E_{oB}$
   c) $2 \cdot E_{oA}$
   d) $2 \cdot E_{oB}$
   e) $3 \cdot E_{oA}$
   f) $3 \cdot E_{oB}$

6) A twist to Q 5. Now the two waves have identical phases. The linear superposition of these two waves results in a wave whose amplitude magnitude is ...
   a) $E_{oA}$
   b) $E_{oB}$
   c) $2 \cdot E_{oA}$
   d) $2 \cdot E_{oB}$
   e) $3 \cdot E_{oA}$
   f) $3 \cdot E_{oB}$

7) Back to Q 6. The recorded intensity resulting from the superposition of the two identical-phase waves A and B is ...
   a) $2 \cdot E_{oA}^2$
   b) $2 \cdot E_{oB}^2$
   c) $4 \cdot E_{oA}^2$
   d) $4 \cdot E_{oB}^2$
   e) $9 \cdot E_{oA}^2$
   f) $9 \cdot E_{oB}^2$

8) Coherence between two waves requires all of the following conditions, <u>except one. Which one?</u>
   a) equal amplitudes
   b) the same frequencies
   c) identical wavelengths
   d) a fixed phase difference

9) Two waves, 1 and 2, intersect at a common point. Their field profiles at this point are as shown:

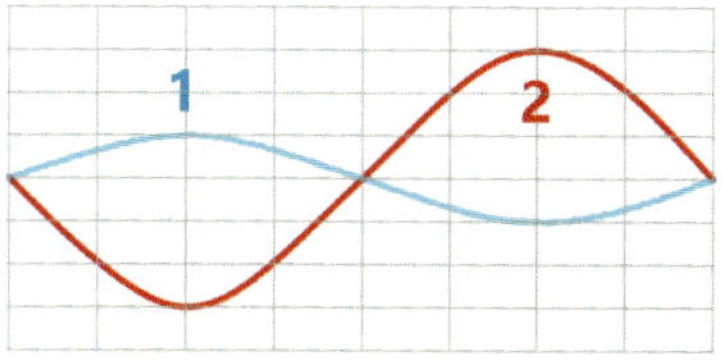


In which of the following graphs does the black line best describe the combined superposition field?

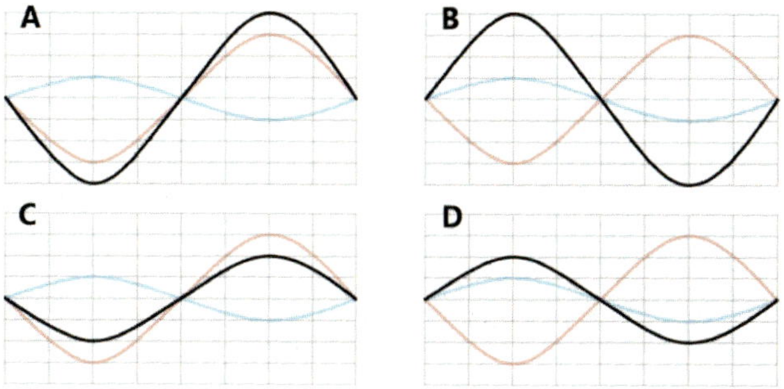


   a) A  b) B  c) C  d) D

10) Two beams, A and B, originating from two distinct, otherwise identical, HeNe laser sources operating at 633 nm are directed at a common point O. What is the expected interference?

a) constructive interference due to their common wavelength
b) destructive interference due to their two distinct origins
c) no interference due to their two distinct origins
d) no interference due to their different wavelengths

11) Back to Q 10. To achieve interference at point O:

a) use just one laser source, then employ amplitude division to produce two coherent beams
b) use just one laser source directed at the desired point of interference
c) use an additional laser source directed at point O
d) use both sources, ensuring that the pathways of the two laser sources are collimated
e) use two crossed linear polarizers, each placed along the pathway of each laser

12) Two equal-amplitude magnitude ($E_o$) coherent waves (wavelength $\lambda$) are interfering at point O. Wave A has traveled an optical path length equal to $502 \cdot \lambda$, while wave B has traveled an optical path equal to $505 \cdot \lambda$. At point O, the two waves produce ...

a) constructive interference of amp. mag. $2 \cdot E_o$
b) constructive interference of amp. mag. $4 \cdot E_o$
c) destructive interference of amp. mag. $1 \cdot E_o$
d) destructive interference of amp. mag. $2 \cdot E_o$

13) Back to Q 12. What is the phase difference between waves A and B at point O?

a) $2\pi$
b) $3\pi$
c) $4\pi$
d) $6\pi$

14) What optical path difference is required such that a phase difference equal to $4\pi$ develops?

a) $4\pi$
b) $4\lambda$
c) $2\lambda$
d) $\pi\lambda$

15) Two perfectly coherent beams A and B intersect at point O. Which of the following path lengths can produce constructive interference (two correct)?

a) beam A: $5/2\ \lambda$; beam B: $6/2\ \lambda$
b) beam A: $5/2\ \lambda$; beam B: $7/2\ \lambda$
c) beam A: $5/2\ \lambda$; beam B: $8/2\ \lambda$
d) beam A: $5/2\ \lambda$; beam B: $9/2\ \lambda$

16) Back to Q 15. Despite all checks pointing to perfect coherence, i.e., temporal, spatial, and initial phase, the beams do not produce interference fringes; instead, a constant field intensity corresponding to the simple sum of the two individual intensities is observed. Possible explanations are that:

a) The universe is conspiring against interference.
b) The two beams are linearly polarized at perpendicular axes.
c) There is a half-wave plate in front of one of the two beams.
d) An opaque plate is blocking one of the two beams.

17) Back to Q 15. Dawf is messing with the setup. He places a half-wave plate in front of beam A. What does he effectively achieve?

a) an addition of $\lambda/2$ in the pathway of beam A
b) an addition of $\lambda$ in the pathway of beam A
c) destruction of the coherence of beam A
d) destruction of the coherence of beam B

18) Back to Q 15. Two perfectly coherent beams A and B intersect at point O. Which of the following path length differences can produce destructive interference?

a) zero
b) an odd integer of half-wavelengths
c) an integer of wavelengths
d) an integer of half-wavelengths

19) Coherent microwave radiation ($\lambda$ = 60 mm) passes through two narrow slits $S_1$ and $S_2$. A microwave detector at D scans for maximum and minimum intensities across the slits. At which value of the path difference $DS_1 - DS_2$ is a maximum intensity detected?

a) 10 mm
b) 20 mm
c) 40 mm
d) 60 mm
e) 90 mm

Questions Q 20 to Q 34 discuss a type of Young's experiment, in which a single coherent source ($\lambda$) produces two mutually coherent illuminating points at two apertures situated at $S_1$ and $S_2$, spaced by a distance $d_1$. This setup produces interference fringes at an observation screen placed a distance $z_1$ away. The peak fringe intensity is 4 $I_o$, where $I_o$ is the intensity of a single wave originating from either $S_1$ or $S_2$.

20) At the observation screen, the center of a (or any) bright fringe occurs when the optical path lengths from the two apertures differ by (two correct):

a) $\lambda/4$
b) $\lambda/2$
c) $3\lambda/4$
d) $\lambda$
e) $2\lambda$

21) Back to Q 20. The centers of two successive bright fringes correspond to a difference in the optical path lengths from the two apertures by ...

a) $\lambda/4$
b) $\lambda/2$
c) $3\lambda/4$
d) $\lambda$
e) $2\lambda$

22) Dr. Seuss places two linear polarizers in front of two apertures $S_1$ and $S_2$, with their polarization axes perpendicular. What happens to the fringes?

a) They shift by a wavelength ($\lambda$).
b) They shift by a fringe spacing ($\lambda \cdot z/d$).
c) They shift by an amount equal to half a fringe spacing ($½\ \lambda \cdot z/d$).
d) They disappear, producing a uniform light distribution.
e) Their spacing increases by a factor of 2.
f) Their spacing decreases by a factor of 2.
g) Their intensity (brightest to darkest) increases by a factor of 4.

23) Mr. Brown places a half-wave plate in front of $S_1$. What happens to the interference fringe pattern?

a) It shifts by an amount equal to one wavelength.
b) It shifts by an amount equal to one fringe spacing ($\lambda \cdot z/d$).
c) It shifts by an amount equal to half a fringe spacing ($½\ \lambda \cdot z/d$).
d) It disappears, producing a uniform light distribution.
e) Its spacing increases by a factor of 2.
f) Its spacing decreases by a factor of 2.
Its intensity contrast (brightest to darkest) increases by a factor of 4.

24) Let me ask Q 23 again. What is the fringe intensity just across the bisector of the two apertures situated at $S_1$ and $S_2$?

a) maximum (bright as before)
b) minimum (dark)
c) one-half of the brightness
d) one-quarter of the brightness

25) A Wocket is messing with the aperture distance, increasing it by a factor of two ($d_2 = 2 \cdot d_1$). As the two apertures are spaced farther apart, the fringes become ...

a) darker (maximum intensity halves)
b) brighter (maximum intensity doubles)
c) more widely spaced (fringe spacing doubles)
d) more closely spaced (fringe spacing halves)

26) Back to Q 25. What is the fringe intensity just across the bisector of the two apertures situated at $S_1$ and $S_2$?

a) maximum (bright as before)
b) minimum (dark)
c) one-half of the brightness
d) one-quarter of the brightness

27) Back to Q 25. What can compensate for the halved spacing induced by widening the aperture separation (two correct answers)?

a) doubling the observation distance $z$
b) halving the observation distance $z$
c) doubling the wavelength $\lambda$
d) halving the wavelength $\lambda$

28) Yertle places an attenuator in front of $S_1$, resulting in halving its beam intensity (originally $I_o$, now $½ I_o$). The effect of this action is that the fringe peak intensity ...

a) increases to 9 $I_o$
b) remains the same (4 $I_o$)
c) decreases to 9/4 $I_o$
d) decreases to 9/16 $I_o$

29) Back to Q 28. The effect of this action is that the minimum (dark) intensity ...

a) remains zero (complete dark)
b) increases to ¼ $I_o$
c) increases to ½ $I_o$
d) increases to $I_o$

30) Back to Q 28. What is the effect of this action on the fringe spacing and location?

a) The fringe pattern shifts by an amount equal to one wavelength ($\lambda$).
b) The fringe pattern shifts by an amount equal to one fringe spacing ($\lambda \cdot z/d$).
c) There is no shift and no widening (spacing).
d) The fringe spacing increases by a factor of 1½.
e) The fringe spacing decreases by a factor of 1½.

31) Sylvester secretly manages to switch the source; the new wavelength is half of the original wavelength ($\lambda_2 = ½\cdot\lambda_1$). Now, the fringes become ...
   a) darker (maximum intensity halves)
   b) brighter (maximum intensity doubles)
   c) spaced farther apart (fringe spacing doubles)
   d) more closely spaced (fringe spacing halves)

32) Back to Q 31. What is the fringe intensity just across the bisector of the two apertures situated at $S_1$ and $S_2$?
   a) maximum (bright as before)
   b) minimum (dark)
   c) half brightness
   d) one-quarter of the brightness

33) What is the fringe spacing produced on a screen at $z$ = 2 m, if $d$ = 2 mm and $\lambda$ = 0.5 µm?
   a) 0.38 mm
   b) 0.50 mm
   c) 0.66 mm
   d) 1.00 mm
   e) 2.00 mm

34) A twist to Q 33. Blue Fish fills the space between the two apertures and the screen with water ($n$ = 1.33). What is new the fringe spacing?
   a) 0.38 mm
   b) 0.50 mm
   c) 0.66 mm
   d) 1.00 mm
   e) 2.00 mm

Questions Q 35 to Q 42 assume normal incidence on a thin, parallel-surface, transparent glass plate (thickness $d$, refractive index $n$), surrounded by air. The two beams interfering by reflection are 1 (partially reflected off the first surface) and 2 [refracted on the first surface, reflected off the second surface, and refracted by first surface (Figure 4-24)]. The two beams interfering by refraction are 1′ and 2′ (Figure 4-27).

35) If the plate thickness $d$ equals two wavelengths ($d = 2\cdot\lambda$) and the refractive index is $n$ = 1.5, what is the optical path difference between beams 1 and 2?
   a) $3\cdot\lambda$
   b) $4\cdot\lambda$
   c) $5\cdot\lambda$
   d) $6\cdot\lambda$

36) Back to Q 35. What is the value of the phase difference developed between beams 1 and 2?
   a) $11\cdot\lambda$
   b) $11\cdot\pi$
   c) $12\cdot\lambda$
   d) $12\cdot\pi$
   e) $13\cdot\lambda$
   f) $13\cdot\pi$

37) Back to Q 35. What is the interference center at the reflection side (topside) of this glass plate?
   a) constructive interference, bright center
   b) constructive interference, dark center
   c) destructive interference, bright center
   d) destructive interference, dark center

38) Back to Q 35. What is the interference center at the refraction side (underside) of this glass plate?
   a) constructive interference, bright center
   b) constructive interference, dark center
   c) destructive interference, bright center
   d) destructive interference, dark center

39) A phase shift by an amount of $+\pi$ occurs in which of the following cases ($n_{air}$ = 1.0, $n_{water}$ = 1.33, $n_{glass}$ = 1.5) (three correct answers)?
   a) reflection off an air–glass interface
   b) reflection off a glass–air interface
   c) reflection off a water–air interface
   d) reflection off an air–water interface
   e) reflection off a water-glass interface
   f) reflection off a glass–water interface

40) What is the (integer) multiple $\pi$ phase difference (how many $\pi$'s are there?) between beams 2 and 1, and what is the appearance of the central fringe in the reflected (topside) of this glass plate ($n$ = 1.5, $d$ = 0.4 mm) when the illumination is coherent light ($\lambda$ = 0.5 µm)?
   a) phase difference = $47\pi$, dark center
   b) phase difference = $47\pi$, bright center
   c) phase difference = $48\pi$, dark center
   d) phase difference = $48\pi$, bright center
   e) phase difference = $49\pi$, dark center
   f) phase difference = $49\pi$, bright center

41) Back to Q 40. What is the phase difference between transmitted beams 2′ and 1′ as they emerge from the transmitted (underside) of this glass plate?
   a) phase difference = $47\pi$, dark center
   b) phase difference = $47\pi$, bright center
   c) phase difference = $48\pi$, dark center
   d) phase difference = $48\pi$, bright center
   e) phase difference = $49\pi$, dark center
   f) phase difference = $49\pi$, bright center

42) Back to Q 40. The thickness of this plate is just 250 nm. What is the (integer) multiple $\pi$ phase difference between beams 2 and 1, and what is the appearance of the central fringe in the reflected (topside) side of this thin plate?

a) phase difference = $1\pi$, dark center
b) phase difference = $1\pi$, bright center
c) phase difference = $2\pi$, dark center
d) phase difference = $2\pi$, bright center
e) phase difference = $3\pi$, dark center
f) phase difference = $3\pi$, bright center

43) Whale Number One spills a thin layer of water ($n$ = 1.333) on top of a glass plate such that the first reflection is between air and water, while the second reflection is between water and glass. Still using the same illumination, which beams are subject to what phase jump (two correct answers)?

a) beam 1: $+\pi$ phase jump at the air–water interface
b) beam 1: 0 phase jump at the air–water interface
c) beam 2: $+\pi$ phase jump at the water–glass interface
d) beam 2: 0 phase jump at the water–glass interface

44) Back to Q 43. What is the (integer) multiple $\pi$ phase difference (how many $\pi$'s are there?) between beams 2 and 1, and what is the appearance of the central fringe in the reflected topside (given that $d_{\text{water layer}}$ = 188 nm, $n$ = 1.333, $\lambda$ = 500 nm)?

a) phase difference 2 – 1 = $0\pi$, bright center
b) phase difference 2 – 1 = $1\pi$, dark center
c) phase difference 2 – 1 = $2\pi$, bright center
d) phase difference 2 – 1 = $2\pi$, dark center
e) phase difference 2 – 1 = $3\pi$, bright center
f) phase difference 2 – 1 = $3\pi$, dark center

45) Sally is making soap bubbles using a glycerine solution ($n$ = 1.4765). If she is using a red 650 nm light, what is the thinnest soap bubble layer that produces a strong central (bright) interference maximum?

a) 55 nm
b) 110 nm
c) 165 nm
d) 220 nm

46) An antireflection coating of refractive index $n$ = 1.288 is ideal for application on optical glass of which refractive index?

a) $n$ = 1.134
b) $n$ = 1.288
c) $n$ = 1.659
d) $n$ = 2.136

47) An anti-reflection coating ($n$ = 1.288) is designed to eliminate reflection from optical glass when illuminated with red light ($\lambda$ = 649 nm). What should be the minimum thickness of this coating?

a) 126 nm
b) 189 nm
c) 252 nm
d) 378 nm

48) In multiple-beam interference, an increased surface reflectivity leads to an increased (two correct) ...

a) coefficient of reflecting finesse
b) fringe sharpness
c) fringe width
d) layer thickness to produce interference

49) Regular optical index glass ($n$ = 1.5) has reflectivity $R$ = 0.04 (= 4%). What is the coefficient of finesse for multiple-beam interference?

a) 0.04
b) 0.17
c) 0.40
d) 0.96

50) Bare silver has reflectivity $R$ = 0.97 (= 97%) for most of the visible range. What is the coefficient of finesse for multiple-beam interference?

a) 0.97
b) 3.88
c) 431
d) 4311

51) The following Newton ring fringe patterns differ in the radius of curvature of the convex lens. Select the correct descriptions (two correct answers):

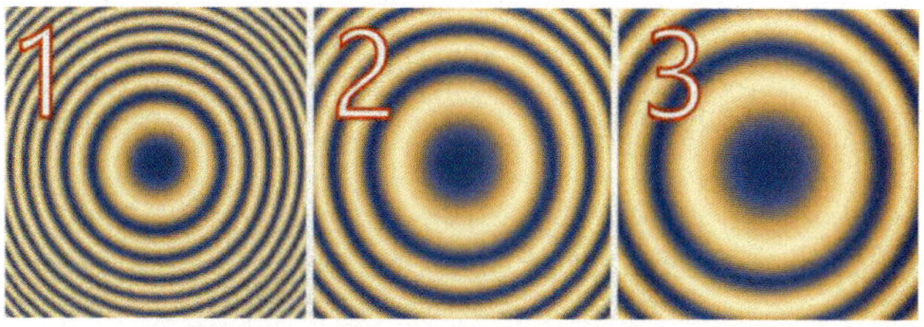


a) smallest to largest radius: 1 – 2 – 3
b) smallest to largest radius: 3 – 2 – 1
c) fringe pattern 3 has about 10× the radius of lens curvature of fringe pattern 1
d) fringe pattern 3 has about 3× the radius of lens curvature of fringe pattern 1
e) fringe pattern 1 has about 10× the radius of lens curvature of fringe pattern 2

52) In this thin-film interference pattern created by white light on a soap bubble, colors appear in a certain order. Select the three correct statements describing this appearance.

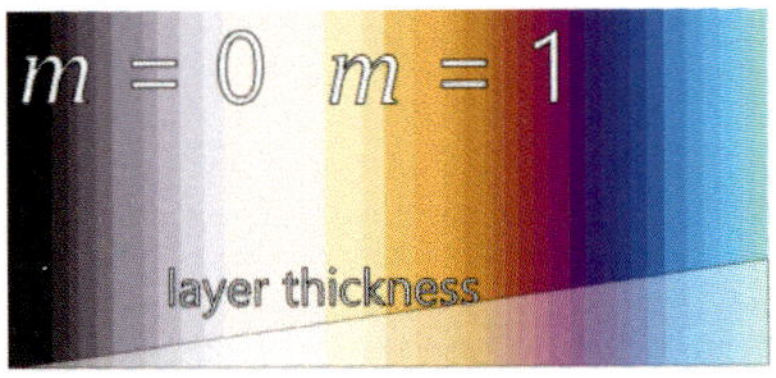


a) Dark appears at zero layer thickness, regardless of the color/wavelength (dark band, $m = 0$).
b) As the layer thickness increases to about $\lambda/4n$, most colors produce constructive interference ($m = 0$ white band to the right of dark band).
c) The red color is seen ($m = 1$) when $d = 2\lambda_{red}/4n$.
d) The red color is seen ($m = 1$) when $d = 3\lambda_{red}/4n$.
e) It is not possible for both the red and the green to produce constructive interference at the same point (layer thickness).

53) A Michelson interferometer operates using HeNe red light ($\lambda = 633$ nm). If the moving mirror is shifted by 30 μm, how many alterations of bright fringes can be counted on the observation screen?

a) 38
b) 95
c) 105
d) 190
e) 380

54) A twist to Q 53. Green Fish pours water ($n = 1.33$) on the optics table. In order to produce 105 alterations of bright fringes, the movable mirror is shifted by ...

a) 25 μm
b) 30 μm
c) 33.3 μm
d) 44.3 μm

55) Mr. Sneelock is preparing a Michelson setup using two colors, a deep blue ($\lambda_B = 400$ nm) and a red ($\lambda_R = 640$ nm). Using the blue light, he moves the moving mirror so that he counts exactly 200 alterations of bright blue fringes. Being careless, he forgets to mark the distance that the moving mirror has to shift. Can you help him?

a) yes
b) maybe
c) no
d) 0.08 m
e) 0.04 mm

56) If Mr. Sneelock had been using the red beam instead, then, over the same distance that the moving mirror was moved, he would have observed how many alterations of bright red fringes?

a) 125
b) 200
c) 250
d) 320
e) 640

57) The interference center in a Michelson interferometer with perfectly parallel mirrors shows a dark center. By how much should one move the moving mirror in order for the center to become bright (minimum change)?

a) $\lambda/8$
b) $\lambda/4$
c) $\lambda/2$
d) $1\,\lambda$
e) $1.5\,\lambda$
f) $2\,\lambda$

58) Marco very, very carefully tweaks the mirror by just 0.16 μm. He joyfully observes that the center brightness in the Michelson interferometer turns black. What is its operating wavelength?

a) 320 nm
b) 480 nm
c) 640 nm
d) 960 nm

59) The Grinch places two linear polarizers along the two branches of a Michelson interferometer, with their polarization axes perpendicular. What happens to the fringes?

a) The Ginch steals them (it's Xmas)!
b) They disappear. The intensity is constantly bright (the former maximum).
c) They disappear. The intensity is constantly bright (about ¼ of the former maximum).
d) They disappear. The intensity is constantly dark (zero).

60) A tilted, yet perfectly flat, mirror in a Michelson interferometer produces a pattern of interference fringes that can be compared to (select two) ...

a) wedged-plate multiple-beam interference
b) Newton's rings
c) an antireflection coating
d) Young's two-slit interference

# 4.5 INTERFERENCE SUMMARY

Interference is a manifestation of the wave nature of light that requires the presence of two coherent sources or paths. It appears as light intensity variations due to electric field phase differences. The phase is the internal clock of the disturbance and determines the wave's momentary magnitude, repeated circularly in time. A phase is expressed by the frequency in the time domain and by the wavevector in space. In harmonic waves, the phase is the argument in the trigonometric function that describes the disturbance.

In order to produce interference by two coherent waves, we use either wavefront division, which produces two coherent sources (Young's two-slit), or amplitude division (thin-film interferometry or Michelson interferometry), which produces two coherent paths.

The physical quantity that is directly detectable in optics is the intensity, which is the time average of the square of the electric field amplitude magnitude. To calculate the intensity in interference, we first add the two interfering individual electric field amplitude vectors (the principle of linear superposition) and then calculate the square of the resultant field amplitude magnitude. Therefore, the intensity can be zero (in destructive interference) or up to four times the initial intensity (in constructive interference).

## Interference Conditions

Expressed in optical path difference,

- the condition for a maximum is $(m)\cdot\lambda$
- the condition for a minimum is $(m+½)\cdot\lambda$, where the integer $m = 0, \pm1, \pm2, \ldots$ is the fringe order.

We can convert the optical path difference to the phase difference using the simple relationship,

$$\frac{\text{phase difference}}{2\pi} = \frac{\text{optical path difference}}{\lambda}$$

Expressed in phase difference,

- the condition for the maximum is $(m)\cdot 2\pi$
- the condition for the minimum is $(m+½)\cdot 2\pi$

To solve an interference problem, we identify the interfering beams and the geometrical factor that causes the optical path difference, which converts to a phase difference.

## Young's Experiment

In two-slit interference, the optical path difference is caused by the different lengths traveled from the two sources (slits) to the observation screen. Exactly across the bisector, the difference is zero, so we observe a maximum (constructive interference). If the screen is placed at a distance $z$ and the sources are separated by $d$, then the optical path difference is $xd/z$. Using the optical path difference,

- the condition for the maximum is $xd/z = (m)\cdot\lambda$. Thus, maxima appear at $\Delta x = m\,\lambda z/d$.

## Thin-Film Interference

In thin-film interference, the optical path difference is caused by the additional optical lengths traveled by the second beam inside the optical plate of thickness $d$ and refractive index $n$. For normal incidence ($\cos\delta = 1$), the optical path difference is $2nd$. The peculiarity with the thin film is that often one of the two reflected beams has an additional phase shift by $+\pi$ (when reflected off a more optically dense medium). Thus, the phase difference between the two interfering reflected beams is $2nd\cdot 2\pi/\lambda - \pi$. Using phase,

- the condition for the maximum is $2nd\cdot 2\pi/\lambda - \pi = (m)\cdot 2\pi$. Maxima appear at $4nd/\lambda = 2m + 1$.

There is no phase shift in the transmitted (refracted) beam interference beneath the plate. Therefore, the phase difference between the two interfering transmitted beams is $2nd\cdot 2\pi/\lambda$. Using phase,

- the condition for maximum is $2nd\cdot 2\pi/\lambda = (m)\cdot 2\pi$. Maxima appear at $4nd/\lambda = 2m$.

## Antireflection Coatings

Antireflection coatings (films) use destructive interference in the reflected beams. The simplest case involves a minimum layer thickness of a quarter of a wavelength of light in that medium. Thus, the wave reflected from the coating–substrate interface (double path) is in a phase opposite to that of the incident wave, and their interference is destructive.

The ideal refractive index $n_f$ of the coating (film) should be $n_f = \sqrt{(n_1 \cdot n_L)}$, where $n_1$ is the refractive index of the initial medium (typically, it is air), and $n_L$ is the refractive index of the lens material. For minimum-order destructive interference ($m = 0$), we obtain a minimum coating thickness of $d = \lambda/4n_1$.

## Michelson Interferometry

In Michelson interference, the optical path difference is caused by the difference in optical path length traveled by the two beams in the two branches, which is $2n \cdot (d_1 - d_2)$. Using the optical path difference,

- the condition for the maximum is $2n \cdot (d_1 - d_2) = m \cdot \lambda$
- maxima appear at $(d_1 - d_2) = m \cdot \lambda / 2n$

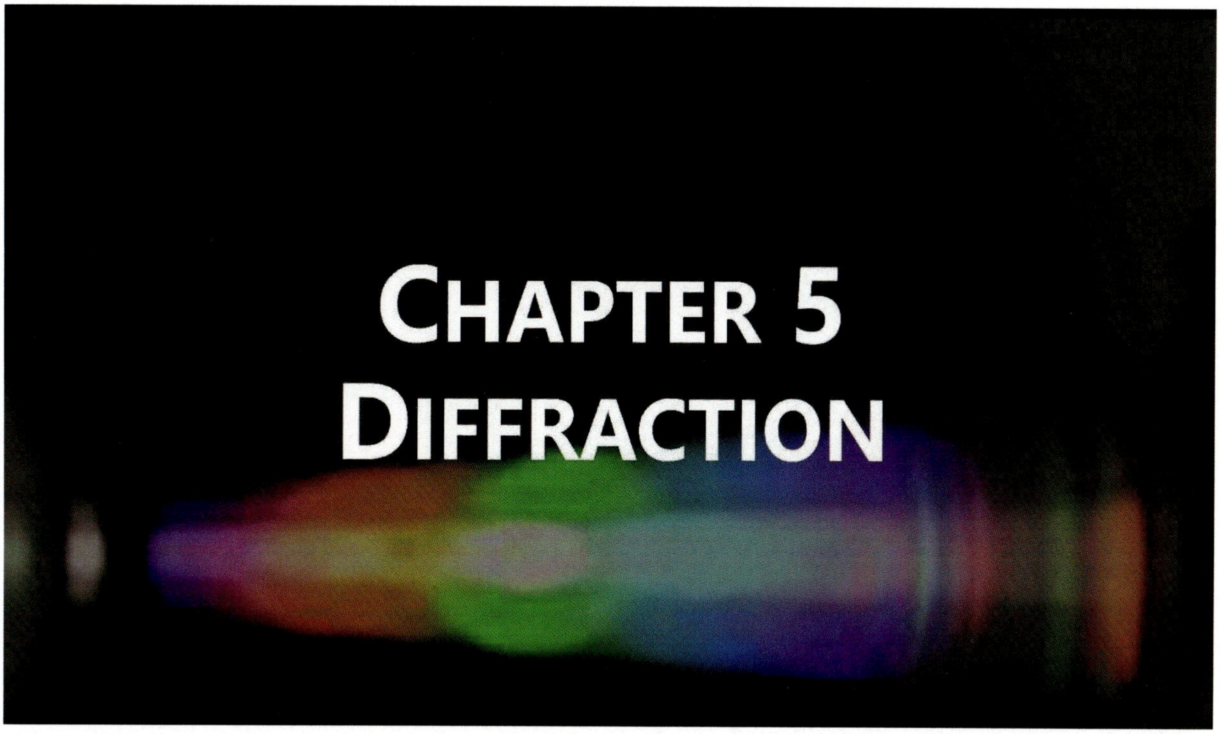

# CHAPTER 5 DIFFRACTION

Geometrical optics deals with light propagation in optical media. Reflection and refraction describe effects resulting from a change in the light propagation speed. This change may occur instantly on an interface, as well as gradually within an inhomogeneous medium in which light is subject to an incremental change in speed (such as within the atmosphere). In effect, geometrical optics ignores certain aspects of the wave nature of light, considering how small the wavelength is: When a propagating beam of light encounters an aperture or obstacle with no other change in the optical medium, it simply continues on its rectilinear propagation through or around this, casting a geometrical shadow of the aperture. The only condition is that the obstacle or aperture is much larger than the wavelength.

Diffraction is noted when the obstacle or aperture is comparable in size to the wavelength, so the interaction of the light wave with the aperture plays an important role. On an observation screen, it is no longer the geometrical shadow that is formed, but the diffraction distribution of luminous intensity resulting from the interaction between the aperture/obstacle and the wave. The first known diffraction observations are attributed to Francesco Maria Grimaldi in his book *Physicomathesis de Lumine, Coloribus, et Iride* (1665):

*...Lumen propagatur seu diffunditur non solum Directe, Refracte, ac Reflexe sed etiam alio quodam quarto modo, Diffracte...*

*[...Light is propagated or diffused not only directly, or refracted, and reflected in the fourth, but even in some other way, a kind of, this is done...]*

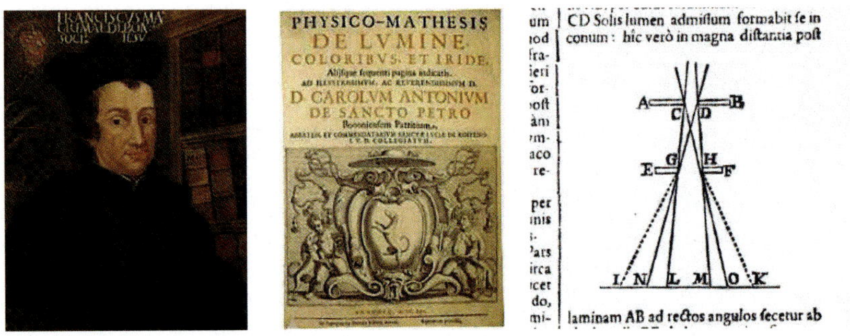


*Figure 5-1: (left) Portrait of Francesco Maria Grimaldi (1618–1663), (center) cover of his book* Physicomathesis de Lumine, Coloribus, et Iride, *and (right) his notes on diffraction.*

According to Huygens, when light is incident on a very small obstacle or aperture that can be approximated by a point, this point becomes a secondary source of spherical waves that are emitted in all directions. This is the most elementary manifestation of diffraction: Light changes its path when passing through an aperture of very small size. This change in direction cannot be described by the laws of geometrical optics.

In this chapter, we investigate the effects in which an aperture, or obstacle, has dimensions comparable with the wavelength of light, which is between 0.3 μm and 0.7 μm. We will implement the Huygens–Fresnel principle, just as we did in the discussion of interference, because, as we will see, diffraction is nothing more than interference by an infinite number of neighboring coherent point sources.

## 5.1 THE GENERALIZED DIFFRACTION PROBLEM

Consider an opaque screen with an aperture that is large compared to the wavelength. Geometrical optics predicts rectilinear propagation past this aperture, independent of the wavelength that leads to the formation (on the observation screen) of a geometrical shadow that has the shape of the aperture shown in Figure 5-2.

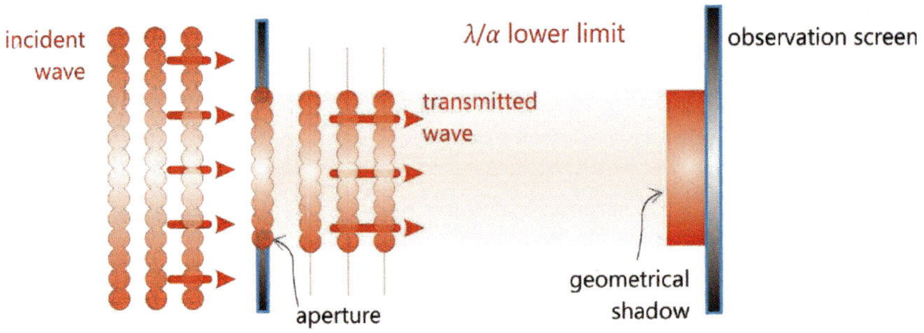


*Figure 5-2: Propagation of a plane wavefront through a large aperture (the $\lambda/\alpha$ lower limit).*

In this configuration, the light will either pass entirely and the transmitted wave will continue with no other change in its path, or the light will not pass at all. The same occurs if we illuminate an obstacle. The shadow has the same shape as the obstacle because it is simply its geometrical projection.

If the aperture is nearly a point, it becomes a source of secondary wave emissions in all directions, according to the Huygens–Fresnel principle (Figure 5-3). In a homogeneous and isotropic medium, the wavefronts become spherical.

The case illustrated in Figure 5-2 is the lower limit of the $\lambda/\alpha$ ratio, while the case illustrated in Figure 5-3 is the upper limit of the $\lambda/\alpha$ ratio. Here $\alpha$ is a representative dimension at the entrance; for a slit, $\alpha$ is its width, and for a circular aperture, $\alpha$ is its diameter. The ratio $\lambda/\alpha$ is a unitless angular quantity expressed in radians.

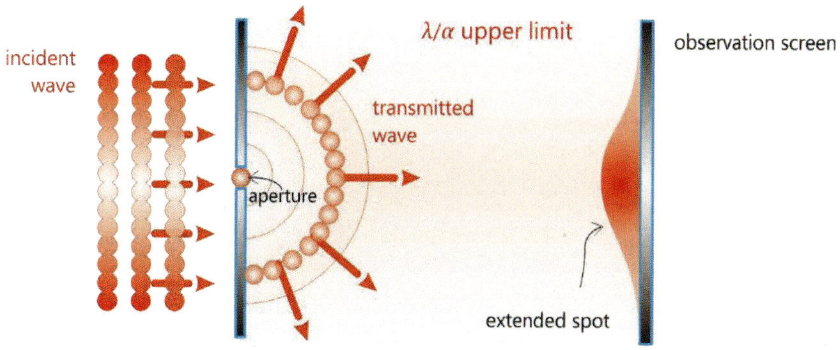


*Figure 5-3: Propagation of a plane wavefront through a small aperture (the $\lambda/\alpha$ upper limit).*

A **diffractive aperture** is defined as a discontinuity in an opaque medium that extends into a small area relative to the order of magnitude of the wavelength $\lambda$. A **diffractive obstacle** is defined as a piece of an opaque medium that extends into a relatively small area (compared to the wavelength) along a cross-section of the incident light.

Diffraction is evident if a wave encounters a diffractive aperture or obstacle. For the visible ($\lambda$ from 0.4 to 0.7 μm), we expect diffraction effects from obstacles or apertures of a few microns up to 500 μm. A diffractive aperture can alter the amplitude of the incident wave if it is an obstacle, a clear aperture, or a partial masking device. The aperture can also alter the phase of the incident wave if the refractive index / optical path varies across its cross-section.

The basic hypothesis employed is that, regardless of its shape, the obstacle or aperture renders all of its point sources of elementary Huygens waves emitted in all directions with the same characteristics as the incident wave. On the observation screen, these waves (specifically, their electric fields) add up as vectors according to the Huygens–Fresnel principle.

Does this sound like interference? Yes, because it is, indeed, interference. In interference, there is interaction between two point sources. In diffraction, there is interaction among an infinite number of point sources that span the entire area of the aperture. The observed diffraction intensity distribution depends on the characteristics of the incident radiation (wavelength, coherence, wavefront form), as well as on the specifics of the obstacle/aperture entrance. All of these contributions define the entrance field.

In the very simple case of an ideal point source, there is emission of spherical waves in all directions. Because of this, we will not observe a bright point, nor will we observe the aperture's geometrical shadow on the observation screen; instead, we will observe an extended distribution of intensity that attenuates symmetrically away from the optical axis (Figure 5-3).

The point aperture constitutes an elementary diffraction problem. When identifying the mathematical expression of the intensity distribution from a nonpoint, an extended aperture can be challenging. To further add to the complexity, the entrance field can no longer be assumed to be a homogeneous, plane wavefront.

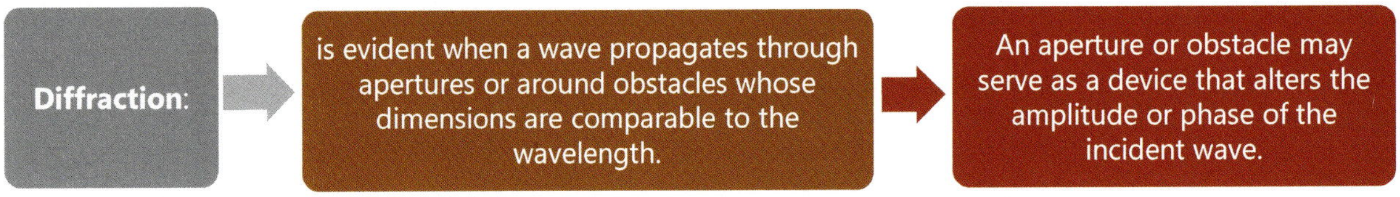


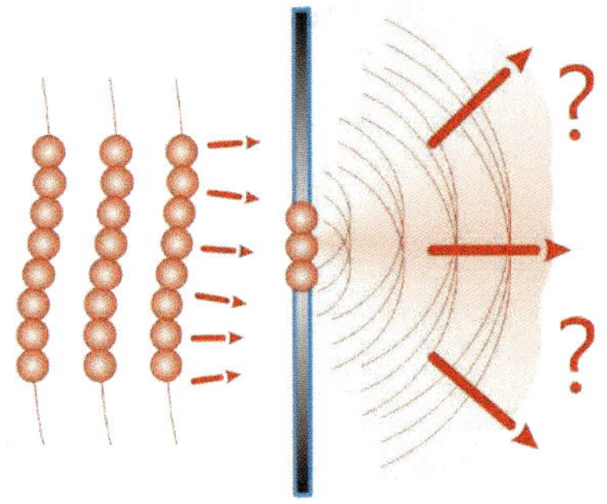


*Figure 5-4: The generalized diffraction problem: A random wavefront passes through a nonpoint aperture.*

It is interesting that Arnold Johannes Wilhelm Sommerfeld defined diffraction as, "*Any deviation of rays from a rectilinear path which cannot be interpreted as reflection or refraction.*"[22] Any reference to light is notably absent from this definition. This is because diffraction effects are encountered in any type of wave, as long as the dimension of the diffractive aperture is comparable to the wavelength. The microwave antennae (in early cell phones) has dimensions of a few centimeters because microwaves have wavelengths that vary from a few millimeters to few centimeters.

[22] Sommerfeld AJW. Zur mathematischen theorie der Beugunsers-cheinenungen. Nachr Kgl Acad Wiss. Gottingen 1894; (4):338-42.

To study the generalized diffraction problem, we consider an aperture of a size comparable with the wavelength, in which there are many, not just one, neighboring point sources. When the aperture is illuminated,[23] the secondary Huygens sources emit in all forward directions. The incident wavefront can have a random form: Its amplitude and phase can change from point to point.

*Figure 5-5: Manifestation of diffraction in sea waves. The aperture is comparable to the wavelength.*

Thus, when solving the general problem of diffraction, in order to find the intensity distribution at a random observation point $P_o$ that is a distance $z$ from the entrance (Figure 5-6), we must follow these steps:

1. Identify the electric fields from all sources originating from every point at the entrance $P_1$.
2. Add all of the fields vectorially.
3. Find the square of the vector sum magnitude. This is the observed intensity at $P_o$.

*Figure 5-6: Geometry of a generalized diffraction intensity distribution.*

This process can be repeated for every other entrance point $P_1$. Since the number of probable point sources is infinite, their sum becomes an integral of the analytical expressions of

[23] We consider coherent light illumination. The theory of noncoherent diffraction is discussed in Born & Wolf, *Principles of Optics*, 6th Edition, Pergamon, 1980, Chapter 11.

the electric fields over the entrance. The general diffraction case is the interference from all of the elementary waves, called secondary wavelets, originating from all points of the entrance at all observation points.

### 5.1.1 Babinet's Principle

The often-cited Huygens–Fresnel principle is only a principle. We accept it on phenomenological grounds based on the acceptable interpretation that it provides to a multitude of wave phenomena. Neither the characteristics of the secondary wavelets, nor the mechanism that explains their creation, is specified.

The Lorentz model explains that, along the light propagation path through an amorphous medium (presented in § 3.1.1), atomic-scale oscillators are set in motion by the incident light and re-emit waves, not only along the original direction, but in all directions. Macroscopically, however, we only note the reflected and refracted waves. This is because within the extended, continuous volume of the medium, the waves traveling in all directions eventually cancel each other out. If the medium is opaque, then at any random point $P_o$ past the medium, the total field is zero: This is how a shadow is formed.

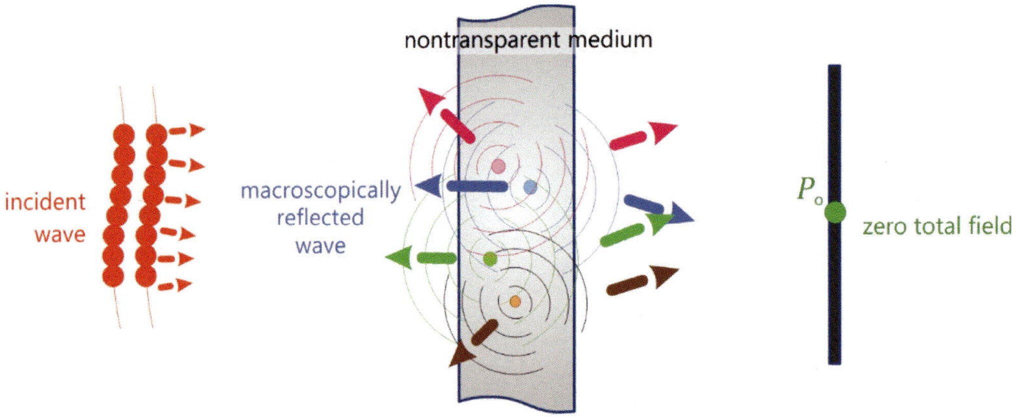


*Figure 5-7: The field past an opaque medium.*

If the medium continuity is interrupted, a diffractive aperture is formed [Figure 5-8 (top)]. The wave that arrives through this aperture to a random point $P_o$ originates mainly from the fields re-emitted from the aperture edge points. A complementary effect occurs when the wave encounters an obstacle [Figure 5-8 (bottom)]. The oscillators inside the material re-emit waves in all directions, but their contribution results (macroscopically) in waves that are not canceled out, particularly near the defining border. The net macroscopic effect is diffraction.

At the specific point $P_o$, the arriving waves have equal amplitude magnitudes and opposite phases with respect to the resultant wave that would be formed if, instead of the obstacle, an aperture of the exact same shape—therefore being complementary to the object—was in place. This is because, if the medium was continuous along the cross-section of the incident light, we would have no light at point $P_o$.

The fields at point $P_o$ can be described by the fields radiated by the secondary wave emitters, which are placed in the exact space that is occupied by the medium, which is now removed; therefore, this space (previously occupied by the medium) functions as an aperture. This is the physical substance of the Huygens–Fresnel principle, which, although phenomenological, provides a reliable explanation of the diffraction effects.

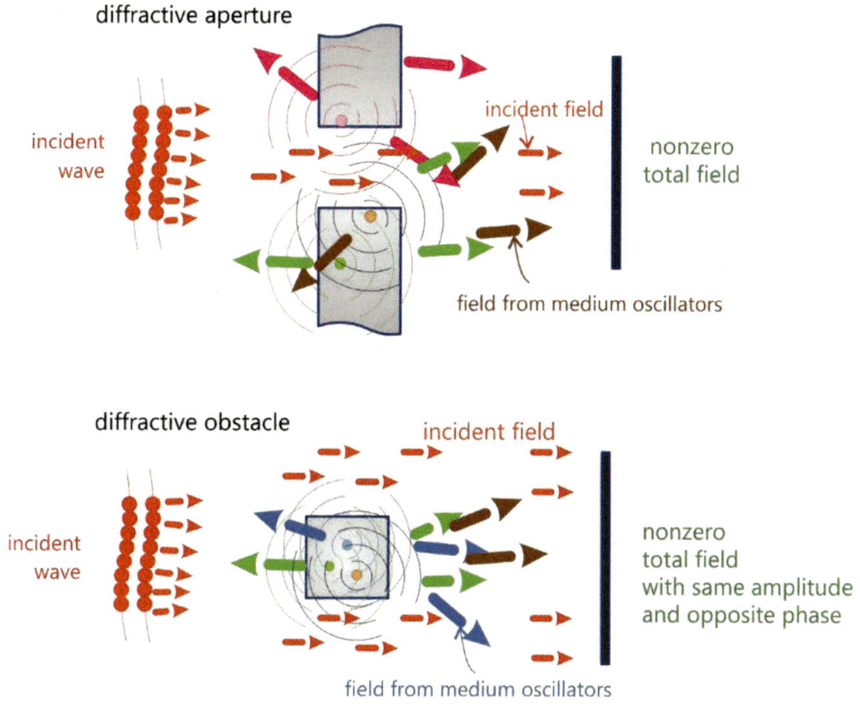


*Figure 5-8: The field (top) past a diffractive aperture and (bottom) past a diffractive obstacle.*

If we consider two complementary apertures, $A_1$ and $A_2$, such that their sum $A = A_1 + A_2$ forms a continuous opaque screen, then there is no field past that medium (Figure 5-9). The field past this completely opaque screen is zero. Thus, it is reasonable to state that, if we add the fields formed from $A_1$ and $A_2$—fields $\mathbf{E}_{A1}(x_o, y_o)$ and $\mathbf{E}_{A2}(x_o, y_o)$, respectively, at every point past aperture $A$—we should obtain zero. Thus, whatever value the field has at a random point $P_o(x_o, y_o)$ past aperture $A_1$, the field from aperture $A_2$, $\mathbf{E}_{A2}(x_o, y_o)$, has the <u>same magnitude</u> and the <u>opposite phase</u>.

This is **Babinet's principle** (named after the French physicist and mathematician Jacques Babinet), which states that complementary diffractive apertures have the same intensity distribution in their respective diffraction pattern formations. The formations are identical, since the observed physical quantity is the fields squared (which are the intensities).

Babinet's principle is a direct consequence of linear superposition implemention in electric fields. Accordingly, we can calculate the diffraction intensity distribution of surface $A_1$ if we subtract the two diffraction pattern formations that originate from surfaces $A$ and $A_2$. For example, assume that surface $A$ is infinite, not finite. Then $A_2$ is the negative of $A_1$. According to the above, for the diffraction pattern formations of surfaces $A_2$ and $A_1$, we should have $\mathbf{E}_{A1}(x_o, y_o) + \mathbf{E}_{A2}(x_o, y_o) = 0$, so these electric fields should have exactly the same amplitudes but opposite phases: $\mathbf{E}_{A1}(x_o, y_o) = -\mathbf{E}_{A2}(x_o, y_o)$. The diffraction intensity distributions are identical.

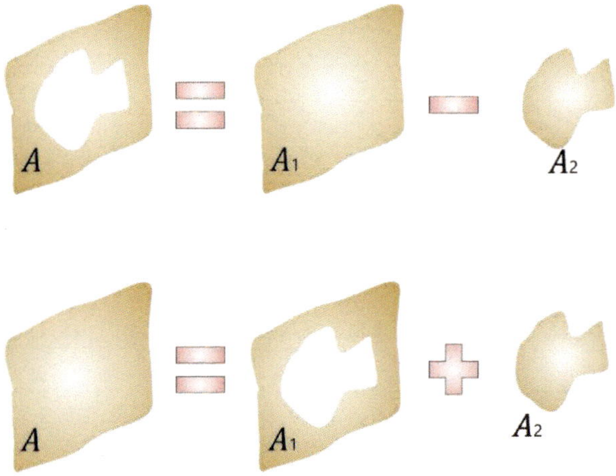


*Figure 5-9: Complementary entry surfaces.*

This principle applies in radiation emission and antennae design. The intensity distribution from a slit of shape $A_1$ has exactly the same amplitude magnitude (and opposite phase) as the diffraction intensity distribution of a thin bar of shape $A_2$.

An extension of Babinet's principle states that if two diffraction intensity distributions of two entry functions are similar, their diffractive properties are similar, too. The diffraction intensity distributions are equal if the the Fresnel number, a unitless quantity denoted as $N$,

Fresnel Number:
$$N = \frac{x^2}{\lambda \cdot z} \tag{5.1}$$

has a fixed value. Here $-z$ is the axial (longitudinal) distance from the entrance distribution. The Fresnel number expresses the diffraction power of an entrance function.

## 5.2 MATHEMATICAL FORMALIZATION

To calculate the magnitude and phase of the electric field at an observation point $P_o(x_o, y_o)$ that results from a diffractive aperture (entrance) point $P_1(x_1, y_1)$, we must possess analytical knowledge of the fields that generate the secondary sources. The field at $P_1$ has a magnitude and phase of $E_1(x_1, y_1) = E_{o1}(x_1, y_1) \times \exp[i\varphi_o(x_1, y_1)]$. We consider a random aperture shape (such as in Figure 5-6) and a longitudinal observation distance $z$. We need to establish the following:

- the magnitude and phase of the field at $P_1$
- the wavelength $\lambda$
- the distance $r_{o1}$ between points $P_1$ and $P_o$: $\sqrt{z^2 + (x_o - x_1)^2 + (y_o - y_1)^2}$

The different wave perturbations that arrive at point $P_o$ from all possible entry points ($P_1$) differ in amplitude and phase because (1) the elementary sources ($P_1$) travel different optical paths from $P_o$ and (2) their paths form different angles of incidence with the observation screen. All of these perturbations are integrated for all possible points $P_1$ over the aperture. Thus, the field $\mathbf{E}(x_o, y_o)$ at point $P_o(x_o, y_o)$ can be expressed as

$$\underbrace{\mathbf{E}(x_o, y_o)}_{\text{field at point of observation}} = \iint_{\text{all aperture points}} \underbrace{h(x_o, x_1; y_o, y_1)}_{\text{inclination factor}} \times \underbrace{\mathbf{E}(x_1, y_1)}_{\text{entrance field}} \, dx_1 dy_1 \tag{5.2}$$

where $h$ is the inclination (slope) factor or diffraction transfer function.

Equation (5.2) is a Fresnel–Kirchhoff integral. For its calculation, we employ the inhomogeneous Helmholtz equation, which provides solutions in terms of Green's functions. We simply borrow the result as presented in Hecht's *Optics*[24] that, for a point-source, the inclination factor $h$ can be expressed as

$$h(x_o, x_1; y_o, y_1) = \frac{1}{\lambda} \frac{\exp\left(i\mathbf{k} \cdot \mathbf{r}_{o1} - \dfrac{\pi}{2}\right)}{z} = \frac{1}{i\lambda} \frac{\exp(i\mathbf{k} \cdot \mathbf{r}_{o1})}{z} \tag{5.3}$$

The calculation must be repeated for every observation point for all elementary emission source points over the entire aperture. The reasoning may be simple, but it is nearly impossible to derive solutions for Eq. (5.2) without certain approximations, as is the case in Fresnel and Fraunhofer diffraction.

[24] Hecht E. *Optics*. 5th Edition, § 10.4, Addison Wesley, 2016.

### 5.2.1 Fresnel Diffraction

We can assume, for example, that the incident light at the diffractive aperture has a constant magnitude $E_o(x_1,y_1)$ and a fixed phase $\varphi_o(x_1,y_1)$. Thus, the expression for the entrance field magnitude simplifies to the terms $E(x_1,y_1) = E_o(x_1,y_1)\times\exp[\,i(\varphi_o(x_1,y_1))\,]$. This approximation is the **near-field diffraction** or **Fresnel approximation**, named after Augustin-Jean Fresnel.

The wavefronts are not necessarily flat, but if the initial source is far away (such that when a wave encounters the diffractive aperture, a small angle similar to the paraxial approximation[25] is formed), then we can assume that only one phase passes through the diffractive aperture. We use Eq. (5.2) for the transfer factor, and add one more step:

$$r_{o1} = \sqrt{z^2 + (x_o - x_1)^2 + (y_o - y_1)^2} \approx z\cdot\left[1 + \frac{1}{2}\left(\frac{x_o - x_1}{z}\right)^2 + \frac{1}{2}\left(\frac{y_o - y_1}{z}\right)^2\right] \tag{5.4}$$

Therefore, the field at the observation point $P_o(x_o,y_o)$ is

$$\mathbf{E}(x_o,y_o) = \frac{1}{i\lambda}\cdot\frac{\exp(ikz)}{z}\cdot\iint_{\text{all aperture points}} \mathbf{E}(x_1,y_1)\cdot\exp\left\{i\frac{k}{2z}\left[(x_o - x_1)^2 + (y_o - y_1)^2\right]\right\}dx_1\,dy_1 \tag{5.5}$$

The squared terms in the exponential part can be separated as follows:

$$\begin{aligned}\mathbf{E}(x_o,y_o) = {} & \frac{1}{i\lambda}\cdot\frac{\exp(ikz)}{z}\cdot\exp\left[i\frac{k}{2z}\left(x_o^{\,2} + y_o^{\,2}\right)\right]\times \\ & \iint_{\text{all aperture points}} \underbrace{\mathbf{E}(x_1,y_1)}_{\text{entrance field}}\cdot\underbrace{\exp\left[i\frac{k}{2z}\left(x_1^2 + y_1^2\right)\right]\cdot\exp\left[-i\frac{2\pi}{\lambda z}\left(x_o x_1 + y_o y_1\right)\right]}_{\text{phase modulation due to propagation}}\,dx_1\,dy_1\end{aligned} \tag{5.6}$$

Thus, the integral becomes more manageable. The vectors to be integrated initially have equal magnitudes and phases. Due to the different optical paths traveled from their respective aperture points, the vectors develop a phase modulation. Note the phase factor, which has a parabolic relationship $(x_o^2 + y_o^2)$ with the transverse displacement at the point of observation $(x_o, y_o)$. We can extend the integration limits of the entry aperture from $-\infty$ to $+\infty$ by setting the field values to zero outside the limits of the entry formation.

Such a problem is solvable schematically using a methodology developed by the French mathematician Marie Alfred Cornu. We construct a vectorial phase diagram (such as the one

[25] *Geometrical Optics* § 8.1.2 The Paraxial Approximation.

introduced in § 4.1.4). The vectors that add up have increasingly larger phases, according to the parabolic relationship ($x_o^2 + y_o^2$). Thus, the relative angle of these phase vectors increases with the square of the transverse displacement from the reference point. Each vector is the contribution of an elementary source from the diffractive aperture. We then obtain the Cornu spiral, or the clothoid (from the Deity of Fate *Μοίρα Κλοθώ*), which has the following shape in Cartesian coordinates ($s$ is the size of the opening):

Abscissa: $$C(x) = \int_0^x \cos\left(\frac{\pi s^2}{2}\right) ds \qquad (5.7)\text{a}$$

Ordinate: $$S(x) = \int_0^x \sin\left(\frac{\pi s^2}{2}\right) ds \qquad (5.7)\text{b}$$

Every point on the spiral corresponds to a parameter $x$, with real and imaginary parts $C(x)$ and $S(x)$. The spiral originates from the inflection point (0,0), the center of symmetry, and is asymptotic at points $+\infty$ ($\sqrt{\pi/2}$, $\sqrt{\pi/2}$) and $-\infty$ ($-\sqrt{\pi/2}$, $\sqrt{\pi/2}$), which are the 'eyes' of the spiral.

In this way, we calculate the field magnitude and phase at point $P_o(x_o, y_o)$, which is a transverse displacement $q$ from the center of symmetry across point $P_{(0,0)}$ and a longitudinal distance $z$ (Figure 5-10). We sum all vectors from the center of symmetry up to the point that corresponds to $B$, the aperture limit of $AB$ of opening size $s$, and repeat the summation for point $A$, the other aperture limit. The difference between vectors $\mathbf{P}(0,0)_B$ and $\mathbf{P}(0,0)_A$ describes vector $\mathbf{AB}$.

If there is no obstacle ($x_1, y_1$), then at the observation screen $P_o(x_o, y_o)$ we join the points filling the interval ($-\infty$, $+\infty$). The magnitude of $\mathbf{AB}$ is the field amplitude at $P_o$. The angle formed with the horizontal axis is the phase. The intensity is the square of field magnitude.

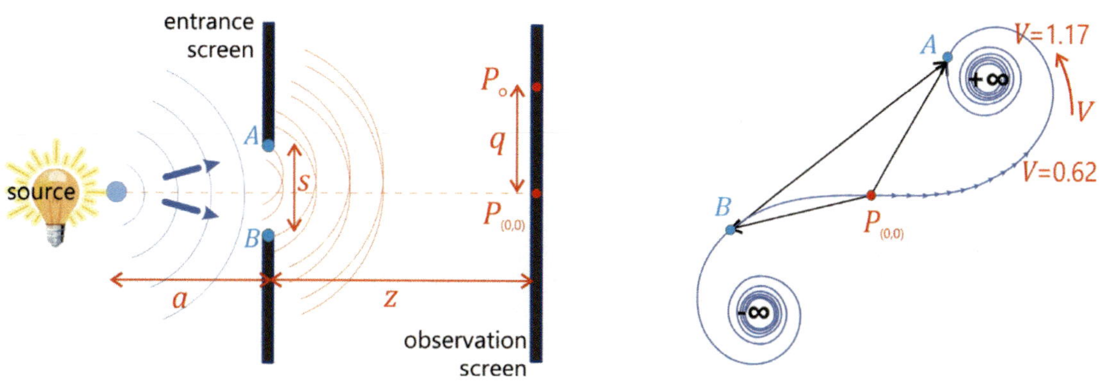


*Figure 5-10: (left) Schematic solution of Fresnel diffraction geometry. (right) The Cornu spiral.*

The correspondence of a point at the diffraction aperture to a point on the spiral is determined by parameter $V$, which depends on the relative transverse displacement, the radius of curvature $\alpha$ of the incident wavefront at the aperture, and the wavelength $\lambda$. Parameter $V$ may express, via the complex number $C(V_x) + iS(V_x)$, the Fresnel diffraction intensity distribution. For example, parameter $V$ for points $A$ and $B$ has the following form:

$$V_A = \left[\frac{s}{2} + q\left(\frac{\alpha}{\alpha + z}\right)\right] \cdot \sqrt{2\frac{\alpha + z}{\alpha z \cdot \lambda}} \quad \text{and} \quad V_B = \left[-\frac{s}{2} + q\left(\frac{\alpha}{\alpha + z}\right)\right] \cdot \sqrt{2\frac{\alpha + z}{\alpha z \cdot \lambda}} \tag{5.8}$$

We can calculate the diffraction intensity distribution in the case of illumination with coherent radiation of a **sharp edge**, setting point $A$ to $+\infty$. Such sharp-edge diffraction patterns are quite often observed; if we illuminate the edges of an object using a coherent source, we can observe, at a relatively close distance around its silhouette, characteristic alternating intensity maxima and minima. The secondary Huygens wavelets from the edges interfere and result in diffraction intensity distributions such as those illustrated in Figure 5-11 (diffraction off a sharp edge of a semi-infinite plane), as well as in Figure 5-12 and Figure 5-13 (diffraction off a slit).

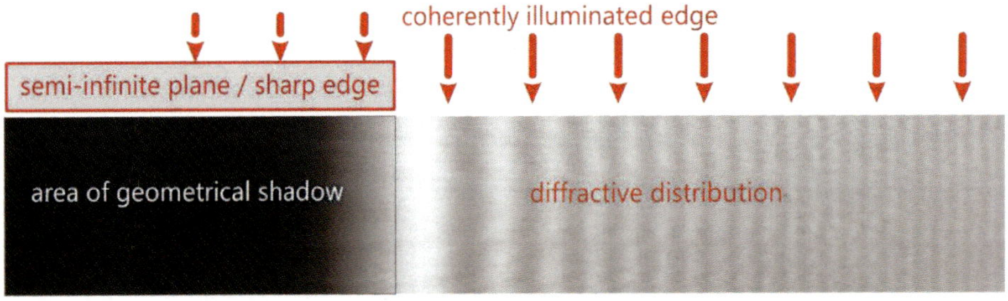


*Figure 5-11: Fresnel diffraction from a sharp edge.*

A variation in the diffraction intensity distribution from the different vector amplitudes results if the vector origin is placed over various points on the spiral. We consider a 1-D entry function that extends to the $(0, -\infty)$ part of the axis. For the point exactly across the edges (axis at 0), the vector tail is at the center of the Cornu spiral. Its head, which corresponds to $\infty$, is one of the two 'eyes' of the spiral. Its magnitude is ½, and its intensity is ¼, of the incident intensity.

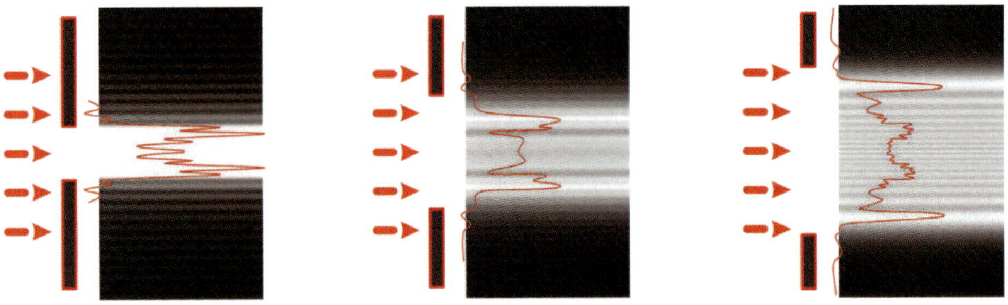

*Figure 5-12: Fresnel diffraction patterns formed by various slit widths.*

The diffraction intensity distribution depends on the entrance field, the aperture size, and the aperture shape $(x_1, y_1)$, as well as on the distance from the observation screen $z$. Observing the development of the diffraction intensity distribution (Figure 5-13), we note that this distribution, while present with significant changes over short distances, gradually takes on a specific distribution with a fixed shape as the distance $z$ to the observation screen increases.

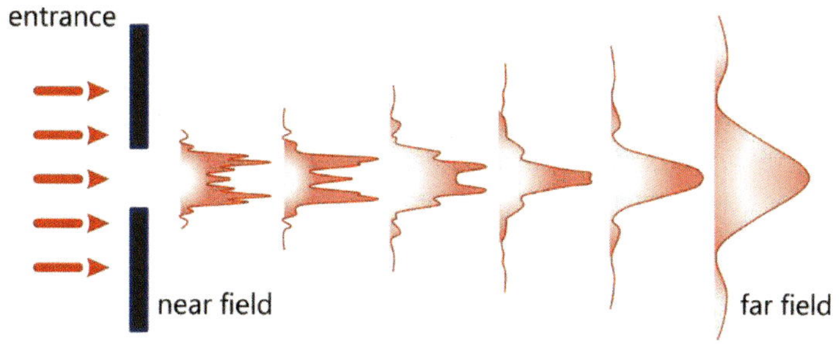


*Figure 5-13: Evolution of a diffraction intensity distribution with increased distance from a simple slit aperture. In the far field, Fresnel diffraction becomes Fraunhofer diffraction.*

## 5.2.2 Fraunhofer Diffraction

The fact that a diffraction intensity distribution attains a fixed shape at long distances from the diffraction aperture leads us to further simplify the mathematical formulas. The following assumptions are made:

- The field amplitude is constant at the entrance.
- The phase is fixed at the entrance.
- The observation screen is at a sufficiently large distance $z$ from the entrance.

This distance $z$ is assumed to be large enough if it satisfies the following condition:

$$z \gg \frac{k}{2}\left(x_1^2 + y_1^2\right)_{\text{MAX}} \approx \frac{\pi}{\lambda} D^2 \tag{5.9}$$

where $D$ is the largest transverse dimension at the entrance. This is the **far-field** or **Fraunhofer approximation**, named after Joseph von Fraunhofer. For example, if the maximum dimension $D$ at the entry aperture is 1 mm and $\lambda = 0.5$ µm, then $z \gg 6$ m.

Although Fraunhofer diffraction is the limit of a more-general Fresnel diffraction, it is a very important case and is mathematically much easier to handle. There is no absolute separation between the two approximations. In effect, the diffraction pattern formation changes progressively from one form to the other.

At such a long distance, the interfering rays from the points of the diffractive aperture can now be assumed to be parallel at every point on the observation screen. Thus, the amplitudes of the secondary waves have nearly equal magnitudes, and their phases follow a linear relationship because they have the same tilt. This is expressed by the approximation that the first phase distribution factor due to propagation in integral Eq. (5.6) essentially equals 1.0, so it can be ignored:

$$\exp\left[i\frac{k}{2z}(x_1+y_1)^2\right] \approx 1.0 \tag{5.10}$$

The diffraction field distribution is now expressed as

$$\mathbf{E}(x_o,y_o)=\frac{1}{i\lambda}\cdot\frac{\exp(ikz)}{z}\cdot\exp\left[i\frac{k}{2z}\left(x_o^{\ 2}+y_o^{\ 2}\right)\right]\times\iint\underbrace{\mathbf{E}(x_1,y_1)}_{\text{entrance field}}\cdot\underbrace{\exp\left[-i\frac{2\pi}{\lambda z}(x_o x_1+y_o y_1)\right]}_{\text{phase modulation due to propagation}}dx_1\,dy_1 \tag{5.11}$$

Ignoring the multiplication factor preceding the integral, and with a change of integration variable at the observation screen (space $x_o$), the 1-D form (space $x_1$) is

$$\mathbf{E}\left(\frac{2\pi x_o}{\lambda z}\right) = \int \underbrace{\mathbf{E}(x_1)}_{\text{entrance field}} \cdot \underbrace{\exp\left[-i\frac{2\pi x_o}{\lambda z}(x_1)\right]}_{\text{phase modulation due to propagation}} dx_1 \tag{5.12}$$

which resembles the well-known mathematical relationship,

$$\mathfrak{F}\{f(t)\} = F(\omega) = \int f(t)\cdot\exp(-i\omega\cdot t)dt \tag{5.13}$$

This relationship is a **Fourier transform**, named after the French mathematician Jean-Baptiste Joseph Fourier. A Fourier transform transforms *f*(*t*), a function expressed in time space, to an expression of the same physical quantity in another space, frequency space (also known as spectral space, since frequency relates to the concept of the spectrum).

*Figure 5-14: A Freudian slip for the Fourier transform.*

For each Fourier transform there is an inverse transform. If we can transform $f(t)$ to its spectral function, $F(\omega) = \mathfrak{F}\{f(t)\}$, then the reverse is also possible; in other words, it is possible to calculate the temporal expression $f(t)$ from the spectral function $F(\omega)$. The function $f(t) = \mathfrak{F}^{-1}\{F(\omega)\}$ is the **inverse Fourier transform** of $F(\omega)$:

$$\mathfrak{F}^{-1}\{F(\omega)\} = F\{F(\omega)\}^{-1} = \frac{1}{2\pi}\int F(\omega)\cdot\exp(i\omega\cdot t)\,dt \qquad (5.14)$$

Here the conjugate variables are time $t$ and angular frequency $\omega$. Their product, the unitless quantity $\omega\cdot t$, expresses the angular phase, and the quantities $t$ and $\omega$ alternate as integration variables in their respective transforms. Comparing Eqs. (5.12) and (5.13), we note that the conjugate variables in the case of far-field diffraction are the entry coordinates $(x_1, y_1)$ and the spatial frequencies $f_x = x_o/\lambda\cdot z$ and $f_y = y_o/\lambda\cdot z$.

*Figure 5-15: Jean-Baptiste Joseph Fourier (1768–1830).*

A very simple Fourier transform involves the function $\cos(\omega_o t)$. We know the time space ($t$) representation of this function quite well: It is a harmonic function with period $T = 2\pi/\omega_o$, uninterrupted to infinity, since it has values for all $t$ from $-\infty$ to $+\infty$. Its shape is shown in Figure 5-16 .

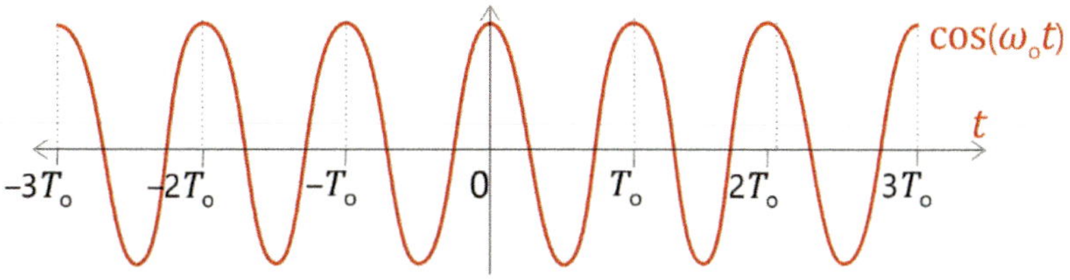


*Figure 5-16: The function $\cos(\omega_o t)$ expressed in time space.*

The expression of this function in frequency space has exactly one frequency:

$$F\{\cos(\omega_o t)\}^{-1} = \int \cos(\omega_o t)\cdot \exp(-i\omega t)\,dt = \frac{1}{2}\int [\exp(i\omega_o t) + \exp(-i\omega_o t)]\cdot \exp(-i\omega t)\,dt$$
$$= \frac{1}{2}\int [\exp(-i(\omega-\omega_o)t)]\,dt + \frac{1}{2}\int [\exp(-i(\omega+\omega_o)t)]\,dt = \pi[\text{delta}(\omega-\omega_o) + \text{delta}(\omega+\omega_o)] \tag{5.15}$$

The last expression consists of two symmetric (with respect to the coordinate origin) functions, called Dirac delta functions, illustrated in Figure 5-17. This description is idealized. It states that the function has only one frequency $\pm\,\omega_o$ ($\omega_o$ for positive $t$ and $-\omega_o$ for negative $t$). The generalized delta function is zero everywhere except for the two spikes at $\pm\,\omega_o$, which are the values of $\omega$ that render its argument to zero.

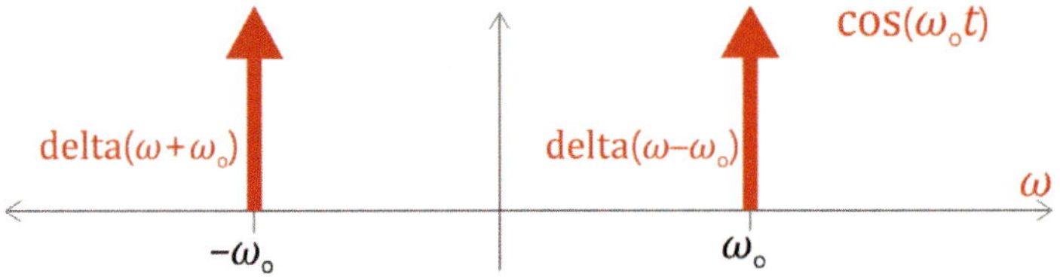


*Figure 5-17: The function* $\cos(\omega_o t)$ *expressed in frequency space.*

The function $\cos(\omega_o t)$ has not changed; it is now simply expressed in a different space: in time space then in its inverse, frequency space. Although we are rather familiar with time-space representations, often the reciprocal spectral expression is more useful. If the function has 2× the frequency, it is compressed in time; however, in frequency space, the separation between the delta spikes doubles. Did we say 'double the frequency?' This is why. Function $\cos(2\omega_o t)$ is depicted in the reciprocal spaces of time and frequency, as shown in Fig. 5-18.

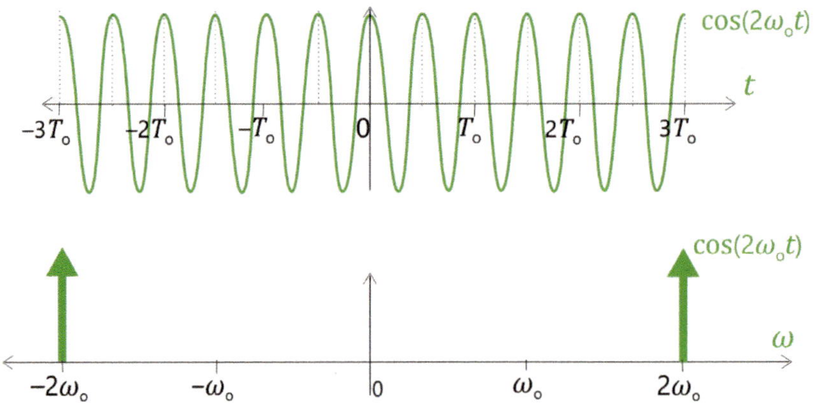


*Figure 5-18: The function* $\cos(2\omega_o t)$ *in both time space (top) and frequency space (bottom).*

This is exactly the same effect as we saw in the interference by two slits (Figure 4-21). The periodic fringe distribution is in an inversely proportional relationship with the source separation $d$. What can be considered now is the field at the observation screen that results from a Fourier transform of the entrance field consisting of the two bright slits.

A generalization of this observation is the basic principle governing Fraunhofer diffraction, which can be stated as follows:

Fourier: For a coherent wave distribution propagating in space, the far-field diffraction pattern formation corresponds to the Fourier transform of the entry (pupil) function.

To calculate the far-field diffraction intensity distribution, we find the Fourier transform of the entry function magnitude $E(x_1, y_1) = E_o(x_1, y_1) \times \exp[\, i\varphi_o(x_1, y_1)\,]$ calculated at the corresponding spatial frequencies $f_x = x_o/\lambda{\cdot}z$ and $f_y = y_o/\lambda{\cdot}z$:

$$\underbrace{E(x_o, y_o)}_{\text{diffraction field magnitude}} = F\{E(x_1, y_1)\}^{-1} = \iint \underbrace{E(x_1, y_1)}_{\text{entrance field}} \times \underbrace{\exp\left[-i2\pi\left(f_x x_1 + f_y y_1\right)\right]}_{\text{phase modulation}} \, dx_1 dy_1 \tag{5.16}$$

The observed intensity $I(x_o, y_o)$ is proportional to the square of the magnitude of the field amplitude:

$$I(x_o, y_o) = E^2(x_o, y_o) \tag{5.17}$$

The pupil function $E(x_1, y_1)$ is determined by the shape (geometry) of the diffractive aperture and the spatial distribution of entrance field. In the simple case of homogeneous illumination, this function is the expression of the geometrical shape of the entrance field.

The **spatial frequencies** correspond to

$$\underbrace{f_x = \frac{x_o}{\lambda z} = \frac{k_x \cdot x_o}{2\pi \cdot z} \approx \frac{k_x \cdot \sin\vartheta_x}{2\pi}}_{\text{spatial frequencies } x} \quad \text{and} \quad \underbrace{f_y = \frac{y_o}{\lambda z} = \frac{k_y \cdot y_o}{2\pi \cdot z} \approx \frac{k_y \cdot \sin\vartheta_y}{2\pi}}_{\text{spatial frequencies } y} \tag{5.18}$$

The conjugate variables are length (such as $x$ and $y$) and spatial frequency (such as $f_x$ and $f_y$). Length is the variable in the spatial domain, using units of length such as meters. Spatial frequency is the variable in the inverse (frequency) domain and is reported in units of inverse length, similar to the wavevector. The product of these two variables, the unitless quantity $f_x \cdot x_1 = (x_o/\lambda \cdot z) \cdot x_1$, is present in the exponential argument of the transformation integrals, just as the product $\omega \cdot t$ is present in the simple transform relationships of Eq. (5.13).

In the expressions of Eq. (5.18), $\vartheta_x$ and $\vartheta_y$ are the angular displacements formed between the optical axis and the observation ray, respectively, in the diffraction intensity distribution. Low frequencies correspond to a wavevector that is close to the optical axis (when $\vartheta$ and $\sin\vartheta$ are very small) and express the coarse content of the optical signal, mainly lightness. On the other hand, high frequencies correspond to a wavevector that is far from the optical axis (when $\vartheta$ is very small) and express the fine content of the optical signal.

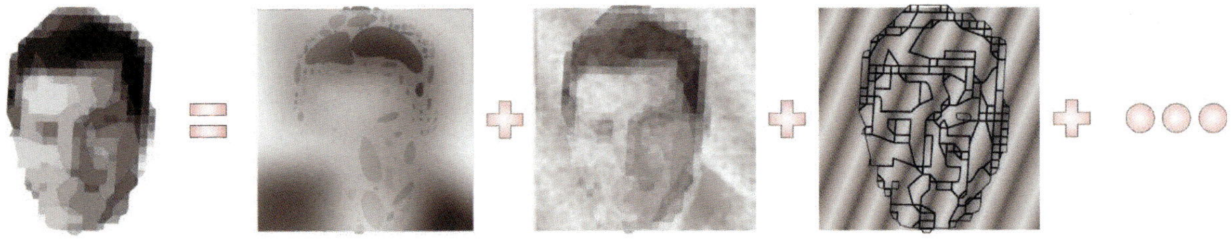

*Figure 5-19: Two-dimensional analysis of an optical signal in spatial frequencies.*

Just as the Fourier transform of a temporal function $f(t)$ describes the harmonic frequencies of which it is composed [its spectrum distribution $F(\omega)$], the Fourier transform of a 2-D spatial function $f(x,y)$, which is the entry distribution $E(x_1,y_1)$, describes its spectral content, which is an equivalent expression in terms of spatial frequencies $F(f_x, f_y)$.

The physical meaning of the spatial frequencies may be understood with an example from acoustics. We know that an audio signal, which is a 1-D signal in time, is composed of high (treble) and low (bass) frequencies. The re-composition of all of these harmonic components reproduces the acoustic signal. However, in any sound reproduction system, even a high-fidelity system, the reproduction quality never matches the quality of the original recording—there is always some loss of signal quality. This is due to the degrading of certain frequencies (tones) and the complete attenuation of others. A high-quality reproduction system preserves a wider range of frequencies, which is reflected in the quality of the reproduced sound.

As in a acoustic signal, there are spatial frequencies in an optical signal, which is a 2-D signal in space. An imaging system (in the simplest case of a converging lens) cuts off some high frequencies due to the lens' confined area. Generally, every entry function in space $(x_1,y_1)$ functions as a filter of the spatial frequencies. The entry distribution cuts off (does not allow propagation of) the spatial frequencies that extend beyond a certain limit.

To implement the far-field Fraunhofer approximation, the observation screen must be at a sufficient distance $z$. We assume a flat wavefront, a fixed amplitude, and coherent illumination of wavelength $\lambda$; the source must be a long distance from the diffractive aperture.

These conditions can practically be realized in a compact lab setting with two converging lenses, where a collimator lens is placed after the source at a distance equal to its focal length in order to create a collimated beam. Thus, the incident intensity is constant, and the wavefront is flat. To satisfy the second condition that at every point on the observation screen the rays are parallel, we must place the observation screen far from the diffractive aperture. We can also compress this distance by using a second converging lens. The diffraction intensity distribution is formed at this lens' focal plane, which is situated at the center of symmetry, the optical axis. In effect, a converging lens carries the Fraunhofer plane on its focal plane (length $f$).

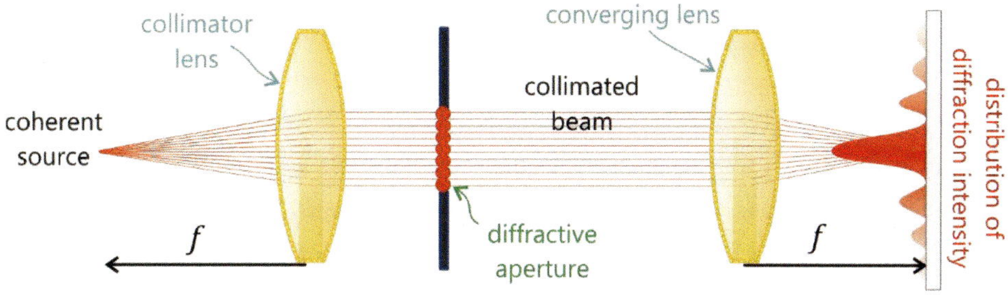


*Figure 5-20: Experimental configuration for Fraunhofer diffraction.*

Here is the procedure we apply to calculate the diffraction intensity distribution on a screen that is far from a diffractive aperture, i.e., the Fraunhofer distribution:

To calculate the diffraction intensity distribution on a screen far from a diffractive aperture (the Fraunhofer distribution): → 1. Mathematically express the entry function [the spatial distribution of the electric field at the diffractive aperture $E(x_1, y_1)$]. → 2. Find the Fourier transform that corresponds to this function, calculated at the spatial frequencies. This is the diffraction field $E(x_o, y_o)$. → 3. Obtain the diffraction intensity distribution $I(x_o, y_o)$, which is the observable quantity, from the square of the diffraction field.

## 5.3 SINGLE-SLIT DIFFRACTION

A diffractive aperture is a single slit along the $-y$ axis (assuming that it extends to infinity), while along the $-x$ axis it has a finite size of width $\alpha$. It is, effectively, a long, rectangular aperture, stretched along one axis. The distribution field at the entrance has a 1-D expression:

$$E(x_1) = E_o \cdot \cos(\varphi_o) \cdot \mathrm{rect}\left(\frac{x_1}{\alpha}\right) \tag{5.19}$$

where rect($x/\alpha$) describes a 1-D step with a value of 0.0 for $-x < -\alpha/2$ and $x > \alpha/2$, and a value of 1.0 for $-\alpha/2 < x < +\alpha/2$.

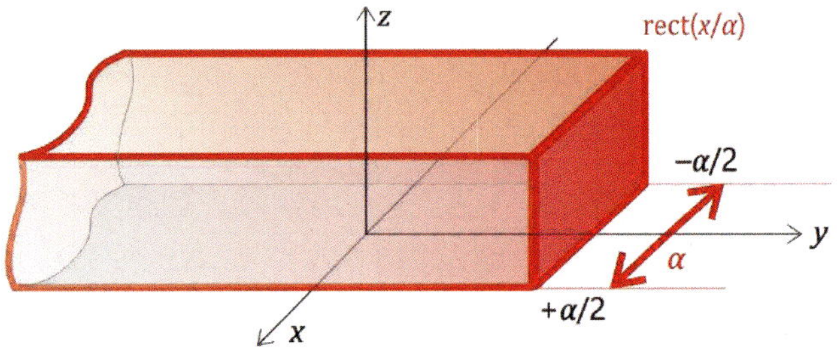


*Figure 5-21: Entrance field distribution: the function* rect($x/\alpha$) *bound along the x axis at* $+\alpha/2$ *and* $-\alpha/2$.

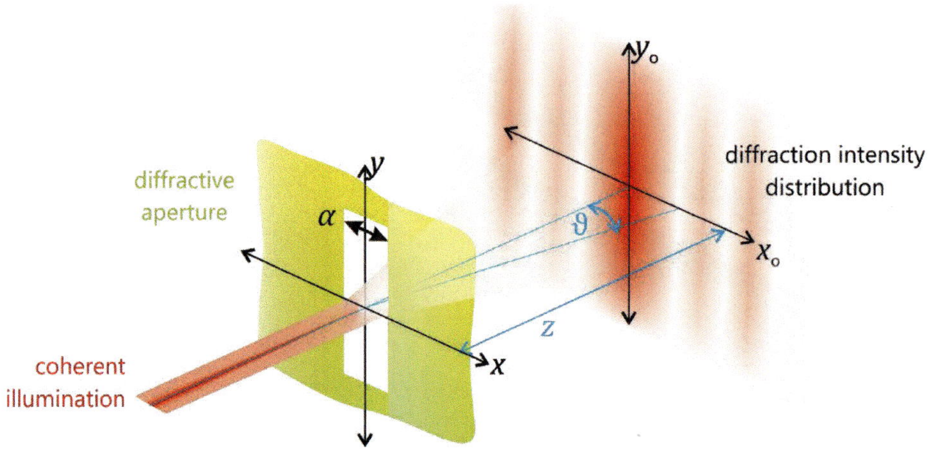


*Figure 5-22: Diffraction intensity distribution from a single slit of size α along the –x axis.*

The Fourier transform of rect($x/\alpha$), calculated at spatial frequencies $f_x = x_o/\lambda \cdot z$, is

$$\underbrace{E(x_o)}_{\text{diffraction}} = F\{E(x_1)\}^{-1} = \int \underbrace{\text{rect}\left(\frac{x_1}{\alpha}\right)}_{\text{entrance field}} \cdot \underbrace{\exp\left[-i2\pi(f_x x_1)\right]}_{\text{phase modulation}} dx_1 = \int_{-\alpha/2}^{\alpha/2} \exp\left[-i2\pi(f_x x_1)\right] dx_1 \quad (5.20)$$

The integral is an expression of the function $\alpha \cdot \text{sinc}(\alpha f_x)$:

$$\underbrace{E(x_o)}_{\text{diffractive field}} = \int_{-\alpha/2}^{\alpha/2} \exp\left[-i2\pi(f_x x_1)\right] dx_1 = \alpha \cdot \text{sinc}\left(\alpha \cdot \frac{x_o}{\lambda \cdot z}\right) \quad (5.21)$$

The sinc function, also known as the cardinal sine function, is defined as

$$\lim_{t=0} \text{sinc}(t) = 1 \quad \text{and} \quad \text{sinc}(t) = \frac{\sin(\pi t)}{\pi t}, \; t \neq 0 \quad (5.22)$$

We introduce the **angular diffraction parameter** $\underline{\alpha}$, which relates the slit size $\alpha$, the wavelength $\lambda$, and the observation distance $z$ according to

$$\underline{\alpha} = \frac{1}{2}ka\frac{x_o}{z} = \pi\frac{a}{\lambda}\frac{x_o}{z} = \pi\frac{a}{\lambda}\sin\vartheta_x \tag{5.23}$$

The diffraction field is then expressed, ignoring multiplicative constants, as

$$\underbrace{E(x_o)}_{\text{diffraction field}} = \text{sinc}\left(\frac{\underline{\alpha}}{\pi}\right) = \frac{\sin(\underline{\alpha})}{\underline{\alpha}} = \frac{\sin\left(\pi\frac{a}{\lambda}\sin\vartheta_x\right)}{\pi\frac{a}{\lambda}\sin\vartheta_x} \tag{5.24}$$

The function $(\sin\underline{\alpha})/\underline{\alpha}$ has a maximum for $\underline{\alpha} = 0$ and is zero for $\underline{\alpha} = \pm m \cdot \pi$ [rad], $m = \pm 1, \pm 2, \ldots$ .

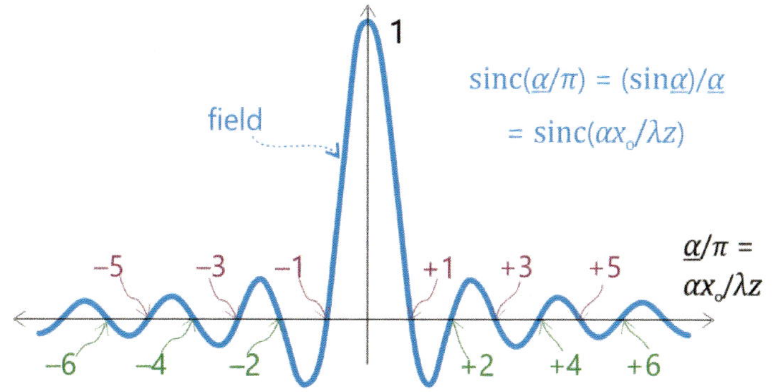


*Figure 5-23: Function $(\sin\underline{\alpha})/\underline{\alpha}$, single-slit diffraction field pattern.*

Between the zero crossings, there are phase changes (amplitude sign alternating from plus + to minus –). The diffraction intensity distribution is

Diffraction Intensity Distribution:

$$I(x_o) = E^2(x_o) = I_o\left(\frac{\sin(\underline{\alpha})}{\underline{\alpha}}\right)^2 \tag{5.25}$$

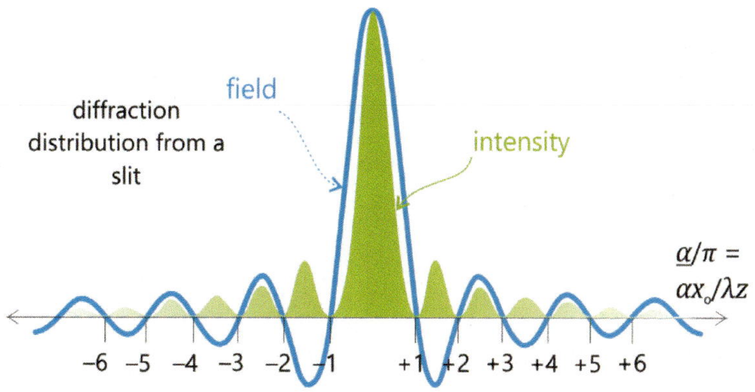


*Figure 5-24: Diffraction intensity distribution pattern from a slit.*

The diffraction intensity distribution appears to spread perpendicularly to the narrow slit (Figure 5-22). There is no modulation along the $-y$ axis, to which the infinitely extended slit dimension is aligned. The intensity maximum is exactly at the center of the distribution. The pattern is concentrated around the **center lobe**, or **main lobe**. Along both sides of the center lobe, there are secondary intensity maxima, or **side lobes** of notably less intensity.

The first minimum defining the center lobe corresponds to the first zero-crossing of the field function, which occurs for $\underline{\alpha} = \pi$ rad and, symmetrically, for $\underline{\alpha} = -\pi$ rad. Thus, its angular width corresponds to $\underline{\alpha} = 2\pi$. The periodic maxima and minima define the side lobes, spaced equally by $\underline{\alpha} = \pi$. This spacing defines their thickness, which is half that of the main lobe.

The diffraction intensity distribution can be described in units of angular diffraction parameter $\underline{\alpha}$, in distances $x_o$ over the screen, or in angular displacement $\vartheta_x$ ($\sin\vartheta_x \approx \vartheta_x$, expressed in radians). The advantage of the latter is its simplicity and independence from the observation distance $z$. Thus, the angular intensity distribution in the diffraction pattern formation has a specific shape and is simply magnified for increased distances from the aperture (Figure 5-25).

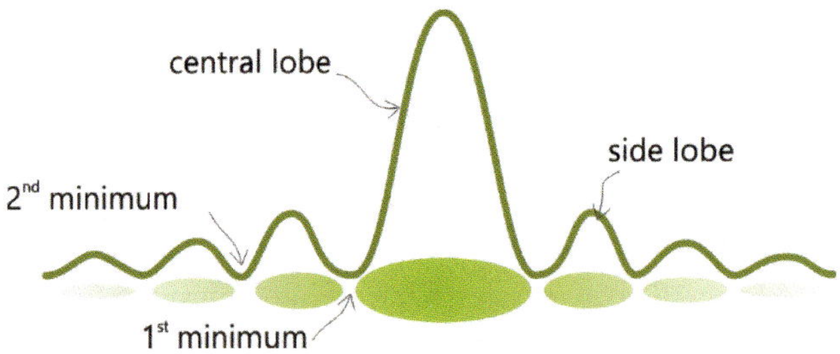


*Figure 5-25: Diffraction intensity distribution profile from a single slit.*

*Table 5-1: Characteristics of a single-slit diffraction intensity distribution.*

| | Angular diffraction parameter $\underline{\alpha}$ | Spatial extent $x_o$ at the screen | Angular extent $\vartheta_x$ ($\sin\vartheta \approx \vartheta$, radians) |
|---|---|---|---|
| extent of the central lobe | $2\pi$ | $2\lambda\ z/\alpha$ | $2\lambda/\alpha$ |
| extent of the side lobe | $\pi$ | $\lambda\ z/\alpha$ | $\lambda/\alpha$ |
| 1st minimum | $\pi$ | $\lambda\ z/\alpha$ | $\lambda/\alpha$ |
| $m^{th}$ minimum | $m\cdot\pi$ | $m\cdot\lambda\ z/\alpha$ | $m\cdot\lambda/\alpha$ |

For expressions involving the observation angle $\vartheta_x$, the characteristic quantity is $\lambda/\alpha$. Angle $\lambda/\alpha$ is a very simple ratio and is characteristic of the angular extent of the side lobe, and of the separation between the side lobes. Minimum intensity corresponds to angles $\vartheta_x = m\cdot\lambda/\alpha$, where the integer $m = \pm1, \pm2, \ldots$ expresses the **diffraction order**. For the observation angle $\vartheta = 0$ ($m = 0$), there is a central maximum: The main lobe is the zeroth diffraction order.

The angular extent $\lambda/\alpha$ of the side lobe depends only on two parameters, the wavelength $\lambda$ and the slit size $\alpha$. The intensity distribution has a smaller angular breath for a smaller ratio $\lambda/\alpha$.

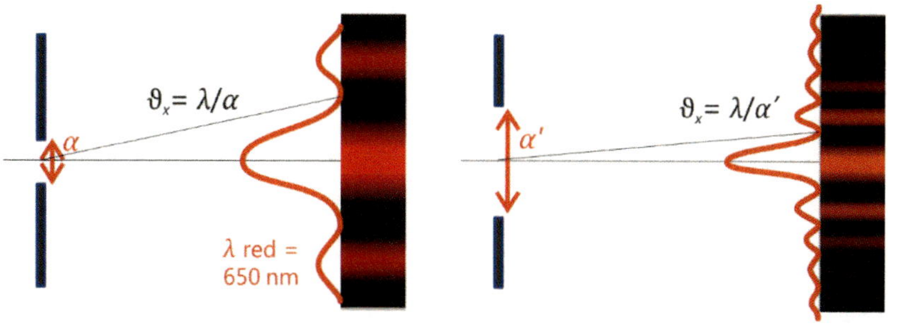


*Figure 5-26: Dependence of the single-slit diffraction intensity distribution on the slit size: (left) small-size slit and (right) large-size slit.*

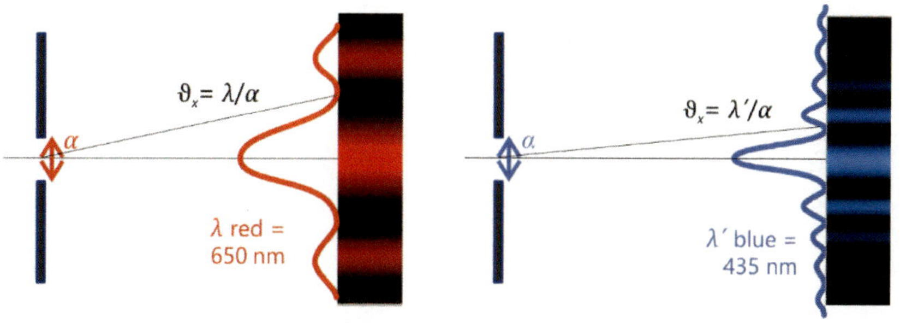


*Figure 5-27: Dependence of the single-slit diffraction intensity distribution on the wavelength: (left) red light and (right) blue light.*

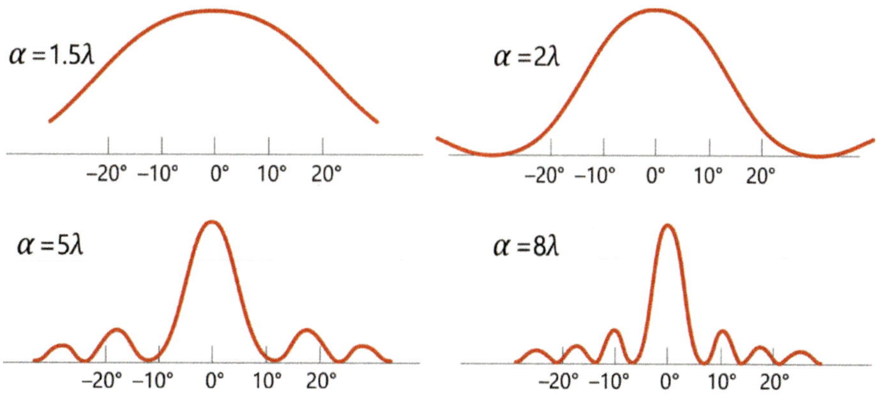


*Figure 5-28: Dependence of the slit diffraction intensity distribution on the slit size (ratio $\lambda/\alpha$). The first zero occurs at an angle equal to the ratio $\lambda/\alpha$ (expressed in radians), which has been converted to degrees in this figure.*

If the wavelength $\lambda$ is much smaller than the slit size $a$ (or $a$ is much larger than $\lambda$), then the characteristic angle $\lambda/\alpha$ is very small. In this case, the diffraction pattern formation at the limit $\lambda/\alpha \ll 1$ becomes the geometrical shadow of the aperture. Conversely, the smaller the aperture $\alpha$ in relation to the wavelength ($\lambda/\alpha \gg 1$), the more closely the aperture resembles a point source, and the more the characteristic angle significantly increases.

What happens if we illuminate the slit with a broadband source? We can find the answer ourselves: With our eyes barely open, our eyelids form a narrow slit, and we can observe a very distant white light source, such as sunlight reflecting from the waves of the sea. We will see a central white spot surrounded by concentric dark lines (the minima), and along both sides we will see a multicolor band.

At the central maximum ($m = 0$), which is the same for all chromatic components, there is no color separation. At the side lobes, the maxima correspond to different angles for different colors, so they separate. The smallest visible wavelength, the blue ($\lambda = 0.4$ µm), appears closer to the center, followed by the rest of the colors with increasing $\lambda$. This is repeated in the higher-order side lobes.

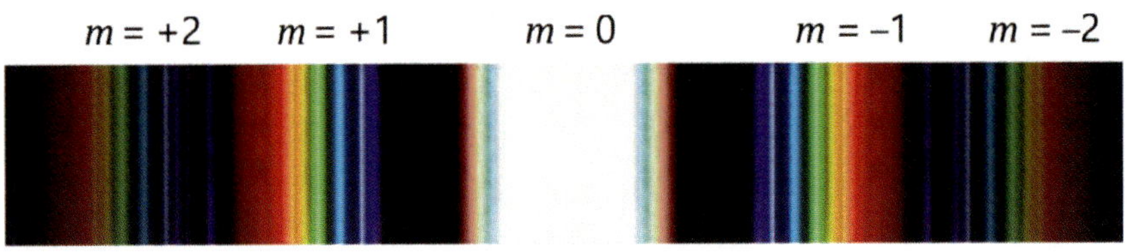


*Figure 5-29: Multicolor diffraction off a simple slit.*

It is interesting to compare the diffraction intensity distribution from a single slit of size $\alpha$ with that from the interference of two apertures separated by $d$ (Young's experiment § 4.2.1).

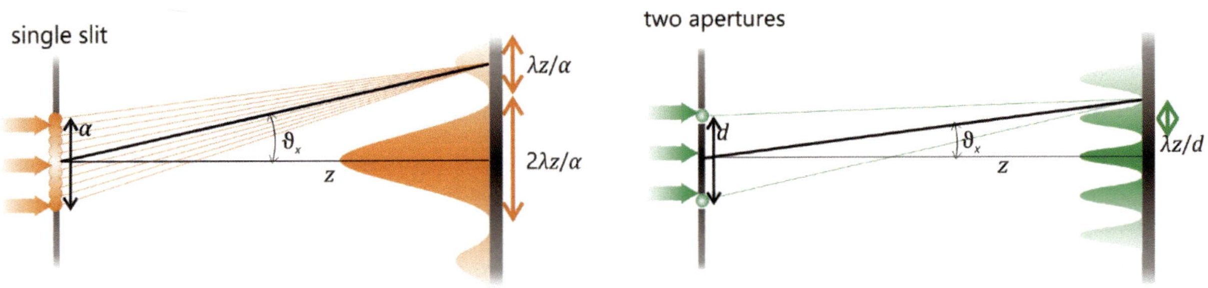


*Figure 5-30: Comparison of the diffraction intensity distributions of (left) a single slit of size α and (right) two small apertures separated by a distance d.*

These two effects have a large degree of similarity and display many common features. Both are manifestations of the wave nature of light, so some degree of coherence is required. The extent of the diffraction pattern formation depends on the degree of coherence of the initial wave. For a low degree of coherence, there are anywhere from a few interference fringes and diffraction lobes (the central lobe and perhaps a few more) to several more orders of interference than a formation by a source with a high degree of coherence, in which many orders of interference appear.

We can regard Young's experiment as a diffraction effect if the two apertures are treated as a diffractive aperture. Thus, at the observation screen, the diffraction field is the Fourier transform of the entry distribution, and the intensity is the square of the field.

Equivalently, the diffraction from a single slit can be regarded as an interference effect if we consider that the wave from the slit results from an infinite number of neighboring waves from each infinitesimal part of the slit, all of which interfere. In the two-slit experiment, there are only two sources, and we apply the principle of linear superposition on the electric fields according to the Fresnel–Huygens principle. In slit diffraction, there is an infinite number of such sources. Therefore, at the observation screen, we integrate the contributions from all of these sources, setting the proper marginal conditions dictated by geometry—in other words, the slit thickness.

*Table 5-2: Comparison of diffraction from a single slit and interference from two point sources (with the observation screen at distance z).*

| | Two point sources, separation $d$ | Slit, size $\alpha$ |
|---|---|---|
| mathematical expression | sinusoidal $\cos^2 \frac{\pi}{\lambda}\frac{xd}{z}$ | $(\sin^2 \underline{\alpha})/\underline{\alpha}^2$ where $\underline{\alpha} = ½\, k \cdot \alpha\, x_o/z$ |
| light distribution | equal-sized fringes | equal-sized side lobes, central lobe with double extent |
| minima separation | $\lambda \cdot z/d$ | $\lambda \cdot z/\alpha$, $2\lambda \cdot z/\alpha$ for central |
| intensity distribution | equal-intensity fringes | intense central lobe, decreasing intensity for side lobes |

In the case of a simple slit of size $\alpha$, an alternative way to calculate the diffraction intensity distribution is to divide the extent of the slit into two equal parts (Figure 5-31). We treat this as interference: two parts = two small apertures separated by $d = \alpha/2$. For the first intensity minimum, the optical path difference must be $\lambda/2$:

First Minimum:
$$\frac{\alpha}{2} \cdot \sin\vartheta = \frac{\lambda}{2} \Rightarrow \frac{\alpha}{2} \cdot \frac{x}{z} = \frac{\lambda}{2} \Rightarrow x = \frac{\lambda z}{\alpha} \tag{5.26}$$

Indeed, in the two-slit experiment, the first minima appear for $\lambda z/2d = \lambda z/\alpha$, which is exactly the same expression as the first minimum from the analytical expression of the function $\mathrm{sinc}(\alpha x_o/\lambda z)$. This treatment may be further expanded. We can divide the entry aperture not in two parts, but in $2m$ parts, and again set the condition that, to obtain the $m^{th}$ minimum, the optical path difference between these elementary sections must be $\lambda/2$.

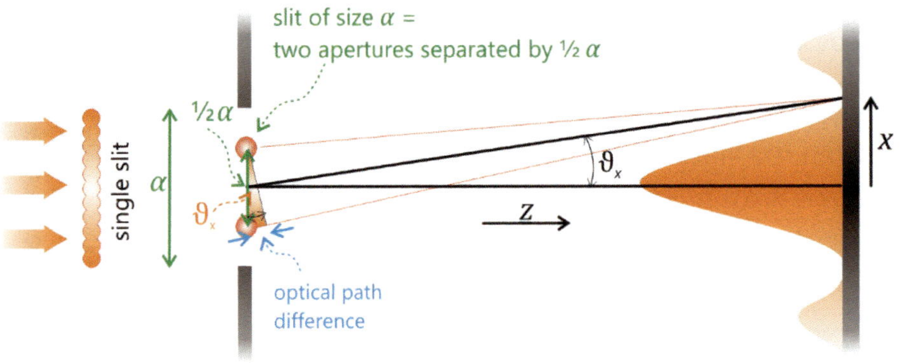


*Figure 5-31: Alternative way to calculate the minima in single-slit diffraction.*

When dividing the aperture into four parts (diffraction order 4), the condition for zero is $x_4 = 4\lambda z/\alpha$, and the condition for the minima of diffraction order $m$ is $x_m = m \cdot \lambda z/\alpha$.

## 5.3.1 Rectangular Aperture Diffraction

Now the diffractive aperture is a rectangle: Along the –$x$ axis its width is $\alpha$, and along the –$y$ axis its width is $\beta$. The aperture therefore can be defined by two slits with dimensions $\alpha$ and $\beta$ along the mutually perpendicular axes. The entrance distribution field has a 2-D form:

$$E(x_1,y_1) = E(x_1)\times E(y_1) = E_o \cdot \exp(i\varphi_o)\cdot \mathrm{rect}\left(\frac{x_1}{\alpha}\right)\times \mathrm{rect}\left(\frac{y_1}{\beta}\right) \qquad (5.27)$$

We compare Eq. (5.27) with Eq. (5.19) and see that they are essentially expressions of two independent but overlapping rectangular slits along the axes –$x$ and –$y$.

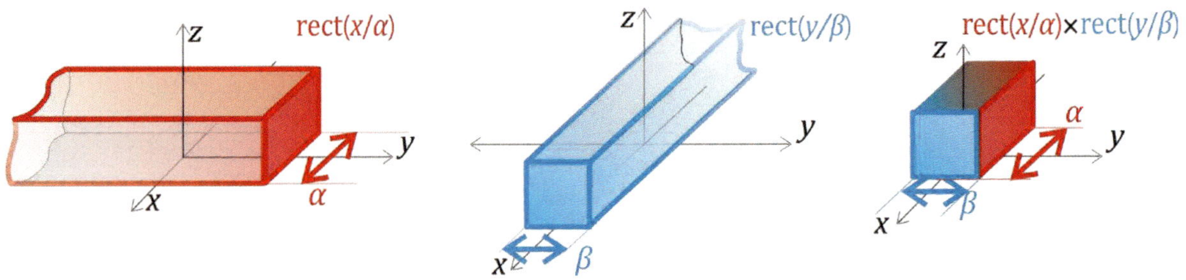


*Figure 5-32: Rectangular aperture represented as a superposition of two slits.*

According to the orthogonality theorem, the Fourier transform of the quantity $E(x_1, y_1)$ in Eq. (5.27) is the product of the corresponding independent (orthogonal) transforms for a single slit [ $E(x_1)$ and $E(y_1)$ ] along the respective axes:

$$\underbrace{E(x_o, y_o)}_{\text{diffraction}} = F\{E(x_1, y_1)\}^{-1} = F\{E(x_1)\}^{-1} \times F\{E(y_1)\}^{-1} = \text{sinc}\left(\frac{\underline{\alpha}}{\pi}\right) \times \text{sinc}\left(\frac{\underline{\beta}}{\pi}\right) \tag{5.28}$$

The characteristic angular diffraction parameters, $\underline{\alpha}$ and $\underline{\beta}$, are

$$\underline{\alpha} = \frac{1}{2} k\alpha \frac{x_o}{z} = \pi \frac{\alpha}{\lambda} \frac{x_o}{z} = \pi \frac{\alpha}{\lambda} \sin \vartheta_x \quad \text{and} \quad \underline{\beta} = \frac{1}{2} k\beta \frac{y_o}{z} = \pi \frac{\beta}{\lambda} \frac{y_o}{z} = \pi \frac{\beta}{\lambda} \sin \vartheta_y \tag{5.29}$$

The diffraction intensity distribution (the square of the diffraction field) is

$$I(x_o, y_o) = E^2(x_o, y_o) = \left(\frac{\sin(\underline{\alpha})}{\underline{\alpha}}\right)^2 \cdot \left(\frac{\sin(\underline{\beta})}{\underline{\beta}}\right)^2 \tag{5.30}$$

This is a 2-D pattern with a symmetry that reflects the diffractive aperture with inverse proportionality. Similar to the diffraction pattern formation from a slit, the extent of the central lobe is 2× that of the side lobes. For a slit of size $\alpha$ along the axis –$x$, the central lobe size is $2\lambda z/\alpha$ and spans the –$y$ axis. The central lobe that corresponds to a slit of size $\beta$ along the –$y$ axis has size $2z\lambda/\beta$ and spans the –$x$ axis. The larger the slit size, for example, $\alpha > \beta$, the smaller the corresponding central lobe size, $2\lambda/\alpha < 2\lambda/\beta$. Thus, the central lobe is elongated along the narrower-slit-size extent of the rectangular diffractive aperture.

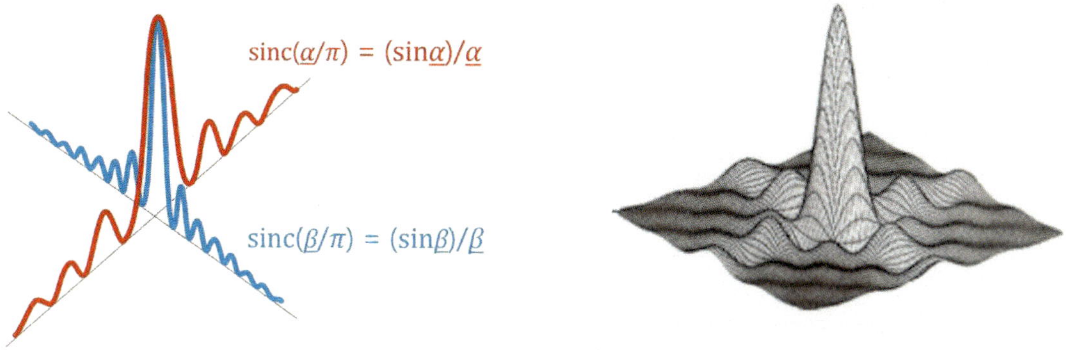


*Figure 5-33: Diffraction intensity distribution from a rectangular aperture $\alpha \times \beta$ = two overlapping slits.*

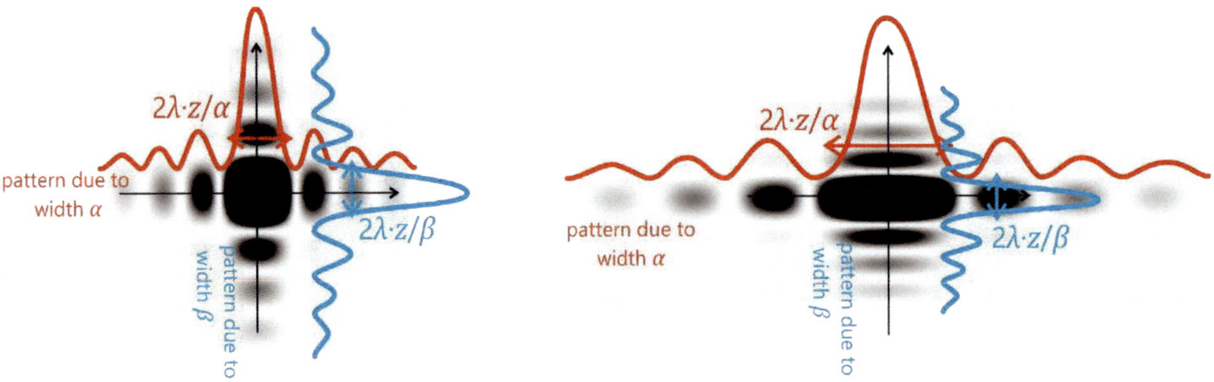


*Figure 5-34: Diffraction intensity distribution from a rectangular aperture $\alpha \times \beta$: (left) $\beta \approx \alpha$ and (right) $\beta > \alpha$.*

*Figure 5-35: Similar to the display resulting from a single-slit diffractive aperture, multicolor lobes can also be displayed from a rectangular diffractive aperture.*

# 5.4 CIRCULAR APERTURE DIFFRACTION

Here, the diffractive aperture is a disk of diameter $D$. Because of the rotational symmetry with respect to the optical axis, the coordinate system is polar ($r, \vartheta$) instead of Cartesian. The origin ($r = 0$) is the center the disk, which is the center of symmetry. The entrance field is given as

$$E(r_1) = E_o \cdot \exp(i\varphi_o) \cdot \text{circ}\left(\frac{r_1}{D/2}\right) \tag{5.31}$$

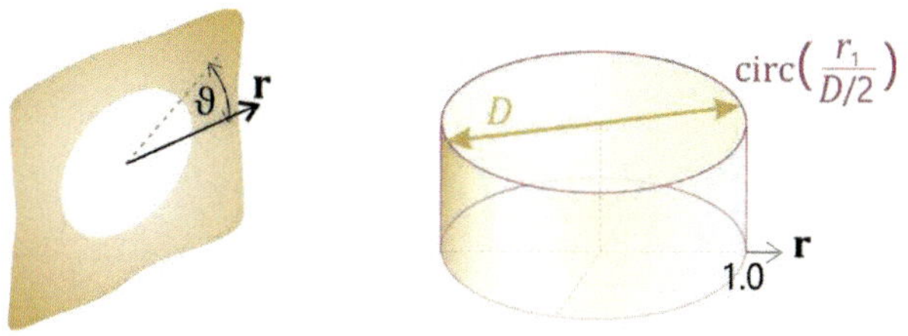


*Figure 5-36: Entrance field for the function* circ[$r/(D/2)$].

The function circ[$r/(D/2)$] has values of 0 for $r > D/2$ and 1.0 for $r < D/2$. The Fourier transform of this function, which is the electric field diffraction intensity distribution, is

$$\underbrace{E(r_o)}_{\text{diffraction}} = F\{E(r_1)\}^{-1} = \frac{J_1\left(\pi D r_o / \lambda z\right)}{\pi D r_o / \lambda z} = \frac{J_1(\rho)}{\rho} \tag{5.32}$$

The diffraction intensity distribution, which is the square of the diffraction field, is

$$I(r_o) = E^2(r_o) = \left(\frac{J_1(\rho)}{\rho}\right)^2 \tag{5.33}$$

The quantity $\rho = \pi \cdot D r_o / \lambda z$ is the **radial angular diffraction parameter** for a circular aperture with diameter $D$, wavelength $\lambda$, and an observation screen at distance $z$, over which we measure radial distances $r_o$:

$$\rho = \frac{1}{2} k \frac{D r_o}{z} = \pi \frac{D}{\lambda} \cdot \frac{r_o}{z} = \pi \frac{D}{\lambda} \cdot \sin\vartheta \tag{5.34}$$

The function $J_1$ [see Eq. (5.32)] is the first-order Bessel function, which is defined as

$$J_1(x) = \frac{1}{x}\int x\, J_o(x)\, dx \qquad \text{and} \qquad J_o(x) = \int_0^{2\pi} \exp(ix\cos\vartheta)\, d\vartheta \tag{5.35}$$

where $J_o$ is the zeroth-order Bessel function.

The diffraction intensity distribution [Eq. (5.33)] is known as the **Airy pattern**, whose central lobe is the Airy disk, named after Sir George Biddell Airy. This expression has radial symmetry around the central maximum $I_o$ for $\rho = 0$. In analogy with Eq. (5.22), the maximum is

$$\lim_{\rho=0} \frac{J_1(\rho)}{\rho} = 1 \tag{5.36}$$

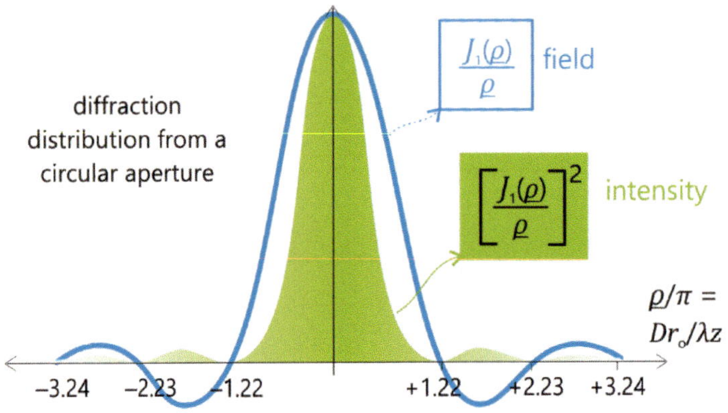


*Figure 5-37: Diffraction distributions from a circular aperture: electric field (blue) and intensity (green). Compare with Figure 5-24 and Figure 5-33.*

The first zero-intensity (the minimum-intensity ring) occurs for $\rho = 1.22{\cdot}\pi$ and defines the **Airy disk radius**. This is the radius of the central bright circular spot, or disc, and corresponds to the central lobe. For expressions involving the observation angle $\vartheta_r$, the characteristic quantity is the apparent angle $\vartheta_o$ drawn to the radius of the central disk from the center of the circular aperture:

$$\rho = \pi\frac{D}{\lambda}\cdot\sin\vartheta_o = 1.22\cdot\pi \qquad \Rightarrow \qquad \sin\vartheta_o \approx \vartheta_o = 1.22\cdot\frac{\lambda}{D} \tag{5.37}$$

The first zero-intensity (generally, the minimum intensity) in angle units is at $1.22{\cdot}\lambda/D$. It is proportional to the wavelength $\lambda$ and inversely proportional to the diameter $D$ of the diffractive aperture, independent of the observation distance $z$. The diameter of the first zero-intensity (the first minimum intensity/dark ring) in the Airy disk is

Diameter of the First Zero-Intensity (in air): $$2\cdot 1.22\,\frac{\lambda z}{D} \tag{5.38}$$

If, in the space between the circular aperture and the observation screen, there is a medium with refractive index $n$, the Airy disk first minimum-intensity diameter appears at

Diameter of First Zero-Intensity (in medium with index $n$): $$2\cdot 1.22\,\frac{\lambda z}{n\cdot D} \qquad (5.39)$$

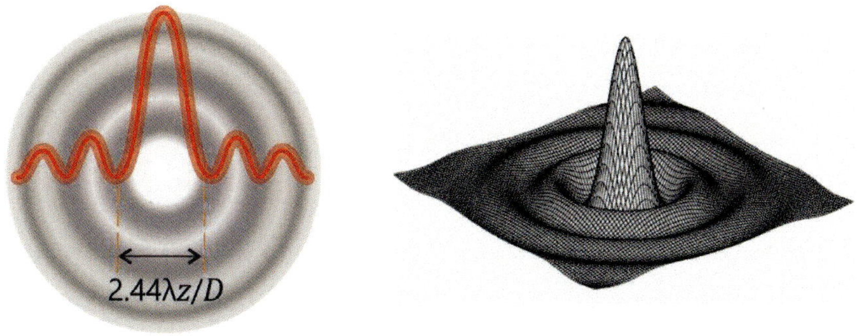


*Figure 5-38: Airy disk diffraction intensity distribution from a circular aperture.*

The minimum-intensity (zero-intensity) ring radius is determined by the values of the radial angular diffraction parameter $\underline{\rho}$ that are the solutions of Eq. (5.32).

In addition to the first parameter, which occurs for $\underline{\rho}_1 = 1.22\,\pi$ (first zero), we also have $\underline{\rho}_2 = 2.233\,\pi$ (second zero), $\underline{\rho}_3 = 3.238\,\pi$ (third zero), etc. The dark rings therefore are not equispaced, contrary to the minima in a single-slit or rectangular aperture, which are equispaced. The secondary maxima around the central disk correspond to much lower intensities and attenuate very quickly away from the central maximum. The central disk encloses approximately 85% of the diffracted light energy.

*Table 5-3: Comparative diffraction characteristics: single slit versus circular aperture.*

| | Single slit, size $\alpha$ | Circular aperture, diameter $D$ |
|---|---|---|
| diffraction intensity distribution | 1-D, Cartesian coordinates<br>$(\sin^2 \underline{\alpha})/\underline{\alpha}^2$, $\underline{\alpha} = ½\, k\cdot \alpha\, x_o/z$ | Polar coordinates<br>$J_1^2(\underline{\rho})\,/\underline{\rho}^2$, $\underline{\rho} = ½\, k\cdot D\, r_o/z$ |
| characteristics | side lobes of equal spacing; central lobe, doubled extent | rings nearly, but not exactly, equispaced |
| central lobe thickness | $2\,\lambda\cdot z/\alpha$ | $2\cdot 1.22\,\lambda\cdot z/D = 2\cdot 1.22\,\lambda\cdot f/\#$ |
| central lobe angular extent | $2\,\lambda/\alpha$ | $2\cdot 1.22\,\lambda/D$ |

# 5.5 IMAGE QUALITY ASSESSMENT

## 5.5.1 Diffraction-Limited Optics

Geometrical optics precisely predicts the image location and size for any given object location and size. However, it offers no information concerning the image sharpness. Image sharpness depends on how small the smallest point image is, which is determined by the wave nature of light, as manifested by diffraction.

The concept of a point is often used in imaging. The problem is that points do not exist. We cannot ignore the wave nature of light, no matter how convenient it is to do so. The reality of the wave nature of light is that the smallest image formation is not a point, but, instead, a specific, finite light intensity distribution, whose extent is determined by diffraction. Thus, every aperture affects the quality of the image.

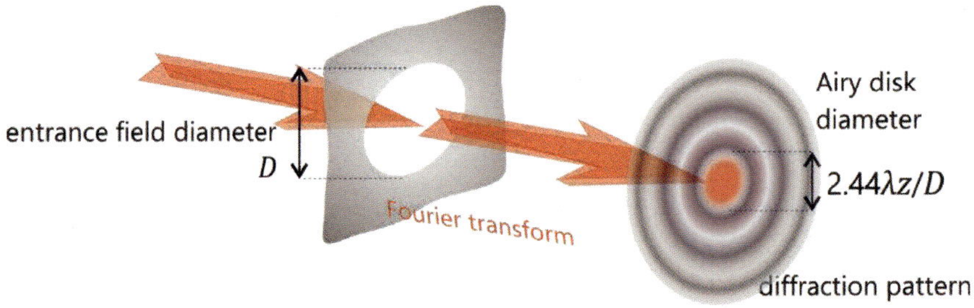


*Figure 5-39: Focal distribution of a 'perfect' lens = the Airy disk.*

Consider the circular aperture of the simplest imaging system, a lens. We illuminate this aperture with a uniform, plane (collimated beam) coherent wavefront. Thus, the lens edge functions as a circular diffractive aperture. The image is none other than the Fourier transform of this aperture. At a distance equal to the system's focal length, we have the formation of the Airy disk, which corresponds to the reciprocal of $D$, its diameter.

The circular aperture (like any aperture) filters the spatial frequencies of an infinitely extended incident wavefront. The smaller the aperture [Figure 5-40 (left)], the fewer spatial frequencies pass through. The Airy disk is broad and not very confined. On the other hand, a larger aperture [Figure 5-40 (right)] allows more spatial frequencies to pass, so the image is more specific. The Airy disk is, accordingly, significantly more confined. Even in an ideal, aberration-free optical system illuminated with a perfectly plane, incident wavefront, the smallest possible point in the image formation cannot have a smaller span than that of the Airy disk.

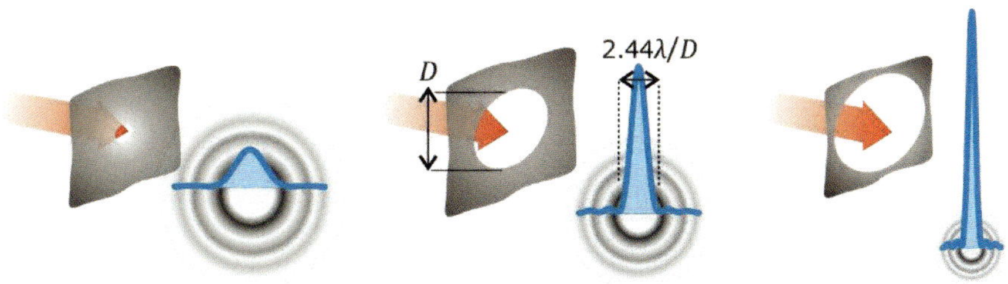


*Figure 5-40: Dependence of the Airy disk size on the entry aperture diameter.*

In the case of imaging through a turbulent medium, or through an imaging system with aberrations, the lower limit of the image span is not the Airy disk, but is a compromised minimum focal spot (Figure 5-41).

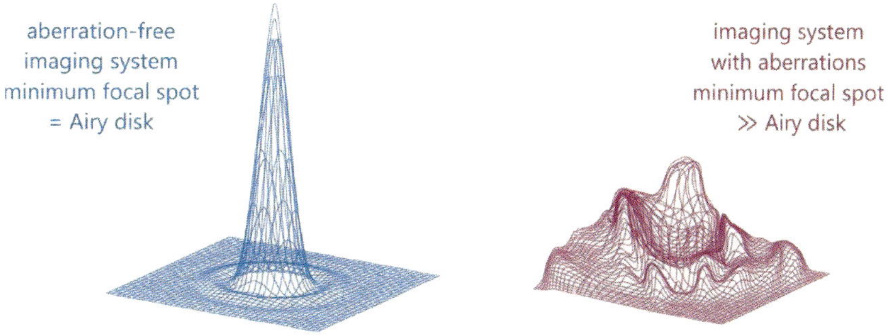


*Figure 5-41: Minimum focal spot in (left) the ideal case and (right) the case of aberrations.*

The wave nature of light, manifested as diffraction, is therefore the ultimate limit that determines the finite extent of the focal point in an imaging system.[26] An optical system designed such that its focal point is not limited by other parameters (such as turbid media or aberrations) is termed **diffraction-limited**.

**Diffraction-limited imaging** occurs in systems with very few optical aberrations.

The Airy disk is the image of a point object.

The minimum size at the image (focal) plane is determined by diffraction.

In a diffraction-limited system with a circular aperture, an Airy spot is formed from a point image. This occurs under the most ideal of conditions: coherent and uniform illumination, absence of aberrations, and absence of medium turbidity and optical noise (stray light).

[26] The only known exception is engineered materials with negative refractive indices (yes, they do exist)!

## 5.5.2 Resolution Limit

Why is a diffraction-limited system important? Because if the optical system can form such a small point from an object point, then it also produces a sharp, crisp image from a real, extended object. It is about image quality.

We want to be able to distinguish the small details in such a crisp image. For example, say we are observing the details of a cell through a microscope. Optical **resolution** describes the ability of an imaging system to discern (resolve) detail. Two closely spaced but independent radiating points should be imaged to two distinct points. The concept of resolution applies to any imaging system, such as a telescope, a microscope, a photography camera, and, of course, the human eye.

The minimum separation (angular or spatial) between the closest distinguishable image points is the **resolution limit**. The lower the resolution limit, the better. The reciprocal of the resolution limit is the **resolving power** or **resolving ability**. The higher the resolving power, the better.

Resolution Limit

- ✔ The **separation** (expressed either angularly or spatially) between the closest distinguishable image points.
- ✔ The lower, the better.
- ✔ Is expressed in angle (arcminute, milliradian) or length (millimeter) units.

Resolving Power (Ability)

- ✔ The **reciprocal of the separation** between the closest distinguishable points imaged through an optical instrument.
- ✔ The higher, the better.
- ✔ Is expressed in inverse angle ($\text{arcmin}^{-1}$) or inverse length (lines/millimeter, cycles/degree) units.

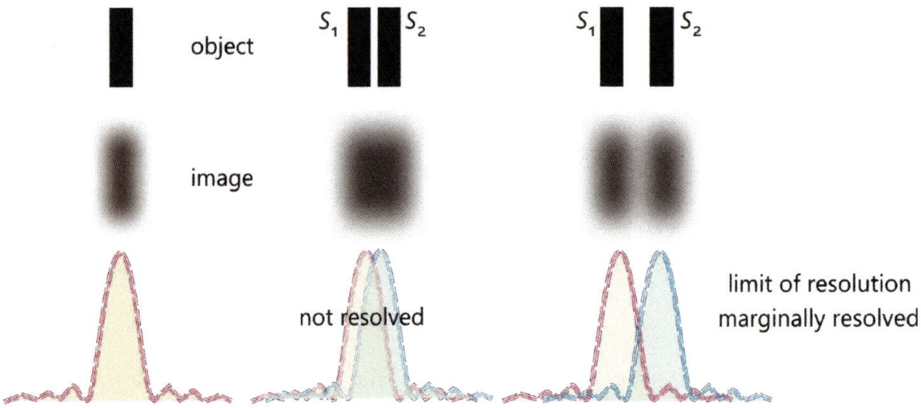


*Figure 5-42: (left) A single object imaged to an Airy disk, (center) two unresolvable spots, and (right) two marginally distinguishable spots, exhibiting the Rayleigh criterion.*

Consider two neighboring, independent, radiating points $S_1$ and $S_2$ that represent two stars imaged through a diffraction-limited telescope (Figure 5-42). At the focal plane, the images are two independent Airy disks. In order to be distinguished as two, their separation must be at least such that the center (bright maximum) of one Airy disk coincides with the first minimum (dark) of the other disk. Then their peak-to-peak separation is at least the radius of the central Airy lobe. This is the **Rayleigh criterion**. We realize therefore that the resolution limit in an aberration-free optical system is, *ceteris paribus* (meaning, other things being equal), governed by diffraction.

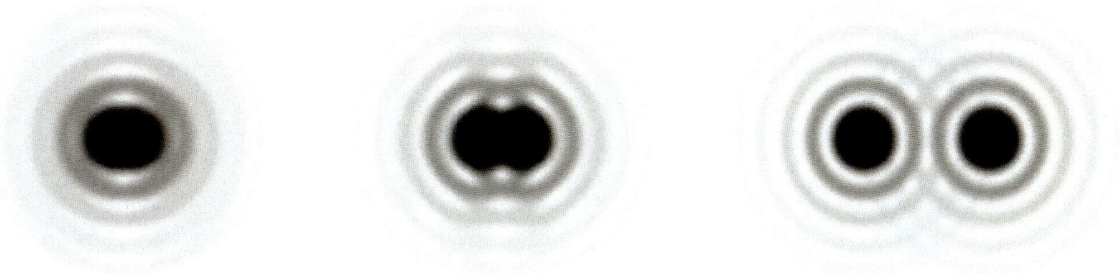

*Figure 5-43: (left) Unresolved, (center) marginally resolved, and (right) resolved Airy patterns.*

Example ☞: Calculation of the Airy disk size. Assume lens diameter $D$ = 1 cm, focal length $f$ = 5 cm, and $\lambda$ = 0.55 μm (mean visible wavelength).

In this case, the radius of the Airy disk is = $1.22 \cdot \lambda \cdot f/D$ = 3.36 μm. This is, indeed, a very small size, 3 thousandths of a millimeter.

It is often preferable to express the resolution limit in angle units. The angular extent corresponding to the Airy disk, which is the smallest angle that two objects may approach such that their images can still be resolved, is derived from Eq. (5.38):

$$\vartheta_{\text{MIN}} = 1.22 \cdot \frac{\lambda}{D} \tag{5.40}$$

where $D$ is the imaging system's aperture stop diameter. If the propagation medium has a refractive index of $n \neq 1$, then, instead of using only the wavelength $\lambda$, we use the quantity $\lambda/n$.

Example ☞: Calculation of the angular size of the example above.

Airy disk angular radius = $1.22 \cdot \lambda/D$ = 0.067 mrad = 0.0038° = 0.23′ (arcmin).

Can we magnify to get a better resolution?

When two image spots overlap, it does not help to further magnify the formed image. The spots will still overlap! They are either resolved or unresolved.

## 5.5.3 Diffraction from a Circular Aperture and Its Effects on Vision

We would like to investigate the resolution limit of the human eye. This resolution limit derives from the relationship $\vartheta_{MIN} = 1.22\ (\lambda/nD)$, where $n = 1.336$ is the refractive index of the aqueous humor, the liquid that fills the eye (vitreous humor, which fills the area of the eye past the crystalline lens, has almost the same refractive index). The entrance pupil of the human eye is about 2 mm in diameter for daylight; therefore, for $\lambda = 0.55\ \mu m$, the human eye can discern objects that are separated by a minimum angle given by

$$\vartheta_{MIN} = 1.22\frac{\lambda}{n \cdot D} = 1.22\frac{0.55\ \mu m}{1.336 \cdot 2\ mm} = 0.25\times10^{-3}\ rad = 0.25\ mrad = 0.86\ arcmin \qquad (5.41)$$

In vision, what really matters is the relative minimum resolvable angular separation in a spatial pattern. We observe two black dots and indeed see them as two. Then we move the dots farther away. The two dots now form a smaller angle. At some point, we can barely tell if they are one or two. This is the resolution limit. Naturally, the smaller this angle, the better.

The human eye thus has a resolution limit of 0.86 arcmin. We typically round this to the value of 1 arcmin (1/60 of a degree): 1 arcmin is the angle subtended by a piece of medium-thickness business card (0.25 mm) held at arm's length. It is a very small angle.

The resolving power of the human eye is its **visual acuity**, which is the reciprocal (expressed in $arcmin^{-1}$) of the **minimum angle of resolution** (MAR) expressed in arcminutes.[27] The 20/20 vision line is what a normal-sighted person can read from 20 ft away and corresponds to patterns (black–white lines) separated by 1 arcmin and therefore by 1.745 mm at 20 ft. Visual acuity can also be reported as the logarithm of the MAR, or the logMAR. In the case of 20/20 vision, logMAR = 0.0, since $\log_{10}(1) = 0$.

The circular aperture serving as an aperture stop (the anatomical pupil) has a variable size. It is affected, among other factors, by the ambient illumination. While we use 2 mm in daylight, the circular aperture can be as large as 8 mm at night. This might lead us to think that the resolution limit of the eye is not 1 arcmin but 0.25 arcmin: If we take the reciprocal to find the resolving power and visual acuity, it appears that the eye has 4× more resolving power with a large pupil. This is a theoretical approach, however. At these large pupil diameters and under very low ambient illumination, other parameters have a much greater affect on vision and limit the achievable resolution. In this case, aberrations are greatly increased, and the retinal spatial sensitivity (rod vision) is reduced. These factors, among others, effectively do not allow such a resolution to be realized.

[27] *Visual Optics* Chapter 3. Visual Acuity.

## 5.5.4 Quantification of Image Quality: the PSF and MTF Functions

The aperture can be a rectangle, a slit, a circle, or it can have a random shape, in which case it would form its respective diffraction pattern. Other conditions that affect the point-image distribution can exist, such as the presence of aberrations. Regardless of the case, the 3-D image spot corresponding to a single point of an object is the **point spread function** (PSF). The $-z$ dimension is the intensity, while the $-x$ and $-y$ dimensions correspond to the cross-sectional shape at the image-forming plane.

A more general term for the PSF is a system's impulse response. The degree of spreading (blurring) in the PSF base is a measure of the quality of an optical system. In Figure 5-44, two cases are illustrated. The system on top shows the diffraction-limited PSF of a circular aperture, which has the form of an Airy disk. The PSF is confined to the $x$–$y$ plane by a narrow disk, while its center peak maximum is intense. Obviously, this is not the case for an optical system with aberrations (lower system), where the PSF is spread over the $x$–$y$ focal plane.

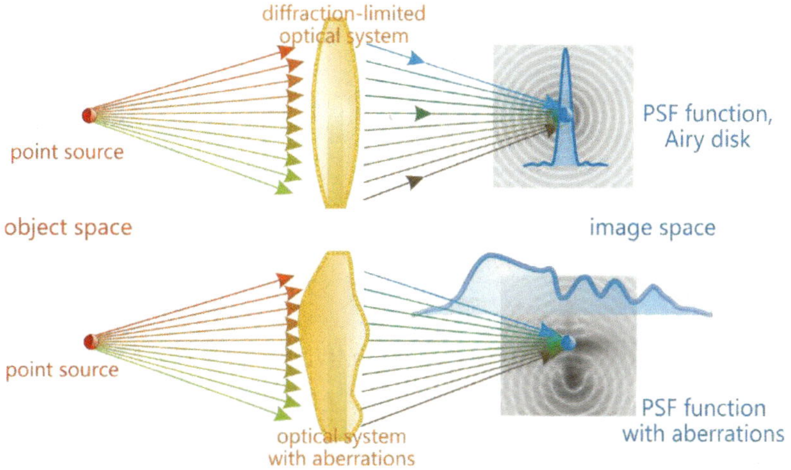


*Figure 5-44: The PSF can be perceived as the visualization of the image corresponding to an object point. (top) In a diffraction-limited optical system, the PSF can be an Airy disk. (bottom) In a real system with aberrations, the PSF is a light distribution with a lower peak intensity and a greater spread at the base than the Airy disk.*

The image of any object can be derived if we consider the object as a superposition of a very large number of points. Each object point has a varying brightness and a PSF. To obtain an image of the entire object, we reconstruct each PSF, shift to the corresponding image location, and scale according to the brightness of the object point. The mathematical expression that describes this shift action is called **convolution** and is denoted by $\otimes$.

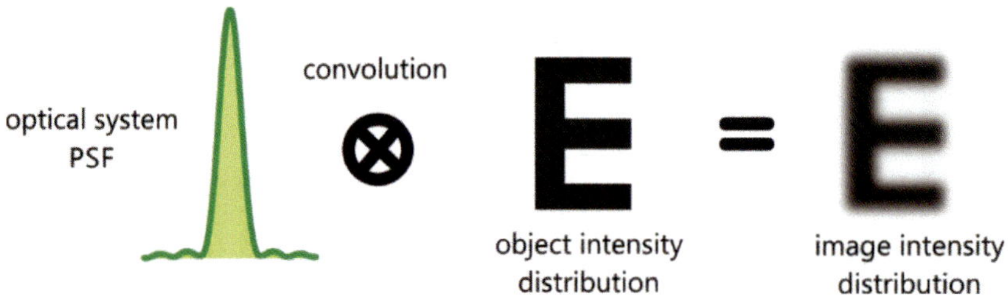


*Figure 5-45: An image can be computed mathematically by a convolution operation between the optical system PSF and the object intensity distribution.*

We can consider the PSF, in effect, as a piece of sandpaper. If we 'rub' the PSF against the object intensity distribution, we get the convolution that is formed in the image shape. The coarser the sandpaper, the blurrier the image. Thus, from a broad PSF we expect a less sharp image, while from a narrow PSF we expect a sharper image.

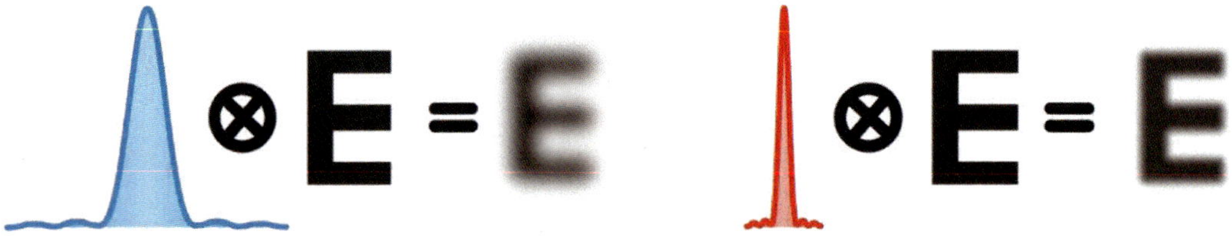

*Figure 5-46: The effect of convolution on the same object is dependent on the shape of the PSF. A broad PSF forms a less clear image.*

The mathematical expression of convolution can take the following form:

$$\underbrace{I(x,y)}_{\text{image intensity distribution}} = PSF(x,y) \underbrace{\otimes}_{\text{convolution}} \underbrace{O(x,y)}_{\text{object intensity distribution}} \tag{5.42}$$

*Figure 5-47: When aberrations are present, the PSF is very broad, often lacking a clear central peak. The image produced is blurry.*

Exactly because the shape of the PSF governs the image intensity distribution, an estimate of the image quality can be derived by a metric based on the PSF. One metric is, of course, the angular extent of its confinement at the base, whose radius is the resolution limit. However, there are other metrics. Here are two examples: In Figure 5-47, the peak of the PSF is compromised (suppressed) due to a defocus aberration. In Figure 5-41 (right), the PSF is compromised due to imaging through a turbulent atmosphere.

The metric that expresses this is the **Strehl ratio**, named after the German physicist, mathematician, and astronomer Karl Wilhelm Andreas Strehl. The Strehl ratio is defined as the ratio of the peak (maximum attainable) PSF intensity corresponding to a real optical system (therefore, an aberrated system) to that of a diffraction-limited optical system of the same aperture diameter. Based on the above, the Strehl ratio cannot be less than 0.0, which is the worst case, and also cannot be more than 1.0, which is the ideal case. The Strehl ratio is not a metric of resolution, but rather of the extent to which the imaging system is aberration free.

Example ☞: In Figure 5-41 (right), the PSF peak is 100 units (arbitrary), while in Figure 5-41 (left), the PSF peak is 45 units. The Strehl ratio is 0.45.

Another function that helps to determine the imaging quality is the **modulation transfer function** (MTF), which expresses the ability of an optical system, such as a lens, to transfer the contrast from an object to its image.

Imagine we are trying to take a photograph of a zebra, whose distinctive black stripes alternate with white stripes of equal width. In the photograph [Figure 5-48 (right)], we note that the line pairs lose sharpness such that the thinner (and denser) pairs may become indiscernible. Beyond a certain line density, we might not see the difference between a black stripe and a white stripe; they are both equally gray. There is a limit to how dense the lines can be and still be observable in the photograph. Thus, the MTF expresses the sharpness with which a specific density of line pairs is imaged.

*Figure 5-48: (left) The stripes of the object are imaged (right) with varying degrees of contrast.*

The **units of spatial frequency** express how close two lines can be, or how many line pairs fit per unit length. The spatial frequency has dimensions of inverse length. In optics, a common metric is **line pairs/millimeter**. A similar metric is dots per inch (DPI), which is used in printers or screens. For example, 300 DPI = 300/2.5″ = 59 line pairs/mm (two dots are needed to make up a line pair). The dots used in printers and scanners express the individual elements and, by extension, the resolution power. In digital screens or digital sensors, the individual elements are called **pixels** (for picture elements), but the metric DPI is still in use.

**Contrast** is an expression of the difference between the brightest (maximum) intensity and the darkest (minimum) intensity in any given light distribution, object, or image. To express contrast we use the modulation index $\mathcal{V}$, a unitless quantity that is defined exactly as in the case of interference fringe visibility (also discussed in § 4.1.3).

Now observe a sinusoidal object and its image. In the object [Figure 5-49 (left)], the intensity varies from a maximum $T_{\text{MAX}}$ (white bands) to a minimum value $T_{\text{MIN}}$ (dark bands) in a sinusoidal fashion (sine chart). The density line pair determines the object spatial frequency: Denser lines correspond to higher spatial frequencies.

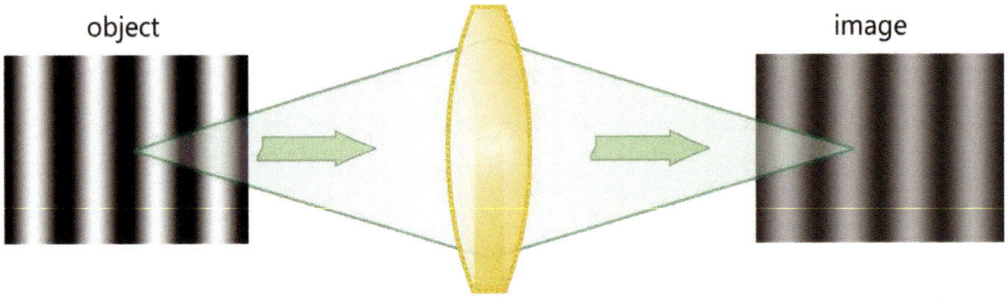


*Figure 5-49: (left) A sinusoidal object and (right) its corresponding image with reduced sharpness.*

Through any imaging system (in the simplest case, a converging lens), we obtain the image [Figure 5-49 (right)]. In the image, there exists again a maximum intensity $I_{\text{MAX}}$ (white bands) and a minimum intensity $I_{\text{MIN}}$ (dark bands). This is how a specific spatial frequency is imaged. In an ideal imaging system, the image appears exactly as the object appears. This does not happen because, in effect, the image of a point is not a point. The (previously) bright spots are light gray, and the (previously) dark spots are dark gray. Therefore, the contrast of the image is less than that of the object. Similar to the relationship developed in interference fringes [Eq. (4.20)], the modulation ratios are

Object Modulation:
$$\mathcal{V}_{\text{OB}} = \frac{T_{\text{MAX}} - T_{\text{MIN}}}{T_{\text{MAX}} + T_{\text{MIN}}} \tag{5.43}$$

Image Modulation:
$$\mathcal{V}_{\text{IM}} = \frac{I_{\text{MAX}} - I_{\text{MIN}}}{I_{\text{MAX}} + I_{\text{MIN}}} \tag{5.44}$$

The ratio of the two modulations is the **modulation transfer function** (MTF):

Modulation Transfer Function:
$$MTF = \frac{\mathcal{V}_{\text{IM}}}{\mathcal{V}_{\text{OB}}} \tag{5.45}$$

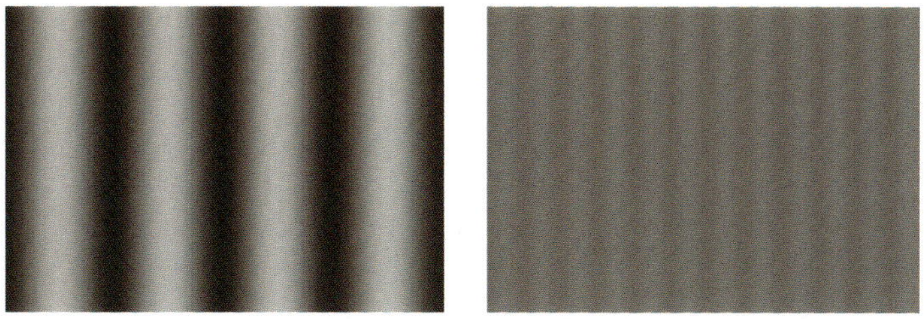

*Figure 5-50: Image with (left) high-contrast, low spatial frequencies and (right) reduced-contrast, high spatial frequencies.*

The MTF value is always less than 1.0. For low spatial frequencies (broad line pairs) the value is close to 1.0, but for high spatial frequencies (dense line pairs) the MTF value is smaller. We note that the value of the MTF depends on the spatial frequency. We can no longer discern image line pairs when the contrast between them is entirely lost, which occurs when $I_{MAX} = I_{MIN}$. For these frequencies, the value of the MTF is zero. The highest spatial frequency, which is the densest line pair that can identified, is termed the **cut-off frequency**.

In a lab setting, we use bar charts or sine charts, for which the line pairs (a succession of dark and white stripes) condense gradually. In doing so, we simultaneously image many spatial frequencies and can observe their sharpness in the image. As the spatial frequency progressively increases, the sharpness of the line pairs in the image decreases to a point where we cannot distinguish any contrast between them. In the cross-sections in Figure 5-51, we note the declining image sharpness and therefore the lower MTF values for the increasing spatial frequencies.

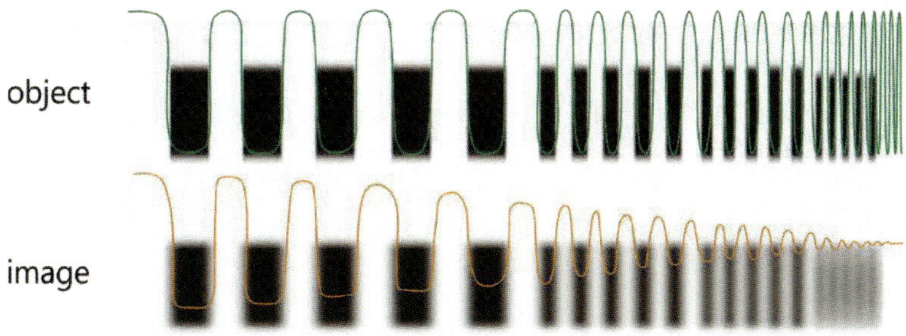


*Figure 5-51: The high spatial frequencies (increasing progressively from left to right) become indiscernible due to the reduced contrast in the image (bottom) compared to the contrast in the object (top).*

We set the maximum value of the spatial frequency conveniently to equal 1.0. This way, all of the other frequencies correspond to a fraction of this cut-off frequency. Figure 5-52 illustrates the variation in MTF with the fraction of the maximum spatial frequency.

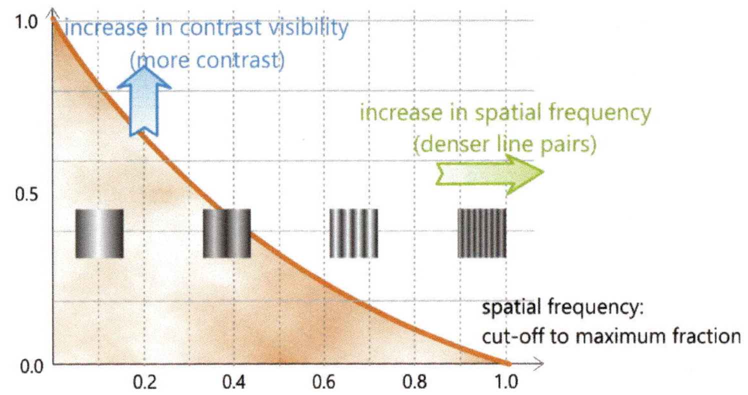


*Figure 5-52: Radial profile of the MTF with the fraction of the maximum spatial frequency.*

We can estimate the cut-off frequency using the Rayleigh criterion. Two neighboring points are no longer discernible when ½ of the width (or the radius for a circular aperture) of the central lobe for slit-diffraction pattern formation is at the maximum of its immediate neighboring lobe. For a circular aperture,

Minimum Distance of Discernible Line Pairs = ½ central lobe thickness: $1.22\ \lambda{\cdot}z/D$ (5.46)

where $\lambda$ is the wavelength, $z$ is the distance to focus, and $D$ is the lens aperture. The spatial cut-off frequency, the reciprocal of the minimum distance, for a circular aperture and a slit are

$$\underset{\text{circular aperture}}{k_{\text{cut-off}}} = \frac{D}{1.22\ \lambda z} \quad \text{and} \quad \underset{\text{slit}}{k_{\text{cut-off}}} = \frac{D}{\lambda z} \tag{5.47}$$

The spatial cut-off frequency can be a metric of the resolving power. The higher this frequency, the better, as denser or a greater number of line pairs can be observed. Typically, we use the simpler equation [Eq. (5.47) (right)] to calculate the cut-off frequency.

Example ☞: Calculation of the cut-off frequency for a lens with aperture diameter $D$ = 1 cm and focal length $f$ = 5 cm. Consider $\lambda$ = 0.55 μm (mean visible wavelength).

The distance to focus is the lens focal length; therefore, $z = f$ = 5 cm. The cut-off frequency is $D/(\lambda \cdot f)$ = 363 lines/mm = 182 line pairs/mm (considering that it takes 2 lines per mm for a line pair per mm).

While in optics we use line pairs/millimeter, in vision, we also use the unit of **cycles/degree**. This unit is defined as a pair consisting of a black stripe and a white stripe (1 cycle = 2 lines = 1 line pair) in a degree. This unit is used because in visual optics resolution is an angular matter. Specifically for the retina of the human eye, an image of 1 mm across corresponds to 3.345°. Therefore,

$$1\ \frac{\text{cycle}}{\text{degree}} = \frac{1 \text{ line pair (2 lines)}}{\frac{1}{3.345}\text{ mm}} = 6.69\ \frac{\text{lines}}{\text{mm}} = 3.345\ \frac{\text{line pairs}}{\text{mm}} \tag{5.48}$$

The resolving power of the human eye with a 2 mm-diameter pupil is about 100 line pairs/mm, which converts to 30 cycles/degree.[28]

The diffraction-limited expressions of the MTF for a slit and a circular aperture are

$$\underbrace{MTF(x)}_{\text{slit}} = 1 - x \quad \text{and} \quad \underbrace{MTF(x)}_{\text{circular aperture}} = \frac{2}{\pi}\left[\cos^{-1}(x) - x\sqrt{1-x^2}\right] \tag{5.49}$$

where $x = k/k_{\text{cut-off}}$ is the normalized, unitless fraction of the maximum spatial frequency.

Figure 5-53 illustrates the variation in the MTF for a diffraction-limited slit, a diffraction-limited circular aperture, and a circular cross-section aperture stop in an actual imaging system that has optical aberrations.

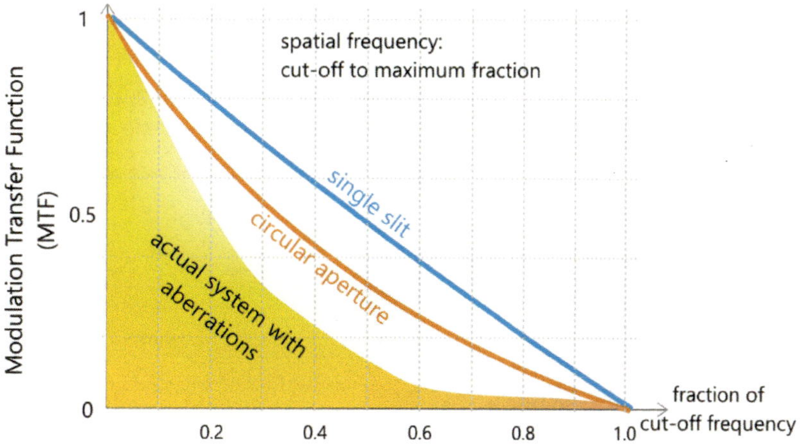


*Figure 5-53: Radial profile of the MTF for a single slit, a diffraction-limited circular aperture, and a circular cross-section aperture stop in an actual imaging system that has optical aberrations.*

A unique advantage of the MTF is that, under certain conditions, the MTF of a system can be calculated by cascading the MTFs of the component systems. In other words, the MTF of an optical system at any frequency can be computed by multiplying the value of the MTFs of each of the system's components at that same spatial frequency.

This is useful in calculating the effects of camera systems, for example, lens + film (or sensor). In general, however, one cannot simply multiply the incoherent MTFs of all the lenses in a multi-element system and obtain the MTF for the system, even if the lenses are all diffraction-

[28] *Visual Optics* § 3.4.3.3 Spatial Frequency and Cycles per Degree.

limited. The system MTF is determined from the cascaded pupil function.[29] Therefore, the element (lens) with the lowest numerical aperture will limit the system cut-off frequency, and the element (lens) with the most aberrations will most influence the shape of the MTF.

Why is the shape of the MTF important? Similar to the PSF, the shape of the MTF can offer a good indication of the expected image quality. In the example of Figure 5-53, the MTF is compromised due to imaging via aberrations. Effectively, the cut-off frequency is reduced to 0.7 of the maximum fraction of the diffraction-limited aperture. More importantly, however, the **area under the curve** of the MTF is reduced. This area under the curve can also be used as a metric of image quality. Its definition is the ratio of the area under the curve of an actual (therefore, aberrated) system to that of the corresponding diffraction-limited pattern from the same aperture. Based on the above, the ratio of the area under the curve cannot be less than 0 and cannot be more than 1.0, which is the ideal case. This ratio is equivalent to the Strehl ratio introduced with the PSF.

The MTF relates to the PSF via a Fourier transform [Eq. (5.13)]. The more expanded the MTF is in the frequency domain, the more condensed the PSF is in the spatial domain.

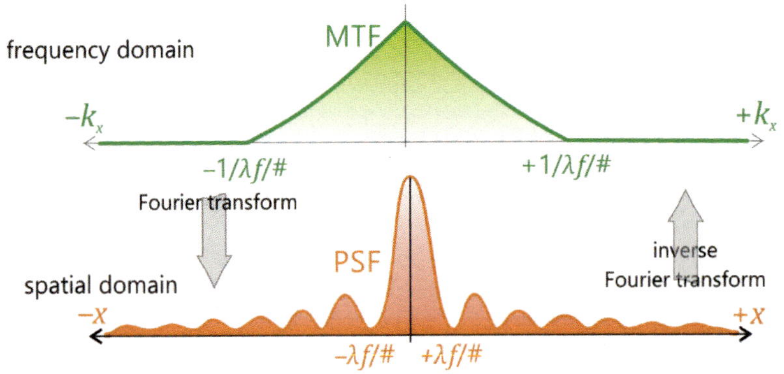


*Figure 5-54: The MTF and PSF functions are Fourier transform pairs.*

With this methodology, we can study the quality of an actual imaging system. In an object, the intensity distribution corresponds to either the object reflectivity or the object transmissivity, represented by an entrance function, which may vary significantly from object to object; however, this function can always be analyzed into a set of harmonic components and, inversely, be composed of the spatial frequencies.

[29] In the case of an incoherent optical imaging system, the Fourier transform of the PSF is the optical transfer function (OTF). The OTF is the auto-correlation of the pupil function, which itself is a complex function. For simplicity, here we only describe the MTF, which is the magnitude of the complex OTF.

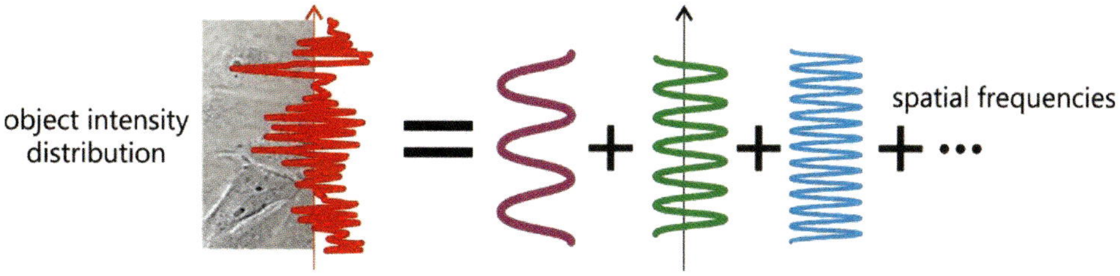


*Figure 5-55: Object intensity breakdown at various spatial frequencies.*

For each of the spatial frequencies, there is a different MTF value, so each frequency is imaged by the optical system with a different contrast. Some low frequencies propagate unaffected and are not eliminated, such as, for example, spatial frequencies a and b in Figure 5-56.

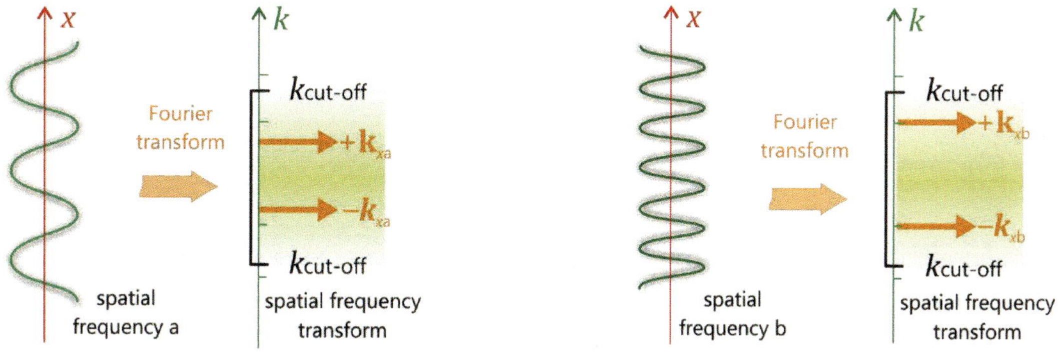


*Figure 5-56: Fourier transforms of spatial frequencies. All frequencies (left and right), represented by the $\pm k_x$ vector magnitudes, propagate through (are not cut off by) the imaging system.*

However, there are certain high frequencies that are cut off; these correspond to values greater than the $k$ cut-off, as illustrated in Figure 5-57.

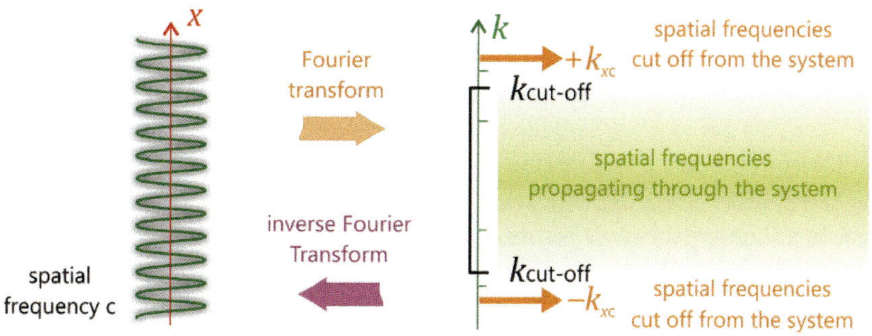


*Figure 5-57: Fourier transform of spatial frequencies. These frequencies, represented by the $\pm k_x$ vector magnitudes, are cut off from the imaging system.*

So, if there is cut-off for the high frequency ($\pm k_{xc}$), then the image is composed only of frequencies $\pm k_{xa}$ and $\pm k_{xb}$. The image thus appears less sharp because it loses part of its high-frequency content. Figure 5-56 and Figure 5-57 offer an illustrative expression that describes the action of an optical imaging system in terms of spatial frequencies—there is always an upper limit to the frequency of the content to be imaged.

In practice, optical resolution charts are used to test the resolution, contrast, distortion, and MTF of imaging systems. U.S. Air Force (USAF) resolution charts are used as a standard for such testing. Each element on the chart comprises two patterns (two sets of three lines) of vertical and horizontal bars separated by spaces of equal width. The lines are five times as long as they are wide. The groups of lines and spaces gradually decrease in size by factors of $\sqrt{2}, \sqrt[3]{2}, \sqrt[6]{2}$. The detail on these transparency slides can be as fine as 0.78 μm (641 line pairs/mm).

The resolution is specified as the group and element of the finest bars for which the difference between the black and the white in the image can be determined. The resolution can thus be expressed as vertical or horizontal.

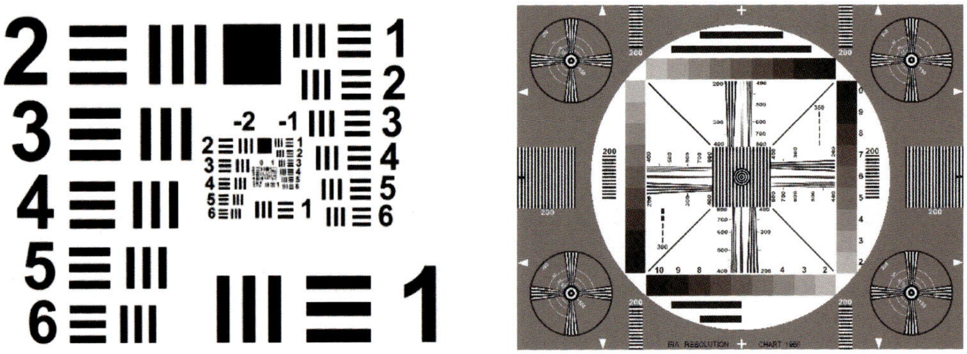

*Figure 5-58: Resolution charts: (left) The 1951 USAF resolution chart and (right) the 1956 EIA resolution chart, which is also used for grayscale contrast testing.*

**PSF:**

- An indication of the smallest possible image that a system can form.
- Can serve as a metric for the resolution limit.

**MTF:**

- An indication of the tightest possible line density that a system can image.
- Can serve as a metric for the resolution power.

# 5.6 DIFFRACTION BY MORE THAN ONE APERTURE

In Young's interference experiment, we considered two infinitesimally small sources. In a more realistic case, the sources have a specific, measurable size. We consider two slits of the same size $\alpha$, whose centers are separated by distance $d$.

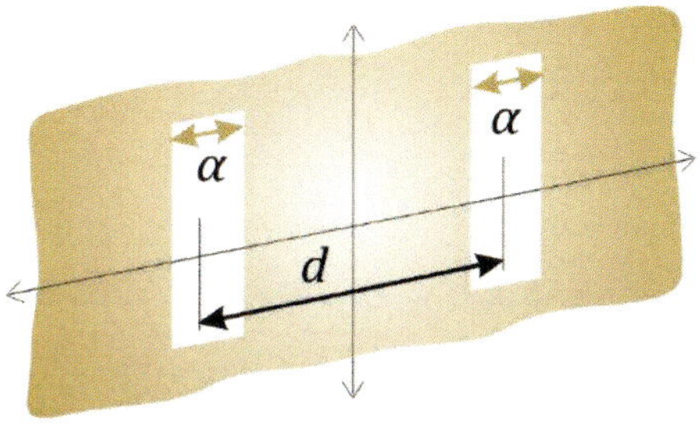


*Figure 5-59: Two similar diffractive apertures: slits of size α separated by distance d.*

The diffraction pattern resulting from one of such slits is well known (as seen in § 5.3). It has a central bright lobe formed on the optical axis, with the first zero at angle $\lambda/\alpha$. The configuration is dependent only on the slit size and the wavelength.

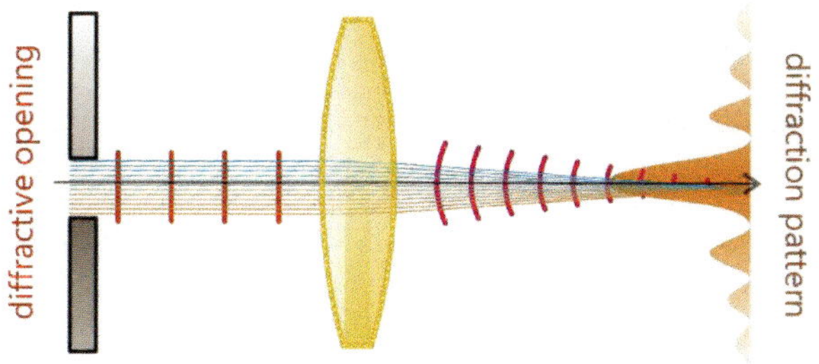


*Figure 5-60: Diffraction intensity distribution of a slit exactly centered on the optical axis.*

Now, instead of one slit, we have two similar slits that are not centered on the optical axis. What is the diffraction intensity distribution of such an entry field? We will prove that each of the two slits forms exactly the same diffraction pattern at the same location as the pattern that would be formed if the slit was on the optical axis. The intensity pattern formed by a displaced slit (with no other change, such as tilt or rotation) is identical (with respect to the optical axis) to the pattern formed when the slit is not displaced.

When a collimated beam is imaged by a converging lens, a ray parallel to the optical axis focuses on the focal point located on the principal optical axis. Let's apply some math here.

*Figure 5-61: Intensity distributions of similar slits displaced by a distance d/2 from the optical axis.*

At slit $P_1$, we assume an incident plane wavefront of fixed amplitude of magnitude $E_o$. A slit of size $\alpha$ centered at $x = 0$ is described by the function rect($x/\alpha$), which may be re-written as rect($x/\alpha$) ⊗ delta(0), where delta is the Dirac delta function, and ⊗ denotes convolution. The result of the convolution can be described by the overlap of the two functions at every location $x$. This overlap is obtained by pulling one function (rect), while keeping the other (delta) in place. A slit at $x = +d/2$ can be expressed as a convolution: rect($x/\alpha$) ⊗ delta($x - d/2$). Thus, at the location $+d/2$, the function rect($x/\alpha$) appears. The field at a slit of width $\alpha$, shifted laterally to the optical axis by $d/2$, is

$$E(x_1) = E_o \cdot \underbrace{\exp(i\varphi_o)\cdot \mathrm{rect}\left(\frac{x_1}{\alpha}\right)}_{\text{slit of thickness } \alpha \text{ (on-axis)}} \otimes \underbrace{\left[\mathrm{delta}\left(x - \frac{d}{2}\right)\right]}_{\text{slit displacement by } +d/2} \tag{5.50}$$

The key question is: What is the diffraction field of this shifted slit? Specifically, what is the effect of the shift ($d/2$)? To find the Fourier transform of the field function, we use the shift theorem of Fourier transformations:

$$\underbrace{E\left(x_{o\,P_1}\right)}_{\text{diffraction from slit 1}} = F\left\{E\left(x_{1\,P_1}\right)\right\}^{-1} = \underbrace{E_P}_{\substack{\text{on-axis diffraction field}\\ \text{from a single slit}}} \cdot \underbrace{\exp\left(+i\frac{\pi d}{\lambda}\cdot\frac{x}{z}\right)}_{\substack{\text{phase factor (ray tilt)}\\ \text{due to shift by } -d/2}} \tag{5.51}$$

Here $E_P$ is the diffraction intensity distribution field of a slit with characteristics of $P_1$ [Eq. (5.19) and Figure 5-23], located exactly on the optical axis (zero shift). The field is now multiplied by a phase factor that expresses a tilt with respect to the optical axis. This is the effect of the slit displacement ($-d/2$): The diffraction field has a slope (with respect to the optical axis) that is determined by the shift.

The diffraction intensity distribution is the magnitude of the field squared. The square of the slope factor is simply 1.0, so it does not affect the intensity pattern. If at the entrance we only had slit $P_1$, the diffraction pattern would be no different from the intensity distribution of a slit on the optical axis. Accordingly, the diffraction field intensity distribution from slit $P_2$ is

$$\underbrace{E\left(x_{\mathrm{o\,P_2}}\right)}_{\text{diffraction from slit 2}} = F\left\{E\left(x_{1\,\mathrm{P_2}}\right)\right\}^{-1} = \underbrace{E_{\mathrm{P}}}_{\substack{\text{on-axis diffraction field}\\ \text{from a single slit}}} \cdot \underbrace{\exp\left(-i\frac{\pi d}{\lambda}\cdot\frac{x}{z}\right)}_{\substack{\text{phase factor (ray tilt)}\\ \text{due to shift by } +d/2}} \tag{5.52}$$

which corresponds to a displacement (shift) by $+d/2$. This distribution has a tilt that is <u>opposite to that of</u> $P_1$.

The two independent diffraction field intensity distributions are both formed on the same spot on the optical axis $z$ and differ only with respect to their phase due to the different tilts of the image-forming rays; the intensity that results from a single slit has exactly the same pattern as the pattern corresponding to a slit of size $\alpha$.

Here comes the fun part. On the observation screen there are two overlapping coherent waves, the two field distributions $E_{P1}$ and $E_{P2}$. The total disturbance results from a <u>linear superposition of these two fields</u>, exactly as in the case of two point sources interfering in Young's interference experiment:

$$E_{\mathrm{TOT}} = \underbrace{E\left(x_{\mathrm{o\,P_1}}\right)}_{\text{diffraction, slit 1}} + \underbrace{E\left(x_{\mathrm{o\,P_2}}\right)}_{\text{diffraction, slit 2}} = E_{\mathrm{P}}\cdot\left[\exp\left(+i\frac{\pi d}{\lambda}\cdot\frac{x}{z}\right)+\exp\left(-i\frac{\pi d}{\lambda}\cdot\frac{x}{z}\right)\right] = 2E_{\mathrm{P}}\cdot\cos\left(\frac{\pi d}{\lambda}\cdot\sin\vartheta_x\right) \tag{5.53}$$

where $x/z = \sin\vartheta_x$. Distances $x$ are measured on the observation screen, situated at a distance $z$ from the slits along the optical axis. The total field forms a sinusoidal function with a periodic step $\lambda\cdot z/d$ (if expressed by its angular extent, then it is simply $\lambda/d$). This is an <u>interference factor</u> that depends only on the source (slit) separation $d$.

In summary: The total diffraction field is obtained by (1) calculating the field of each of the entrance slits, (2) implementing a Fourier transform to find the diffraction field, and (3) implementing linear superposition between the two fields.

We can also take a different approach. We can first add the entrance fields, then implement Fourier transform on the combined field, and we will find the exact same result. Let's prove this. At the entrance there are two slits of size $\alpha$, spaced by $d$. At the locations $+d/2$ and $-d/2$, the spikes produced by function $\mathrm{rect}(x/\alpha)$ appear. Therefore, the total entrance field is

$$\underset{\text{two slits}}{E(x)} = E_{\mathrm{o}}\cdot \underbrace{\exp(i\varphi_{\mathrm{o}})\cdot\mathrm{rect}\left(\frac{x}{\alpha}\right)}_{\text{slit of thickness } \alpha \text{ (on-axis)}} \otimes\left[\mathrm{delta}\left(x-\frac{d}{2}\right)+\mathrm{delta}\left(x+\frac{d}{2}\right)\right] \tag{5.54}$$

which is illustrated in Figure 5-62. Compared to Eq. (5.19), there is an additional factor, the function $[\mathrm{delta}(x-d/2) + \mathrm{delta}(x+d/2)]$.

The Fourier transform [following the relationships in Eq. (5.15)] of this term is the cosine function $\cos(\pi d\, f_x)$. As for the remaining part, its Fourier transform is the typical single-slit diffraction intensity distribution. According to the convolution theorem, the Fourier transform of these two convolved fields is the product of their independent Fourier transforms.

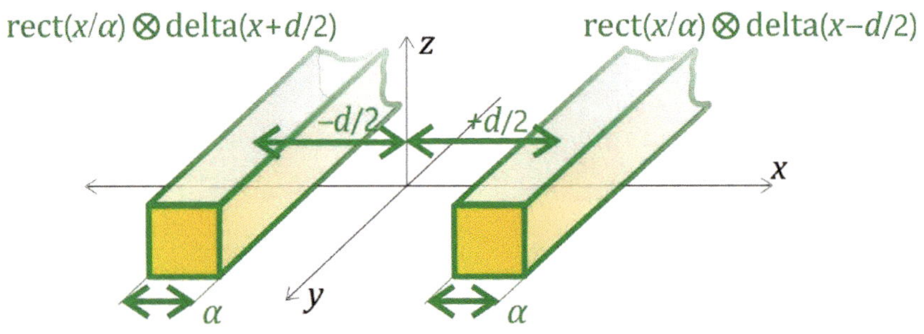


*Figure 5-62: Mathematical modeling of two slits of size α separated by distance d.*

Thus, the distribution of the intensity due to diffraction from two slits is the product of the diffraction pattern formation owing to only one slit, multiplied by the interference factor owing to their separation, $\cos^2[\pi d/\lambda \cdot \sin(\vartheta_x)]$:

$$I_{\text{two slits}} = 4I_o \times \underbrace{\left(\frac{\sin(\underline{\alpha})}{\underline{\alpha}}\right)^2}_{\text{diffraction factor}} \times \underbrace{\cos^2\left(\frac{\pi d}{\lambda}\sin\vartheta_x\right)}_{\text{interference factor}} \tag{5.55}$$

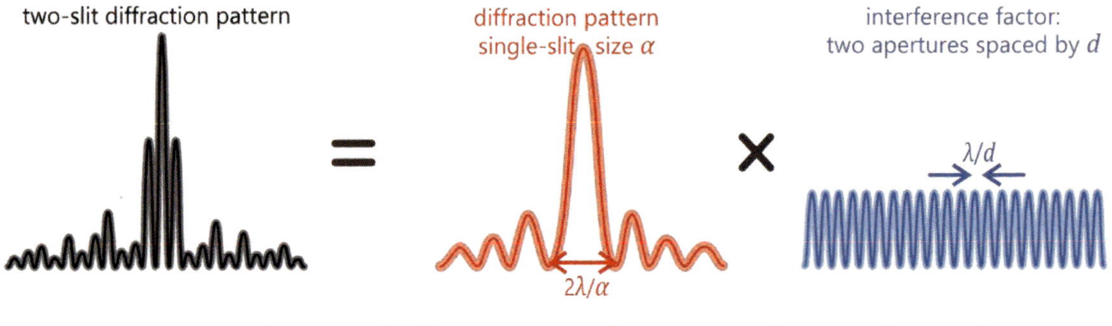


*Figure 5-63: Combination of diffraction pattern formations with an interference factor.*

The combined diffraction intensity distribution can be considered as a pattern of amplitude-modulated carrier frequencies. The amplitude-modulated distribution (envelope) is the independent diffraction pattern of a single slit, whose main parameter, the central bright lobe, is inversely proportional to the slit size $\alpha$. The carrier frequency has a periodic modulation that is inversely proportional to the two-slit spacing $d$, just as in the interference by two aperture point sources.

The formation of diffraction patterns draws on aspects of both diffraction and interference. It has the characteristics of diffraction from an aperture (assuming the same shape, size, and orientation) and simultaneously displays line pairs (the interference factor)—a periodic distribution that is inversely proportional to the aperture spacing. The latter is independent of the geometry of the diffractive apertures.

To find the diffraction intensity distribution of two similar apertures shifted symmetrically from the optical axis: → Find the diffraction pattern formed by only one aperture, ignoring the transverse shift. → Calculate the interference factor (fringes) from two points located at the shift centers. → Multiply the two formations. The diffraction pattern formation is the amplitude modulated by the interference fringes.

For the two slits with thickness $\alpha$ separated by $d$, obviously, $d > \alpha$; otherwise, the slits would overlap. The number of line pairs that appear in each lobe depends on the relationship between $d$ and $\alpha$. Figure 5-64 describes the case where $d = 2\alpha$. The first minimum of the diffraction factor, the central lobe edge, appears at angle $\vartheta_x = \lambda/\alpha$. Simultaneously, the interference factor is zero every $\vartheta_x = \lambda/d = ½\, \lambda/\alpha$. It is noted that all angles are measured from the optical axis. For $\vartheta_x = 0$ there are maxima for both of the parameters, which we set to 1.0. The diffraction intensity distribution is obtained by multiplying the two factors, point by point.

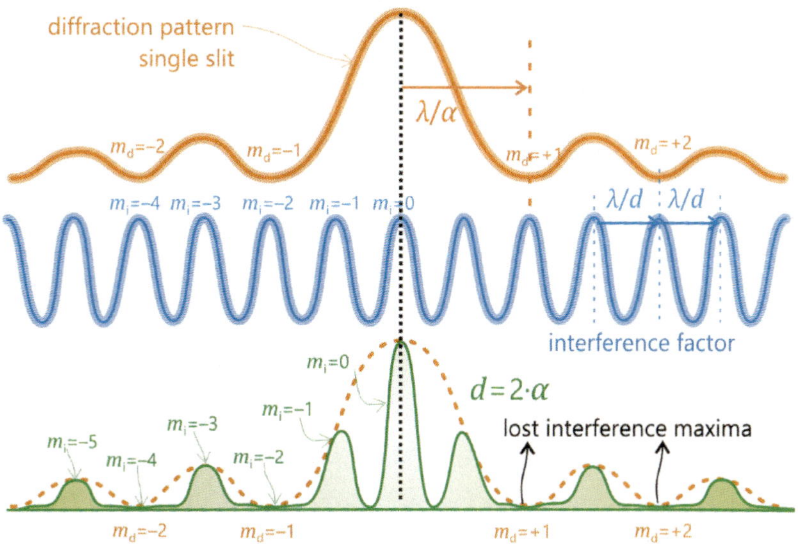


*Figure 5-64: Amplitude-modulated intensity distribution of the interference factor by the diffraction factor for d = 2α.*

Within the main central lobe there are three interference fringes, and in each side lobe there is only one fringe. This is because some maxima of the interference factor vanish due to the fact that specific maxima (orders $m$ = ±2, ±4, ...) of the interference factor coincide with the minima (zero) of the diffraction factor. In general, if $d = k{\cdot}\alpha$, where $k$ is an integer, the eliminated maxima (bright) of the interference factor correspond to the orders $m = \pm k, \pm 2k, \ldots$ . The main lobe contains $(2{\cdot}k - 1)$ bright fringes, and each side lobe contains $(k - 1)$ bright fringes.

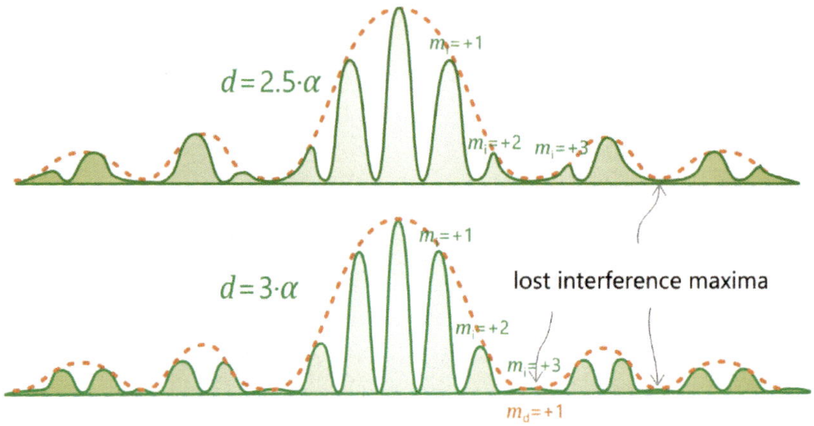


*Figure 5-65: Amplitude-modulated two-slit diffraction intensity distribution for (top) d = 2.5α and (bottom) d = 3α.*

When the ratio of the slit spacing to the thickness is a semi-integer number, the zeros in the central diffraction coincide with the minimum (dark fringe) of the interference factor. This means that no fringes vanish. However, the point between the first and second side lobes (the second zero of the diffraction factor) coincides with the maximum (bright fringe) of the interference factor and causes a fringe to vanish. The same occurs between the third and fourth lobe, and so on. We can state a general condition of fringe elimination as follows:

Minimum Diffraction Factor = the maximum (bright fringe) interference factor:

$$m_{\text{diffraction}} \cdot \frac{\lambda}{\alpha} = \frac{\lambda}{d} + m_{\text{interference}} \cdot \frac{\lambda}{d} \qquad (5.56)$$

The diffraction orders $m_{\text{diffraction}}$ and interference orders $m_{\text{interference}}$ are, in principle, different integers.

The smaller the slit size or the greater their separation ($d/\alpha \gg 1$), the larger the number of interference fringes within the central lobe ($\approx 2 \cdot d/\alpha$). In the vicinity of the optical axis, there are many fringes with nearly equal intensities. Therefore, particularly near the optical axis, the distribution resembles one in which the diffraction factor does not even exist and appears as interference by two point slits.

### 5.6.1 Two Circular Apertures

Up to now, we have presented diffraction by two slits, but we emphasized that the formulation is independent of the diffractive aperture shape as long as the two apertures are similar—of the same shape and same size. We now study the diffraction intensity distribution of two similar circular apertures. The two apertures have exactly the same diameter $D$ and are separated by distance $d$.

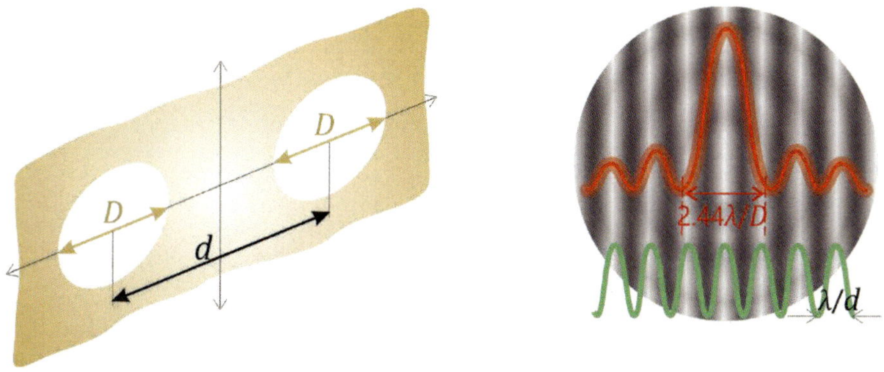


*Figure 5-66: (left) Entrance field of two circular apertures of diameter D separated by distance d. (right) Output of the diffraction intensity distribution of two coherent circular apertures. This distribution corresponds to d = 2D.*

The diffraction intensity distribution results from a carrier frequency with a periodicity of $d/\lambda$ (angular extent $\lambda/d$) that is amplitude-modulated by the diffractive Airy disk formation per each individual circular aperture. The angular extent of the central lobe is $2.44\cdot\lambda/D$.

Using the single circular aperture pattern, as in Eq. (5.32), the expression for two circular apertures now becomes

$$I_{\text{two circular apertures}} = 4I_o \cdot \underbrace{\left( \frac{J_1\left(\pi D r_o / \lambda z\right)}{\pi D r_o / \lambda z} \right)^2}_{\text{diffraction factor}} \times \underbrace{\cos^2\left(\frac{\pi d}{\lambda}\sin\vartheta_x\right)}_{\text{interference factor}} \tag{5.57}$$

The same technique is applied to more than two of such similar diffractive apertures. This is the case when the diffractive entry aperture is composed of periodic repetitions of a specific shape. The same shape is repeated at exactly the same distance, the transposition step.

We will now explore the example of three equispaced slits of equal size.

## 5.6.2 Diffraction by Three Slits

The diffraction intensity distribution that corresponds to each slit is the envelope that modulates the amplitude of the interference factor. All we now need is to calculate the interference factor that corresponds to three point apertures.

We consider three point apertures separated by $d$ and illuminated coherently with wavelength $\lambda$ (Figure 5-67). Therefore, we treat Young's experiment with three, instead of two, point apertures. Assume that we observe this aperture under angle $\vartheta$ with respect to the optical axis. Because points A, B, and C are equispaced, it is easy to prove that the optical path difference that develops between A & B is exactly the same as the one that develops between B & C, and twice that between A & C. Thus, if a condition of constructive or destructive interference applies for the pair A & B, it also applies for the pair B & C, as well as for all combinations, such as A & C.

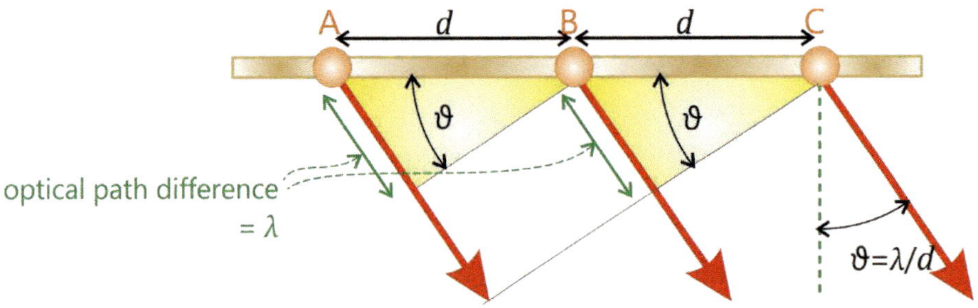


*Figure 5-67: Optical path difference of three point apertures.*

We already know the solution here: The optical path difference is expressed simply by [Eq. (4.37)]:

Optical Path Difference: $$d \cdot \sin(\vartheta) \approx d \cdot \vartheta \text{ [rad]} \quad (5.58)$$

The condition $\sin\vartheta \approx \vartheta$ [rad] is valid for small angles. When expressing the angle as the ratio of the arc to the radius, its value has the units of radians (rad).

To find the angles $\vartheta$ for which there are maxima, all we need to do is to apply the condition of constructive interference: optical path difference = $m \cdot \lambda$:

$$d \cdot \vartheta = m \cdot \lambda \;\Rightarrow\; \vartheta = m \cdot \lambda / d \quad (5.59)$$

where $m$ = 0, ±1, ±2 … . Compared to the pattern resulting from the interference by two points (Figure 4-20), the following can be said about interference by three points:

- The main maxima are equispaced by angles $\lambda/d$ in both cases. They correspond to the same exact locations.
- The intensity of the main maxima within each formation is $4\times I_o$ for two-point interference, while it is $9\times I_o$ for three-point interference. This is because in two-point interference we add two fields ($2\times E_o$), while in three-point interference we add three fields ($3\times E_o$); their squared magnitudes, which yield the intensity, are 4× and 9×, respectively.
- The main maxima are thinner.

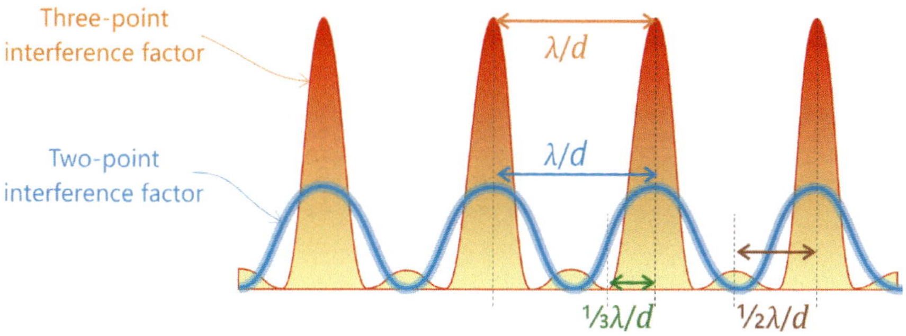


*Figure 5-68: Interference factor comparison between two sources and three sources.*

There is one more fine difference in the case of three-point interference. Halfway between the main maxima, there is a secondary maximum that corresponds to angles with a semi-integer multiple of $\frac{1}{2}\lambda/d$. This is because, at these angles, pairs A & B and B & C are in destructive interference (optical path length = $\frac{1}{2}m\cdot\lambda$), while pair A & C is in constructive interference because the optical path difference between A & C is twice that of pair A & B (optical path length = $m\cdot\lambda$).

Secondary maxima will be further discussed in the section on diffraction gratings (§ 5.7), where we will see that, in general, <u>for $N$ interfering sources, there are $N-2$ secondary maxima</u> and $N-1$ zeros between the main maxima.

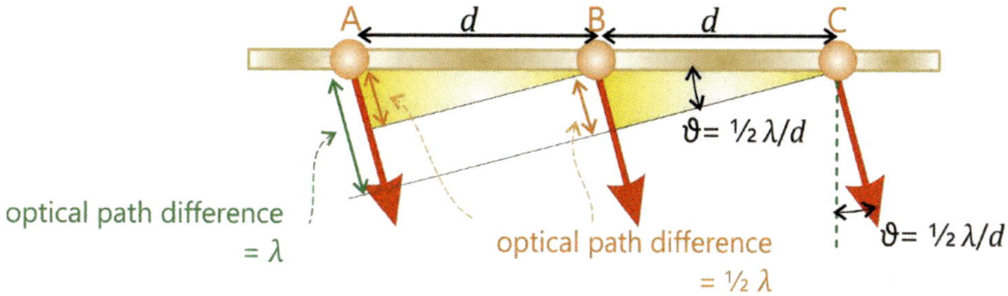


*Figure 5-69: Condition for the appearance of secondary maxima by three sources.*

There is nothing to restrict the number of similar and equispaced diffractive openings from being spread along a specific axis, as has been the case in our discussion so far. The two apertures can be arranged along any line (see, for example, Figure 5-70).

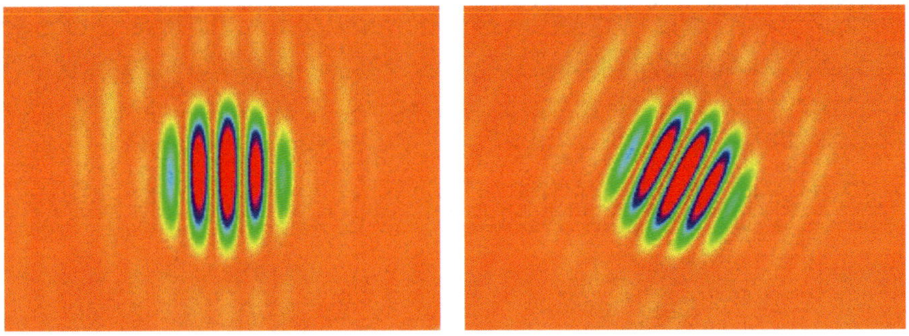

*Figure 5-70: Diffraction pattern of two slits separated along (left) the horizontal axis and (right) the 45° orientation.*

Of course, multiple apertures arranged on a grid (as long as this grid maintains a fixed spacing) can offer intriguing diffraction patterns.

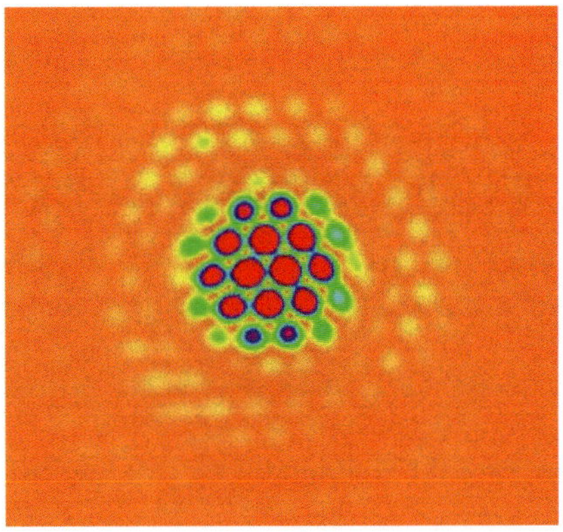

*Figure 5-71: Diffraction pattern of six circular apertures arranged over the corners of a hexagon.*

# 5.7 DIFFRACTION GRATINGS

We just examined diffraction from two and three equispaced and similar sources. The effect is essentially a combination of interference and diffraction. The diffraction part relates to the characteristics of each source, and its spread is inversely proportional to the slit width. For example, the extent of the main lobe is inversely proportional to the slit size $\alpha$. The interference part relates to the source spacing $d$. Its main parameter, the periodic fringe step, is proportional to the wavelength $\lambda$ and inversely proportional to the distance $d$, exactly as in two-point source interference. The diffraction pattern can be considered as an interference-wise amplitude-modulated form of the simple diffraction intensity distribution.

This also applies to a very large number of similar and equispaced diffractive apertures, such as slits. We can find the diffraction pattern formation of each individual slit and then find the interference factor corresponding to the periodic spacing. A **diffraction grating** is fabricated by adding grooves to a glass or metal plate that consist of $N$ parallel slits of size $\alpha$, spaced by $d$.[30] The spacing $d$ is the **grating constant**. It is typical for a diffraction grating to have 500 lines/mm or more, meaning that its grating constant is $d = 2\,\mu\text{m}$.

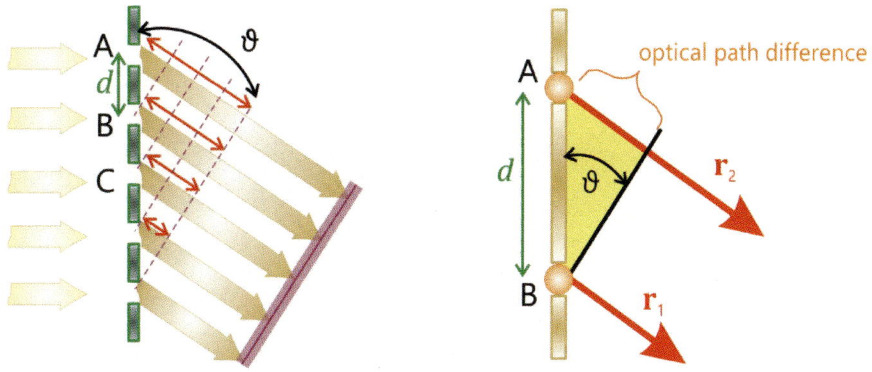


*Figure 5-72: Diffraction grating having equispaced slits of size α with a periodic distribution d.*

When such a grating is illuminated with coherent radiation of wavelength $\lambda$, the diffraction factor corresponds to the function $\mathrm{sinc}^2(\alpha x_o/\lambda z)$. The main diffraction lobe has an angular size of $2\alpha x_o/\lambda z$. This function is a sinusoidal function with step $dx_o/\lambda z$. It acts as an amplitude envelope on the interference factor. Therefore, the diffraction factor depends on the slit geometry, which is determined by the width of size $\alpha$, while the interference factor depends on the grating constant $d$.

[30] The term diffraction grating does not completely reflect the physics, which is a combined effect of diffraction and interference.

To find the interference factor, we consider a plane wavefront propagating along the optical axis. The propagation medium is air (we can also assume a medium with a refractive index $n$). Between two successive slits, the optical path difference is

Optical Path Difference: $$d \cdot \sin\vartheta \tag{5.60}$$

where $\vartheta$ is the observation angle with respect to the optical axis. The phase difference is

Phase Difference: $$\frac{2\pi}{\lambda} \cdot d \cdot \sin\vartheta = kd \cdot \sin\vartheta \tag{5.61}$$

Because the slits are equispaced, the phase difference between the next pair (C & B) is same as that between A & B. Thus, the condition for constructive interference (main or principal maxima) is

$$d \cdot \sin\vartheta = m \cdot \lambda \quad \text{or} \quad kd \cdot \sin\vartheta = m \cdot 2\pi \tag{5.62}$$

The above relationships constitute the **grating formula**. The grating formula is exactly the same expression as the expression for the interference maxima for two slits separated by $d$ [such as Eq. (5.59)]. The integers $m = 0, \pm1, \pm2, \ldots$ represent the diffraction order. For observation at angles $\vartheta$, all of the slits that form the grating are in phase, and the principal maxima appear at

$$\sin\vartheta = m \cdot \frac{\lambda}{d} \tag{5.63}$$

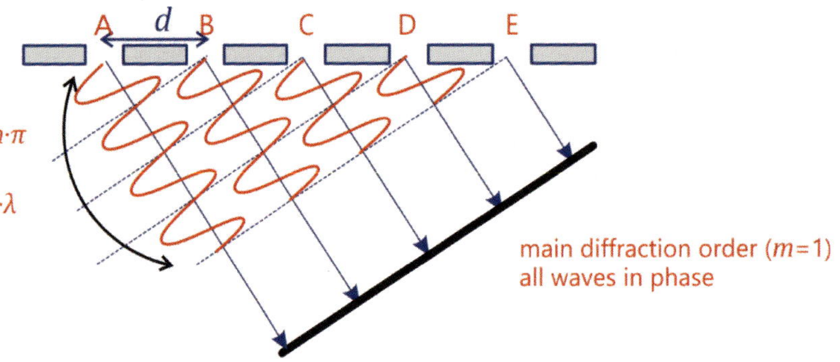


*Figure 5-73: Diffraction grating: condition for the main (principal) maxima.*

The very large number of interfering beams is similar to the case of multiple reflections from parallel plates (as discussed in § 4.2.5). The resultant field by $N$ components is the vector sum of all of the $N$ fields. Thus, the resultant wave has a field amplitude of magnitude

$$E_{\mathrm{TOT}} = E_{\mathrm{A}} \exp\left[-i(N-1)\underline{\gamma}\frac{\sin N\underline{\gamma}}{\sin\underline{\gamma}}\right] \tag{5.64}$$

The square of field amplitude magnitude is the intensity:

$$I_{TOT} = N^2 \cdot I_o \cdot \underbrace{\left[\frac{\sin(\underline{\alpha})}{\underline{\alpha}}\right]^2}_{\text{diffraction factor}} \cdot \underbrace{\left(\frac{\sin N\underline{\gamma}}{N\sin\underline{\gamma}}\right)^2}_{\text{interference factor}} \tag{5.65}$$

where $\underline{\alpha} = \frac{1}{2}\, k\cdot \alpha x_o/z = \pi\ \alpha x_o/\lambda z = \pi(\alpha/\lambda)\cdot\sin\vartheta_x$, which is the slit angular diffraction parameter, and $\underline{\gamma} = \frac{1}{2}k\cdot dx_o/z = \pi dx_o/\lambda z = \pi(d/\lambda)\cdot\sin\vartheta_x$, which is the angular interference parameter.

The diffraction factor is due to the diffraction intensity distribution from each individual slit, while the interference factor is due to the existence of many of such equidistant slits.

We note the striking similarities between Eq. (5.65) and Eq. (5.55), which we can re-arrange as follows:

$$I_{\text{two slits}} = 4I_o \cdot \underbrace{\left[\frac{\sin(\underline{\alpha})}{\underline{\alpha}}\right]^2}_{\text{diffraction factor}} \cdot \underbrace{\cos^2\left(\frac{\pi d}{\lambda}\sin\vartheta_x\right)}_{\text{interference factor}} = 4I_o \cdot \underbrace{\left[\frac{\sin(\underline{\alpha})}{\underline{\alpha}}\right]^2}_{\text{diffraction factor}} \cdot \underbrace{\cos^2(\underline{\gamma})}_{\text{interference factor}} \tag{5.66}$$

We apply $N = 2$ in Eq. (5.55):

$$I_{TOT\,|\,N=2} = N^2 \cdot I_o \cdot \underbrace{\left[\frac{\sin(\underline{\alpha})}{\underline{\alpha}}\right]^2}_{\text{diffraction factor}} \cdot \underbrace{\left(\frac{\sin N\underline{\gamma}}{N\sin\underline{\gamma}}\right)^2}_{\text{interference factor}} = 4I_o \cdot \underbrace{\left[\frac{\sin(\underline{\alpha})}{\underline{\alpha}}\right]^2}_{\text{diffraction factor}} \cdot \underbrace{\left(\frac{\sin 2\underline{\gamma}}{2\sin\underline{\gamma}}\right)^2}_{\text{interference factor}} \tag{5.67}$$

By applying the trigonometric identity $\sin(2\gamma) = 2\cos\gamma\cdot\sin\gamma$, we note that Eqs. (5.65), (5.66), and (5.67) are identical.

For $\underline{\gamma} = m\cdot\pi$, both functions $\sin N\underline{\gamma}$ and $\sin\underline{\gamma}$ are zero. For the fraction $(\sin N\underline{\gamma})/(N\cdot\sin\underline{\gamma})$, in analogy with Eq. (5.22), the limit is

$$\lim_{\underline{\gamma}=m\pi}\frac{\sin N\underline{\gamma}}{N\sin\underline{\gamma}} = \lim_{\underline{\gamma}=m\pi}\frac{N\cos\underline{\gamma}}{N\cos\underline{\gamma}} = 1 \tag{5.68}$$

Indeed, for the values $\gamma = m\pi \rightarrow \pi dx_o/\lambda z = \pi(d/\lambda)\cdot\sin\vartheta_x = m\pi$ (the condition for the principal maxima), the interference factor is maximized: $I^2_{MAX} = N^2$.

Comparing the interference from two slits to the interference from $N$ slits, we find exactly the same conditions for maximum (constructive) interference. A similar comparison can also be made between two-beam interference of a thin plate and thin-film multiple-beam interference. Here, too, the geometrical conditions for the maxima with two interfering beams are identical to those for $N$ beams produced by $N$ slits (see also § 5.6). The difference is that with $N$ slits the fringes are thinner and brighter, according to the Airy factor. The interference maxima appear at the same angles but are significantly finer and brighter.

To find the thickness of the maxima, we search for conditions that zero out the interference factor between the main maxima in the in-between space. This occurs when

$$\sin(N\underline{\gamma}) = 0 \quad \text{and} \quad \sin(\underline{\gamma}) \neq 0 \quad \text{or} \quad N\underline{\gamma} = n\cdot\lambda \quad \text{and} \quad m\cdot\lambda < \underline{\gamma} < (m+1)\cdot\lambda \tag{5.69}$$

Thus, the values of the angular factor $\underline{\gamma}$ for the minima are

Condition for Minima:
$$\underline{\gamma}_N = \left(m + \frac{n}{N}\right)\cdot\pi \tag{5.70}$$

where $m = 0, \pm1, \pm2, \ldots$ is the main diffraction order, and $n = 1, 2, \ldots, N-1$, is the secondary diffraction order. For a diffraction grating of $N$ (as opposed to two) slits, there are $N-1$ minima between the main maxima. There are also $N-2$ secondary maxima.

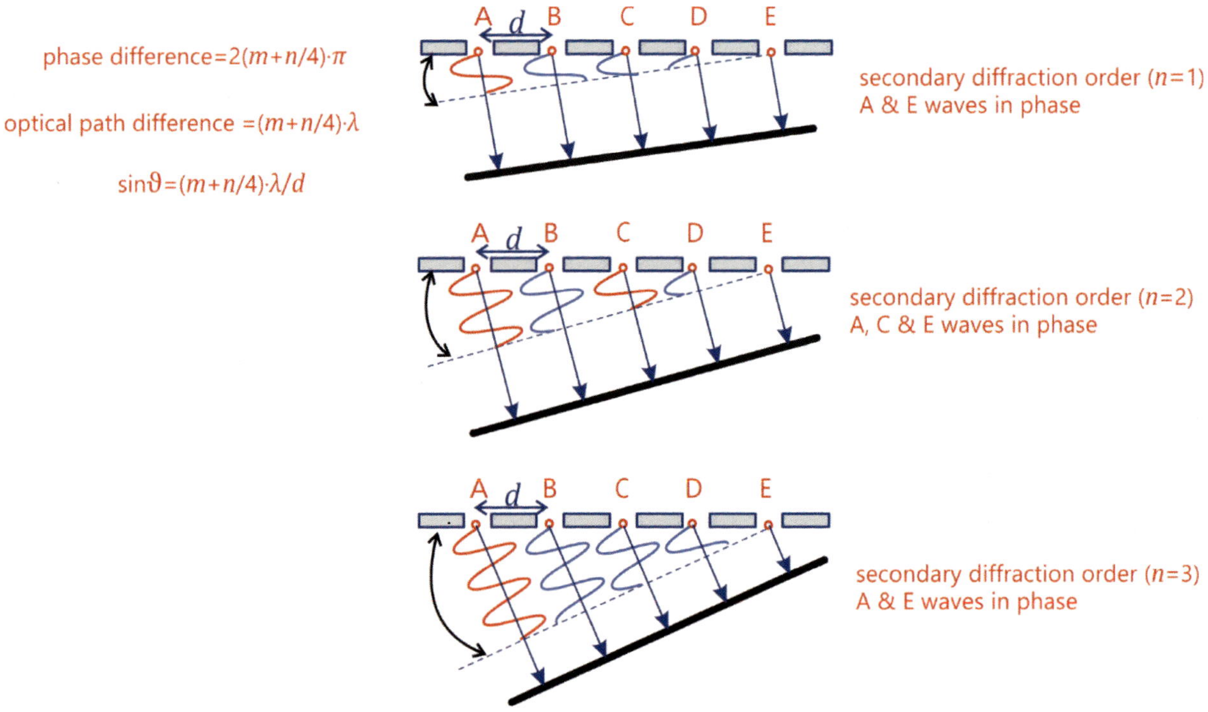


*Figure 5-74: Diffraction grating under the conditions of secondary maxima.*

The angular extent (finesse) of a principal diffraction order is determined by the nearest minima at each side. For the zeroth-order ($m = 0$) main maximum, the first zero intensity is formed when $\underline{\gamma}_n = (n/N)\pi$, where $n = 1$ ($\gamma_1 = \pi/N$). The distance between these symmetric secondary minima defines the angular extent of the principal maximum:

Principal Maximum (angular extent): $$\frac{2}{N}\cdot\frac{\lambda}{d}$$ (5.71)

where $N$ is the total number of slits (grooves) in the grating.

The greater the value of $N$, the smaller the angular extent of the main (principal) maxima. For absolutely monochromatic radiation, an ideal grating with an infinite number of grooves forms a zero-width main maximum, to the extent that the width of the principal maxima is determined only by the spectral range of the incident radiation, such as in the fringes of non-monochromatic interference (see also § 4.2.2.2).

The bright intensities of the secondary maxima are much lower compared to the intensities of the main (principal) maxima. In the example shown in Figure 5-74, where out of five slits we get three secondary maxima, the intensity of the principal diffraction order—simultaneous coherence of all five slits—is proportional to $N^2 = 25$. For $n = 1$, we obtain one secondary diffraction order of coherence of only $N = 2$ slits, so its intensity is proportional to $N^2 = 4$; this is much less than the peak of the main principal diffraction order. As the total number of grating grooves $N$ increases,

- the intensity of the principal maxima ($N^2$) increases drastically,
- the lateral extent of the principal maxima ($1/N$) decreases, and
- the number of the secondary maxima ($N - 1$) increases.

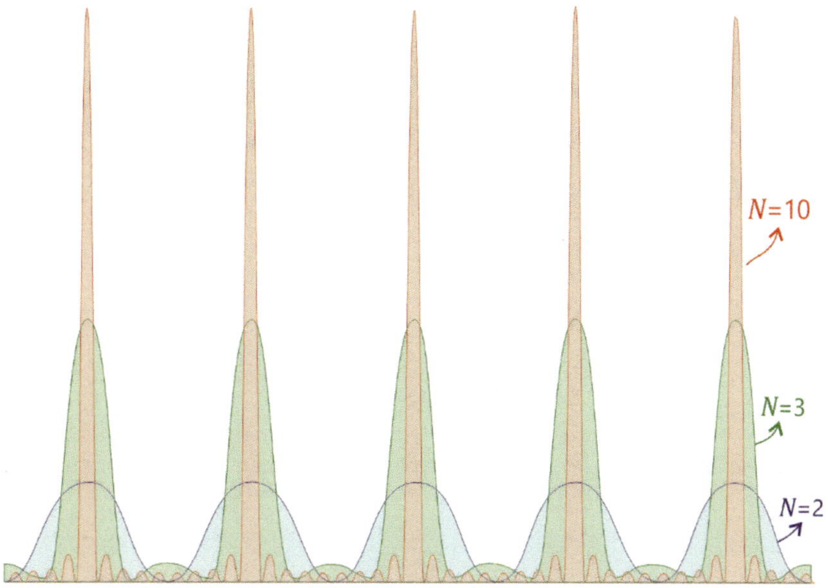


*Figure 5-75: Diffraction intensity distribution of a grating with 2, 3, and 10 grooves. The relative heights of the principal maxima have, in reality, proportionality 4:9:100.*

The angular separation of the main maxima depends on the characteristic angular parameter $\sin\vartheta = \lambda/d$. For the zeroth diffraction order ($m = 0$), the main (central) maximum is formed exactly on the optical axis, independent of the wavelength. The diffraction orders $m$ are formed at the angles $\vartheta_m$ that satisfy the condition $\sin\vartheta_m = m\,\lambda/d$.

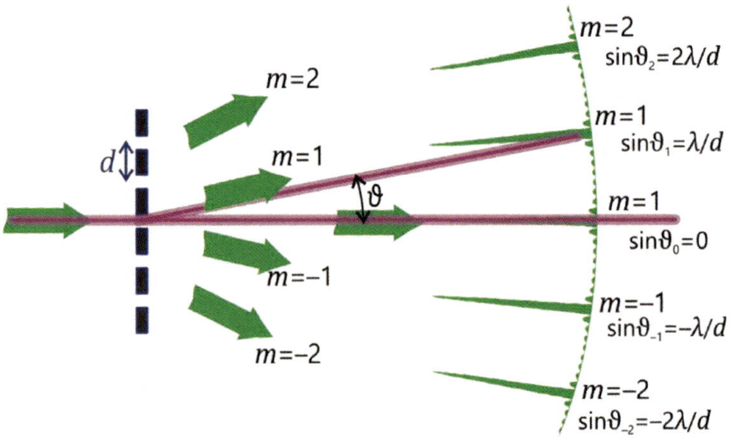


*Figure 5-76: Angular expressions for the principal diffraction orders from −2 to +2.*

We can study the grating action by using vectorial analysis, just as with two-beam interference (§ 4.2.2.1). We associate a wavevector $\mathbf{k}_G = 2\pi/d$ to the grating. The incident light has a wavevector $\mathbf{k}_o = 2\pi/\lambda$. The diffraction orders result from a vector sum of $\mathbf{k}_o$ and $\mathbf{k}_G$:

Vectorial Diffraction Grating: $$\mathbf{k}_m = \mathbf{k}_o \pm \mathbf{k}_G \qquad (5.72)$$

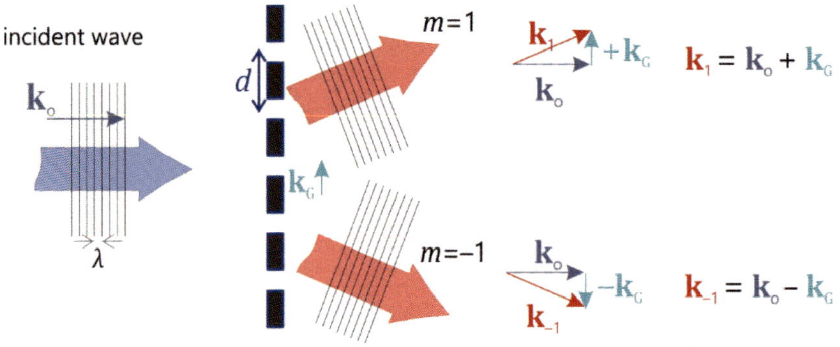


*Figure 5-77: The diffraction of a grating as a wavevector summation.*

The zeroth diffraction order is simply $\mathbf{k}_{m=0} = \mathbf{k}_o$, which corresponds to a wave with the same wavevector as that of the incident wave. The first ($m = \pm1$) diffraction order corresponds to a wavevector $\mathbf{k}_{m=1} = \mathbf{k}_o \pm \mathbf{k}_G$, the second to a wavevector $\mathbf{k}_{m=2} = \mathbf{k}_o \pm 2\times\mathbf{k}_G$, and so on. The magnitude of the transmitted (or reflected, depending on the grating type) wavevector does not change. What changes is the vector direction, which determines the loci of the diffraction maxima.

If the grating is illuminated with a multicolor source, the maxima of the same order for the different wavelengths are formed at different angles, which are expressed by the relationship $\sin\vartheta = m \cdot \lambda / d$. For a specific diffraction order (same $m$), except for the center lobe of the zeroth order for which $m = 0$, various chromatic components appear at different angles. There is therefore source spectral analysis within each diffraction order: The first-order spectrum appears when $m = 1$, the second order when $m = 2$, … . Moreover, within the increasing order, the angular separation of the color components increases.

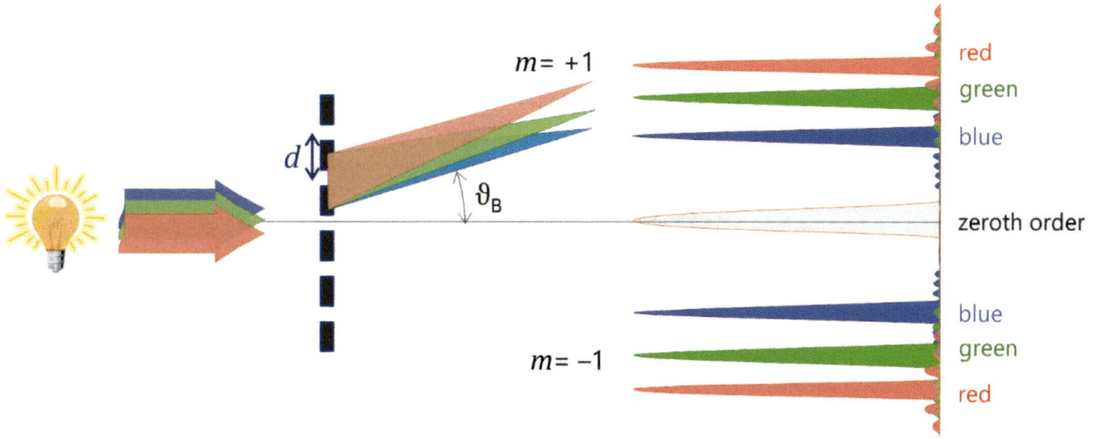


*Figure 5-78: Diffraction grating: chromatic component analysis.*

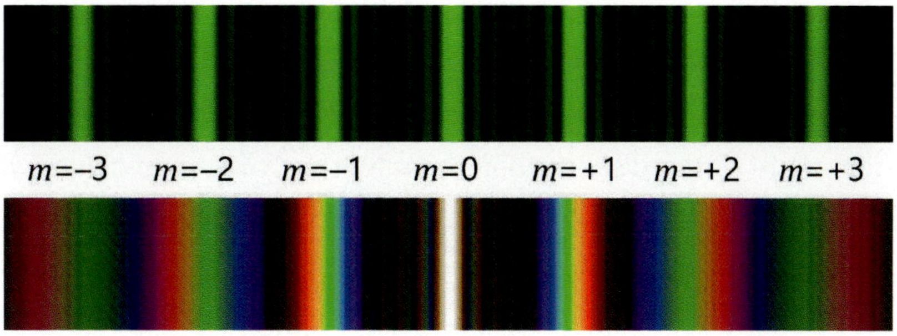


*Figure 5-79: (top) Monochromatic versus (bottom) multicolor diffraction by six grooves.*

It is possible for spectral lines of neighboring diffraction orders to overlap. For example, a red line with $\lambda_{0R} = 700$ nm corresponds to a third-order diffraction angle of $\sin\vartheta = 3 \cdot (700)/d$ (with $d$ expressed in nm), which is exactly the same diffraction angle as the fourth order for the green line $\lambda_{0G} = 525$ nm, corresponding to $\sin\vartheta = 3 \cdot (700)/d = 4 \cdot (525)/d$.

For a diffraction order $m$, two different wavelengths $\lambda_1$ and $\lambda_2$ are diffracted at angles $\vartheta_1$ and $\vartheta_2$, respectively, for which $\sin\vartheta_i = m \cdot \lambda_i / d$ applies. The angular difference is

$$\sin\vartheta_1 - \sin\vartheta_2 = m \cdot \frac{(\lambda_1 - \lambda_2)}{d} = m \cdot \frac{\Delta\lambda}{d} \tag{5.73}$$

Thus, the angular separation of two wavelengths differing by $\Delta\lambda$ is inversely proportional to the grating constant $d$ and proportional to the difference $\Delta\lambda$ and the diffraction order $m$. Analysis of the chromatic components is quantitatively evaluated by the angular separation of the maxima that correspond to two different wavelengths $\lambda_1$ and $\lambda_2$. We define the **angular dispersion** $D_m$ of a grating as the following quantity:

Angular Dispersion: $$D_m = \frac{d\vartheta_m}{d\lambda} \tag{5.74}$$

We apply differential calculus, using the relationship $\sin\vartheta = m\cdot\lambda/d$, and find that the angular dispersion $D_m$ depends on the grating constant $d$, the wavelength $\lambda$, and the diffraction order $m$ according to

$$D_m = \frac{d\vartheta_m}{d\lambda} = \frac{m}{d\cos\vartheta_m} = \frac{m}{d\sqrt{1-\sin^2\vartheta_m}} = \frac{m}{d\sqrt{1-\left(m\cdot\lambda/d\right)^2}} = \frac{1}{\sqrt{\left(d/m\right)^2-(\lambda)^2}} \tag{5.75}$$

The grating resolution ability expresses the smallest difference $\Delta\lambda$ (within an average wavelength range $\lambda_{AVE}$) that may be resolved. There is a limit to how close two wavelengths can be and still be resolved. The **resolving power** of a grating is the unitless quantity

Grating Resolving Power: $$R \equiv \frac{\lambda_{AVE}}{\Delta\lambda} \tag{5.76}$$

Also, in analogy with the resolving power of an optical system, two main diffraction maxima for the diffraction order $m$ are distinguished when the minimum of one coincides with the maximum of the other (Rayleigh criterion, § 5.5.2). The maxima correspond to

$$\pi\frac{d}{\lambda_1}\sin\vartheta_{1\,MAX} = m\cdot\pi \quad \text{and} \quad \pi\frac{d}{\lambda_2}\sin\vartheta_{2\,MAX} = m\cdot\pi \tag{5.77}$$

The first (secondary) minimum for $\lambda_2$ appears when [Eq. (5.70)]:

$$\pi\frac{d}{\lambda_2}\sin\vartheta_{2\,MIN} = \left(m+\frac{1}{N}\right)\cdot\pi \tag{5.78}$$

The resolution dictates that $\vartheta_{2\,MIN} = \vartheta_{1\,MAX}$, so $m_{\lambda 2} - m_{\lambda 1} = d\cdot(\sin\vartheta_{2\,MIN} - \sin\vartheta_{1\,MAX}) - \lambda_2/N \approx \lambda_{AVE}/N \rightarrow m\cdot\Delta\lambda = \lambda_{AVE}/N$. The grating resolving power is proportional to the diffraction order $m$ and to the number $N$ of the grating grooves:

Grating Resolving Power: $$R \equiv \frac{\lambda_{\text{AVE}}}{\Delta\lambda} = m \cdot N \tag{5.79}$$

Example ☞: For the separation of the doublet sodium (Na) lines $D_1$ ($\lambda$ = 0.5890 µm) & $D_2$ ($\lambda'$ = 0.5896 µm), the required minimum resolution ability is $R = \lambda/\Delta\lambda$ = 589.0 nm / 0.6 nm ≈ 1000. Therefore, for the separation of this line in the first diffraction order ($m$ = 1), we need at least $N$ = 1000 grooves, while just 500 grooves are required for the separation of the same doublet lines in the second order.

Equation (5.79) states that the higher the diffraction order, the higher the resolving power. Someone may propose that it is simple (or it is not simple?) to increase the resolution by simply increasing $m$. Therefore, we move to a higher diffraction order. It is useful to ask how many principal orders can be obtained from a grating. This can be calculated by the following: The angle $\sin\vartheta = m\cdot\lambda/d$ may not be greater than $\pm\pi/2$. There is an absolute upper limit to the order of a diffraction grating:

$$\sin\vartheta_{\text{MAX}\,|\,\text{Interference}} = 1.0 \quad\Rightarrow\quad m_{\text{MAX}} \cdot \frac{\lambda}{d} = 1.0 \quad\Rightarrow\quad m_{\text{MAX}} = \frac{d}{\lambda} \tag{5.80}$$

Other than the limiting factor of the observation angle, there is an additional factor that limits diffraction orders. The diffraction intensity distribution is amplitude-modulated by the diffraction factor, so the intensity of the higher diffraction orders is attenuated for increased orders. We can see diffraction orders up to observation angles $2\lambda/\alpha$, which are all contained within the principal diffraction envelope, the diffraction factor corresponding to the slit size. For a much smaller slit size, this angle is quite large (Figure 5-77 and Figure 5-78). Thus, with an increase in the diffraction order $m$, the chromatic components are more separated but have reduced intensity.

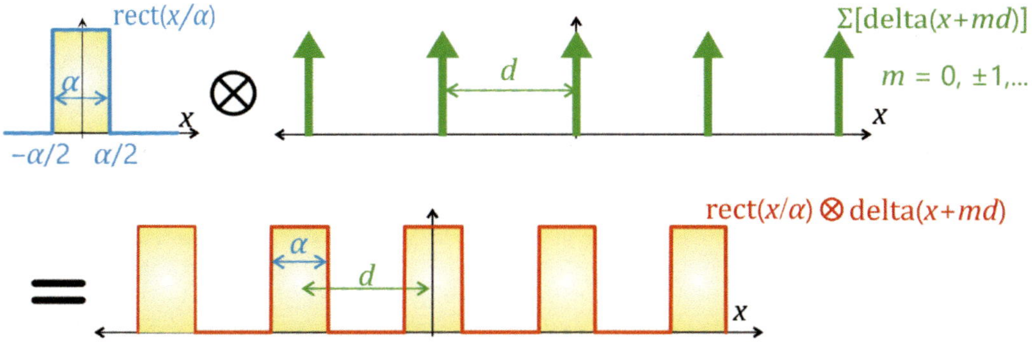


*Figure 5-80: Diffraction grating shown as a periodic convolution function* rect *with step d.*

We can accept this conclusion if we consider the diffraction pattern as the Fourier transform of the field distribution just after the diffraction grating. For a periodic grating amplitude, the field after the grating is expressed as a convolution of the rect function with a step equal to $d$ (Figure 5-80), so the diffraction pattern formation results from the product of the corresponding Fourier transformations.

The Fourier transform of the function that describes slit size $a$, rect($x/\alpha$), is sinc($\alpha x_o/\lambda z$) (§ 5.3). This transform is multiplied by the Fourier transform of the periodic displacement delta($x+md$), which is also a periodic delta function ($x_o+m\lambda z/d$), with step $\lambda z/d$, which is the spatial frequency corresponding to the grating periodic distribution $d$. Their product is the diffraction field, as shown in the two figures below.

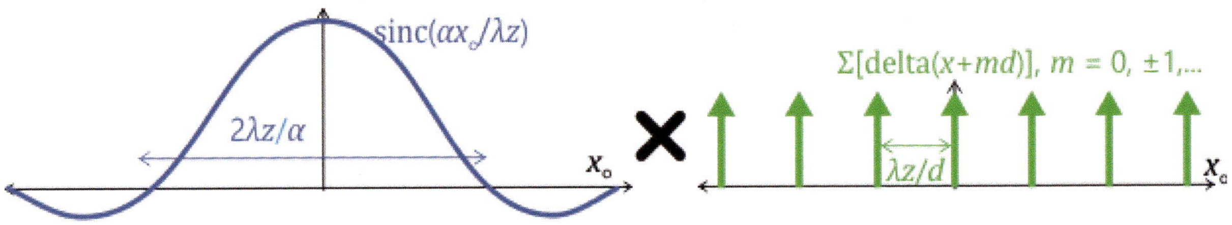


*Figure 5-81: Fourier transform components for the slit function and periodic function.*

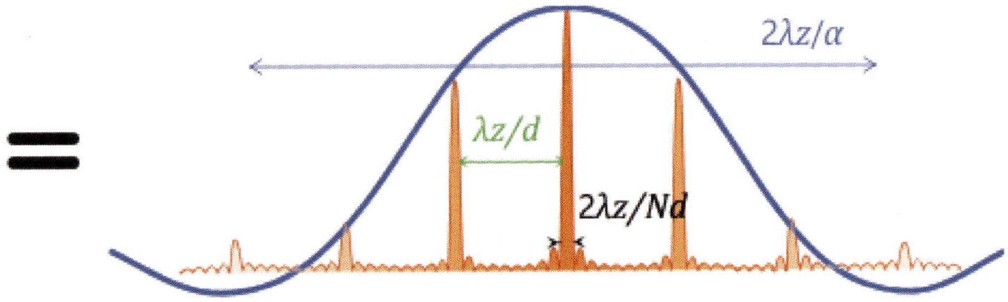


*Figure 5-82: The product of the Fourier transforms of the two components (slit function and periodic function).*

We note that the observable diffraction orders are those under the envelope of the first lobe that corresponds to the diffraction pattern formation of a single slit. The most intense diffraction order is determined by the reciprocal of the grating groove size:

$$\sin\vartheta_{\text{MAX}\,|\,\text{Diffraction}} = \frac{\lambda}{\alpha} \quad\Rightarrow\quad m_{\text{MAX}}\cdot\frac{\lambda}{d} = \frac{\lambda}{\alpha} \quad\Rightarrow\quad m_{\text{MAX}} = \frac{d}{\alpha} \tag{5.81}$$

There are, of course, higher diffraction orders within the secondary lobes, but these have a much lower intensity and probably are not observable. Unlike Eq. (5.81), Eq. (5.80) is absolute because it is impossible to have any beam past the 90°angle, either in reflection or refraction.

One more observation from Figure 5-82 is as follows: The thickness of each diffractive maximum corresponds to the thickness of the delta function; in other words, it is infinitesimal. This is because we assumed an infinite diffraction grating. A real grating with $N$ grooves has a thickness of $N{\cdot}d$, so the grating function is multiplied by $\text{sinc}(Nd \cdot x_o/\lambda z)$ (not illustrated). Thus, the diffractive maxima are not delta spikes but lobes of size $\lambda z/Nd$. The nonzero thickness of the main maxima are expressions of the finite grating area.

A diffraction grating may reflect or transmit light. Therefore, there are reflection and transmission gratings. In a transmission grating, the zeroth diffraction order follows the same path as the incident beam, and the diffraction orders are symmetrically spaced around it. The diffraction angle is measured with respect to the transmitted beam.

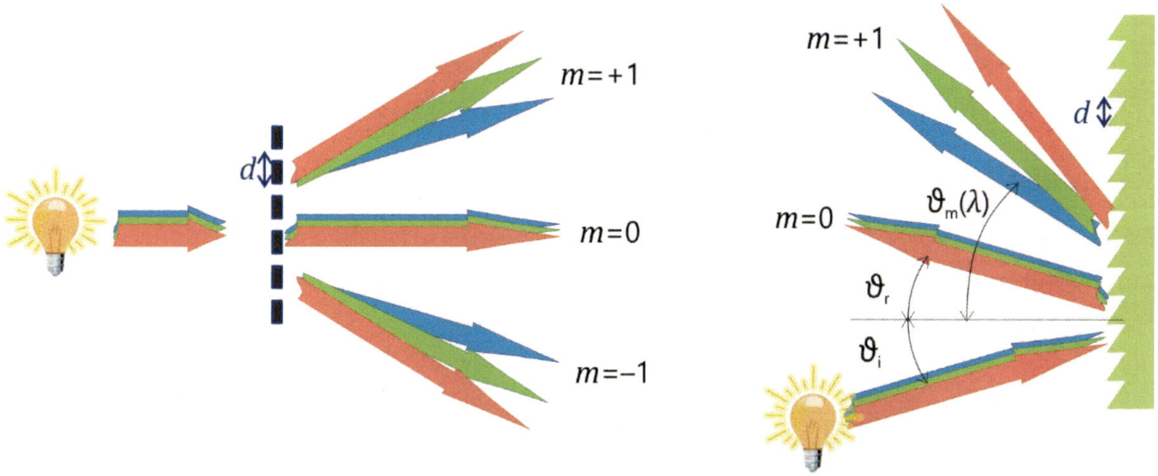


*Figure 5-83: (left) Transmission grating and (right) reflection grating.*

Examples of a reflection grating are the familiar compact disk (CD) and the digital versatile disk (DVD). The information contained in the CD or DVD (music or data) is encoded in equispaced grooves. This periodic distribution enables the CD/DVD to function as a diffraction grating, resulting in an effect that is easy to observe.

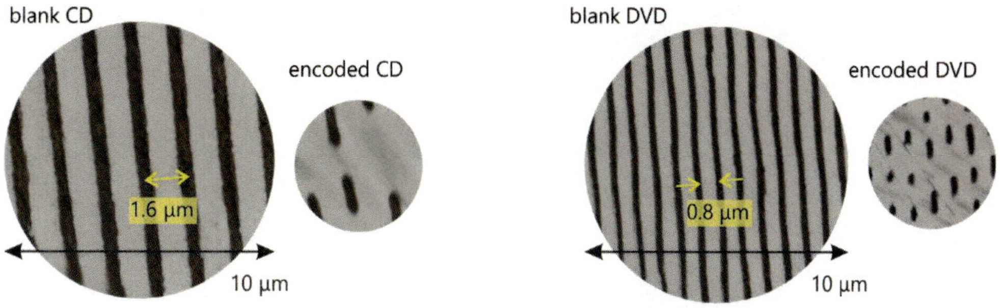


*Figure 5-84: Grooves in (left) a CD and (right) a DVD.*

It is not only the grooves, but also the digitally encoded information inside the grooves that give off diffraction effects. The spacing of the encoded bits inside the groove is fixed, although not exactly periodic, as there may be successive ON bits (1's) and OFF bits (0's), not in any specific order. If we illuminate a CD with a select radiation, it is possible to observe, inside every diffraction order, diffractive maxima that correspond to these bits.

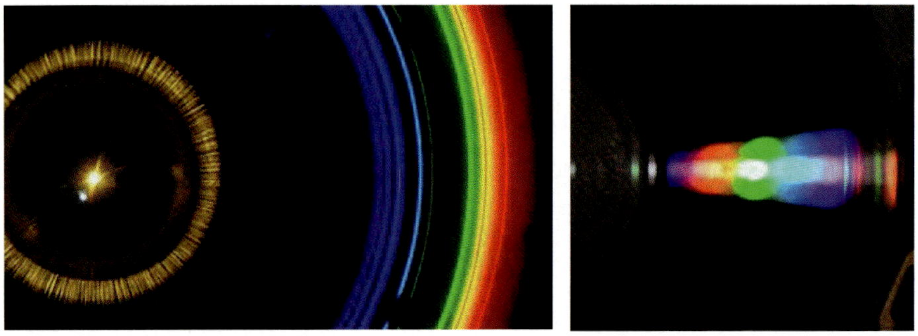

*Figure 5-85: Diffraction off a CD: Streaks of colors appear when white light falls on a CD.*

In a reflection grating, the zeroth diffraction order is reflected according to the law of reflection, so no color separation appears. Diffraction orders usually appear on only one side of the reflected beam, and for a diffraction angle with respect to the reflected angle $\theta_i$,

$$\sin\vartheta_m(\lambda) = m\frac{d}{\lambda} - \sin\theta_i \;\Rightarrow\; m\frac{d}{\lambda} = \sin\vartheta_m(\lambda) + \sin\theta_i \;\Rightarrow\; \vartheta_m(\lambda) = \sin^{-1}\left(m\frac{d}{\lambda} - \sin\theta_i\right) \quad (5.82)$$

A grating with a periodic distribution of reflectivity or transmissivity, in effect, modulates the amplitude of an incident wave: For an incident wave of fixed amplitude magnitude $E_o$, there is $E_o$ at the locations of the slits (grooves) and zero amplitude at the location of the part between the slits, with no phase change. Such gratings are **amplitude gratings**.

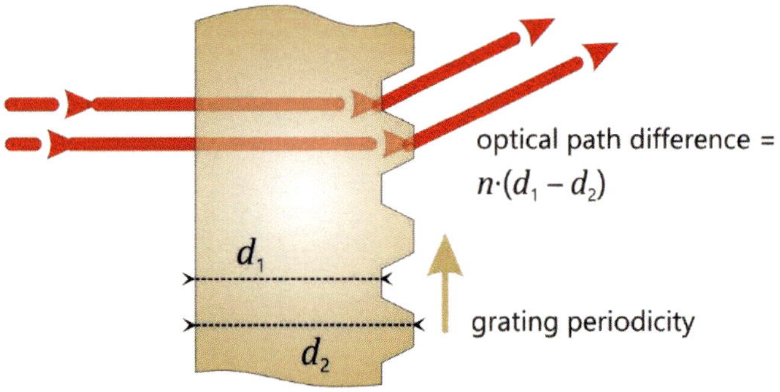


*Figure 5-86: A phase grating is a transparent plate in which one side has periodic changes to the cross-sectional thickness.*

A transparent grating can periodically alter the phase of the incident wave if its optical path length changes periodically from $nd_1$ to $nd_2$. Light exits with a constant amplitude, but with a phase modulation that varies periodically from $\varphi_1 = knd_1$ to $\varphi_2 = knd_2$. This is a **phase grating**. If the refractive index is fixed, but the thickness varies, then the periodic phase difference is $\delta\varphi = kn(d_1 - d_2)$. Phase gratings with periodic refractive index changes are formed via the photorefractive effect, which is achieved with the interference of two coherent beams inside the volume of specific crystals.[31]

## 5.7.1 Monochromator

Diffraction gratings are used in devices designed to separate various chromatic components. The **monochromator** (or spectrograph) is one of these devices, a typical example of which is the Czerny–Turner monochromator.[32] Light analysis is achieved with a reflective diffraction grating, the central element of the device. Multicolor light enters the device through an entrance slit of a controlled size (the slit width), at a characteristic angle that is determined by the geometry (the numerical aperture). The light is collimated by either a lens or a mirror and therefore forms a collimated beam of all of the chromatic components incident on the diffraction grating.

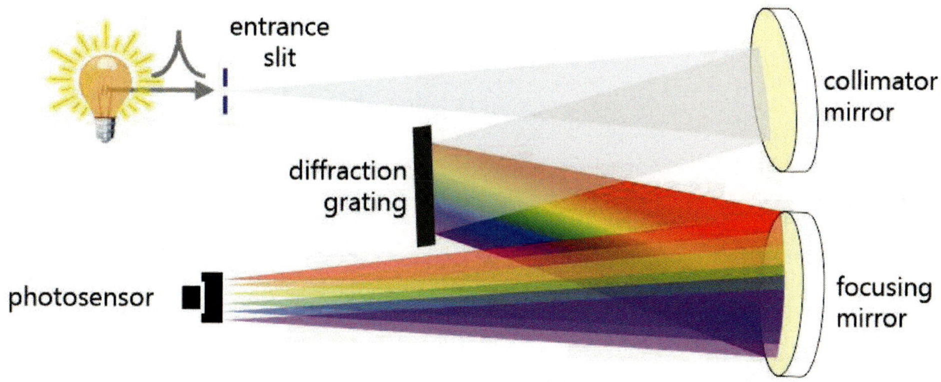


*Figure 5-87: Schematic diagram of a Czerny–Turner monochromator.*

The diffraction grating separates the chromatic components into their spectral content. The gratings in the monochromator are designed to favor one specific diffraction order for a given angle of incidence. Although it is theoretically possible to obtain many diffraction orders within an angle $\pi$, we usually use only one diffraction order, within which the chromatic

[31] Asimellis G, Khoury J, Woods CL. Experimental demonstration of the holographic incoherent-erasure joint-transform correlator. Opt Eng. 1997; 36(9):2392-2399 [doi: 10.1117/1.601493].

[32] Although they work on the same principle, a monochromator isolates a narrowband light beam from the source, while a spectrograph is designed to measure the spectrum of the light source.

components of the incident light are analyzed. These beams are focused on an observation screen or a photonic sensor using a focusing mirror.

### 5.7.2 X-ray Diffraction in Crystals

For an arrangement to function as a diffraction grating, it must have a constant order of magnitude of the wavelength of the radiation with which it interacts. For example, a diffraction grating with 600 lines/mm corresponds to a grating constant $d = 1667$ nm—approximately 3× the yellow wavelength (555 nm). A diffraction grating for X rays, with a wavelength on the order of 0.1 nm, requires a correspondingly small periodicity. In 1912, the German physicist Max von Laude proposed that a crystal lattice, which is composed of a highly arranged atomic structure, can form a physical 3-D diffraction grating. This is possible because the average distances between the atoms inside a crystal lattice have exactly this order of magnitude.

For example, the unit cube cell in sodium chloride (NaCl) has edges of $\alpha_0 = 0.5627$ nm. This lattice has a cubic symmetry with an alternating succession of sodium ions (small spheres) and chlorine ions (large spheres).

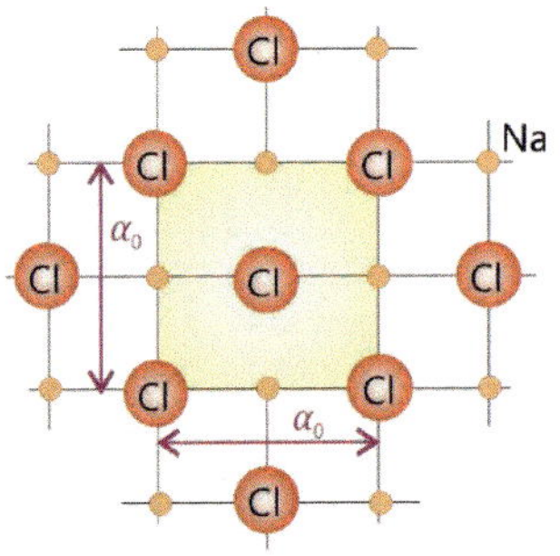


*Figure 5-88: Sodium chloride unit cell.*

This is a unit cell. A unit cell has a repeated (therefore, periodic) structure that can function as the groove in a grating, be it a reflection or a transmission grating. In comparison to the size of a unit cell, effectively, the grating extends infinitely. A cubic lattice can have a multitude of such gratings. Each grating is obtained by different combinations of the elementary diffracting centers, for instance, the chlorine ions. Specific crystallographic groups with a characteristic grating constant $d$ correspond to the vectors of the inverse lattice.

Now consider an X-ray beam incident on these planes (shown as blue dashed lines in Figure 5-89 and Figure 5-90) with an angle $\vartheta$ with respect to the surface of the crystal [in crystallography this angle is the complementary of the usual angle of incidence (Figure 5-90), measured not from the perpendicular to the plane, but from the plane of reflection]. Only part of the incident beam is reflected off these planes; this constitutes the quantitative problem.

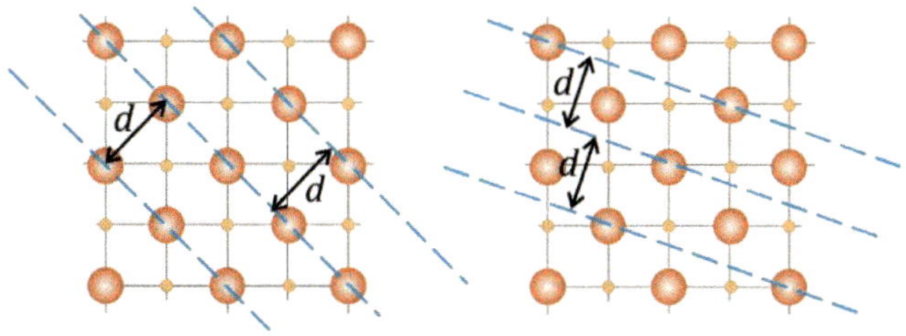


*Figure 5-89: Crystallographic planes in sodium chloride.*

Here we ignore the quantitative problem and investigate only the qualitative problem: We investigate the angles at which the beam emerges. These are the angles of principal interference maxima between the reflected beams from two successive crystallographic planes. The part of the beam that is reflected off the lower plane travels a longer path than the beam that is reflected off the upper plane, and this constitutes an optical path difference. We treat this as interference between two parallel, optically active surfaces (Figure 4-24). The optically active surfaces are separated by $d$, while in the medium between the surfaces, the refractive index is 1.0. Employing Eq. (4.55), it follows that the optical path difference is

Optical Path Difference (2 – 1): $$2nd \cdot \cos(90° - \vartheta) = 2d \cdot \sin(\vartheta) \tag{5.83}$$

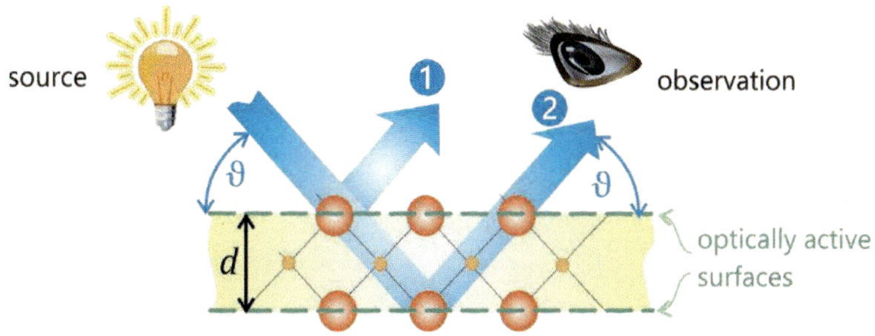


*Figure 5-90: X-ray reflection by two parallel crystallographic planes separated by a distance d.*

Exactly as in the visible frequencies, there is constructive interference when the optical path difference between two successive planes is a multiple integer of the wavelength. We apply the condition for constructive interference:

Constructive Interference Condition:

$$\text{optical path difference} = 2d \cdot \sin(\vartheta) = m \cdot \lambda \tag{5.84}$$

The above formula is known as **Bragg's law**, named after William Henry Bragg and his son William Lawrence Bragg. Bragg's law allows the measurement of spacing between the crystal planes and therefore allows the study of crystalline structures in detail. The Braggs earned the 1915 Nobel Prize for their studies in the measurement of the crystallographic plane spacing in NaCl and ZnS crystals, as well as in diamond using X-ray diffraction.

# 5.8 DIFFRACTION QUIZ

1) As a wave propagates, which of the following can be affected by diffraction?
   a) the direction of propagation
   b) the frequency
   c) the wavelength
   d) the photon energy

2) Diffraction effects can be observed in all except one of the following:
   a) sea waves
   b) light
   c) sound
   d) sandstorms

3) Max punches a tiny hole in his shirt collar. He keeps the button as well as the newly created buttonhole, which are both illuminated by coherent, collimated red light. On a screen placed a distance from the button and buttonhole, the two diffraction fields (two correct answers) ...
   a) are identical in every aspect (intensity & field)
   b) have identical phase fields
   c) have opposite phase fields
   d) have identical intensities

4) To calculate the Fraunhofer field intensity, the steps involve (1) computation of the electric field at the diffractive aperture and (2) computation of the square of the field distribution to calculate the intensity of the diffraction pattern. What step is not listed, and in what order does it belong?
   a) the computation of the magnetic field, before step (1)
   b) the Fourier transform of the light intensity, after step (2)
   c) the Fourier transform of the electric field, after step (1)
   d) the summation of field and intensity, after step (2)

5) The intensity distribution pattern following a vertically oriented narrow slit has fringes of maximum (bright) and minimum (dark) intensity separated along the ...
   a) vertical axis
   b) horizontal axis
   c) direction of propagation of light

The following questions Q 6 to Q 17 pertain to diffraction created by a narrow slit with aperture width $\alpha$ (illuminated by coherent light with wavelength $\lambda$) that forms an intensity distribution on a distant screen. The central lobe and side lobe widths are measured between the dark fringes on either side.

6) What are the central lobe <u>angular</u> width and side lobe angular width in relation to the aperture width $\alpha$ and wavelength $\lambda$ (angular width in radians)?
   a) central lobe: 0.5 $\lambda/\alpha$; side lobe: 0.5 $\lambda/\alpha$
   b) central lobe: 1 $\lambda/\alpha$; side lobe: 0.5 $\lambda/\alpha$
   c) central lobe: 2 $\lambda/\alpha$; side lobe: 1 $\lambda/\alpha$
   d) central lobe: 2 $\lambda/\alpha$; side lobe: 2 $\lambda/\alpha$

7) Which of the following decreases the angular width of the central bright lobe (select two)?
   a) decreasing the slit width
   b) increasing the slit width
   c) increasing the light wavelength
   d) decreasing the light wavelength

8) Whale Number One spills water ($n$ = 1.333), filling the space between the slit and the screen. Which of the following changes occur (select three)?
   a) the bright center lobe disappears
   b) the bright center lobe narrows
   c) the bright center lobe widens
   d) the bright side lobe narrows
   e) the bright side lobe widens
   f) the spacing between side lobes decreases
   g) the spacing between side lobes increases

9) The bright central lobe is 1.00 cm wide. What is the bright side lobe width (length, in cm)?
   a) 0.50 cm
   b) 1.00 cm
   c) 1.50 cm
   d) 2.00 cm

10) Determine the first-order ($m$ = 1) dark fringe (minimum) from across the bisector of the slit on the screen when $\lambda$ = 633 nm, $\alpha$ = 0.3 mm = 300 μm, and the distant screen is 2.00 m away.
   a) 0.42 mm
   b) 2.11 mm

c) 4.22 mm
d) 6.33 mm
e) 8.44 mm

11) Back to Q 10. What appears at 8.44 mm?
a) the first-order ($m$ = +1) side-lobe maximum
b) the second-order ($m$ = +2) dark minimum
c) the second-order ($m$ = +2) side-lobe maximum
d) the third-order ($m$ = +3) side-lobe maximum

12) Back to Q 10. Where is the first-order ($m$ = +1) side-lobe maximum?
a) 0.42 mm
b) 2.11 mm
c) 4.22 mm
d) 6.33 mm
e) 8.44 mm

13) When using $\lambda$ = 633 nm, the first minimum occurs 3.6° from across the bisector of the slit on the screen. What is the slit aperture width $\alpha$ [μm]?
a) 100 μm
b) 10 μm
c) 1 μm
d) 0.1 μm

14) Back to Q 13. Yertle the Turtle somehow manages to double the slit aperture width. Now, what falls on the exact same spot as the first minimum in the unaltered case?
a) the first ($m$ = 1) dark minimum
b) the first ($m$ = 1) bright side lobe
c) the second ($m$ = 2) dark minimum
d) the second ($m$ =2) bright side lobe

15) When $\lambda$ = 650 nm and $\alpha$ = 3.25 μm, what is the angular subtense of the bright central diffraction lobe?
a) 0.04 rad
b) 0.10 rad
c) 0.20 rad
d) 0.40 rad
e) 4.00 rad

16) When using coherent red ($\lambda_R$ = 630 nm) illumination, the first-order ($m$ = +1) bright side-lobe maximum appears at 1.9 mm from the center-lobe maximum. Foot Guy magically switches to blue ($\lambda_B$ = 400 nm) illumination. Now where is the first-order bright side-lobe maximum?
a) 0.8 mm
b) 1.2 mm
c) 1.3 mm
d) 1.6 mm
e) 3.0 mm

17) The slit is now illuminated with white light instead of coherent monochromatic light. What do we know about the blue ($\lambda_B$ = 400 nm) and the red ($\lambda_R$ = 640 nm) components (two correct answers)?
a) The central ($m$ = 0) maximum is identical for both the blue and the red.
b) The first-order ($m$ = +1) dark minimum is identical for both the blue and the red.
c) The first-order ($m$ = +1) bright maximum is closer to the center for the blue.
d) The first-order ($m$ = +1) bright maximum is closer to the center for the red.

Questions Q 18 to Q 23 pertain to diffraction created by a rectangular-shaped aperture with width $\alpha$ along the $-x$ axis (horizontal) and height $\beta$ along the $-y$ axis (vertical), illuminated with coherent light of wavelength $\lambda$, that forms an intensity distribution at a distant screen.

18) At the limit where $\beta$ is infinitely large, the diffraction pattern formed on a distant screen along the $-z$ axis appears with the fringe minima separated along the ...
a) vertical direction ($-y$ axis)
b) horizontal direction ($-x$ axis)
c) direction of propagation of light ($-z$ axis)

19) A rectangular-shaped aperture has $\alpha$ = 65.0 μm and $\beta$ = 32.5 μm. Select the appropriate statements describing the diffraction intensity distribution of this aperture (two correct answers).
a) The horizontal and the vertical diffraction intensity distributions share the same center maximum.
b) The horizontal and the vertical diffraction intensity distributions have the same dark-fringe (minimum) distances from the center.
c) The dark-fringe minima along the horizontal axis are spaced twice as far as the minima along the vertical axis.
d) The dark-fringe minima along the horizontal axis are spaced half as far as the minima along the vertical axis.

20) The diffraction intensity distribution of a rectangular aperture ($\alpha \neq \beta$) can be made to have equal spacing between the dark fringes by ...

a) rotating the aperture by 45°
b) using a different observation screen distance
c) using white light
d) not possible, Captain Sully!

21) A rectangular aperture ($\alpha \neq \beta$) with $\alpha$ = 0.3 mm forms a diffraction intensity distribution with a first-order ($m$ = 1) dark minimum 3.33 mm from the center-lobe maximum along the horizontal. If the first-order ($m$ = 1) dark minimum along the vertical is 5.0 mm from the center, what is the vertical aperture size?

a) 5.0 mm
b) 3.3 mm
c) 2.2 mm
d) 1.1 mm
e) 0.2 mm

22) Back to Q 21. If the distant screen is at $z$ = 2 m, what is the wavelength that illuminates that rectangular aperture?

a) 0.400 μm
b) 0.500 μm
c) 0.600 μm
d) 0.700 μm

23) A square (□) aperture ($\alpha = \beta$) with $\alpha$ = 0.3 mm forms a diffraction intensity distribution with a first-order ($m$ = +1) dark minimum 3.33 mm from the center-lobe maximum along the horizontal. Where is the first-order ($m$ = +1) dark minimum along the vertical ?

a) 5.00 mm
b) 3.33 mm
c) 2.22 mm
d) 1.11 mm

24) A circular aperture (○) of diameter $D$ and a square aperture (□) of width $\alpha$ may have the same position as the first-order ($m$ = +1) dark minima if ...

a) $\lambda_{\square} = \lambda_{\bigcirc}$ and $D = 1.22 \cdot \alpha$
b) $\lambda_{\square} = \lambda_{\bigcirc}$ and $D = 1.22/\alpha$
c) $D = \alpha$ and $\lambda_{\square} = 1.22 \cdot \lambda_{\bigcirc}$
d) $D = \alpha$ and $\lambda_{\square} = \lambda_{\bigcirc}/1.22$

25) Assuming that the results of Q 24 achieve the formation of the first-order dark minimum at the same position, what else can be at the same screen position between the diffraction intensity distribution of a circular aperture (○) of diameter $D$ and a square aperture (□) of width $\alpha$?

a) the location of the bright central maximum
b) the location of the second-order ($m$ = +2) and higher-order dark minima
c) the brightness of the central maximum
d) the brightness of the side lobes

Questions Q 26 to Q 30 discuss diffraction created by a circular aperture with diameter $D$, illuminated with coherent light of wavelength $\lambda$, that forms an intensity distribution on a distant screen. Recall that any angle computed as a ratio of lengths is expressed in radians. One radian is $180/\pi$ degrees (°), and one degree is 60 minutes of an arc (arcmin). One thousandth of a radian is denoted as 1 mrad.

26) Coherent illumination of $\lambda$ = 0.500 μm enters a round pupil with an aperture diameter of 1 mm. What is the first-order diffraction ($m$ = +1) minimum angular subtense from the center if the medium between the pupil and the distant screen is air ($n$ = 1.00)?

a) 0.61 mrad
b) 0.61 rad
c) 0.0175°
d) 2.0°
e) 1.573 arcmin

27) Back to Q 26. Weird Fish fills the space between the aperture and the screen with water ($n$ = 1.33). What is the first-order ($m$ = 1) minimum angular subtense from the center?

a) 0.61 mrad
b) 0.61 rad
c) 0.0175°
d) 2.0°
e) 1.573 arcmin

28) Back to Q 26. Zebra widens the opening of the aperture to 2 mm. What is the first-order ($m$ = 1) minimum angular subtense from the center if the medium between the pupil and the distant screen is air ($n$ = 1.00)?

a) 0.61 mrad
b) 0.61 rad
c) 0.0175°
d) 2.0°
e) 1.573 arcmin

29) In an optical system that is free of aberrations, which of the following circular aperture diameters produces the most concentrated (i.e., smallest angular subtense) central diffraction lobe (everything else being equal)?

a) 2 mm
b) 3 mm
c) 4 mm
d) 6 mm

30) A circular aperture diffraction pattern is obtained on a screen using yellow light ($\lambda_Y$ = 580 nm). If instead we use violet light ($\lambda_V$ = 420 nm) without any other changes, what happens to the diffraction pattern (two correct answers)?
   a) It disappears.
   b) The dark minima are spaced farther apart.
   c) The central lobe becomes broader.
   d) The dark minima are more closely spaced.
   e) The central lobe becomes narrower.

31) In an optical system that is free of aberrations, the bright central-lobe maximum most likely resembles which function used to describe image quality?
   a) the point spread function (PSF)
   b) the modulation transfer function (MTF)
   c) the object intensity distribution (OID)
   d) the image intensity distribution (IID)

32) Which of the following point spread function (PSF) attributes are most desired in a high-resolution optical system (two correct answers)?
   a) a small (narrow) angular width
   b) a large (wide) angular width
   c) a small (low) central maximum
   d) a large (high) central maximum

33) What aspect of the circular aperture diffraction pattern best defines the lateral extent of the PSF?
   a) the central-lobe angular extent
   b) the central-lobe intensity maximum
   c) the side-lobe angular extent
   d) the side-lobe intensity maximum

34) The cut-off frequency value in the modulation transfer function (MTF) is related to what parameter of the diffraction pattern from the same aperture?
   a) the angular extent of the first-order ($m$ = 1) minimum
   b) the reciprocal of the angular extent of the first-order ($m$ = 1) minimum
   c) the Fourier transform of the aperture
   d) the Strehl ratio of the PSF

35) The presence of aberrations in a system results in what changes with respect to the PSF and MTF functions (two correct answers)?
   a) a reduction in the area under the MTF curve
   b) a reduction in the angular size of the PSF
   c) an increase in the angular extent of the PSF
   d) an increase in the cut-off frequency of the MTF

36) When imaging the 1951 USAF target, certain line groups on it are no longer distinguishable in the image. The spacing of these line groups is __________ the cut-off frequency.
   a) a quarter of
   b) half
   c) just about
   d) twice

37) When observing the diffraction intensity pattern formed by two point objects in image space, the points are said to be just resolved when ...
   a) the central maximum of one coincides with the central maximum of the other
   b) the central maximum of one does not coincide with the central maximum of the other
   c) the central maximum of one does not coincide with the first minimum of the other
   d) the central maximum of one coincides with the first minimum of the other

Questions Q 38 to Q 48 pertain to diffraction created by a multitude of similar apertures (all having the same width, i.e., $\alpha$ if slits, $D$ if circular apertures). If more than two, the apertures are equispaced as well: Their center-to-center separation equals a fixed distance $d$. The interference fringe spacing is measured between two successive maxima or two successive minima. The diffraction center-lobe (the central maximum of the single slit or circular disk diffraction) width is measured between the two minima (zeros) on either side.

38) A rectangular slit of width $\alpha$ illuminated with coherent wavelength $\lambda$ forms its first-order ($m_d$ = +1) diffraction minimum at an angle [rad] of ...
   a) $2\ \lambda/\alpha$
   b) $1\ \lambda/\alpha$
   c) $0.5\ \lambda/\alpha$
   d) $0.25\ \lambda/\alpha$

39) Two infinitesimally small apertures separated by a distance $d$ and illuminated with light of coherent wavelength $\lambda$ form their first-order ($m_i$ = +1) interference fringe maximum at an angle [rad] of ...
   a) $2\ \lambda/d$
   b) $1\ \lambda/d$
   c) $0.5\ \lambda/d$
   d) $0.25\ \lambda/d$

40) In the diffraction pattern from similar apertures (such as two slits of equal width), what is the relationship between $d$ and $\alpha$?

a) $d < \alpha$
b) $d = \alpha$
c) $d > \alpha$

41) Assume that two slits, each of width 0.10 mm, are separated by 0.40 mm. If the interference fringe spacing is 2.5 mm, what is the angular width of the central diffraction lobe?

a) 1.25 mm
b) 2.5 mm
c) 5.0 mm
d) 10.0 mm
e) 20.0 mm

42) Back to Q 41. What is formed at ±1.25 mm alongside the center maximum?

a) diffraction lobe first-order ($m_d = \pm 1$) minima
b) diffraction lobe second-order ($m_d = \pm 2$) minima
c) interference fringe first-order ($m_i = \pm 1$) minima
d) interference fringe first-order ($m_i = \pm 1$) maxima
e) interference fringe fourth-order ($m_i = \pm 4$) minima
f) interference fringe fourth-order ($m_i = \pm 4$) maxima

43) Back to Q 41. What is formed at ±10.0 mm alongside the center diffraction lobe maximum (two correct answers)?

a) diffraction lobe first-order ($m_d = \pm 1$) minima
b) diffraction lobe second-order ($m_d = \pm 2$) minima
c) interference fringe first-order ($m_i = \pm 1$) minima
d) interference fringe first-order ($m_i = \pm 1$) maxima
e) interference fringe fourth-order ($m_i = \pm 4$) minima
f) interference fringe fourth-order ($m_i = \pm 4$) maxima

44) Back to Q 41. How many interference maxima fit within the center maximum?

a) 4
b) 7
c) 8
d) 9

45) When $d$ = 0.90 mm, 5 fringes of interference maxima fit within the center maximum of the single-slit diffraction pattern. What is the width $\alpha$ of each slit?

a) 0.10 mm
b) 0.20 mm
c) 0.30 mm
d) 0.40 mm
e) 0.50 mm

46) Back to Q 45. What happens to the third-order ($m_i = \pm 3$) interference maxima under the center maximum of the single-slit diffraction pattern?

a) They fall exactly on the diffraction lobe first-order ($m_d = \pm 1$) minima, so they disappear.
b) They form at full height.
c) They form at half height.
d) They shift to outside the diffraction lobe.

47) When $d = 2.5\ \alpha$, what coincides with the diffraction lobe first-order ($m_d = \pm 1$) minima?

a) second-order ($m_i = \pm 2$) interference minima
b) second-order ($m_i = \pm 2$) interference maxima
c) third-order ($m_i = \pm 3$) interference minima
d) third-order ($m_i = \pm 3$) interference maxima

48) When $d = 2.5\ \alpha$ (in general, an odd-number multiple of half $\alpha$) ...

a) there are no lost interference maxima
b) there are interference maxima lost at an even ($m_d$ = 2, 4, ...) number of the diffraction order minima
c) there are interference maxima lost at an odd ($m_d$ =1, 3, ...) number of the diffraction order minima

Questions Q 49 to Q 60 pertain to diffraction gratings. $N$ refers to the total number of grooves/lines, also reported as density ($N$/mm or $N$/cm). The grating constant $d$ is the reciprocal of the groove density.

49) Green light ($\lambda_G$ = 550 nm) illuminates a diffraction grating with 4000 lines/mm; the diffraction fringes are observed on a screen 1 m away. Which of the following increases the fringe spacing on the screen?

a) replacing green with violet ($\lambda_V$ = 420 nm)
b) using a grating with 6000 lines/mm
c) moving the observation screen closer to the grating

50) A grating is composed of a total of six slits. How many secondary maxima and secondary minima are present between two successive main diffraction peaks?

a) five minima; four maxima
b) four minima; five maxima
c) four minima; four maxima
d) five minima; five maxima

51) In a grating diffraction pattern, there are three secondary maxima between two successive main diffraction peaks. How many slits are present in this grating?
a) three
b) four
c) five
d) six

52) A grating composed of ten total slits differs from a grating composed of only two (same size and spacing) slits by (three correct answers) ...
a) the angle(s) at which the principal maxima are observed
b) the number of secondary maxima between two successive principal maxima
c) the intensity of the principal maxima
d) the angular spread of each individual principal maximum
e) the requirement for constructive interference between two successive slits

53) A diffraction grating has 4000 lines/cm. What is the groove constant?
a) 0.25 μm
b) 2.5 μm
c) 4.0 μm
d) 25 μm

54) Back to Q 53. The angle between the central maximum and the second-order maximum is 30°. What is the illumination wavelength?
a) 420 nm
b) 490 nm
c) 570 nm
d) 625 nm

55) Monochromatic light of wavelength 0.550 μm is incident on a grating whose line spacing is 2.0 μm. What is the angle between the zeroth-order and first-order maxima?
a) 8.25°
b) 14.42°
c) 15.96°
d) 17.41°
e) 32.0°
f) 33.37°

56) Back to Q 55. What is the angle between the zeroth-order and second-order maxima?
a) 8.25°
b) 14.42°
c) 15.96°
d) 17.41°
e) 32.0°
f) 33.37°

57) Back to Q 55. What is the angle between the first-order and second-order maxima?
a) 8.25°
b) 14.42°
c) 15.96°
d) 17.41°
e) 32.0°
f) 33.36°

58) Polychromatic light is incident on a grating. If the violet ($\lambda_V$ = 420 nm) component has a first-order peak at 12.12°, the first-order peak for the red ($\lambda_B$ = 635 nm) component appears at what angle?
a) 39.41°
b) 24.83°
c) 22.35°
d) 18.51°

59) Back to Q 58. What is the line density for this grating?
a) 500 lines/cm
b) 2000 lines/cm
c) 5000 lines/cm
d) 10,000 lines/cm

60) A grating of 500 total grooves has what resolving power at the second diffraction order ($m_d$ = +2)?
a) 250
b) 500
c) 1000
d) 2000

# 5.9 DIFFRACTION SUMMARY

Diffraction is a manifestation of the wave nature of light that is observed when coherent light illuminates a small aperture or obstacle whose dimensions are comparable to the wavelength. It is, essentially, an interference, not by two, but by an infinite number of elementary Huygens secondary sources that correspond to the aperture or the obstacle.

Certain approximations are made for the two main types of diffraction: Fresnel, near field, and Fraunhofer, far field. In the latter diffraction type, we assume a constant field amplitude at the entrance, a fixed phase at the entrance, and, more importantly, an observation screen at a sufficiently large distance from the entrance. An important mathematical statement is that the far-field diffraction pattern formation corresponds to the Fourier transform of the entry (pupil) function.

## Fraunhofer Diffraction

To calculate the diffraction intensity distribution on a screen far from a diffractive aperture, we:

1. Express mathematically the entry function [the spatial distribution of the magnitude and phase of the electric field at the diffractive aperture $E_1(x_1, y_1)$], which is the pupil function.
2. Find the Fourier transform that corresponds to this function, calculated at the spatial frequencies. This is the diffraction field $E(x_o, y_o)$.
3. Obtain the diffraction intensity distribution $I(x_o, y_o)$ (the observable quantity) from the square of the diffraction field.

## Single-Slit Diffraction

The diffraction intensity distribution of a slit of size $\alpha$ illuminated by coherent light of wavelength $\lambda$ on a screen at a distance $z$ has a central maximum surrounded by weaker side lobes. The angular and spatial extent of the main-lobe central maximum are $2\lambda/\alpha$ and $2\lambda \cdot z/\alpha$, respectively. The angular and spatial extent of the side lobes are $\lambda/\alpha$ and $\lambda \cdot z/\alpha$, respectively.

## Circular Aperture Diffraction

The diffraction intensity distribution of a circular aperture of diameter $D$ illuminated by coherent light of wavelength $\lambda$ on a screen at a distance $z$ has a central maximum surrounded by weaker side lobes. The angular and spatial extent of the main-lobe central maximum are $2.44\lambda/D$ and $2.44\lambda \cdot z/D$, respectively. The central shape of this formation is known as the Airy disk.

In well-corrected optical systems, the smallest possible image spot, formed of a point object, has the shape not of a point, but of an Airy disk. Such systems are called diffraction-limited. A system with a larger aperture stop (diameter $D$) results in a finer image spot.

## Rayleigh Criterion of Resolution

The minimum separation (angular or spatial) between the closest distinguishable image points is the resolution limit. The reciprocal of the resolution limit is the resolving power or resolving ability. For two points to be discernible, their separation has to be at least such that the center (bright maximum) of one Airy disk coincides with the first minimum (dark) of the other disk. Then their peak-to-peak separation is at least the radius of the central Airy lobe. This is the Rayleigh criterion. Thus, the smallest angular separation between two points is $1.22 \cdot \lambda/D$.

This criterion has profound implications in the resolution limit of the visual system of the eye—a limit that determines visual acuity. For the average human eye in daylight conditions (pupil diameter 2 mm), the resolution limit is 1/60° or 1 arcmin.

Another implication of diffraction is the quality of the image. In a diffraction-limited system, the smallest image point has the shape of the Airy function. In general, the smallest image point forms a distribution described by the point spread function (PSF). The narrower and more confined the PSF, the sharper the image.

Likewise, the image contrast is affected as well. The parts containing certain spatial frequencies (fine detail) lose contrast to the point of not being imaged. This is the cut-off frequency, which determines the closest spacing of the black-and-white alternating stripes that are distinguishable in the image. The typical resolving power of the human eye with a 2 mm diameter pupil is approximately 100 line pairs/mm, which converts to about 30 cycles/degree.

## Diffraction by More than One Aperture

To find the diffraction intensity distribution of two similar apertures shifted symmetrically from the optical axis, we:

1. Find the diffraction pattern formation formed by only one shift, ignoring the transverse shift.
2. Calculate the interference factor (fringes) from two points located at the shift centers.
3. Multiply the two formations. The diffraction pattern is amplitude-modulated by the interference fringes.

The combined diffraction intensity distribution is a pattern of the amplitude-modulated carrier frequency. The amplitude-modulated distribution (envelope) is the independent diffraction pattern of a single slit whose main parameter, the bright central lobe, is inversely proportional to the slit size $\alpha$. The carrier frequency has a periodic modulation that is inversely proportional to the two-slit spacing $d$, just as in the interference by two aperture point sources.

A diffraction grating consists of a large number of equally spaced, identical slits of spacing $d$. The condition for the intensity maxima for near-normal incidence is $d \cdot \sin\vartheta = m \cdot \lambda$.

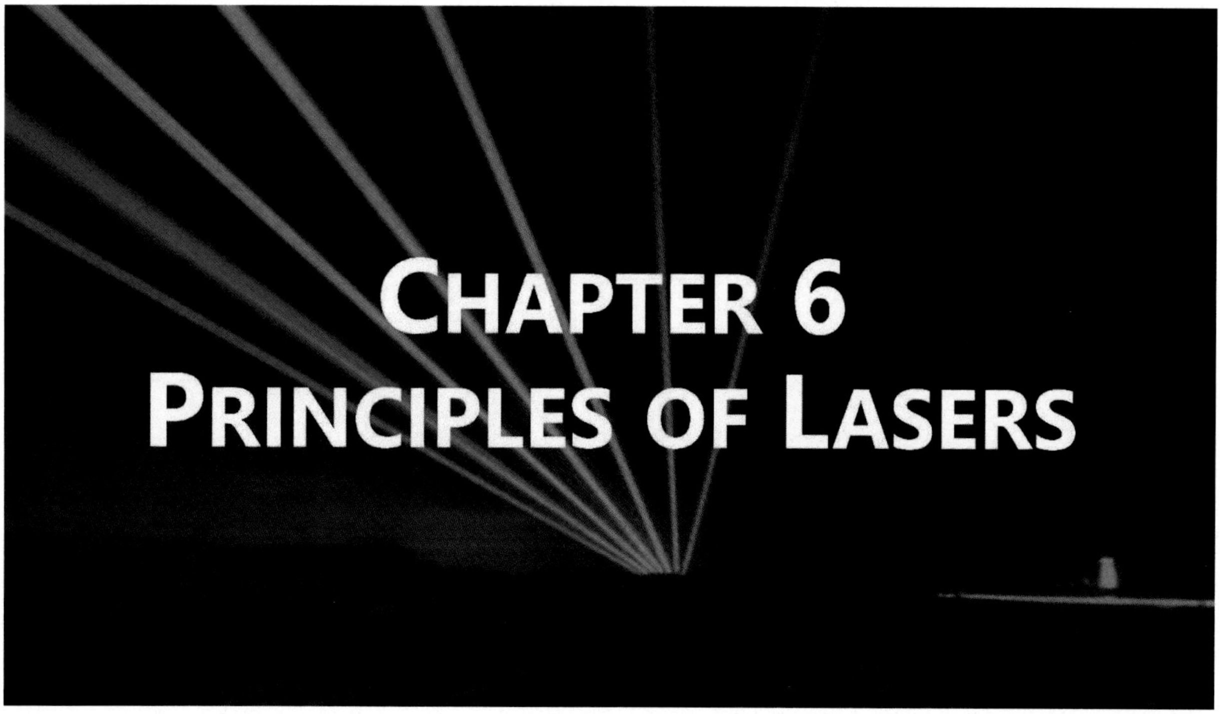

# CHAPTER 6
# PRINCIPLES OF LASERS

Laser is light. However, comparing the monochromatic light of a simple lamp to laser light is akin to comparing the kids in a schoolyard to a military march! The kids' random and unpredictable movement is entirely different from the highly synchronized movement of the soldiers. A laser is light with a very high degree of inherent organization.

This highly structured form of laser radiation is responsible for certain specific properties, such as high monochromaticity, directionality, and high pulse energy. These properties make the laser uniquely suitable for applications in which ordinary light would not or could not be applied. The laser is used in a very broad range of applications, including industrial (high-precision cuts, metal welding, semiconductor etching), medical (cornea keratomileusis for the correction of refractive error, as well as other therapeutic and cosmetic surgical applications), and optical telecommunications (lasers are used as both sources and signal modulators), in addition to simple applications such as label scanning, which is the function of the compact disc (CD).

To comprehend the basic mechanism of a laser, we need some understanding of the nature of light, as well as an understanding of the atomic structure and quantum theory. Then we deal with the appropriate energy transitions inside an atom. A surprising mechanism of energy exchange is the stimulated emission of radiation, a transition that is the 'heart' of the laser. This emission is a purely quantum effect—a legacy of Einstein's attempt to interpret black-body radiation. So, yes, despite the fact that the laser was born in the 1960s, it would not have been possible without the groundbreaking contribution of Einstein's ideas.

## 6.1 THE ATOMIC STRUCTURE

Democritus was the first to propose that matter cannot be split into increasingly smaller pieces indefinitely, but is instead composed of tiny particles he named **atoms** ($\alpha$- for not & -$\tau\acute{\varepsilon}\mu\nu\omega$ for cut) because they cannot not be any further divided. Today we know that the atom is the smallest unit into which matter may be split with no electric charge separation.

The atomic structure describes the atomic particle (electrons and protons) arrangement inside the atom, the type of forces between them, and the relative motion of the electrons. Initially, with the Rutherford experiment (1910), we realized that all of the positive charge is condensed at the center (the nucleus), whose radius is some tens of thousands times smaller than that of the atom. If we compare an atom to a football stadium, a few electrons 'loiter' in the stands, while the nucleus is the size of a dice at the center of the field. The atom is mostly empty space!

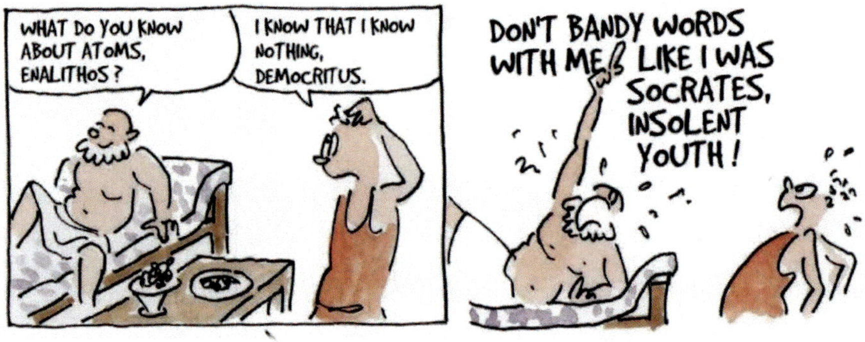


*Figure 6-1: Democritus was the first to propose the concept of the atom (© www.fiami.ch).*

The different atoms are tabulated in the periodic table as elements. Each element has a different integer **atomic number** $Z$, which expresses the number of protons in the nucleus of an atom; this number equals the number of electrons in a neutral atom. In a first approximation, the electrons orbit the nucleus, just as the planets orbit the sun according to the planetary laws of Johannes Kepler.

However, this model cannot not explain atomic stability: No light is emitted by the orbiting electrons. An accelerating charge (orbiting involves acceleration) emits electromagnetic radiation. So why do the orbiting electrons not do the same?

A second failure of the model was based on the linearity of the emission spectra. The Rutherford model could not explain why a hydrogen atom, for example, emits only at specific frequencies.

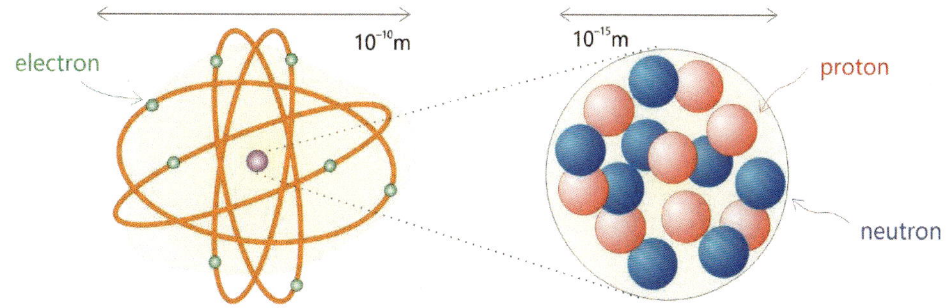


*Figure 6-2: The Rutherford atom. Most of it is empty space.*

*Figure 6-3: Atomic hydrogen-like single-electron orbital, also called the 6s-orbital. The image is a 3-D rendering of the spatial density distribution of* $|\Psi|^2$*, with the color depicting the phase of* $\Psi$*. The spatial distribution is smooth and vanishes for long radii. The cloud is a more realistic representation of an orbital than the more common solid-body approximations. (Image by Geek3 from Wikipedia.)*

A steady-state nonradiating atom excludes gradual and continuous changes in the electron states. This is possible only if we consider that the energy in an atom can take only specific, discrete values. The existence of discrete energy states is a consequence of the wavelike behavior of matter as described by quantum theory.

During its first stages, quantum theory employed the model proposed by Niels Henrik David Bohr: a minification of the solar system, with electrons in geometrical orbits around the nucleus, in agreement with the laws of classical Physics. No spiral orbits were permitted, nor could electrons 'fall' into the nucleus, emitting radiation along the way, because electrons had to emit energy in quanta—in prescribed quantities. Thus, electron orbital stability was explained: Each orbit corresponds to a specific energy level, with no intermediate levels.

Today we know that Bohr's theory is simply a first approximation. The electron is not 'a small planet around the sun,' but a particle with a specific energy associated with a wave function probability according to the theory developed by Werner Heisenberg, Erwin Rudolf Josef Schrödinger, and Paul Adrien Maurice Dirac. The electron state is described by a probability function. Strictly prescribed geometrical orbits do not exist, but we can predict the probability that an electron will be at a certain point in space and time. This probability is expressed by the solutions of an associated wave equation.

Stationary solutions result if we consider that the electron, under the influence of the nucleus, behaves like a standing wave. These solutions are the system **eigenstates**. The values of energy $E_n$ that correspond to these solutions provide the **energy states** (levels) of the atom. Each level is described by a set of three quantum numbers: $n$, $l$, and $m$. The first number $n$ is termed the **principal quantum number** and determines the electron energy:

$$E_n = -2\frac{me^4\pi^2Z^2}{h^2n^2} = E_{Z=1,\,n=1}\cdot\frac{Z^2}{n^2} \tag{6.1}$$

where $m$ is the mass, $Z$ is the atomic number, and $e$ is the electron charge. The energy that an electron can have—as well as the total energy in an atom—does not take just any value, but takes only specific values. This noncontinuous energy in an atom is called **quantization** of the energy levels. The principal quantum number takes the integer values from $n = 1$ to $n = 7$.

For $Z = 1$ (hydrogen) and $n = 1$, the lowest energy level corresponds to $E_{Z=1,n=1} = -13.6$ electron volts (eV). The photon energies that can be emitted from a hydrogen atom are described by

$$\Delta E_n = E_{Z=1,\,n=1}\cdot\left(\frac{1}{n_{\text{final}}^2} - \frac{1}{n_{\text{initial}}^2}\right) \tag{6.2}$$

The hydrogen spectrum is composed of many different series. The Balmer series, named after the Swiss mathematician Johann Jakob Balmer,[33] corresponds primarily to the visible part of the emission and is produced for $n_{\text{final}} = 2$. The Paschen series, named after the German physicist Friedrich Paschen,[34] corresponds to $n_{\text{final}} = 3$ (lines in the infrared); the Lyman series, named after the American physicist Theodore Lyman,[35] corresponds to $n_{\text{final}} = 1$ (lines in the ultraviolet). The greater the value of $n$, the farther the electron distribution is from the nucleus, and the greater the mean value of the radial orbit. For the hydrogen atom at its fundamental state, the (smallest possible) radius is $\alpha_o = 0.529\,\text{Å}$, while for an atom with atomic number $Z$ for an orbit with a principal quantum number $n$, the radius $r_n$ is given by

$$r_n = \alpha_o\cdot\frac{n^2}{Z} = 0.529\frac{n^2}{Z} \quad [\text{Å}] \tag{6.3}$$

The first three energy levels in a hydrogen atom ($Z = 1$) can be described as shown in the figure.

[33] Balmer JJ. Notiz über die Spectrallinien des Wasserstoffs [Note on the spectral lines of hydrogen]. Annalen der Physik und Chemie. 1885; 25:80-7.

[34] Paschen F. Zur Kenntnis ultraroter Linienspektra. I. (Normalwellenlängen bis 27000 Å.-E.) Annalen der Physik. 1908; 332(13):537-70.

[35] Lyman T. An Extension of the Spectrum in the Extreme Ultra-Violet. Nature. 1914; 93(2323):241.

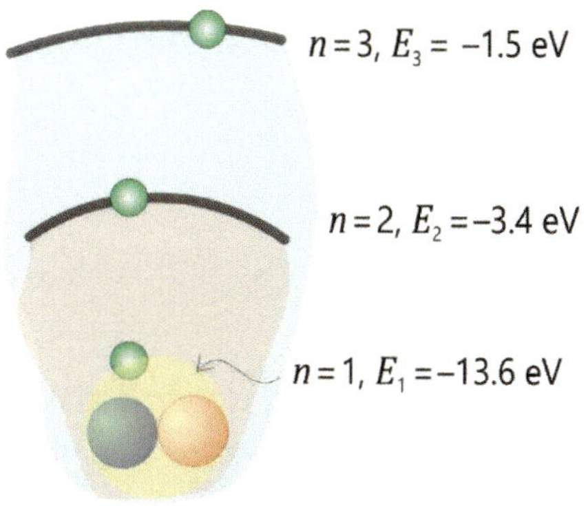


*Figure 6-4: Energy levels in the hydrogen atom.*

The secondary quantum number $l$, also called the orbital quantum number, or the **azimuthal quantum number**, determines the angular momentum $\mathbf{l}$ (whose magnitude is denoted by $\underline{l}$):

$$\underline{l} = \sqrt{l \cdot (l-1)} \cdot \frac{h}{2\pi} = \sqrt{l \cdot (l-1)} \cdot \hbar \tag{6.4}$$

The angular momentum is zero for spherical distributions and increases for an elliptical orbit. Thus, $l$ determines the distribution shape. For each $n$, $l$ takes the $n$ integer values:

$$\text{Azimuthal Quantum Number:} \quad l = 0, 1, 2, \ldots, n-1 \tag{6.5}$$

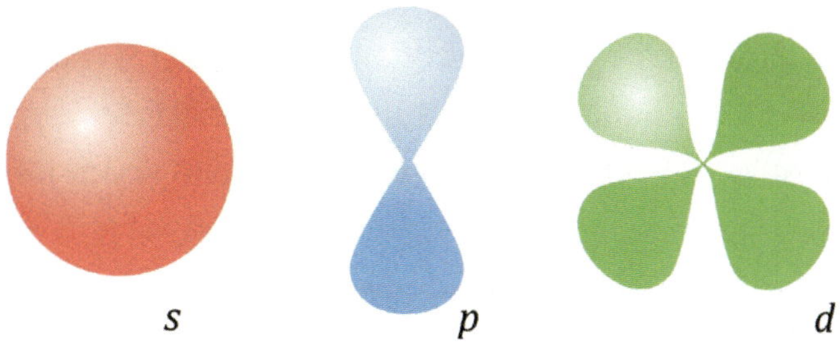


*Figure 6-5: Simple distributions with n = 1 and l = 0, 1, 2.*

Figure 6-5 presents the quantized orbits for $n = 1$, where $s$ (sharp), $p$ (principal), $d$ (diffuse), and $f$ (fundamental) stand for angular momentum: $s \mapsto l = 0$, $p \mapsto l = 1$, $d \mapsto l = 2$, and $f \mapsto l = 3$. For example, $2p$ indicates that $n = 2$ and $l = 1$, while $1s$ indicates that $n = 1$ and $l = 0$.

The third quantum number $m$, the **magnetic quantum number**, determines the projection of the angular momentum onto the $-z$ axis:

$$l_z = m\frac{h}{2\pi} = m \cdot \hbar \tag{6.6}$$

The orientation of the elliptical orbit is not random, but instead correlates to a set of values of specific, possible, magnetic quantum numbers. For every $l$ value, $m$ takes the $2l+1$ integer values from $-l$ to $+l$:

Magnetic Quantum Number: $m = -l,\ \ -l+1,\ \ -l+2, \ldots,\ \ +l-2,\ \ +l-1,\ \ +l$ (6.7)

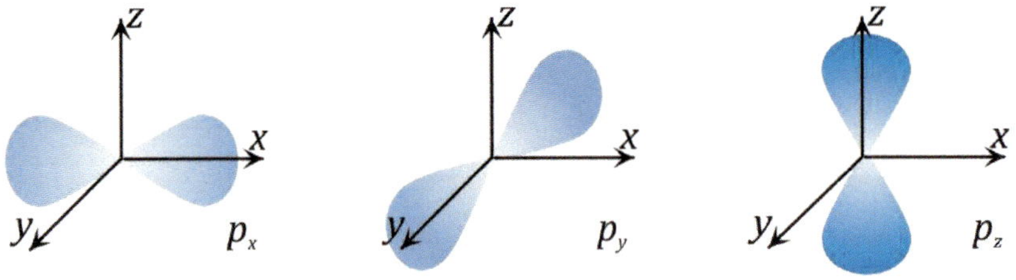


*Figure 6-6: The p-orbital orientation in space.*

Thus, for every value of the principal quantum number $n$, there are more than one orbits that differ in their spatial distribution of probabilities. Specifically, there are $n^2$ different eigenstates for every $n$. The geometric illustrations for these orbits are circular or elliptical and have different orientations in space. For a specific $n$, there is only one spherical distribution ($l = 0$), while the remaining values of $l$ correspond to elliptical distributions that become increasingly complicated. The most modern theory does not even engage geometric illustrations and orbits, but uses only probability wave functions. However, in a first approximation, it helps to keep this orbital representation in mind.

The exclusion principle proposed by Wolfgang Ernst Pauli states that only two electrons can have the same set of quantum numbers ($n$, $l$, $m$). These electrons are distinguished by the **magnetic quantum number** of the **spin** $m_s$ (simply called spin) which takes the values $-½$ and $+½$. Thus, each electron in an atom at any given time is described by a single set of four numbers ($n$, $l$, $m$, $m_s$). This is the quantum state.

In the elementary particle world, there are either bosons, which can share quantum states, or fermions, which do not share quantum states due to the Pauli exclusion principle. **Fermions** are named after the Italian physicist Enrico Fermi, and **bosons** are named after the East Indian physicist Satyendra Nath Bose. A typical fermion is the electron, while a typical boson is the photon. Thus, photons may favor sharing the same quantum state by grouping (§ 1.4.6) into a photon state. To the contrary, in an electron state, there is only one electron! Any atomic level with energy $E_n$ (the main quantum number of $n$) can be occupied by a specific number of electrons. Level $n = 1$ has room for only two electrons (1, 0, 0, ±½), and level $n = 2$ has room for eight electrons [(2, 0, 0, ±½), (2, 1, 0, ±½), (2, 1, −1, ±½), and (2, 1, +1, ±½)]. In general, the level capacity is $2{\cdot}n^2$. Even within a given energy level (principal quantum number $n$), the $2{\cdot}n^2$ electrons have slightly different energies. This is the **fine structure**.

The total electron angular momentum $\mathbf{j}$ is the vector sum of the angular momentum vector $\mathbf{l}$ and the spin vector $\mathbf{s}$:

$$\text{Electron Angular Momentum:} \quad \mathbf{j} = \mathbf{l} + \mathbf{s} \tag{6.8}$$

Similar to atoms and molecules, energy states are quantized. In addition, molecules can oscillate and/or rotate. The oscillation and rotation energy states in a molecule are also quantized and are characterized by their respective quantum numbers.

*Figure 6-7: Attendees at the Fifth Solvay Conference on Electrons and Photons (1927), where the world's most notable physicists met to discuss the newly formulated quantum theory. Seventeen of the twenty-nine attendees were (or became) Nobel Prize winners, including Marie Curie, who alone among them won Nobel Prizes in two separate scientific disciplines. The leading figures were Albert Einstein and Niels Bohr.*

## 6.1.1 Permissible Transitions

Even an eigenstate, which is a temporally fixed solution of the electron wave function, is not fixed over time. Eigenstates oscillate, and it is possible (for many reasons) for an electron to shift states and transition to another state. The concept of stability is relative compared to other probable states. In quantum mechanics, the expected value of a physical quantity is calculated by the integral of its respective wave function. For example, the probability of a state $\Psi_i(x, y, z, t)$ being observed can be expressed in terms of a probability amplitude $\alpha_k$:

$$|\alpha_k|^2 = \int \Psi_i^*(x,y,z,t) \cdot \Psi_i(x,y,z,t)\, dx\, dy\, dz \tag{6.9}$$

where * indicates the complex conjugate of a function. If the probability amplitude $\alpha_k$ is not time-dependent, then the state is termed an eigenstate and its energy is well-defined.

It is possible—due to oscillation, rotation, or electron exchanges in an atom or molecule—to have a **transition** from one state to another. The calculation of the transition probability in quantum mechanics involves a **transition matrix**. The matrix elements $M_{if}$ are dependent on the coupling between the two states and are calculated by the spatial integrals of the wave functions of the final state, the interaction potential $V$, and the wave function of the initial state. This is the transition moment integral:

$$M_{if} = \int \Psi_f^*(x,y,z,t) \cdot V \cdot \Psi_i(x,y,z,t)\, dx\, dy\, dz \tag{6.10}$$

The **transition probability** $\lambda_{if}$ depends on both the coupling strength between the states and the number of possible transitions (density $\rho_f$). The latter plays the role of providing statistical weight. We express the transition probability $\lambda_{if}$ in units of inverse time using the following relationship:

Transition Probability:
$$\lambda_{if} = \frac{2\pi}{\hbar} \cdot \left|M_{if}\right|^2 \cdot \rho_f \tag{6.11}$$

This transition probability is also termed the decay probability and is related to the reciprocal of the mean transition lifetime $\tau$:

Mean Transition Lifetime:
$$\tau_{if} = 1/\lambda_{if} \tag{6.12}$$

There are **permissible** transitions if the corresponding transition probability is nonzero. If the probability is zero (or very small), the transition is **forbidden**. Even between two permissible transitions of same nature, their transition probabilities may be significantly different. This is noted as large differences between the intensities of the spectral emission lines.

In a class of transitions termed **electric dipole transitions**, an electron transits from a state with an azimuthal number of $l_i$ to a final state with an azimuthal number of $l_f$. The transition is permitted if the change in electron orbital angular momentum is by just one unit $\hbar$, meaning that the change in azimuthal quantum number is by just the number 1:

$$\Delta l = l_i - l_f = \pm 1 \tag{6.13}$$

This is described by the Fermi golden rule: The change in electron orbital angular momentum (and therefore in the atomic system) by quantity $\hbar$ exactly corresponds to the angular momentum of the emitted or absorbed photon. This can be considered an extension of the conservation of momentum of the entire system: atom + photon. Thus, a transition from state $2p$ to state $1s$ is permitted, while a transition from state $2s$ to state $1s$ is not. The latter is a forbidden transition; in other words, the transition probability is very small.

Another rule that formally constrains the possible transitions is the **selection rule** or **transition rule**, expressed as

$$\Delta m = 0 \text{ or } \pm 1 \tag{6.14}$$

Transitions for which $\Delta m = 0$ are termed $m$-even or linear, while transitions for which $\Delta m = \pm 1$ are termed $m$-odd and relate to rotating dipoles. In each transition, the magnetic spin quantum number ($m_s$) does not change.

In addition to electric dipole transitions, there are also magnetic dipole transitions, as well as higher-order electric transitions such as electric quadrupole transitions, each having different selection rules. Certain forbidden transitions may be permissible as magnetic dipole or electric quadrupole transitions. However, these transition probabilities are many orders of magnitude smaller than those of a corresponding electric dipole transition.

## 6.1.2 Occupancies...

The energy levels in an atom are like the steps in a stairwell. The nucleus is at the floor; just as the gravitational field pulls a ball to steps of smaller potential energy, the electromagnetic field of the nucleus pulls the electrons to levels of lower energy. Imagine such a stairwell and many balls thrown from above. The balls first fill the bottom steps of the stairwell, and the top steps are filled later.

When filling the energy states with electrons, the levels of lower energy are filled first, so that the system has the smallest possible energy and therefore the highest stability. This state is termed **fundamental**, while if an electron, for some reason, leaves its state to transition to another state of higher energy, the state is termed **excited**.

The Boltzmann distribution describes the energy-level relative populations in a state of thermal equilibrium. Figure 6-8 represents the prediction of Ludwig Boltzmann. The length of each line indicates the energy-level population (for energies from $E_0$ to $E_3$). The 'take-home message' is this: No energy level in thermal equilibrium can have a larger population than another level to which a lower energy corresponds.

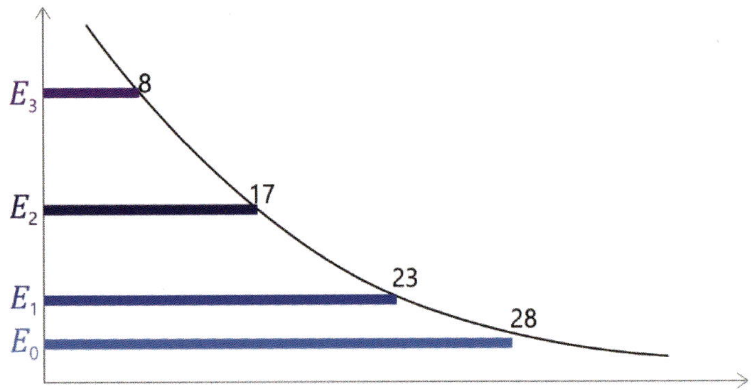


*Figure 6-8: Relative populations in energy levels.*

In thermal equilibrium, the atom tends to be in a state in which the electron-energy distribution ensures the lowest possible energy. Thus, the levels of lower energy have a greater probability of being occupied than other levels of higher energy. The Boltzmann distribution is expressed as follows: If $N_1$ is the number of atoms in a level with energy $E_1$, and $N_2$ is the number of atoms in a level with energy $E_2$, then in a state of thermal equilibrium, the ratio $N_2/N_1$ is the following:

$$\frac{N_1}{N_2} = \exp\left(-\frac{E_2 - E_1}{kT}\right) \tag{6.15}$$

where $T$ is the absolute temperature (expressed in kelvins K), and energy $E$ is expressed in electronvolts (eV). Just as in every similar relationship in quantum mechanics, the temperature is accompanied by the Boltzmann constant $k = 1.38 \times 10^{-23}$ J/K.

### 6.1.3 Radiative Processes

Let's return to the analogy of the stairwell with the balls. We raise a ball from the ground level (1) to the higher level (2). To do so, we must 'pay a toll' for the transfer—an energy amount that corresponds to the potential energy difference between the two steps. If we let the ball go, it falls back from level 2 to ground level 1, releasing that same amount of energy in the form of kinetic energy and eventually in the form of heat.

*Figure 6-9: Energy-level mechanical analog.*

Radiation emission involves processes of energy exchange between matter and light (as discussed in § 1.5.2) that result in light emission. Light emission is due to an electron (the ball in the above analogy) transition from an energy state $E_2$ to another state $E_1$ that is slightly lower. For the emission of a specific photon, the difference between the two atomic state energy levels corresponds to the photon frequency $\nu_{12}$:

Photon Energy: $$h \cdot \nu_{12} = E_2 - E_1 \quad (6.16)$$

### 6.1.3.1 Spontaneous Emission

Consider two isolated energy states in an atom, states 1 and 2, between which an electric dipole transition is permitted. Ground level 1 has energy $E_1$, while the upper level 2 has energy $E_2$ ($> E_1$). Electric dipole transitions are permitted between the two states.

We assume that the atom has one electron at level 2 and no electrons at level 1. This is an excited state, not the fundamental state, and is metastable: The system has a tendency to relax (decay) spontaneously when the electron returns to level 1 under no external influence. Thus, the system returns to the fundamental state, which is stationary since it corresponds to a lower energy. The life expectancy of the excited state depends, among other things, on the conditions of pressure and temperature, and does not exceed a few nanoseconds.

Upon relaxation, the energy difference $E_2 - E_1$ is released in the form of the emitted photon. This radiative process is termed **spontaneous emission**. The longer the mean lifetime of the electron in the excited state, the lower the decay probability in the unit of time. The transition probability for spontaneous emission in the unit of time was expressed by Einstein as

Spontaneous Emission Probability: $$A_{21} \cdot dt \quad (6.17)$$

where $A_{21}$ is the **coefficient of spontaneous emission**. The quantity $\tau_{spont} = 1/A_{21}$ has dimensions of time and is called the **transition lifetime** for the spontaneous emission. Coefficient $A$ has dimensions of inverse time and is inversely proportional to the lifetime of the excited state. It relates to the mean lifetime and [per Eq. (6.12)] to the atomic structure, and is entirely independent of any field. Typical values for the lifetime are on the order of magnitude of $10^{-7}$ s.

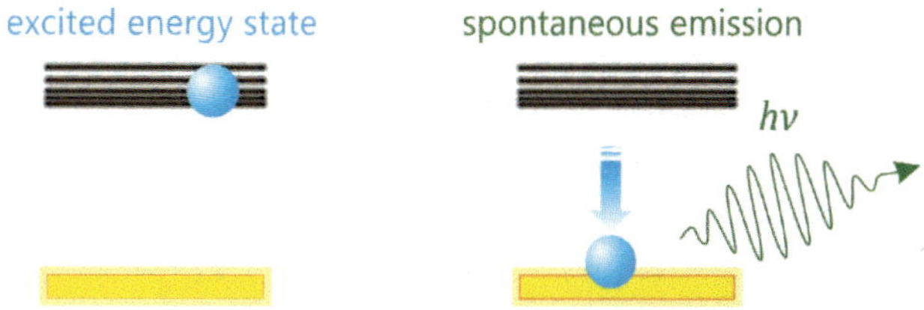


*Figure 6-10: Spontaneous emission.*

Light emitted from warm bodies (such as a heated piece of iron or an incandescent lamp) and gas-discharge sources (such as lightning) is mainly due to spontaneous emission from excited atomic and/or molecular states. The direction of the emitted photon in a gas or a homogenous medium can be entirely random, since there is no mechanism to determine or restrict a favorable direction of emission. In a solid source, the direction of the emitted photons can be dictated by the orientation of the excited atoms and the symmetry of their arrangement, if such a condition exists.

### 6.1.3.2 Stimulated Emission

Albert Einstein realized that, in addition to spontaneous emission, another form of emission is possible. Consider an atom configuration in the excited state. The transition probability is such that an electric dipole transition is permitted between the two states.

A photon of energy $h \cdot \nu = E_2 - E$, which is exactly the energy difference between the two atomic energy levels, is incident on the atom. It is possible for the photon to trigger, or stimulate, the atom to decay. The energy difference $E_2 - E_1$ is released in the form of a photon that has exactly the same frequency (energy), same direction (momentum), and same phase as the initial photon. The emitted photon is in the same photonic state as the incident photon—they have the same frequency and direction.

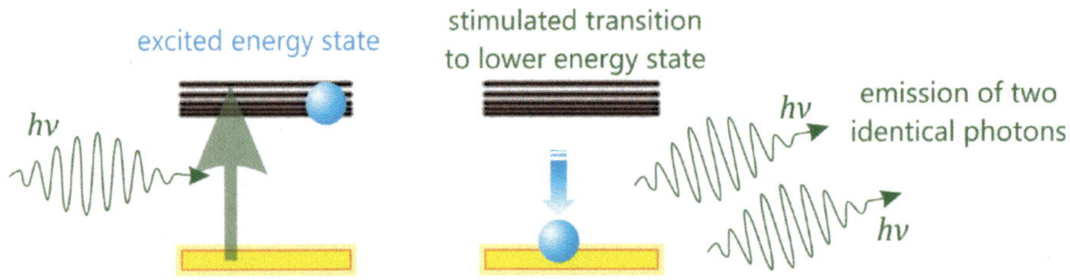


*Figure 6-11: Stimulated emission.*

This photon 'cloning' process is known as **stimulated emission**, or induced emission, of radiation. Photons are coherently multiplied by this process. The probability of stimulated emission occurring in a given unit of time is

Stimulated Emission Probability: $$B_{21} \cdot \rho \cdot dt \tag{6.18}$$

where $B_{21}$ is the **Einstein coefficient of stimulated emission** and $\rho$ is the radiative energy density in a given unit of volume. Einstein arrived at this discovery when attempting to model black-body radiation [*Zur Quantentheorie der Strahlung* (*On the Quantum Theory of Radiation*)] in 1917.

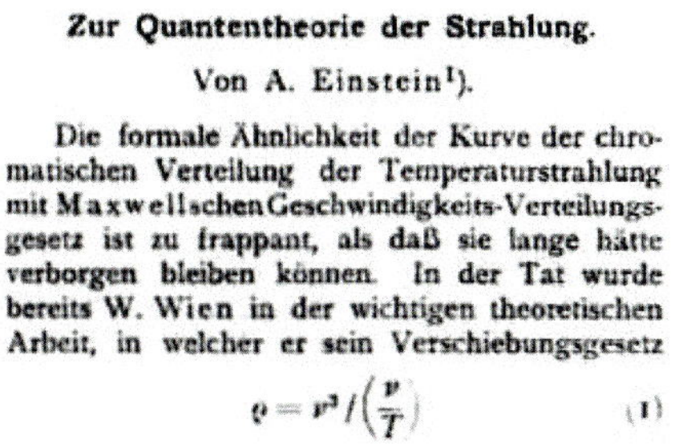
**Zur Quantentheorie der Strahlung.**

Von A. Einstein[1]).

Die formale Ähnlichkeit der Kurve der chromatischen Verteilung der Temperaturstrahlung mit Maxwellschen Geschwindigkeits-Verteilungsgesetz ist zu frappant, als daß sie lange hätte verborgen bleiben können. In der Tat wurde bereits W. Wien in der wichtigen theoretischen Arbeit, in welcher er sein Verschiebungsgesetz

$$\varrho = \nu^3 f\left(\frac{\nu}{T}\right) \quad (1)$$

*Figure 6-12: (left) Einstein's introduction of the concept of stimulated emission of radiation. (right) The discovery of stimulated emission was fundamental to the development of the laser. And, yes, back then this might have been a thought experiment (© www.fiami.ch).*

The probability of stimulated emission occurring is proportional to the density of the already existing radiation in place and is a consequence of the statistical nature of the photon, which follows Bose–Einstein statistics (§ 1.4.6). The coefficients $B$ have units of $J^{-1} \cdot m^3 \cdot s^{-1}$.

Stimulated emission was initially termed negative absorption and was validated by Rudolf Walter Ladenburg in 1928.

### 6.1.3.3 Quantum (Resonant) Absorption

Radiation absorption describes the processes of energy exchange between matter and light (§ 1.5.2) that result in light 'disappearance.' Of course, no energy disappears; photon energy converts to another form that is released inside the medium. An example of such a process is nonresonant absorption (§ 3.3.2.1), in which an electromagnetic wave—even just a single photon with a frequency that does not coincide with any of the material resonant frequencies—can cause the electron cloud in an atom to start oscillating. Thus, the photon energy is transformed to an electron cloud oscillation. This oscillation can be treated with small disturbances around a state of equilibrium (first-order perturbation theory). Sequentially, the oscillating dipoles either re-emit the absorbed energy with the same dipole oscillation frequency (scattering), or the energy is transferred to lattice oscillations, eventually being converted to heat.

The process described here involves radiation absorption in which the photon frequency corresponds to the energy difference between two atomic states. The transition probability is such that an electric dipole transition is permitted between the two states.

Consider the fundamental state. This state is stationary, which means that if there is no external influence, the atom remains in that state. When a photon with energy $h \cdot \nu = E_2 - E_1$ interacts with the atom, because the incident energy equals the energy difference between the two atomic levels, it is possible for the atom to become excited. The photon is absorbed, and the ground-state electron transitions to the upper level. This is **quantum absorption** or **resonant absorption**.

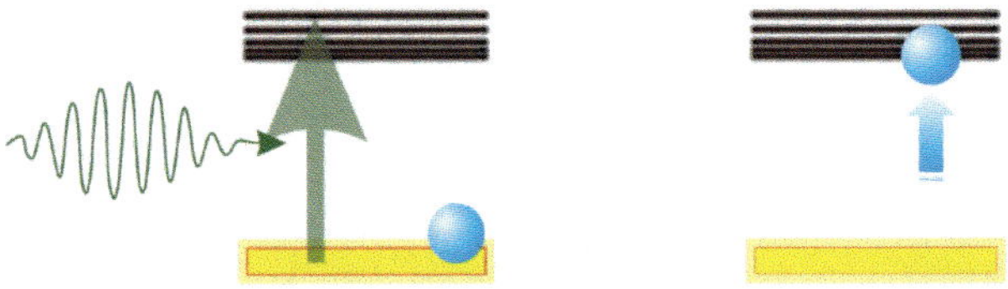


*Figure 6-13: Quantum (resonant) absorption.*

The probability of photon absorption occurring in a given unit of time is

Absorption Probability: $$B_{12} = \rho \cdot dt \qquad (6.19)$$

where $B_{12}$ is the respective $B$ **Einstein coefficient of quantum absorption** with the same dimensions and physical content as the coefficient of stimulated emission.

The basic points of Einstein's groundbreaking work are the following:

- Since the processes of quantum absorption and stimulated emission are reciprocal effects, the probabilities $B_{12}$ and $B_{21}$ are equal: $B_{12} = B_{21} = B$. It is often stated that the transition cross-section for stimulated emission and re-absorption are equal.[36]
- The calculation of these coefficients involves black-body spectral density radiation. The coefficients $A$ and $B$ for a specific pair of energy levels are related by

$$A = \left( 8\pi \cdot h \frac{\nu^3}{c^3} \right) \cdot B \qquad (6.20)$$

where $\nu$ is the photon frequency. The coefficient of spontaneous emission increases with the cube of the corresponding energy difference between the two states. We note that the ratio of the coefficients, $A/B$, has units of density radiation.

The quantum effect of stimulated emission not only provides an interpretation for ordinary effects such as the color of a heated body, but is also responsible for the function of the laser. For stimulated emission to produce laser emission, all that is needed is a material in a state in which more atoms (or molecules) are in a higher energy state than in the ground state.

[36] This equal-transition cross-section for stimulated emission and re-absorption does not hold for transitions between manifold levels with nondegenerate Stark levels, a case frequently occurring in solid-state gain media.

# 6.2 THE LASER CONCEPT

Although Einstein did not invent the laser, he provided the theoretical background of stimulated emission, which is the foundation of laser operation. The two photons encounter two excited atoms and, with subsequent stimulated emission, create four photons. With a sufficiently large population of excited atoms, a large population of identical photons can be created in the form of a laser beam. LASER stands for light amplification by stimulated emission of radiation.

## 6.2.1 Building the Laser Beam: Atomic Rate Equations

Radiation of energy density $\rho$ and frequency $\nu$ interacts with a material along the direction $-z$. We consider two energy levels $E_1$ and $E_2$ such that the only permissible transition corresponds to the frequency $h \cdot \nu = E_2 - E_1$. The material in a given unit of volume has $N_1$ atoms in the ground state 1 (occupied by an electron at $E_1$) and $N_2$ atoms in the excited state 2 (occupied by an electron at $E_2$). As radiation propagates through the material, three exchanges take place between the radiation and matter: spontaneous emission, stimulated emission, and quantum absorption, each of which is described in the next three subsections.

### *6.2.1.1 Spontaneous Emission*

At any random time, a photon is emitted from an excited atom in a random direction and in a random phase. This process is spontaneous and completely independent of any incident field. The rate of change of the radiation energy density $d\rho/dt$ is

$$d\rho/dt = A \cdot \rho \tag{6.21}$$

The emitted photon has no phase correlation to any incident field. The emitted photon direction is random, so the radiation is isotropic ($4\pi$ solid angle). The photon gain along any given direction, including that of the initial radiation, is very small and can be ignored.

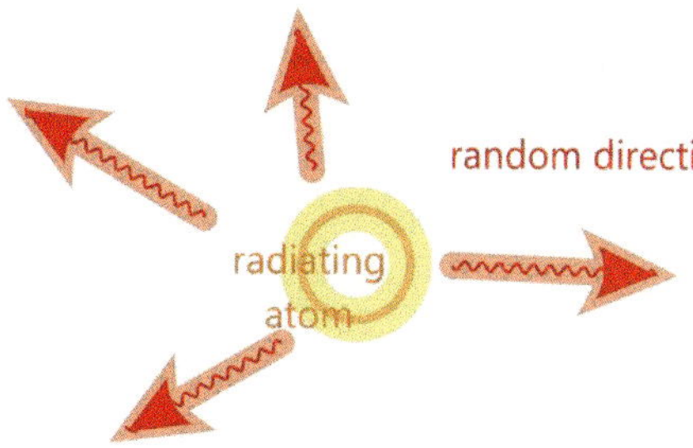


*Figure 6-14: Spontaneous emission of radiation.*

### 6.2.1.2 Stimulated Emission

A photon interacting with an atom in its excited state triggers emission of an additional photon with the same frequency, same phase, and same direction. Unlike spontaneous emission, stimulated emission amplifies the initial radiation. This is due to the perfect match between the stimulated emission photon and the incident radiation. In spontaneous emission, other than the frequency correlation, there is no other correlation, in phase or direction, with the radiation already in place. Essentially, amplification is possible by spontaneous emission because of an avalanche of photon cloning. The gain in density $\rho$ is

$$\frac{d\rho}{dt} = B \cdot \rho \cdot N_2 \cdot (h\nu_{21}) \tag{6.22}$$

### 6.2.1.3 Quantum Absorption

A photon interacting with an atom in its fundamental state is absorbed. The change in radiation density along the direction $-z$ is

$$\frac{d\rho}{dt} = -B \cdot \rho \cdot N_1 \cdot (h\nu_{21}) \tag{6.23}$$

The radiation transcending through the material gains in energy density as follows:

$$\frac{d\rho}{dt} = B \cdot \rho \cdot N_2 \cdot (h\nu_{21}) - B \cdot \rho \cdot N_1 \cdot (h\nu_{21}) = B \cdot \rho \cdot (N_2 - N_1) \cdot (h\nu_{21}) \tag{6.24}$$

Equation (6.24) expresses the photon balance. The quantity $d\rho/dt$ is the rate of energy change inside a given unit of volume inside the medium. It is obvious that a net gain is achieved when the rate of change is positive. The question is, how can $d\rho/dt$ be > 0?

Assume an initial state with only one atom in the fundamental state and only one matching photon. The first possible interaction is photon absorption, followed by spontaneous emission. To have stimulated emission, we need at least one initially excited atom and more atoms in the excited state than in the fundamental state; otherwise, a net gain is not possible.

For radiation amplification ($d\rho/dt > 0$), more atoms must be at the excited level $N_2$ than at the lower level $N_1$: $N_2 > N_1$. Otherwise, if $N_2 < N_1$ the gain is negative, and absorption overpowers stimulated emission. Thus, the radiation is attenuated and the material acts as an absorber.

In a thermal equilibrium state (Boltzmann distribution), the larger populations correspond to the atomic states of the lower energy levels: $N_2 < N_1$. The condition that involves more excited atoms than non-excited atoms is a non-equilibrium condition termed **population inversion**, in which atoms with upper energy levels have a greater population than those with lower energy levels:

$$N_2 > N_1 \text{ for energy levels } E_2 > E_1 \tag{6.25}$$

Population inversion is, by nature, a non-equilibrium state. It requires continuous or pulsed excitation. This state is **metastable** because it is inherently unstable but may have a relatively long lifetime, on the order of $10^{-6}$ s or even $10^{-3}$ s, which is comparable to the lifetimes of atomic transitions, which are typically on the order of $10^{-7}$ s. The prediction that a population inversion is possible is credited to the Russian scientist Valentin Alexandrovich Fabrikant. If it is possible to achieve a population inversion in a medium, this medium can be an **active medium**.

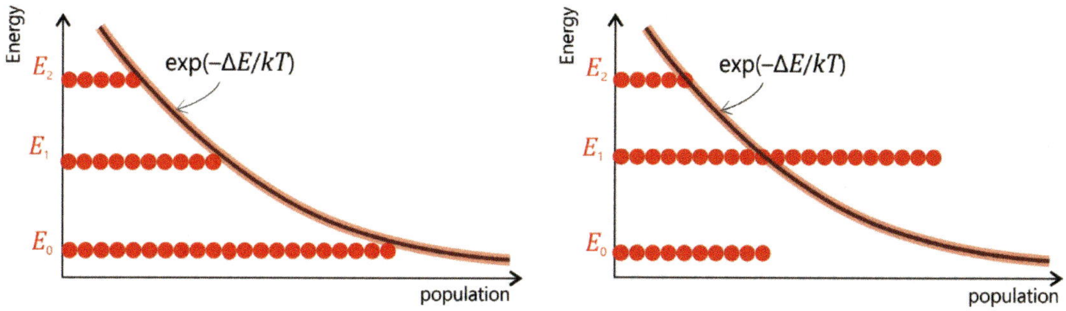


*Figure 6-15: Population inversion between energy levels $E_1$ and $E_0$.*

When the desired population inversion is achieved ('desired' here means that the inversion at least compensates for possible losses, which is the criterion for achieving critical population inversion, critical indicating just enough to commence the lasing action), the laser process can commence with a chain avalanche reaction that forms continuously 'profit-generating' photon generation. The photons multiply with each pass, giving rise to an intense beam of coherent photons moving in the same direction. A new verb describing this action has thus been coined: "to lase."

While it is not possible to predict whether a specific photon will meet an excited atom, over a large population, what matters is statistics. In a population inversion state, the probability of a photon meeting an excited atom is higher than the probability of it meeting an atom in the fundamental state. Thus, statistics ensures that the photon population will increase!

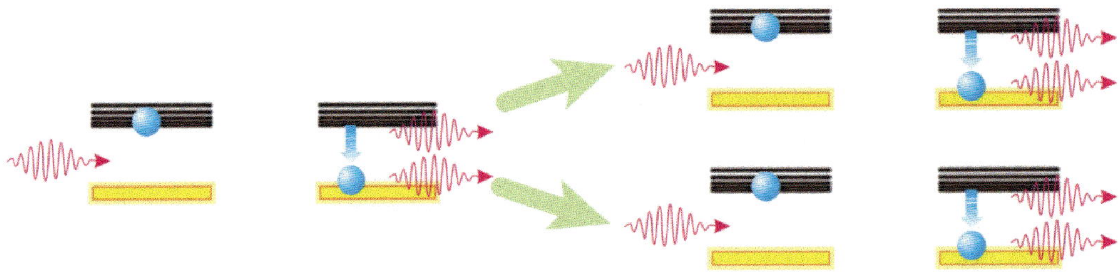

*Figure 6-16: Commencement of the laser process.*

## 6.2.2 The Active Medium

Assume external radiation with frequency $\nu_{12}$ and intensity $I$ incident on an active medium. If $\rho$ is the radiation density, and $u$ is the propagation speed of light in the medium, then intensity $I$ expresses the energy incident on a unit of surface per unit of time:

$$I = u \cdot \rho \tag{6.26}$$

If $\underset{\cdot\cdot}{N}$ is the number of photons in the unit of volume and $\varepsilon_{21}$ is the photon energy, then

$$\rho = \varepsilon_{21} \cdot \underset{\cdot\cdot}{N} = (h\nu_{21}) \cdot \underset{\cdot\cdot}{N} \quad \Rightarrow \quad I = (h\nu_{21}) \cdot \underset{\cdot\cdot}{N} \cdot u = (h\nu_{21}) \cdot N \tag{6.27}$$

where $N$ is the number of photons incident normally on the unit surface in the unit of time.

We emphasize that quantities $I$, $\rho$, and $N$ refer to the specific frequency $\nu_{21}$. As it passes through the active medium, this radiation will be amplified <u>if</u> there is an adequate population inversion. We now combine Eqs. (6.27) and (6.24) to express the photon-generation rate as

$$\frac{d\rho}{dt} = B \cdot \rho \cdot (N_2 - N_1) \cdot (h\nu_{21}) = B \cdot \underset{\cdot\cdot}{N} \cdot (N_2 - N_1) \cdot (h\nu_{21})^2 = \frac{B \cdot h\nu_{21}}{u} \cdot (N_2 - N_1) \cdot I \tag{6.28}$$

The quantity

$$\sigma_{21} = \frac{B \cdot h\nu_{21}}{u} \tag{6.29}$$

is the active **cross-section** of the stimulated emission. This is a dimensionless (pure) number and describes the fraction of photons that contribute to stimulated emission in relation to the incident photons.

There are several ways to achieve a population inversion. The basic steps involve a massive transfer of energy to the non-excited population. Optical means (flash tubes, light from other lasers), electrical means, and even mechanical means may be employed. Some examples of these methods will be discussed in the presentation of the different types of lasers.

This massive energy transfer process is termed **energy pumping** and must be achieved very quickly—faster than the lifetime of the corresponding atomic energy levels, which is about $10^{-7}$ to $10^{-8}$ s. In any case, the process aims to induce—even for a short time—a sufficiently large atom population in the excited state, 'disobeying' the distribution law, and thus to achieve a short-lived population inversion. The source of energy that provides the pumping is simply called the pump.

A laser is characterized by the kind of active medium it uses and the energy differences involved in the lasing action. While there is a multitude of materials that may be utilized in laser action, only a small fraction of energy level sets qualify for lasing. These may be atomic or molecular (oscillating) energy levels, or energy levels in a solid-state arrangement, such as in a semiconductor. Accordingly, the frequencies may vary from microwave up to X ray. The state of matter may be gas, solid, or liquid.

While pumping is an essential condition for a laser to be sustained, pumping alone will not produce lasing. Population inversion and stimulated emission must take place in the presence of optical feedback in an optical resonator / oscillator. This is achieved inside a cavity that acts as a resonator. The **gain envelope** describes the range of frequencies supported by the resonator.

For a given active cross-section $\sigma_{21}$ and population inversion ($N_2 - N_1$), in order to achieve significant amplification for lasing, multiple interactions along an axis of gain are necessary to multiply the gain over one single transition. The photons are 'recycled;' before they are released, each photon can participate in not one, but multiple, simulated emission interactions. This is **optical feedback**. Optical feedback can be achieved if the photons are contained in an optical box, which is an optical cavity that is confined by high-reflectivity mirrors. An aperture on one side allows the laser photons to escape, forming the laser beam.

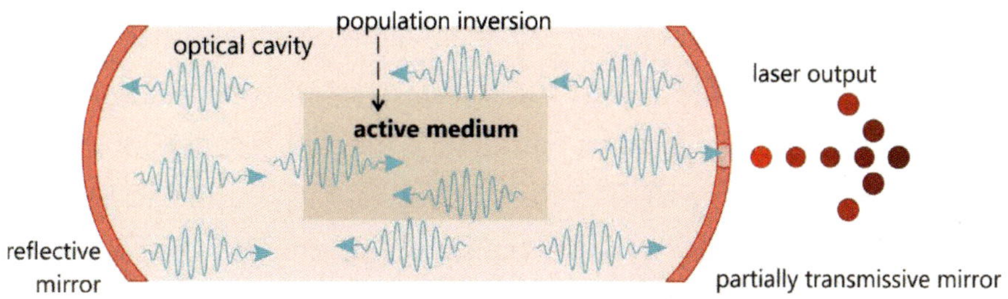


*Figure 6-17: Schematic diagram of the laser model.*

### 6.2.2.1 Optical Resonators and Oscillation Modes

The **optical resonator** (or **oscillator**) is an optical cavity confined by two highly reflecting surfaces. It is designed to sustain multiple transitions of the beam through the active medium; thus, it effectively multiplies the active medium length for significant amplification. In addition to acting as an amplifier, the optical cavity acts as a resonator, which is a filter of specific frequencies. The electric field distributions supported by the resonator are called modes and correspond to a few specific frequencies. Thus, the optical cavity contributes to the laser monochromaticity.

Laser resonators use facing (plane or spherical) reflecting surfaces. Two waves counterpropagating from mirror 1 to mirror 2 and from mirror 2 to mirror 1 interfere constructively along this round trip if they satisfy the interference conditions. In the simplest case, one double-pass includes (a minimum of) one reflection, and the wave arrives at the same origin with the same phase (or a multiple integer of the phase). This is a simple standing wave along the cavity length. The fundamental frequency, or **resonant frequency**, corresponds to a distribution with no intermediate node.

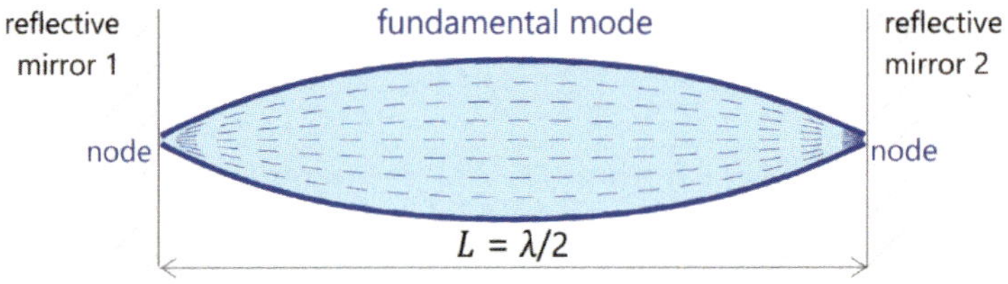


*Figure 6-18: Fundamental-frequency standing wave in a cavity.*

The fundamental wavelength corresponds to $\lambda_o = 2L$, where $L$ is the optical path length between the mirrors (or simply the cavity axial length if it is empty, but it often contains the active medium). Constructive interference occurs when the fundamental state corresponds to an optical path difference equal to one wavelength, or a phase difference equal to $2\pi$.

In addition to the fundamental frequency, there are higher-order harmonics. These harmonics are also standing waves along the cavity axis. Their orders correspond to the number of intermediate nodes (minus one). The first harmonic is the fundamental mode (no intermediate node), the second has one node, the third has two nodes, and so on. In each higher order, the wavelength halves, and the frequency doubles, triples, and so on. The supported oscillation modes have frequencies that are integer multiples of the fundamental (equispaced) modes.

Again using interference reasoning, the harmonics of order $m$ correspond to a path for which the round-trip optical path length equals $m$ wavelengths, or the phase difference equals $m \cdot 2\pi$. A harmonic is also known as an oscillation mode or **propagation mode**. This is a **longitudinal mode**, corresponding to standing-wave formation along the optical axis.

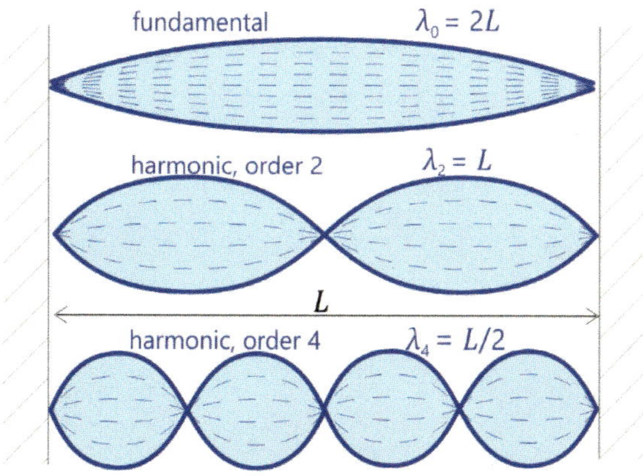


*Figure 6-19: Harmonic frequencies as standing waves in an optical cavity.*

Similar arguments apply for the transverse distribution. Perpendicular to the axis cross-section, additionally, the intensity distribution is not fixed. A fixed cross-section distribution does not satisfy the marginal conditions of, say, a cylinder, and does not exist as a solution of a wave equation within any restricted space. An infinitely extending plane wavefront is a mathematical idealization, even over a nonrestricted space. Thus, in a restricted volume, a transverse electromagnetic wave must arrive at a peripheral node smoothly and not abruptly (the fixed amplitude does comes to a sharp end) at the cavity edges.

The **transverse** oscillation **modes** are solutions of the transverse wave equations that satisfy the marginal conditions. These modes correspond to standing-wave formation / interference perpendicular to the optical axis. They are the $TEM_{XY}$ (transverse electromagnetic) modes, where the indices $_{XY}$ indicate the number of nodes along the $-x$ and $-y$ axes in Cartesian coordinates. The simplest distribution is the normalized, or Gaussian, distribution (§ 6.3.4), which peaks at the center while the intensity falls nearly parabolically toward the periphery. This distribution is the lowest-order oscillation mode, known as the fundamental $TEM_{00}$ mode. This mode is preferable for diffraction-limited output and helps to optimize the efficiency if the radius of the beam in the gain medium matches the radius of the active pump region. In addition to Cartesian coordinates, there are also polar-indexed transverse modes whose indices refer to the polar coordinates.

*Figure 6-20: Electric-field distributions for the first three transverse oscillation modes expressed in polar coordinates.*

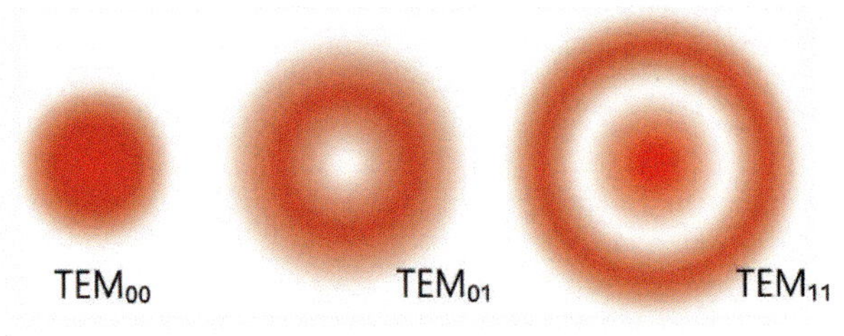


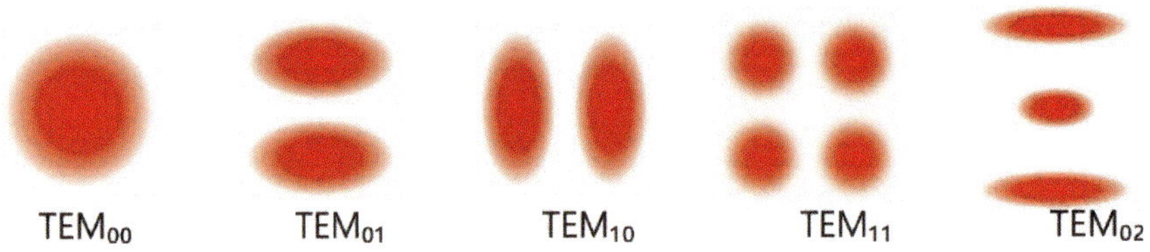


*Figure 6-21: Intensity distributions for transverse oscillation modes expressed in (top) polar coordinates and (bottom) Cartesian coordinates.*

In reality, the laser beam is a superposition of modes, despite the possibility of a prevalence of one mode over another. The questions of which and how many of the oscillation modes prevail, as well as what the fundamental frequency might be, depend on many parameters, such as the shape and dimensions of the resonator cavity (see next page), the wavelength, the optical medium, the initial conditions, etc.

The spectral density of the oscillation modes in a laser resonator per unit of frequency is

$$\frac{d}{d\nu}N(\nu) = \frac{8\pi}{c^3}\nu^2 \tag{6.30}$$

This is exactly the mode density employed in the body-black spectral distribution [Eq. (1.26)].

The design of the resonator (including the optical elements, angles of incidence, and distances between the components) determines the beam radius of the fundamental mode at all locations along the beam, as well as other important properties. The characteristics of an optical resonator are:

(a) In one mirror there is a semi-transparent point that allows the laser beam to exit.

(b) Its dimensions, although often much greater than the wavelength of radiation, are multiple integers of the wavelength $\lambda$; the resonator is tuned to the radiation. The fundamental frequency is determined by the cavity specifics (length, optical medium, shape). The cavity is designed for the wavelengths that satisfy the interference conditions.

(c) It is open, meaning that, practically, it is composed of two opposed mirrors, while the other sides are not confined. The mirrors can simply be the polished sides of a crystal, the refractive surfaces at Brewster's angle (Figure 6-39), or conventional plane or spherical mirrors.

(d) It is axial: Inside the cavity, amplification is favored only for waves that travel nearly parallel to its axis. For these axial modes, the losses are quite low to permit the lasing action. All other possible modes decay nearly completely after a few passages through the resonator.

Common resonator configurations, distinguished by the relative positions and shapes of their mirrors, are described next.

A **plane-parallel** (linear) resonator is a set of two opposing, plane (flat) mirrors. This is essentially an interferometer, called a Fabry–Pérot etalon, named after Marie-Paul Charles Fabry and Jean-Baptiste Alfred Pérot[37] (its concept is also credited to Raymond Boulouch). The only difference between a plane-parallel resonator and an etalon is that in the laser resonator the axial dimension (mirror separations) may be large compared to the transverse mirror dimensions. While simple in design, this arrangement is not often used in large-scale lasers because it is extremely alignment-sensitive; the mirrors must be perfectly parallel within a few arcseconds, or else 'walk off' of the intra-cavity beam will result in it spilling out of the sides of the cavity. Small-scale, plane-parallel resonators are commonly used in microchip and microcavity lasers, as well as in semiconductor lasers.

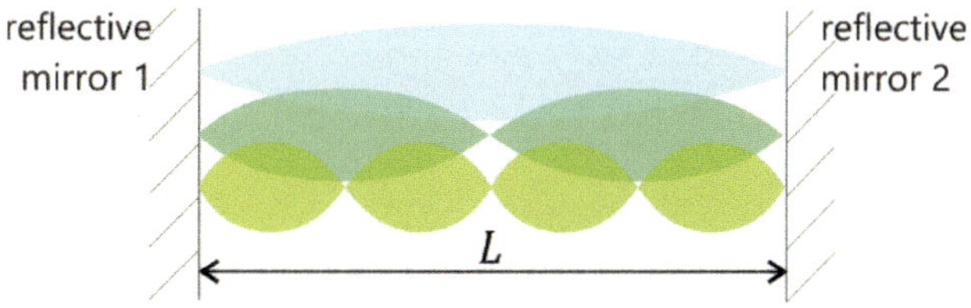


*Figure 6-22: Plane-parallel resonator.*

Practical resonators typically employ curved mirrors to maintain well-defined, laterally confined beams. They are variations of the plane-parallel resonator as described below.

A **concentric** (or **spherical**) resonator [Figure 6-23 (left)] is composed of two spherical mirrors of the same radius of curvature $R$ whose distance $L$ is such that the centers of curvature of the mirrors $C_1$ and $C_2$ coincide ($L = R$). This type of cavity produces a diffraction-limited beam waist right at the center of the cavity, with large beam diameters at the mirrors, filling the entire mirror aperture.

A **confocal** resonator [Figure 6-23 (right)] consists of two spherical mirrors of equal radius of curvature $R$ separated by $L$ such that the focus of one mirror coincides with the center of the other. Because $f = R/2$, the center of curvature $R$ of one mirror actually coincides with the surface of the other: $L = R$. This design produces the smallest possible beam diameter at the

[37] Fabry C, Pérot A. Sur la constitution des raies jaunes de sodium. Comptes Rendus Acad Sci Fr. 1900; 130:653-5.

mirrors for a given cavity length and is often used in lasers where transverse-mode pattern purity is important.

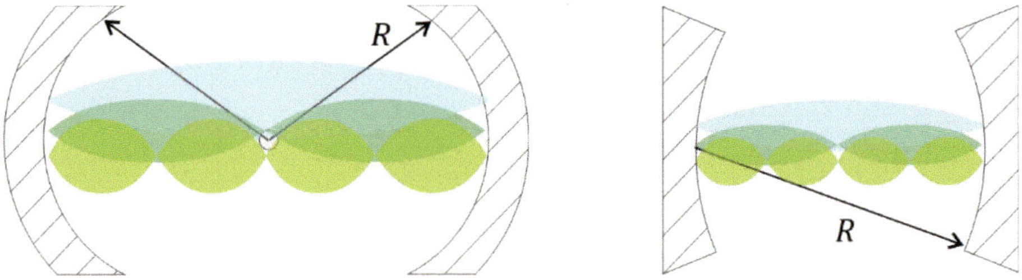


*Figure 6-23: (left) Concentric or spherical resonator. (right) Confocal resonator.*

Other types of resonators include the linear [Figure 6-24 (left)] and the ring resonators [Figure 6-24 (right)]. For practical reasons, many laser resonators comprise more than one mirror pair; three- and four-mirror arrangements are common, producing a folded cavity.

*Figure 6-24: (left) Linear resonator and (right) ring resonator.*

Resonators are classified either as stable or metastable. A resonator is assumed to be metastable if one random ray, after a large number of reflections between the two mirrors, 'walks off' the resonator, as shown in Figure 6-25 (left). To the contrary, in a stable resonator, the ray remains bound. The mirror curvature coincides with the wavefront curvature within the cavity, as shown in Figure 6-25 (right). The output laser beam is formed of photons that are parallel to the cavity axis.

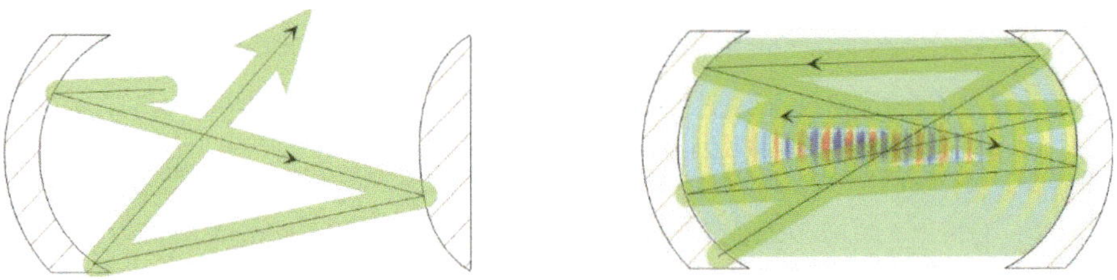

*Figure 6-25: (left) Metastable and (right) stable resonator.*

#### 6.2.2.2 The Optical Resonator in Action

The concept of a resonator is not unique to lasers. Sound systems employed this concept long before optics did; a deep water well is an example of a resonator. In a properly designed acoustic resonator, the audio oscillations amplify and filter the acoustic output.

Since an optical resonator is based on multiple beam reflections off of the confining mirrors, in effect its operating principle is that of multiple-beam interference. Based on this argument, the optical path length for a double path ($2L$) must be a multiple integer of the wavelength. Therefore, the fundamental frequency corresponds to a wavelength $\lambda_o = 2L$ (with the appropriate adjustment for the value of the optical path length instead of $L$ in the obvious case that the cavity is not empty, but contains the active medium).

The theoretical analysis of interference in a Fabry–Pérot resonator is similar to that presented in § 4.2.5. Comparing the interference effects between two beams and multiple-beam interference, we realize that the conditions for the maxima are exactly the same: Maxima appear under the same conditions, regardless of whether we consider only two waves or multiple beams. The factor that determines the appearance of the maxima is the phase difference $\varphi_\delta$ for a double path, which, for the axial modes—the angle propagation in relation to the optical axis $\delta = 0$—is expressed by

$$\varphi_\delta = \frac{4\pi \cdot nL}{\lambda} \tag{6.31}$$

where $n \cdot L$ is the cavity optical length. Constructive interference within the cavity occurs when

Condition for Constructive Interference: $$\varphi_\delta = \frac{4\pi \cdot nL}{\lambda} = 2m \cdot \pi \tag{6.32}$$

where the integer $m$ expresses the interference order ($m$ = 0, 1, 2, ...). Equivalently, this relationship expressed in wavelength units, using the optical path difference, is similar to the multiple-beam interference conditions:

Condition for Constructive Interference (in wavelength): $$2n \cdot L = m \cdot \lambda_m \tag{6.33}$$

Resonant Wavelengths and Frequencies: $$\lambda_m = \frac{2 \cdot nL}{m} \quad \text{and} \quad \nu_m = m \cdot \frac{L}{2c} \tag{6.34}$$

The oscillation modes in a resonator are equispaced only if they are expressed in frequency. Their periodicity is the frequency difference between consecutive maxima. This is the resonator **free spectral range** (FSR). In general, the free spectral range expresses the range of frequencies (or wavelengths) in which there is no neighboring order overlap. Using the above relationships, we can express the FSR for the Fabry–Pérot resonator in frequencies or wavelengths:

$$FSR = \frac{c}{2nL} \text{ (frequency)} \qquad FSR = \frac{2nL}{m} \text{ (wavelength)} \tag{6.35}$$

The frequency expressions are more helpful because the oscillation modes in a resonator are equispaced only if they are expressed in frequencies: The oscillation modes depend exclusively on the cavity optical length ($n \cdot L$). However, the acuity of the maxima is affected by the number of waves participating in the interference, or equivalently, by the cavity mirror reflectivity $R$.

While the intensity of two interfering beams is described by a cosine function squared, for multiple beams, the variation in relative intensity displays maxima that are described by an Airy relationship (Figure 6-26), just as in multiple-beam interference by reflection (Figure 4-44).

Figure 6-26 expresses the Airy function for different values of the mirror reflectivity $R$. This is equivalent to stating the wave population that contributes to the interference effect. We note that the higher the reflectivity,

- the thinner the maxima of the Airy function, and
- the greater the contrast between the maximum and minimum intensities.

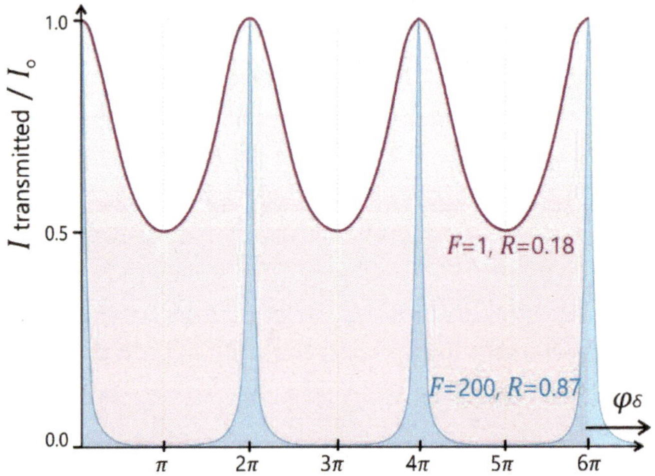


*Figure 6-26: Airy function, in which the variation in relative intensity results from multiple cavity reflections.*

The property of 'being thin' and the contrast between the maximum and minimum both depend on the **coefficient of finesse** $F$, a unitless quantity similar to that defined in interference:

$$F \equiv \left(\frac{2\rho}{1-\rho^2}\right)^2 = \frac{4R}{(1-R)^2} \tag{6.36}$$

We also use an equivalent quantity, the **coefficient of reflecting finesse** or simply, the finesse $\mathfrak{F}$:

$$\mathfrak{F} \equiv \frac{\pi\sqrt{R}}{1-R} \tag{6.37}$$

The larger the value of the reflectivity, the larger the value of the coefficient of finesse. In addition, when the resolution increases, the minimum value of the Airy function decreases (it becomes zero for values of phase difference $\varphi_\delta = \pm\pi, \pm3\pi, \ldots$). The horizontal axis may be also expressed in wavelength, based on the following conversion relationship:

$$\frac{\text{phase difference}}{\pi} = \frac{\text{optical path difference}}{\lambda} \tag{4.17}$$

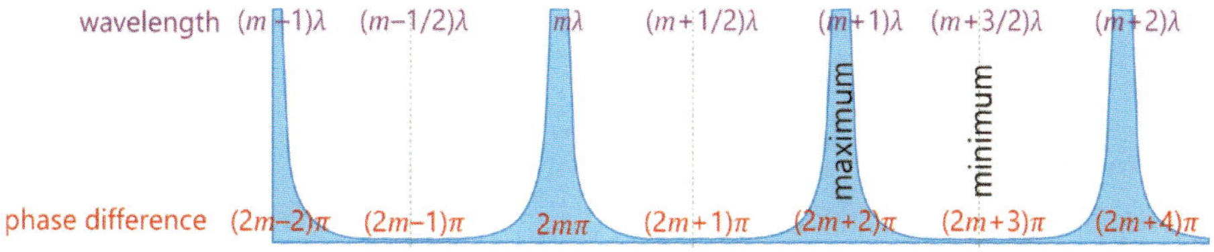


*Figure 6-27: Phase difference and wavelength for the maxima and minima.*

If the wavelengths differ slightly by $\Delta\lambda$ compared to the condition for constructive interference [Eq. (6.33)], the intensity decreases drastically. Assume that there are two wavelengths within the cavity, $\lambda_m$ and $\lambda_{(m+1)} = \lambda_m - \Delta\lambda$. A pattern expressed by the Airy relation corresponds to each wavelength. For wavelength $\lambda_m$ (Figure 6-28), the condition for the maximum is

$$2n \cdot L = m \cdot \lambda_m$$

while for wavelength $\lambda_{m+1}$, the condition for the maximum is

$$2n \cdot L = (m + 1) \cdot \lambda_{m+1}$$

If we plot two successive maxima of orders $m$ and $m+1$ for these two wavelengths (Figure 6-28), we note that (per the Rayleigh criterion) the order of the maxima that corresponds to two wavelengths that are distinguishable when their separation is at least equal to the FWHM of the principal maximum, provided by Eq. (4.97):

$$FWHM = \frac{4}{\sqrt{F}} = \frac{2\pi}{\mathfrak{F}} \text{ (expressed in phase)} \qquad FWHM = \frac{\lambda_m}{\mathfrak{F}} \text{ (expressed in wavelength)} \tag{6.38}$$

The above relationship is a very a simple one that is often encountered in optical resonators. As mentioned above, another parameter is the **finesse**, a factor expressing the reflective 'fineness'.

Finesse: $$\mathfrak{F} = \frac{FSR}{FWHM} \tag{6.39}$$

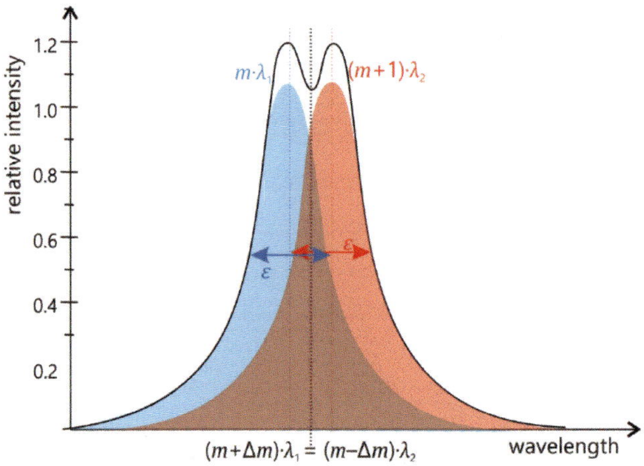


*Figure 6-28: Resolution ability as a result of two neighboring maxima for different wavelengths.*

Finesse is directly related to the quality factor distribution (the quality factor, termed the Q-factor, is involved in Q-switching—a laser technique presented in § 6.3.1). We define the **resolving power** of a resonator as the following unitless quantity:

Resolving Power: $$R = \frac{\lambda_m}{\Delta\lambda} = m\cdot\mathfrak{F} \tag{6.40}$$

The resolving power increases with the interference order and with finesse. A typical resolving power is on the order $10^6$, which means that $\Delta\lambda$ for the supported laser wavelengths may be on the order of 0.001 Å. The concepts of both resolution ability and free spectral range are also employed in prisms and diffraction gratings, in addition to, of course, laser cavities.

Here is the problem: It appears that only specific oscillation modes with wavelengths that are multiple integers of the fundamental $\lambda_m$ can be supported in the cavity. However, only photons with a specific wavelength $\lambda_{12}$ can be amplified by the active medium. Two different physical conditions exist, both of which are fundamental: The wavelength is supported in the cavity and the wavelength is amplified by the active medium. Is there a match? The answer is yes, so this describes the concept of the **gain envelope**.

Recall that, even in a high-quality resonator, for any $\lambda$, there is a minimum extent $\delta\lambda$, and not absolute monochromaticity. The quantity $\delta\lambda$ may be considered as a wavelength uncertainty and can be approximated by the FWHM:

$$\delta\lambda \approx \frac{2}{\sqrt{F}} \cdot \frac{\lambda_m}{m} = \frac{1}{\mathfrak{F}} \cdot \frac{\lambda_m}{m} \tag{6.41}$$

Thus, the resonator supports modes with wavelengths:

$$\lambda_m \pm \delta\lambda \approx \frac{2nL}{m}\left(1 + \frac{2}{\sqrt{F}\,\pi m}\right) \tag{6.42}$$

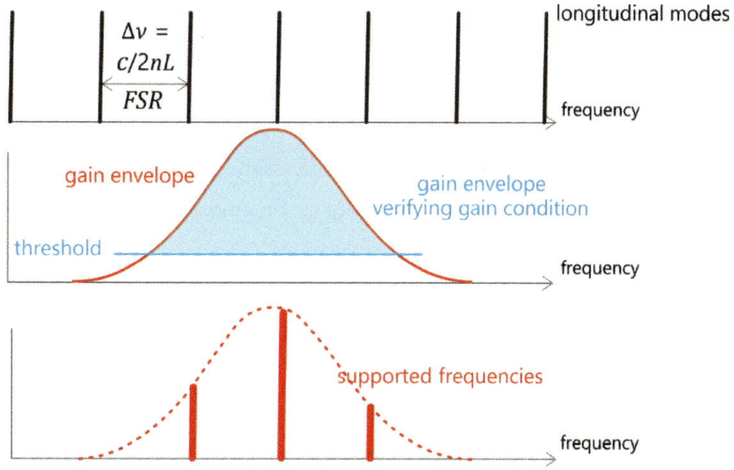


*Figure 6-29: Supported frequencies in a cavity. We can describe their shape as: laser frequencies = cavity × gain envelope. Only some of the longitudinal cavity modes are eventually supported by the gain condition.*

The above relationships describe the support envelope for the resonator modes. By increasing the interference order, we can proportionally increase the resonator resolving power. This can be easily achieved with a simple increase in the cavity optical length. However, simultaneously, the free spectral range reduces, so a compromise is necessary.

There is an additional factor that contributes to the wavelength uncertainty. The atomic emission lines $\lambda_{12}$ have an extent, just as every emission line has a finite extent, expressed in either wavelength or frequency. The line extent relates to the degree of color specificity, or monochromaticity, of the laser. The narrower the line, the more monochromatic the laser. In Eq. (6.16), we stated that the photon energy is exactly defined as

Photon Energy:
$$h \cdot \nu_{12} = E_2 - E_1 \tag{6.43}$$

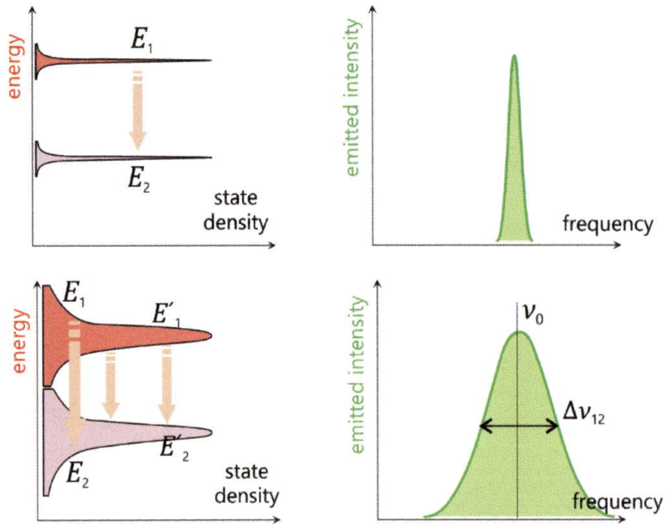


*Figure 6-30: Emission line broadening and its associated degeneration (uncertainty) of energy levels.*

Confined energies of the initial and final states lead to a well-defined photon energy and, as a consequence, a well-defined frequency / wavelength. However, the atomic energy levels are not fine lines. They, too, have some spread, and it is exactly the uncertainty (range) corresponding to the energy values for the initial and final states that is responsible for **emission line broadening** (Figure 6-31). Recall that the mean transition lifetime [Eq. (6.12)] is not infinitesimally small. The span of the intensity curve $I(\nu)$ in an atomic emission line (to one-half of the maximum value) is inversely proportional to the mean transition lifetime. This span is expressed in frequency terms as

$$\Delta\nu_{12} = \frac{1}{\pi}\left(\frac{1}{2\tau}\right) \tag{6.44}$$

The relative intensity $I(\nu)$ is described by a Lorentzian distribution:

$$I(\nu) = I_o \cdot \frac{\Delta\nu_{12}}{2\pi\left[(\nu-\nu_o)^2 + \left(\frac{\Delta\nu_{12}}{2}\right)^2\right]} \tag{6.45}$$

This is the simplest case of intrinsic line broadening. The order of magnitude is $\delta\nu = 10^8$ Hz, or $\delta\lambda = 10^{-4}$ Å. Other effects further contribute to the increase in energy line spread, such as atomic collisions (collisional broadening), temperature gradients (thermal broadening), interaction with other atomic fields (the Stark effect), and broadening due to the Doppler effect. It is expected, even in thinner emission lines, to have the extent $\delta\lambda$ span from 0.01 to 0.1 Å. Thus, the spectral linewidth may extend up to few nanometers. In the case where it is possible to have multiple emission lines, the cavity retains only those frequencies that it can support. When properly selecting the optical length of the cavity, it is possible to select a frequency near the gain envelope maximum.

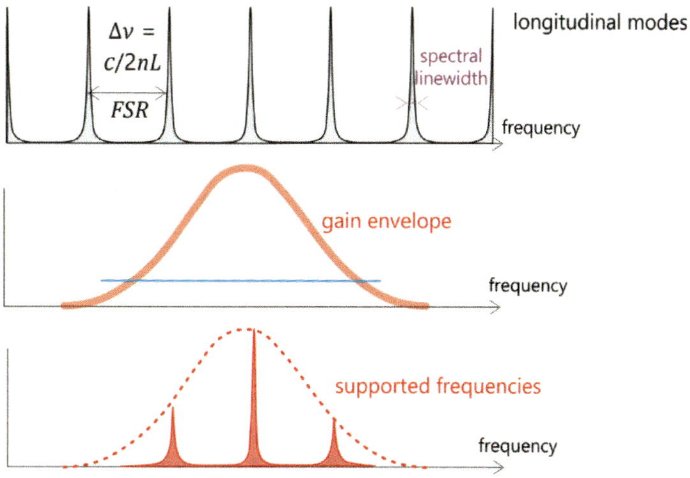


*Figure 6-31: Supported frequencies in a cavity, including line broadening. We can describe their shapes as: laser frequencies = cavity × gain envelope.*

Laser emission is not exactly a single frequency: It is a superposition of (equispaced) frequencies, with an extent determined by the optical oscillator and the atomic-level structure. We can impose high monochromaticity if the fundamental frequency, which equals the distance between the harmonics, is quite large (and $n \cdot L$ is small). Then, however, the optical gain is poor.

### 6.2.2.3 Resonator Gain

We presented the optical resonator as a frequency selector. Some additional conditions must be satisfied in order for the resonator to function as an amplifier and not just a filter. For example, the amplifier **gain** must overcome the total losses. Referring to Figure 6-32, we follow the ray path: 1 ⇢ 2 ⇢ 1 ⇢ output. Mirror reflectivities 1 and 2 are $R_1$ and $R_2$, respectively. We consider an initial intensity $I_o$. At mirror 2, after an optical path length $L$, the beam intensity is

$$I_1 = I_o \cdot \underbrace{\exp\left[\left(B - \alpha_e\right) \cdot L\right]}_{\text{gain over cavity length } L} \tag{6.46}$$

where $B$ is the gain coefficient per unit length, and $\alpha_e$ is the absorption coefficient, which expresses the loss per unit of length. This relationship has the form of the Beer–Lambert law [Eq. (3.37)], with the difference that, in addition to absorption, there is also a gain coefficient.

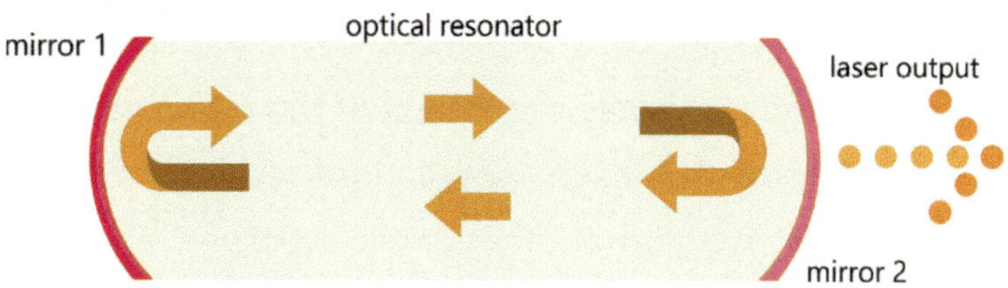


*Figure 6-32: Path of a beam in an optical resonator, including the laser output.*

After reflection off mirror 2, the intensity of the reflected beam becomes

$$I_2 = \underbrace{R_2}_{\text{losses due to reflection}} \cdot I_1 = R_2 \cdot I_o \times \exp\left[(B-\alpha_e)\cdot L\right] \tag{6.47}$$

At mirror 1, after traveling an optical path length $L$, the intensity of the reflected beam becomes

$$I_2' = I_2 \cdot \underbrace{\exp\left[(B-\alpha_e)\cdot L\right]}_{\text{gain over cavity length } L} = R_2 \cdot I_o \times \exp\left[2(B-\alpha_e)\cdot L\right] \tag{6.48}$$

After reflection off mirror 1, the intensity of the reflected beam becomes

$$I_3 = \underbrace{R_1}_{\text{losses due to reflection}} \cdot I_2' = R_1 \cdot R_2 \cdot I_o \times \exp\left[2(B-\alpha_e)\cdot L\right] \tag{6.49}$$

Thus, the amplifier gain $G$, which is the ratio of the intensity of a beam after a full cycle to its initial intensity, is expressed by

Resonator Gain:
$$G = \frac{I_{\text{full cycle}}}{I_o} = R_1 \cdot R_2 \cdot I_o \times \exp\left[2(B-\alpha_e)\cdot L\right] \tag{6.50}$$

For laser action to occur, the round-trip gain must be ≥ 1; therefore,

$$G = R_1 \cdot R_2 \cdot I_o \cdot \exp\left[2(B-\alpha_e)\cdot L\right] \geq 1.0 \tag{6.51}$$

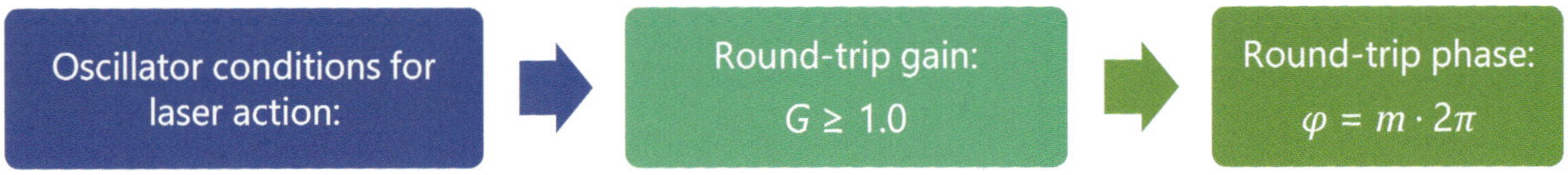


## 6.2.3 Three- and Four-Level Lasers

An important element for achieving laser efficiency is the selection of suitable energy levels. We emphasize that, with only two levels, laser action is impossible. This is because the transition modes are the same for both directions (1⇢2 and 2⇢1). Thus, even if, in principle, there is a population inversion between the two levels, regardless of the pump action or pump rate, as soon as the excited atoms relax, the two states' populations reach equilibrium. Then, for every photon generated by stimulated emission, another one is absorbed, and the total gain in photon population is zero. For transitions between only two levels, simultaneous pump absorption and signal amplification cannot occur.

To achieve a sustainable population inversion, the energy levels that participate in the process of energy pumping and the levels that participate in the radiative processes must be different. This requires at least three planes, while improved efficiency is achieved with the participation of four levels.

In a **three-level system**, the laser transition ends in the **ground state**. The unpumped gain medium exhibits strong absorption for the laser transition. A population inversion and the consequent net laser gain result only when more than half of the atoms are pumped into the upper pump level (level 3); the threshold pump power is thus fairly high. The upper pump level may be manifold (broadband energy level) when it is properly selected for efficient energy pumping. For example, in optical or mechanical pumping, the energy expenditure can be better utilized. If that upper pump level has a relatively short lifetime, this short lifetime helps to enable rapid and direct electron relaxation when the atoms transition to the upper laser level (level 2), ready to lase to level 1. A relatively large energy span for level 3 ensures a relatively short lifetime; this is yet one more advantage of a three-level system. The first-ever laser, the ruby laser ($Cr^{3+}$:$Al_2O_3$) (§ 6.4.2.5), is an example of a three-level laser.

In a **four-level system**, two levels (0 and 3) participate in the pumping process and are entirely disengaged from those involved in the laser action (levels 1 and 2).

The fact that the lower laser level (level 1) is different from the pump ground level (level 0) drastically reduces the pump population requirements because it only takes a few more ($\Delta n$) electrons at level 3 than the required population for inversion $\Delta n$. Thus, a four-level system has a very low pump threshold. To the contrary, in a three-level system, a transfer of $N_1/2 + \Delta n$ electrons is required: $N_1/2$ to equate the populations and $\Delta n$ to achieve the inversion.

With a four-level system, the pumping process can be efficient because the expendable energy is funneled in a broadband state (or states) that transits quickly to laser level 2, while laser level 1 is still vacant, ready to be filled by the stimulated-emission-produced electrons. Once electrons arrive (due to the laser action), laser level 1 is quickly depopulated by multi-photon transitions to energy level 0. If the depopulation rate from level 1 to level 0 is quite fast, the occupation in level 1 is always less than that of level 2, so it is possible to support population inversion for much longer time intervals; in addition, re-absorption of the laser radiation is avoided. This means that there is limited absorption of the gain medium in the unpumped state, and the gain usually rises linearly with the absorbed pump power.

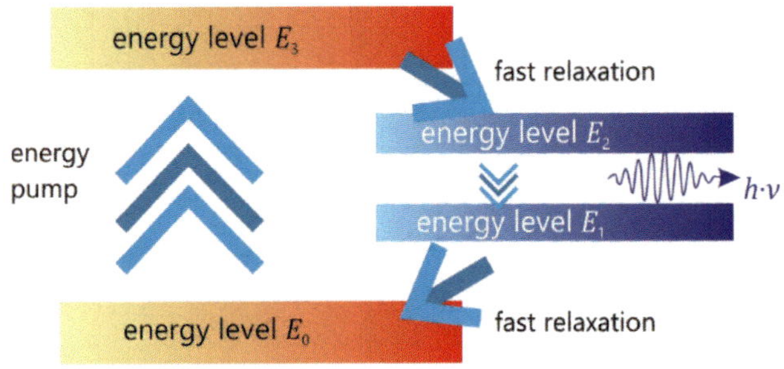


*Figure 6-33: A four-level laser model.*

Example ☞: The natural cycle of water (Figure 6-34) is a system with a gross similarity to a four-level laser. The sea is level 0. Evaporation is the pumping mechanism, with the energy coming from the sun. The upper cloud level is level 3. Vapors condense to form the clouds, which are the upper laser level 2, and then there is rain, which is the lasing action. Water drain (run-off) from the earth/ground (level 1) returns via a fast decay to the sea (level 0).

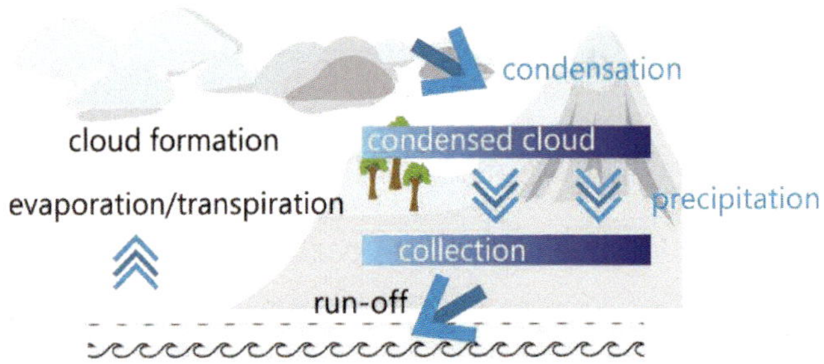


*Figure 6-34: The water cycle in nature.*

## 6.2.4 Laser Fundamentals

What makes the laser so special? The critical features of optical beams are temporal, pertaining to the beams' ability to be modulated (pulsed), spectral, pertaining to their very narrow wavelength range (color), and spatial, pertaining to their ability to be confined in a very small divergence beam and achieve very strong optical intensity at a focal point. Lasers uniquely combine all three of these general specifications. The particular characteristics of laser radiation are: monochromaticity, coherence, directionality, brightness, and polarization, as described in the following subsections.

### *6.2.4.1 Monochromaticity*

While not absolutely monochromatic, laser light is far more monochromatic than conventional light sources. Laser light is a wave of a central frequency $\nu$ with a very restricted extent, called the **linewidth** $\delta\nu$. This restriction is not only due to all of the photons originating from a specific pair of energy levels (in contrast with photons of a common light source that result from radiative decay from a multitude of atomic or molecular states), but is mainly due to the resonant cavity (optical oscillator) amplifying specific resonant frequencies $\nu$ that are compatible with Eq. (6.34) and are subject to sustainable gain.

Thus, laser light does not have exactly one frequency. It is possible for laser light to be composed of a series of equispaced frequencies. The degree of monochromaticity may be expressed in terms of $\delta\nu$ (frequency) or $\delta\lambda$ (wavelength):

$$\delta\nu = \frac{d\nu}{d\lambda}\delta\lambda = \left|\frac{d\,{}^{c}\!/_{\lambda}}{d\lambda}\right|\delta\lambda = \frac{c}{\lambda^2}\delta\lambda \tag{6.52}$$

### 6.2.4.2 Coherence

Coherence is perhaps the most important laser property. A high degree of coherence is a property that makes many interferometric applications possible. Coherence derives directly from the mechanism of stimulated emission: All of the photons are exact copies of the initial photon with regard to frequency, direction, and phase.

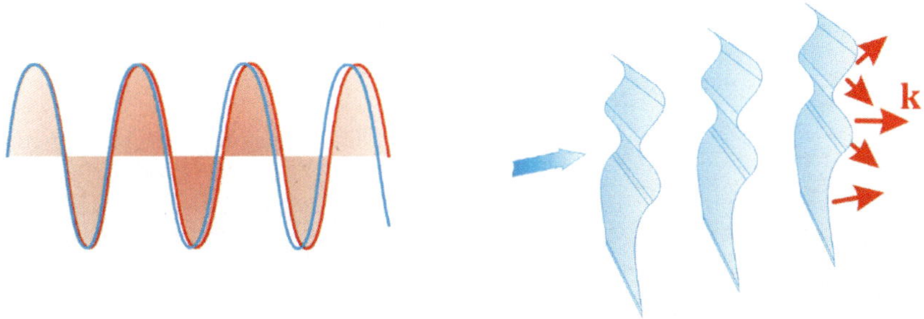


*Figure 6-35: (left) Partial spatial coherence and (right) partial temporal coherence.*

The concept of coherence can be expressed as either a spatial or a temporal property (as discussed in § 4.1.1). Spatial coherence refers to wave synchronization in space, so an ideal spatial coherence exists when, for every two wavefront points, the phase difference between the corresponding fields is preserved (zero for the same wavefront), even if the wavefront points are distant. Temporal coherence refers to wave synchronization in time, so an ideal temporal coherence exists when the phase difference between two fields is preserved for every time $t$ over a large time interval $\tau$. Temporal coherence is an expression of monochromaticity.

The degree of temporal coherence depends directly on the monochromaticity, which may, in certain lasers, be very high. However, there is quite a range. Certain gas lasers have a very long temporal coherence length of tens of meters, while other lasers, such as diode lasers, have a temporal coherence length on the order of few millimeters. The degree of temporal coherence largely depends on the multitude of supported longitudinal modes.

The degree of spatial coherence depends on the multitude and distribution of transverse modes inside the optical cavity. The $TEM_{00}$ mode provides the highest possible spatial coherence.

#### 6.2.4.3 Directionality

Directionality can be quantified as the angle formed between the two confining rays in a beam; as an angle, it is reported in radians. For example, a laser beam with high directionality (1 mrad) may propagate over a long distance with no measurable change in its cross-sectional diameter. Such photon alignment, reported as collimation, is nearly impossible with an ordinary source. Perhaps even more important than this (and which may be quite impressive in laser shows), directionality (together with the related property of high spatial coherence) is directly associated with the ability of the laser beam to focus to a very small point.

The minimum diameter $d$ associated with directionality $D$ for a laser beam focused by a lens of focal length $f$ can be expressed as

$$D \approx f \cdot d\varphi \tag{6.53}$$

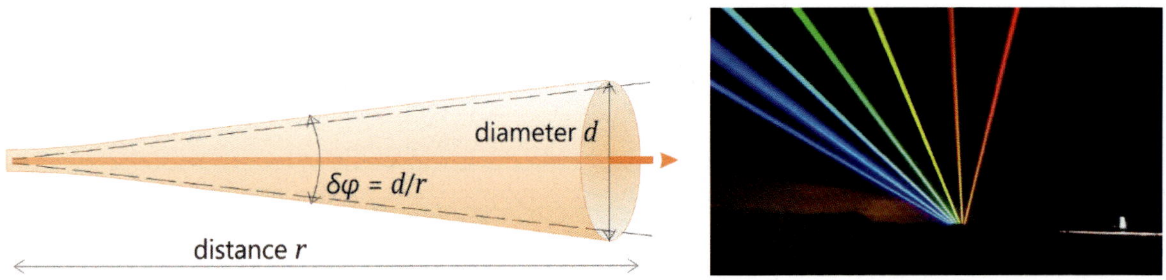


*Figure 6-36: Laser directionality.*

The smaller the value of $\delta\varphi$, the higher the directionality. Directionality is influenced by the characteristics of the optical cavity (transverse mode distribution): The multiple reflections produce a well-collimated beam because only photons traveling parallel to the cavity walls are reflected off both mirrors. Directionality is optimal for the fundamental mode $TEM_{00}$, which is the Gaussian beam distribution, and can be exceptionally good in gas lasers (with $\delta\varphi$ down to a fraction of a milliradian), while in solid-state lasers, this angle increases to a few milliradians—and it increases even more in semiconductor lasers, which are the worst performers in terms of directionality.

The physics behind high directionality—low angular laser beam divergence—is none other than that of high spatial coherence. Physics (the wave nature of light and diffraction) is also responsible for the limit on directionality: There is no such thing as zero divergence. Even for a perfectly collimated but spatially confined wavefront, the wave nature of light is responsible for a measurable angular divergence. Thus, due to the high degree of coherence in the laser beams, the divergence can be confined only by diffraction and might be small enough to have no effect on directionality. More on this in § 6.3.4.

#### 6.2.4.4 Brightness

Laser light can be very bright. Brightness is the common word for the photometric quantity of luminance $L$, which is the luminous flux emanating from a surface per the unit of solid angle of observation and the unit of the projection area. Brightness is reported in candelas per meter squared (cd/m$^2$), a unit called a nit.[38] It is noteworthy that the luminous intensity of solar light on the surface of the earth ($1.6\times10^9$ cd/m$^2$) is approximately the luminous intensity of a simple laser of just 1 mW of power, with a diameter of 1 mm!

The high brightness of the laser is related to its directionality because the energy is confined in a very small divergence cone. For a given beam configuration, the laser brightness is related to the output power. Pulsed lasers can produce anywhere from a few watts of power per pulse (solid-state lasers) to thousands of watts per pulse in solid-state lasers applied in metal welding. Continuous-wave (CW) lasers offer power ranging from a few milliwatts in a HeNe laser to many kilowatts in a $CO_2$ laser.

This raises the subject of **laser danger**, an issue that is particularly important due to possible damage to the eyes. It is not only the output power/brightness that constitutes this danger, but also the high directionality of the laser beam.

A strong light bulb that emits isotropically affects the eye only by a fraction of the angular extent with which it is perceived. A 100 W light bulb (100 W of electrical power, of which about 30 W is light, the rest being mostly heat) at 1 m distance spreads light energy over a surface area of $4\pi$ m$^2$, and within the eye pupil (2 mm diameter), less than 10 μW of that energy enters the eye. To the contrary, with a 2 mm-diameter laser beam of just 100 mW output (this is typical for a laser pointer, folks!), all of the laser energy may enter the eye (Figure 6-37). Thus, a laser that is 1000 times weaker than a typical home light bulb can affect the eye with 4000× more energy.

[38] *Introduction to Optics* § 2.2.1 Photometric Quantities.

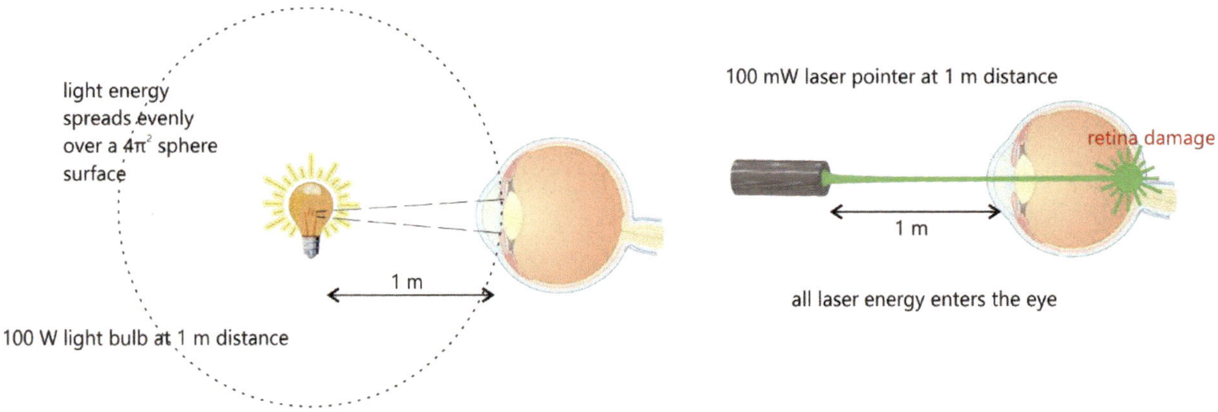


*Figure 6-37: A laser beam should NEVER enter the eye.*

According to the Center for Devices and Radiological Health (CDRH) classification system, Class I Laser products include devices that are considered to be safe during normal use. Their low power or enclosed beam is incapable of causing injury—with the possible exception that they may belong to a higher class during maintenance or service.

Class II laser products require a caution warning label. Staring into the beam of such lasers is an eye hazard. A typical example is the red laser in consumer label scanners. These are typically lasers in the visible spectrum, with CW power of 1 mW.

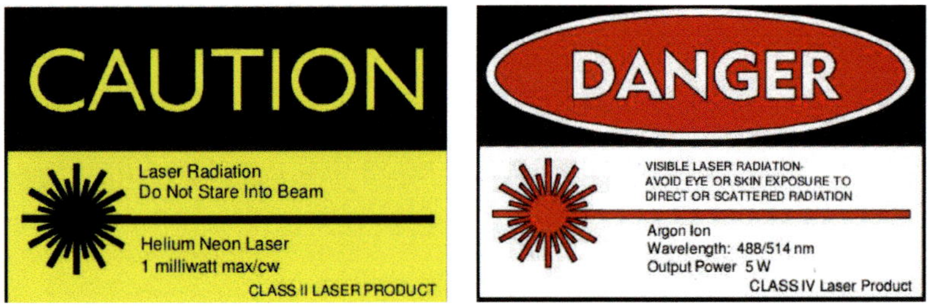


*Figure 6-38: Center for Devices and Radiological Health (CDRH) laser warning signs. (left) CAUTION for Class II and Class III with expanded beam lasers and (right) DANGER for Class IIIa with small beam, Class IIIb, and Class IV, for which safety goggles are absolutely required.*

Class III laser products (and higher classes) require danger warning signs and protective eye goggles. Direct exposure to the beam is an ocular hazard. These lasers may be in the visible or the UV/IR. Continuous-wave power may be up to 500 mW.

Class IV laser product exposure, both direct and indirect (scatter, reflection, diffusion), is an eye and skin (and even a fire) hazard. Continuous-wave power may be up to 0.5 W. In addition to warning signs and mandatory eye protection, required safety measures include laboratory door interlocks, entryway warning lights indicating laser operation, laser protective barriers, and curbs on the optical table.

#### 6.2.4.5 Polarization

Many lasers produce linearly polarized light (see § 2.2) either by employing an optical-polarizing element inside the cavity, or simply because of the cavity geometry. Specific oscillation modes have linear polarization, but in a superposition of many of such modes, the specific orientation may be compromised. Thus, it is possible for a laser beam to be simply unpolarized. We emphasize here that the term 'unpolarized' does not refer to the production mode but rather to the fact that such a beam may have a prevailing polarization state.

The polarization property has a very practical application when elimination of reflection is desired, as in the case of multiple reflections in laser resonators. Selecting parallel polarization and Brewster's angles of incidence, the reflectivity is zero, so the losses due to multiple reflections inside the active medium are significantly reduced.

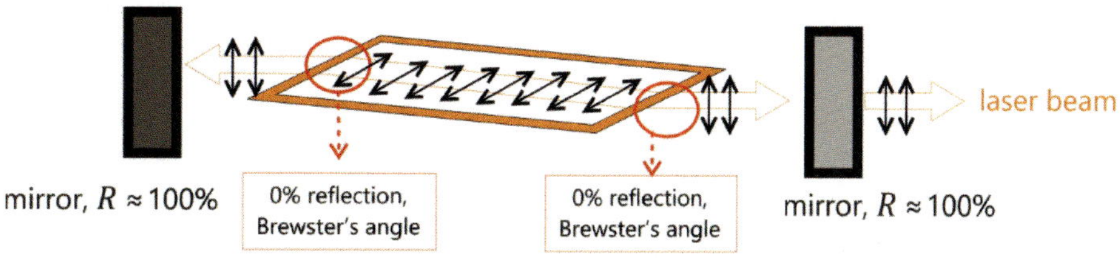


*Figure 6-39: Brewster's windows in a laser resonator.*

# 6.3 Laser Techniques

Laser pulsing is a broad term describing techniques that modulate the laser output from its natural CW mode to a rapid succession of short, powerful pulses. These techniques are involved in a wide range of laser applications that address different requirements. In fact, only a few lasers are pulsed because a continuous mode is simply not possible. Laser pulses can be as short as a few nanoseconds (1 ns = $10^{-9}$ s) or even a few femtoseconds (1 fs = $10^{-15}$ s).

The key parameter in pulsed laser output is the peak pulse power and the pulse repetition rate, also termed the laser speed in certain commercial systems. The peak pulse power (W) is proportional to the pulse energy and inversely proportional to the pulse duration. The repetition rate, reported in hertz (Hz), reflects the number of pulses per second—not, of course, the frequency of the emitted radiation. These parameters are determined by the pumping capacity and other engineering aspects of the laser device, but their values cannot be less than the reciprocal of the pulse width.

Laser pulsing is driven by specific power requirements. In certain applications, it is the peak power rather than the total dissipated energy that matters, as in, for example, the excimer and femtosecond lasers applied in modern ocular corneal surgery. A very high peak pulse power is necessary for special nonlinear optical applications.

## 6.3.1 Q-Switching

The technique of Q-switching can produce very powerful pulses, on the order of megawatts. The principle of operation is simple: During the time when no lasing occurs, the pump action continues, and the population inversion accumulates. Then laser action is permitted, and the 'loaded' system discharges. The challenge is not only to store energy in the form of a population inversion, but also to manage a laser 'halt' by not permitting stimulated emission during this accumulation period. There are mechanical and optical approaches for this, such as disengaging one of the two cavity mirrors.

The technique is termed quality factor distribution, or Q-switching. Here, Q is the oscillator quality coefficient. An oscillator with a high Q value has very low losses. It is obvious that an oscillator with, for example, poor reflectivity or a de-activated mirror has a low Q value. If the mirror is activated, the Q value switches and the lasing action produces a high output for a very short time. This is a Q-switched pulse.

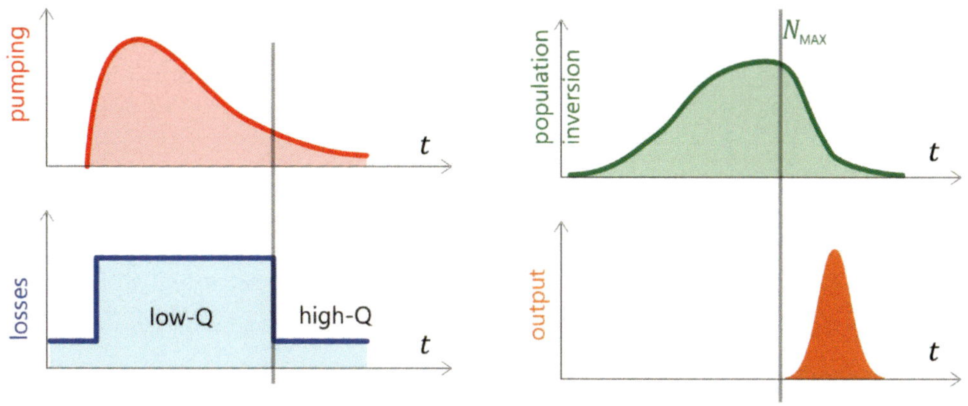


*Figure 6-40: Laser pumping, loss, population inversion, and output pulses in Q-switching.*

The faster the release of the stored energy, the more intense the pulse. Such a short time duration is in the range of a few nanoseconds. Q-switching can be achieved mechanically by moving a specific mirror very quickly, or by employing acousto-optic or electro-optic modulators that control the constants of the medium with an acoustic or electric signal, respectively. Other techniques involve dyes or saturable-transparency-modulating absorbers.

## 6.3.2 Mode-Locking

In any laser resonator, there exist simultaneously many longitudinal (axial) modes whose relative prevalence may vary over time. A pulse is composed of many harmonics. As the time duration of the pulse becomes smaller, its spectral content expands. To obtain a sharp pulse, a large number of harmonics must simultaneously be in phase, or synchronized.

If the various modes are forced to be in phase at specific time intervals, the laser can produce a very intense pulse. This is the **mode-locking** technique. For mode-locking to be efficient, the resonator must be able to support a relatively large number of modes. A specific mechanism that forces the modes to maintain a specific phase difference is a another key requirement. The first condition is ensured in a resonator of large length, and the second condition can be met by using a modulation mechanism that controls the resonator losses with a frequency equal to $c/2L$.

In a mode-locked laser, the lines in the spectrum are equidistant because a saturable absorber or active mode-locker enforces this condition. An exact equidistance of the lines in the Fourier spectrum arises from the periodicity of the generated pulse train, which is ensured by the action of the saturable absorber or active mode-locker; such a device prevents progressive broadening in the pulses.

Mode-locking can be achieved either actively (active mode-locking) if the energy required for the modulator is provided externally, or passively (passive mode-locking) if there is no external intervention.

An example of active mode-locking involves the insertion of a modulator element into the laser cavity, which may be an electro-optical or acousto-optical crystal. After the creation of the first pulses, one of the pulses becomes in phase via the nonlinear effect (for example, by using saturable dye absorption, or localized modulation of the refractive index due to strong absorption). In many cases, a combination of Q-switching and mode-locking is implemented.

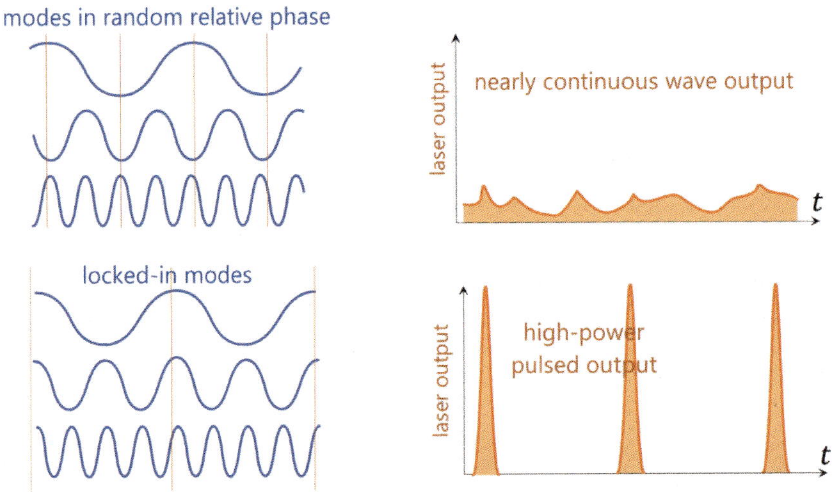


*Figure 6-41: Principle of operation of mode-locking: (top row) modes in random phase and (lower row) locked-in phase modes. The left column illustrates the mode distribution, and right column illustrates the laser output.*

With mode-locking, very intense laser pulses are possible, with a very small pulse duration on the order of picoseconds (1 ps = $10^{-12}$ s). The maximum peak power increases with the number of the lockable axial modes. In a Nd:Glass (neodymium-doped glass) laser with a cavity length of 30 cm, the available longitudinal modes are $10^3$ to $10^4$. It is noted that the total laser energy in both cases (Q-switching and mode-locking) neither increases nor decreases, but is simply redistributed over time.

In the case of extremely short pulses, a pulse of such short temporal length has a spectral spread over a considerable bandwidth. This is in contrast to the very narrow bandwidths typical of continuous-wave lasers. Pulsed lasers are used when peak power is prioritized over temporal coherence.

## 6.3.3 Second-Harmonic Generation

Second-harmonic generation is applied in order to convert the laser output wavelength to half of its value. This technique fulfills the requirement for a laser to be capable of operating at a different wavelength (e.g., the green, instead of the infrared). For example, the Nd:YAG 1.064 μm output is converted to 0.532 μm. This frequency/wavelength transformation can be achieved with energy coupling between two beams. In terms of photonics, we obtain one photon from two separate photons, and the energy of the resulting photon is twice that of the contributing photons (photon suturing). This is a **nonlinear** optical effect.

The principle of linear superposition (§ 4.1) states that the total disturbance (field) of two waves is the vector sum of their respective fields:

Linear Superposition Principle: $$\mathbf{E} = \mathbf{E}_1 + \mathbf{E}_2 + \mathbf{E}_3 + \ldots \quad (6.54)$$

In the case where it is possible to violate linearity, there are additional terms:

Nonlinear Optics: $$\mathbf{E} = \alpha_1(\mathbf{E}_1 + \mathbf{E}_2 + \ldots) + \alpha_2(\mathbf{E}_1 + \mathbf{E}_2 + \ldots)^2 + \ldots \quad (6.55)$$

where the term $\alpha_2$ corresponds to the degree of nonlinearity and is called a nonlinear optical coefficient. The presence of the nonlinear term is a violation of linearity and induces a polarization nonlinearity. The induced electric dipole displacement is not in a linear relationship with the cause, the electric field. The second-order term allows for the appearance of a term that expresses beam coupling:

Beam-Coupling Parameter: $$2\alpha_2 \cdot E_1 \cdot E_2 \quad (6.56)$$

If each wave is expressed in the complex harmonic form [Eq. (1.7)], whose exponential is

$$\exp[i(\omega \cdot t - \mathbf{k} \cdot \mathbf{r})]$$

the resulting wave has a frequency that is the sum of the two individual frequencies:

Second-Harmonic Contribution: $$\mathbf{E} = \mathbf{E}_o \cdot \exp[i(\omega_1 + \omega_2) \cdot t - (\mathbf{k}_1 + \mathbf{k}_2) \cdot \mathbf{r})] \quad (6.57)$$

Beam coupling can be achieved with select nonlinear crystals, such as lithium niobate. If the conditions are such that the same beam interacts with itself, and the emerging beams are in phase, then it is possible for the coupling to lead to a nonlinear interaction, and a new wave with twice the frequency can emerge. This is the process of second-harmonic generation (Figure 6-42).

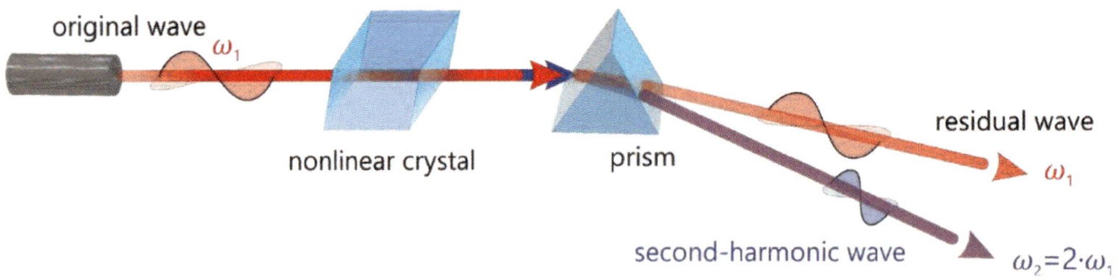


*Figure 6-42: Second-harmonic generation.*

## 6.3.4 The Gaussian Beam

In most laser applications, the laser beam is spatially modulated. The initial beam, as formed from the laser device, can have a small initial cross-section diameter; however, an expanded, yet collimated, beam may be desired.

To understand how a beam (and specifically a laser beam) can be spatially transformed, it is necessary to consider the basic propagation properties of the individual transverse modes that are formed in the optical resonator. No transverse pattern has a fixed cross-section amplitude (and intensity)—even in the simplest of all modes, the $TEM_{00}$ mode.

This simplest distribution is the normalized or Gaussian form, which corresponds to the lowest-order transverse mode, the fundamental distribution of intensity. This distribution has axial symmetry, peaking at the center, while the intensity falls nearly parabolically toward the periphery in a fashion resembling a symmetric bell curve (Figure 6-43).

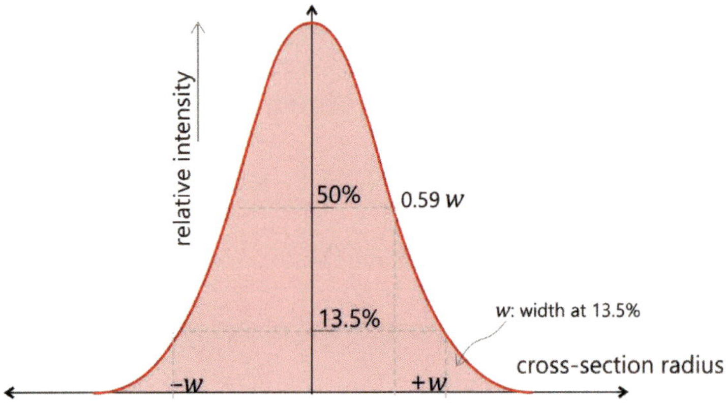


*Figure 6-43: Radial distribution of the relative intensity in a Gaussian beam.*

The transverse intensity pattern can be expressed as

$$I(r) = I_o \cdot \exp\left(-\frac{2r^2}{w_o^2}\right) \tag{6.58}$$

The transverse radius $w_o$ determines the reduced intensity points of 13.5% (= $1/e^2$) of the maximum (center) value. A parameter that is often used is the **beam waist** of the radial cross-section $2w$, which defines the limits of the transverse cross-section as at least 13.5% of its maximum value. The half-intensity point corresponds to a radius of $r = 0.59w$.

Let's see how such a beam propagates in free space or through an optical system. Even if a spatially confined beam had a 'perfect' plane wavefront with rays absolutely parallel at some point in space along the $-z$ axis, it is more than certain that diffraction would eventually result in the development of wavefront curvature and angular divergence (beam opening). Consider (at point $z = 0$) a collimated Gaussian beam (with an infinite radius of curvature and a plane wavefront) with a radial cross-section $w_o$ and a wavelength $\lambda$. For various points along the axis of propagation $z$, the radius of curvature $R(z)$ and the radial cross-section of the beam $w(z)$ are described by[39]

$$R(z) = z \cdot \left(1 + \frac{\pi \cdot w_o^2}{\lambda z}\right) \quad \text{and} \quad w(z) = w_o \sqrt{1 + \left(\frac{\lambda z}{\pi \cdot w_o^2}\right)^2} \tag{6.59}$$

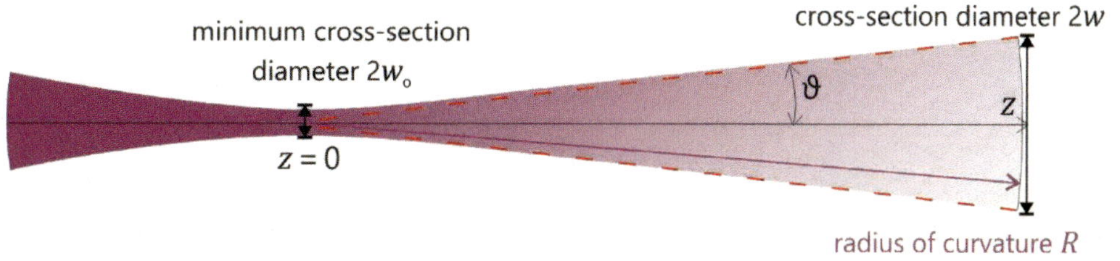


*Figure 6-44: Radial cross-section showing the radius of curvature and the angular divergence of a Gaussian beam.*

At the limit $z = 0$ (known as the near-field distribution), there is a minimum transverse diameter $2w_o$, and the radius of curvature is

$$\lim_{z \to 0} R(z) = \frac{\pi \cdot w_o^2}{\lambda} = z_R \tag{6.60}$$

The quantity $z_R$ has dimensions of length and is termed the **Rayleigh range**. In an 'ideal' point source, we expect that $w_o \dashrightarrow 0$ and $z_R \dashrightarrow 0$. This is the limit where a Gaussian beam becomes the geometrical idealization of a ray.

[39] Boyd GD, Gordon JP. Confocal multimode resonator for millimeter through optical wavelength masers. Bell Syst Tech J. 1961; 40:489-508.

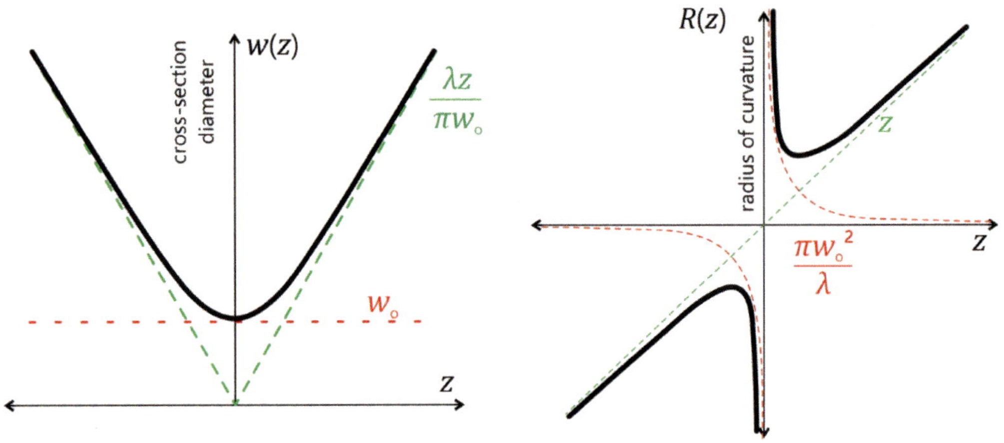


*Figure 6-45: (left) Radial cross-section and (right) radius of curvature of a Gaussian beam.*

At the limit $z \rightarrow \infty$, which is the far-field distribution, the radius of curvature and the radial cross-section approach

$$\lim_{z\to\infty} R(z) = z \quad \text{and} \quad \lim_{z\to\infty} w(z) = \frac{\lambda z}{\pi \cdot w_o} \cdot z \tag{6.61}$$

The angular divergence of the cone that describes the beam approaches the value

$$\lim_{z\to\infty} \vartheta(z) = \frac{w(z)}{z} = \frac{\lambda z}{\pi \cdot w_o} \tag{6.62}$$

It is easy to identify the similarities between Eq. (6.62) and Eq. (5.38), the latter of which describes the angular extent due to the diffraction of a circular aperture of radius $D$. In both cases, the angular extent is proportional to the wavelength $\lambda$ and inversely proportional to the cross-section diameter. We note that the geometric configuration of the beam after a considerable propagation distance depends only on the wavelength and its minimum cross-section—exactly as in far-field diffraction.

The longitudinal propagation of a Gaussian beam through an optical system is still subject to the rules of geometrical optics. According to a basic property of Fourier transforms, an intensity distribution always remains Gaussian. However, the transverse cross-section and the radius of curvature of the wavefront change due to diffraction, according to the relationships in Eq. (6.59), which also apply for negative $z$ (reverse propagation). This can be viewed as if the beam at point $z < 0$, with a negative radius of curvature (converging), is transformed by a converging lens and propagates toward its focal point (Figure 6-46).

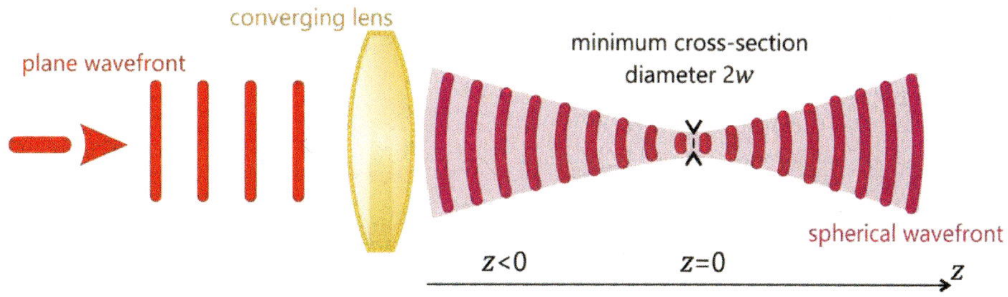


*Figure 6-46: Spatial transformation of a Gaussian beam by a lens.*

In this specific example, if we assume that the Gaussian beam with a plane wavefront (collimated along the principal optical axis) passes through the circular aperture whose diameter $D$ corresponds to the lens with focal length $f$, then at the lens focal point, the minimum radial cross-section is

$$w_o \approx \frac{\lambda f}{\pi D} \tag{6.63}$$

Geometrical optics predicts that the point image will be formed at distance $f$.[40] Its cross-section will be zero or, equivalently, will have a zero Rayleigh range ($z_R \to 0$). If $x$ is the object location and $x'$ is the image location with respect to the lens, the familiar imaging relationships are still valid, with a small corrective factor:

Without Gaussian Correction:

$$\frac{1}{x} + \frac{1}{f} = \frac{1}{x'} \tag{6.64}$$

With Gaussian Correction:

$$\frac{1}{x + \dfrac{z_R^2}{x+f}} + \frac{1}{f} = \frac{1}{x'} \tag{6.65}$$

If we assume a point source (meaning that it has a zero-minimum cross-section), then $w_o = 0$ and therefore $z_R = 0$. Thus, at the limit where the Gaussian distribution is ignored, the minimum cross-section point is at the exact same location as the point of image formation, as predicted by the simpler rules that apply in geometrical optics.

Comparing the minimum diameter [Eq. (6.63)] with the Airy disk central lobe thickness [Eq. (5.38)], we note that the propagation of a Gaussian beam is a continuous Fourier transform of itself! The Fourier transform of a Gaussian distribution always remains Gaussian. The statement that the Fourier transform of a Gaussian distribution has a Gaussian distribution is a

[40] *Geometrical Optics* § 4.1 Lens Imaging Relationship.

general property of the transverse propagation modes. These modes include the fundamental $TEM_{00}$ mode, which has is a Gaussian distribution.

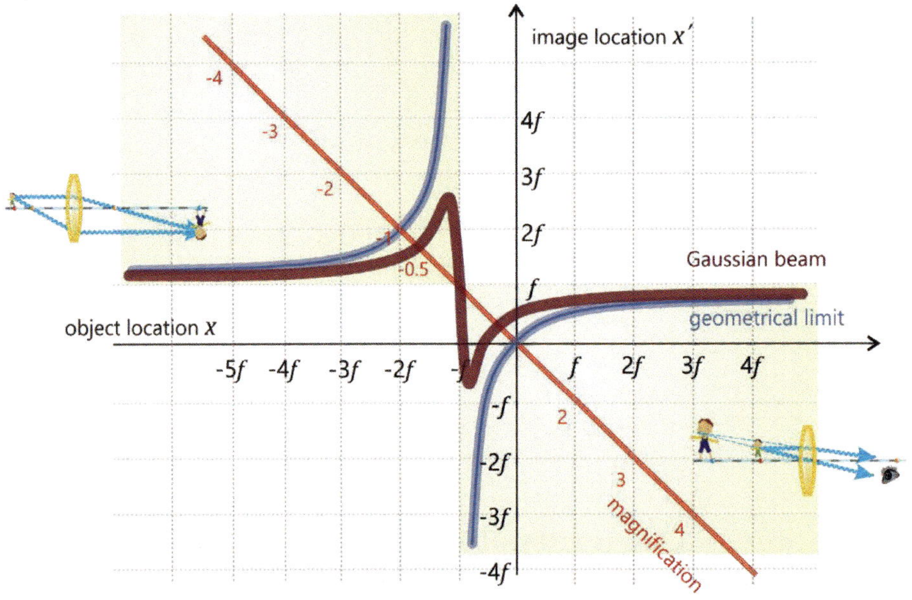


*Figure 6-47: Parametric curve of image location versus object location for a Gaussian beam, compared to the geometrical optics limit.*

Thus, both diffraction and geometrical optics can describe the propagation of a Gaussian beam. To the limit that the Gaussian distribution is ignored, the point of minimum distribution (focusing) is the point predicted by geometrical optics, but the minimum radial cross-section, the focal spot size, and the depth of field are all governed by diffraction.

All of these parameters are affected by lens aberrations or by any wavefront flatness deviations. Here we examine only the diffraction-limited (§ 5.5.1) case.

Depth of field and depth of focus depend on the longitudinal extent along the radial cross-section in which the beam waist remains nearly constant.[41] To find the extent, we apply the second relationship in Eq. (6.59) for the radial cross-section $w(z)$, setting a typical tolerance of 5% in diameter increase. We then find the axial points corresponding to the cross-section radius $1.05 \cdot w_o$. These points correspond to an axial separation of

$$\Delta z = \pm \frac{0.32\ \pi}{\lambda} w_o^{\ 2} = \pm \frac{0.32\ \pi}{\lambda \cdot \pi} \left( \frac{\lambda \cdot f}{D} \right)^2 \tag{6.66}$$

[41] *Visual Optics* § 6.1.4 Depth of Field and Depth of Focus.

# 6.4 THE LASER SPECTRUM

## 6.4.1 As Far Back as 1905...

Laser development dates back to 1905, with the proposition of light quanta and the subsequent introduction of the concept of stimulated emission by Albert Einstein. The next milestone was in 1939 when the possibility of implementing stimulated emission for the amplification of electromagnetic radiation was suggested by the Russian physicist Valentin Fabrikant (at that time stimulated emission was called 'negative absorption in gas discharge') in his doctorate thesis. A 1951 patent application (made jointly with Fatima A. Butaeva and Mikhail M. Vudynsky) for the amplification of electromagnetic radiation (including ultraviolet, visible, infrared, and radio spectral regions) can be considered the laser's 'birth certificate.'[42]

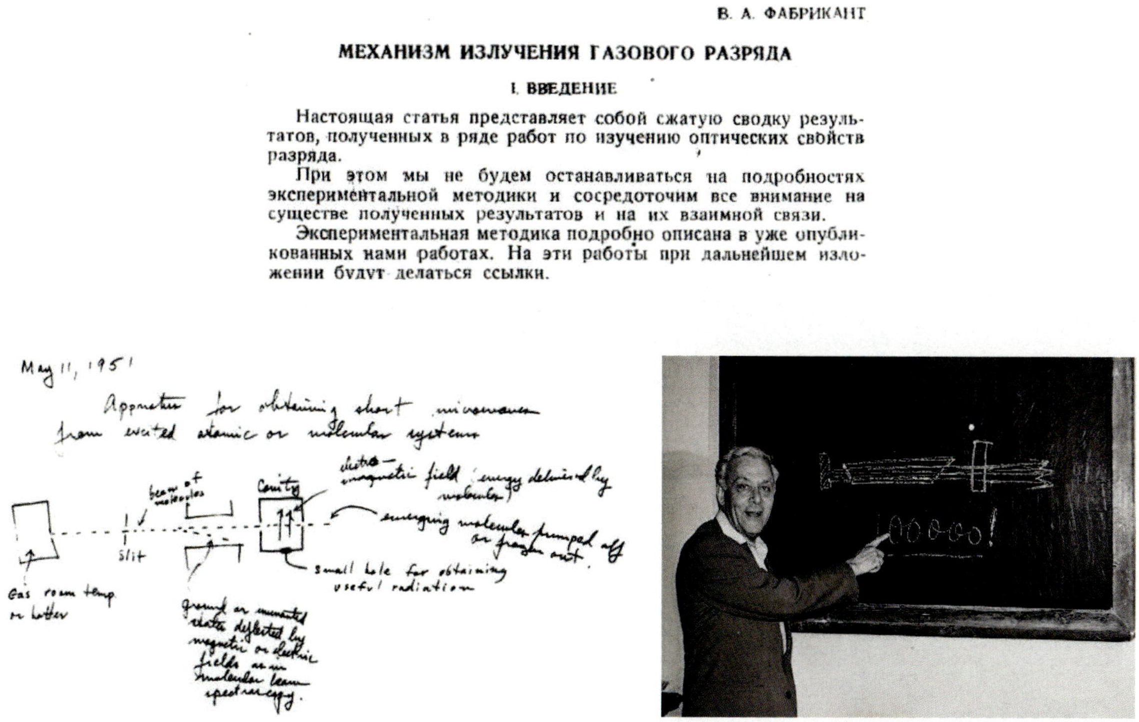

В. А. ФАБРИКАНТ

МЕХАНИЗМ ИЗЛУЧЕНИЯ ГАЗОВОГО РАЗРЯДА

I. ВВЕДЕНИЕ

Настоящая статья представляет собой сжатую сводку результатов, полученных в ряде работ по изучению оптических свойств разряда.

При этом мы не будем останавливаться на подробностях экспериментальной методики и сосредоточим все внимание на существе полученных результатов и на их взаимной связи.

Экспериментальная методика подробно описана в уже опубликованных нами работах. На эти работы при дальнейшем изложении будут делаться ссылки.



*Figure 6-48: (top) A page from Fabrikant's 1940 publication of his 1939 Doctorate thesis, (bottom left) notes from his early laser theory development, and (bottom right) Fabrikant delivering a lecture in 1960.*

[42] The patent was entitled: "A method of amplification of electromagnetic radiation (in the ultraviolet, visual, infrared, and radio frequency ranges) having as its distinct feature the transmission of the radiation to be amplified through the medium in which an excessive, compared to the equilibrium, concentration of atoms or other particles and their systems at upper energy levels corresponding to the excited states of the said medium is produced by an additional radiation or by other means."

Initially, the concept of stimulated emission was realized for operation in the microwave regime. A molecular oscillator was suggested by Nikolay G. Basov and Aleksandr M. Prokhorov at the Lebedev Physical Institute of the Soviet Academy of Sciences in the 1950s. At nearly the same time, Charles H. Townes and Arthur L. Schawlow at Bell Laboratories, and Gordon Gould at Columbia University constructed a microwave amplification radiation device, named the **maser** (microwave amplification by stimulated emission of radiation), which operated at $\lambda$ = 1.25 cm.

It is not a coincidence that the first form of a laser was in the microwave regime: Because the value of the spontaneous emission is proportional to the cube of the emitted frequency value, spontaneous emission in microwaves is very low compared to stimulated emission and quantum absorption in microwaves. In addition, population inversion may be easier to achieve, since the population inversion ratio is in an exponential relation to the energy difference. Thus, in the small microwave-energy differences (compared to energy differences in the visible), the population ratio threshold is easier to reach. Population inversion in the first maser was applied in ammonia molecules composed of a nitrogen atom at the top of a pyramid and three hydrogen atoms at the pyramid base. The two lower ammonia energy levels result from the separation of the oscillating energy levels having a double minimum. Population inversion can be achieved with an electrically induced separation of the higher-energy-level molecules from the lower-energy-level molecules.

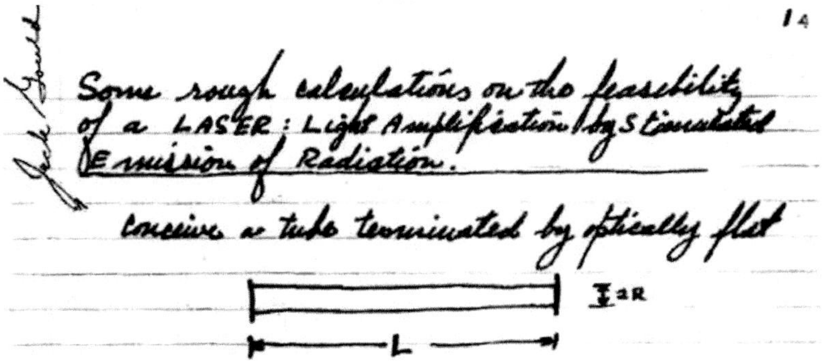


*Figure 6-49: Hello, LASER! The first use of the word LASER in notes by Gordon Gould, notarized by Jack Gould (see vertical script); no known relation between the two men.*

The idea of extending the maser principle (in microwaves) to frequencies in the visible light is claimed by many. There was the Townes, Schawlow, and Gould faction (credited with the word LASER), as well as the Basov and Prokhorov faction, who shared the 1964 Nobel Prize with Townes for their fundamental work in the field of quantum electronics, which led to the construction of oscillators and amplifiers based on the maser–laser principle. On 16 May 1960, at the Hughes Laboratory in Malibu, California, Theodore Harold (Ted) Maiman operated the world's first laser, which was based on a ruby crystal doped with chromium ions.

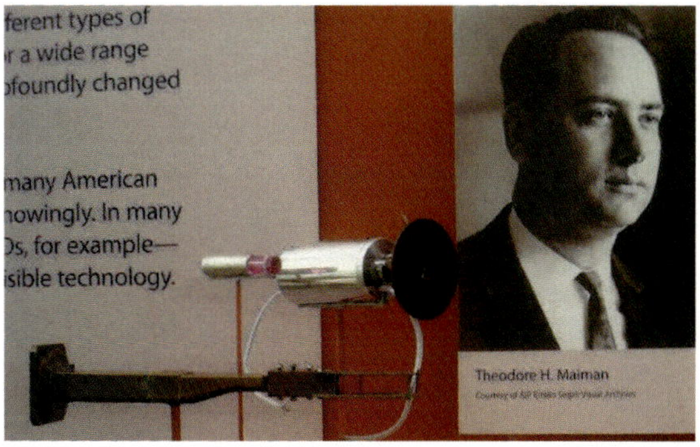


*Figure 6-50: The first laser and its ruby crystal. In the background is a portrait of Theodore Maiman. (Photo from the National Museum of American History, Washington, D.C.)*

The first laser was very small. A 1 $cm^3$ cross-section ruby rod, whose opposing surfaces were silver-coated, was the active medium and the oscillator. Optical pumping was provided by a coil flash-discharge tube.

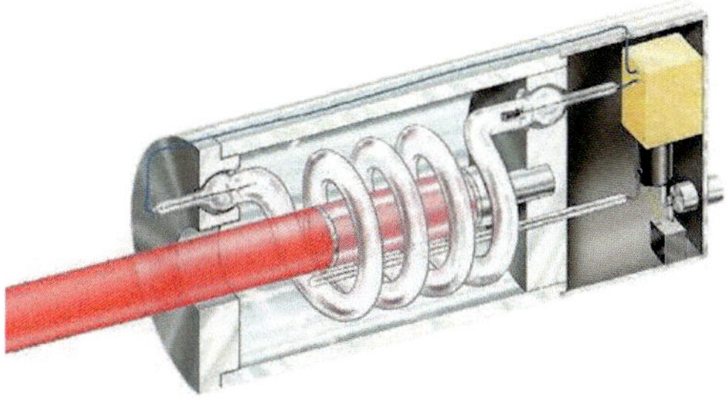

*Figure 6-51: The architecture of the original ruby laser.*

Soon, more lasers followed: in 1961 the HeNe laser, in 1962 the semiconductor laser, and in 1964 the Nd:YAG laser. A frenetic bout of laser development soon followed.

It is noteworthy that it took nearly ten years for the transition from microwave to optical frequencies. The reason for this was two-fold. The first factor was the delay in the development of optical cavities, which are fundamental for laser action.

The second and most important factor, however, was that population inversion for systems with energies that correspond to the visual spectrum was very challenging and, of course, more demanding than for microwave frequencies because the atom population required for inversion is in an exponential relation to the energy difference, which, in the visual frequencies, is quite a bit larger.

## 6.4.2 Laser System Classification

Lasers can be classified into several categories. Five classification criteria include: (1) the active medium, (2) the pump method, (3) the wavelength of emitted radiation, (4) the emitted power, and (5) the laser function itself.

1. With respect to the **active medium**, lasers are classified according to the following:

   (a) Solid state, including:

   - *Rare-earth lasers.* These are solid-state gain media employing rare-earth-doped laser crystals and glasses such as trivalent $Nd^{3+}$, $Er^{3+}$, $Eu^{3+}$, etc. In most cases, the rare-earth ions replace the other ions of similar size and of the same valence (charge state) as in the host medium; for example, a $Nd^{3+}$ ion in Nd:YAG (yttrium aluminum garnet) replaces an yttrium ($Y^3$+) ion.

   - *Semiconductor lasers.* These are lasers based on semiconductor gain media such as GaAlAs. Stimulated emission occurs at an inter-band transition under the condition of a high carrier density in the conduction band. Most semiconductor lasers are laser diodes pumped with an electrical current in a region where an n-doped and a p-doped semiconductor material meet.

   - *Color center lasers.* These lasers can be excellent sources of tunable radiation. Tunable refers to a laser whose operating wavelength can be altered in a controlled manner. The color center, also known as the F-center (or Farbe center, from the German *Farbe* for color), is a crystallographic defect in which an anionic vacancy in a crystal is filled by one or more unpaired electrons. Electrons in such a vacancy tend to absorb light in the visible spectrum such that a material that is usually transparent becomes colored.

   (b) Liquid state. The main laser in this class is the tunable dye laser. In the dye laser, the liquid material (dye) consists of diluted organic molecules, such as rhodamine B, rhodamine 6G, and sodium fluorescein. The output from liquid-state lasers can vary from the near-UV to the near-IR. The $H_2O$ vapor laser is another laser in this class.

   (c) Gas state. These lasers use an electric current discharged in a gas medium to produce a laser beam. Gas lasers are used in applications that require long coherence lengths, very high beam quality, or single-mode operation. In this class, we have:

   - *Neutral atom lasers.* These include HeNe and Ar.

   - *Ion lasers.* These include $Ar^+$ and $Kr^+$.

- *Molecular lasers.* These include simple molecular lasers ($CO_2$) and excited dimers (excimer lasers), including KrF, XeCl, ArF, etc.

2. With respect to the **pump technique**, lasers are differentiated according to:

   (a) Electron beam (e-beam). In this type of laser, the lasing medium consists of very high-speed electrons moving freely through a magnetic structure and ionizing the gaseous active medium to provide a population inversion.

   (b) Electric discharge. In this class, the population inversion is achieved via a strong electric field on the order of a few kilovolts in a low-pressure gas chamber.

   (c) Optical pump. The pump is achieved in most cases with a flash lamp controlled by a triggering circuit. Energy from the lamp (which may require high voltage on the order of kilovolts) provides the population inversion. Flash lamps were the earliest energy source for lasers, including the first-ever laser. A second laser may be used as an optical pump (as in dye lasers). The pump laser's narrow spectrum is suitable for more-efficient energy transfer compared to flash lamps.

   (d) Chemical pump. Pumping results from energy released from an exothermic chemical process. This allows for high output powers that are difficult to reach by other means.

   (e) Thermodynamic pump. Pumping is achieved by adiabatic expansion inside the active medium. The lasers in this class include gas dynamic lasers.

   (f) Nuclear-pumped lasers. Pump energy is provided by a controlled, small-scale nuclear fission.

3. With respect to the **frequency** of **emitted radiation**, lasers can be grouped as infrared, visible, ultraviolet, X-ray, and finally, gamma-ray lasers, which are sometimes called grasers.

4. With respect to the **power** of **emitted radiation**, which can vary from a few milliwatts to a megawatt or even a terawatt, lasers span from low-power to very high-power lasers.

5. With respect to the **mode** of **operation**, lasers can be pulsed or continuous wave. The output from pulsed lasers consists of bursts with a small (to very small) temporal duration (a few nanoseconds, or even 0.1 ps). The purpose of pulsing is to produce very high-power output (energy over time).

### *6.4.2.1 The HeNe Laser*

Helium–neon (HeNe) lasers were among the first to be produced and are quite widespread. The active medium is a mix of helium (He) and neon (Ne) in a 10:1 ratio. The radiated laser results from atomic electron transitions within the neon atom's energy levels. Optical pumping of the helium is achieved via electric collisions caused by discharge or continuous voltage (1.5 ⟼ 10 kV). The optical cavity is composed of a small-bore capillary tube containing the gas mixture under low pressure (~10 mbar); the tube also forms the resonator via a set of opposing reflective mirrors.

The HeNe laser is a four-level system. Population inversion is achieved between levels $2^1s$ He ⇢ $3s$ ($2p^5$ $5s$) Ne or He $2^3s$ ⇢ $2s$ ($2p^5$ $4s$) Ne. The electric discharge ionizes the He atoms at levels $2^3s$ and $2^1s$ (energy ≈ 20.61 eV), which are metastable. Via atomic collision mechanisms, energy is transferred to tuned Ne levels $3s$ (energy ≈ 20.66 eV) and $2s$. The energy-level diagram for the HeNe laser is illustrated in Figure 6-52. Lasing corresponds to 0.543 μm (green), 0.6328 μm (red), 1.152 μm (near-infrared), and 3.39 μm (infrared); filtering and amplification depend on the resonator settings. The resonator is usually constructed to operate in the red.

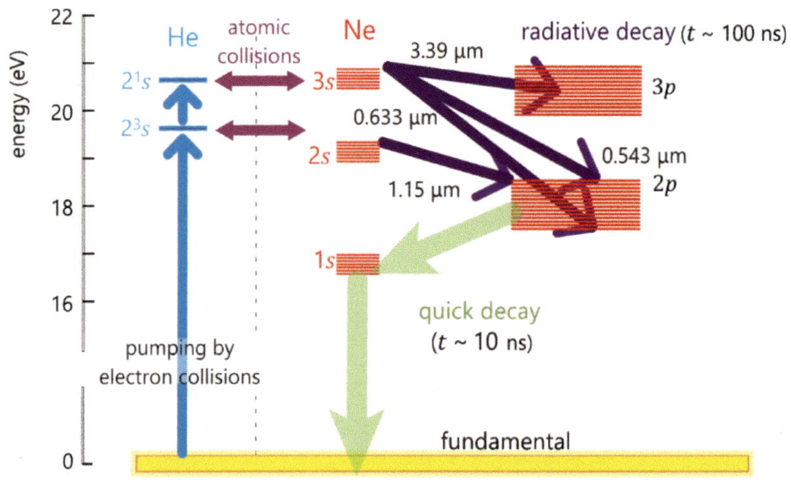


*Figure 6-52: HeNe laser energy-level diagram.*

Characteristics of the HeNe laser include a small spectral range and therefore high coherence, and very high beam quality, but relatively low output power (0.5 to 10 mW). For these reasons, as well as the fact that it is widely accessible and inexpensive, the HeNe laser finds many laboratory applications.

### 6.4.2.2 The $CO_2$ Laser

The $CO_2$ laser is a gas laser capable of delivering a very high output—particularly in pulsed modes—of more than 20 W per unit of length of active cavity. And yes, these lasers are constructed to be large: Some $CO_2$ lasers are tens of meters long with CW power in the range of some tenths of a kilowatt. In a pulsed function, these lasers can reach a power of up to 2000 J per pulse. This is a very efficient system, with approximately 20% wall-plug efficiency (the ratio of the initial pump energy to the energy converted in the laser beam). Due to their high power, they find many industrial applications, mainly in materials processing (metal welding and cutting). $CO_2$ lasers operate in the infrared: 9.6 μm and 10.6 μm. Their optical pumping functions are similar to those of the HeNe laser and are achieved with electric discharge based on the presence of nitrogen gas.

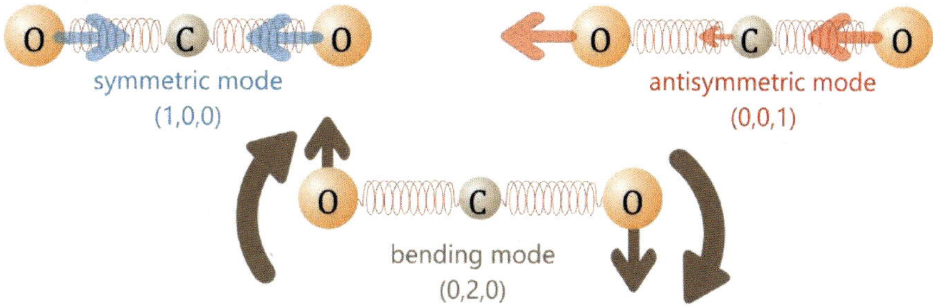


*Figure 6-53: Oscillation modes of the triatomic molecule $CO_2$.*

The involved energy states are oscillation modes of the triatomic $CO_2$ molecule and are described by three quantum numbers that correspond to the excitation states/modes. These are the bending mode ($\nu_2 \approx 2\times10^{13}$ Hz), the symmetric mode ($\nu_1 \approx 4\times10^{13}$ Hz), and the antisymmetric mode ($\nu_3 \approx 7\times10^{13}$ Hz). Within each oscillation mode is a large number of sublevels. Via a mechanism involving electron collisions due to an applied voltage, nitrogen molecules are excited to the $E_1$ level (Figure 6-54).

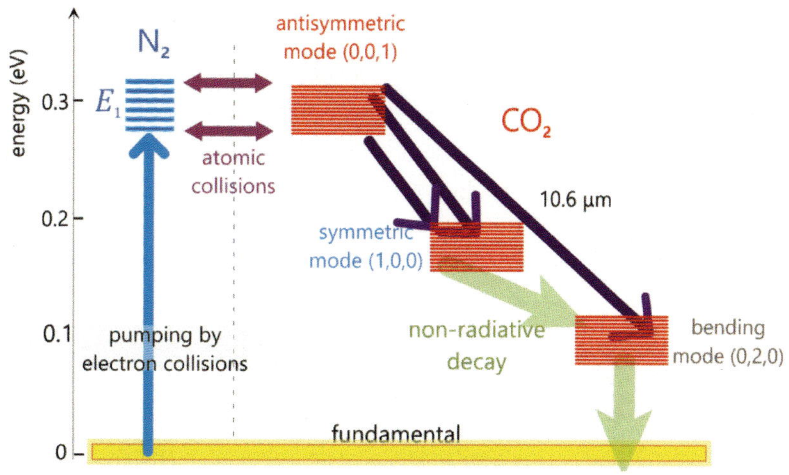


*Figure 6-54: $CO_2$ laser energy-level diagram.*

The energy difference between the $E_1$ level and the antisymmetric oscillation $CO_2$ level (0, 0, 1) is negligible: approximately $2.25\times10^{-3}$ eV. For comparison, the excitation energy that corresponds to these planes is approximately 0.3 eV. Therefore, it is relatively easy to couple energy between excited $N_2$ and $CO_2$ molecules, and that energy then excites the molecules to their antisymmetric mode.

Thus, a population inversion of excited $CO_2$ molecules is formed in level (0, 0, 1), which is also the upper level of a four-level system. This upper $CO_2$ lasing level (0, 0, 1) achieves radiative transitions to levels (1, 0, 0) and (0, 2, 0).

### 6.4.2.3 Argon-Ion Laser

The argon-ion laser is a gas laser. Its radiative procedure results from electronic transitions between excited $Ar^+$ states (just as the radiative procedure of krypton-ion lasers, neon-ion lasers, and xenon-ion lasers results from electronic transitions between excited states of $Kr^+$, $Ne^+$, and $Xe^+$, respectively). The CW radiation in the visible is quite powerful (30~100 W), which is why this laser is used in laser light shows! The argon laser can have many emission lines in the range of 408.9 nm to 686.1 nm. The stronger lines correspond to the green (514.5 nm) and the blue (488 nm), and result in high beam quality. The $Ar^+$ laser requires high pumping energy, so it does not have high efficiency (~10 to 20%). The dissipated heat that is not absorbed is removed by air cooling or water cooling. Laser safety issues arise both from the high output power of such ion lasers and from the high voltage applied to the tube. Krypton-ion lasers typically emit at 647.1 nm, 413.1 nm, or 530.9 nm, but various other lines in the visible, UV, and IR also exist.

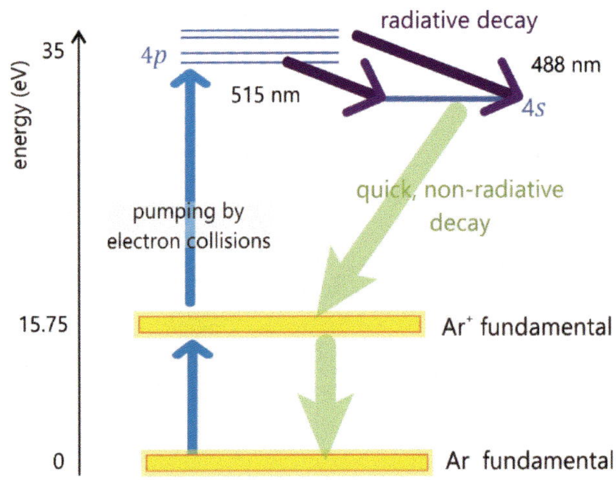


*Figure 6-55: The $Ar^+$ laser energy-level diagram.*

### 6.4.2.4 Excimer Lasers

The word excimer derives from exci<u>ted</u> di<u>mers</u> = excimers. Excimers are molecules composed of two atoms that only bind together in an excited electronic state. In such lasers, optical amplification occurs in a plasma containing excimers (or other molecules) with an anti-binding electronic ground state.

Excimer lasers are pulsed lasers whose main emissions are in the UV. The active medium is composed of bi-atomic noble gas and halogen systems, such as the KrF laser (main emissions at 248 nm), XeCl laser (308 nm), ArF laser (193 nm), and XeF laser (351 nm). These systems present with bound states only when excited, while if the two component atoms are at their ground state, they are mutually repelled, with a relatively long lifetime. After emission, the excimer rapidly dissociates, so re-absorption of the generated radiation is avoided. For these reasons, it is possible to achieve a fairly high gain, even for a moderate excimer concentration.

Excimer lasers are typically pulsed with a repetition rate of up to a few kilohertz and an output power between a few watts and hundreds of watts. This makes these lasers the most powerful laser sources in the UV and supports their use in applications where controlled absorption of a focused beam inside a medium is desired, such as materials processing, micromachining, and (predominantly) tissue photo-ablation, with applications in eye refractive surgery. In this surgery, successive pulses of an excimer laser shape the anterior cornea (the clear covering on the front of the eye) to provide the desired change in its shape and therefore its refractive power, resulting in the correction of myopia, hyperopia, and/or astigmatism.

*Figure 6-56: The excimer laser achieves precision on a small scale, here, having been used for etching "IBM" onto a strand of human hair. Three IBM researchers, Samuel Blum, Rangaswamy Srinivasan, and James J. Wynne, together with the ophthalmologist Stephen Trokel, are credited with the idea of using the 'clean' excision properties of the excimer laser in corneal refractive surgery.*

### 6.4.2.5 Ruby, Nd:YAG, and Nd:Glass Lasers

The ruby crystal laser was the first laser! The active medium is a crystalline lattice of aluminum oxide ($Al_2O_3$) with chromium-ion doping (0.05 % $Cr_2O_3$) or $Cr3^+$:Sapphire crystal. The emitted light corresponds to wavelengths at 694.3 nm and 692.9 nm.

Optical pumping is provided by a flash lamp, which emits a broad, optical-spectrum intense light. Illuminating with a broad-spectrum white light results in poor wall-plug efficiency in a gas active medium with thin absorption lines. Inside the solid, however, the interaction between the $Cr^{3+}$ ion and the lattice ($Al_2O_3$) gives rise to substantial absorption due to degeneration zones, so the broadband white light is selectively, but relatively efficiently, absorbed by the green and the blue lattice absorption bands.

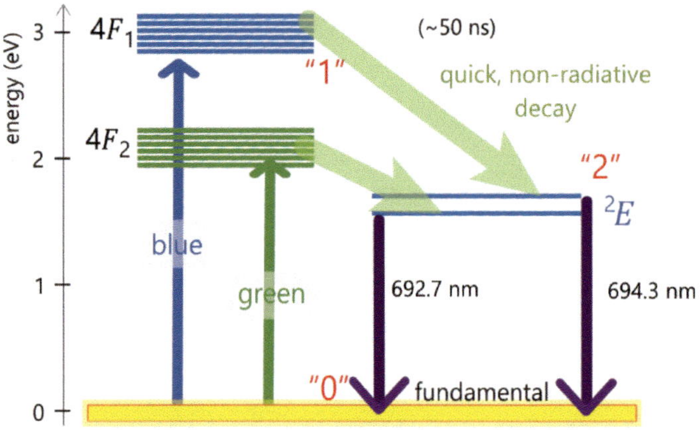


*Figure 6-57: Energy-level diagram for the ruby laser.*

Immediately after absorption, a nonthermal transition to a metastable level occurs with a mean lifetime of 5 ms, which is quite long in atomic standards. If the pump energy overcomes a specific threshold, population inversion can be achieved between the ground level 0 and the metastable level. This is essentially a three-level system (levels 0, 1, and 2) that is possible because level 2 has a relatively long lifetime. The ruby laser functions in pulsed modes with high energies of about 100 J per pulse.

In the neodymium (Nd) laser, the active medium is either a crystal $Y_3Al_4O_{12}$ (yttrium aluminum garnet, known as YAG), in which some $Y^{3+}$ ions have been substituted by $Nd^{3+}$ ions [as well as with other rare-earth ions such as erbium ($Er^{3+}$) and holmium ($Ho^{3+}$)], or a glass doped with $Nd^{3+}$ ions. The most powerful line in these lasers is the near-infrared ($\lambda$ = 1.06 μm). In contrast with the ruby laser, the Nd:YAG laser is a four-level system with very high efficiency: The pump bands at 0.73 μm and 0.8 μm are coupled by a quick non-radiative decay to level $4F_{3/2}$, which has a long lifetime (~0.2–0.3 ms). A quick decay occurs between levels $4I_{11/2}$ and $4I_{9/2}$, which are the ground laser and pump levels, respectively. Radiation with $\lambda$ = 1.064 μm corresponds to the most powerful of the transitions, $4F_{3/2} \dashrightarrow 4I_{11/2}$.

The simplified energy-level diagram of the Nd:YAG laser (Figure 6-58) is similar to that of the Nd:Glass laser because the energy levels of interest are not affected by the crystal structure. Optical pumping in the Nd:YAG laser is achieved with a flash lamp or another semiconductor laser that emits exactly at the spectral absorption bands of the active medium.

The Nd:YAG laser provides a very powerful output. This output can be continuous (~100 W) or pulsed with a pulse duration of few nanoseconds and an energy of 1 J per pulse. The laser output can reach a repetition rate of 10 to 500 Hz. The Nd:YAG laser is used in either a pulsed or continuous fashion for a wide range of applications, such as in materials processing, telemetry, and surgery. The Nd:Glass laser can be used instead of the Nd:YAG if the laser repeatability requirements are low.

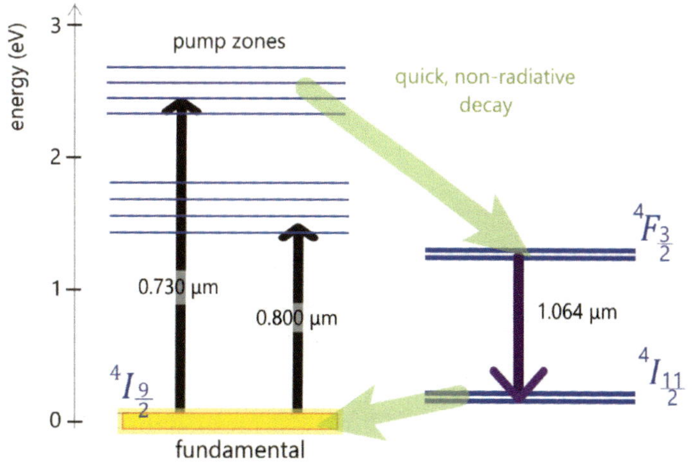


*Figure 6-58: Nd:YAG laser energy-level diagram.*

### 6.4.2.6 Semiconductor Laser

If the $CO_2$ laser is the 'maxi' laser, the semiconductor or diode laser is the 'mini' laser. The active area has a thickness of just 1 µm and is a p–n junction of two adjacent semiconductors: The p-type semiconductor is doped with positively charge carriers, while the n-type is doped with negatively charge carriers.

In an ideal semiconductor, the energy-level diagram is composed of broad bands. Instead of atomic or molecular energy levels, there are crystal energy bands that correspond to the bulk of the crystal. A solid-state energy diagram has the following bands: the lower valence band $E_V$, the upper conduction band $E_C$, and the 'forbidden' band that separates them—the energy gap $E_G$ (with values between 0.1 eV to 3 eV). In fact, these bands are present in any solid. A large energy gap (of more than 3 eV) renders the material an insulator; if there is no energy gap, or if it is very small, the material is a conductor.

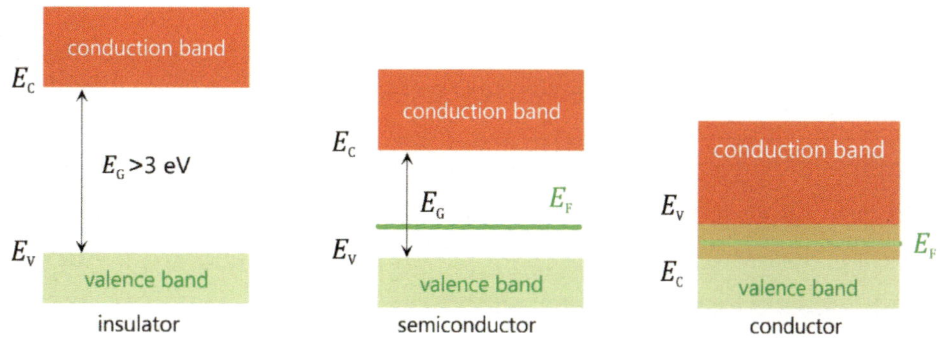


*Figure 6-59: Energy bands in (left) an insulator, (center) a semiconductor, and (right) a conductor. Here, $E_F$ stands for the Fermi level, which is the thermodynamic work required to add one electron to the material.*

At low temperatures, the valence band is almost completely occupied by electrons, while the conduction band is nearly empty. Electron excitation correlates to a transition from the valence band through the energy gap to the conduction band. A pump beam with a photon energy slightly above the bandgap energy can excite electrons into a higher state in the conduction band, from which they quickly decay to states near the bottom of the conduction band. Thus, some availability arises at the upper part of the valence band $E_V$, while the lower parts of the conduction band $E_C$ may constitute an upper laser level.

Pumping can be achieved with application of a forward bias in a p–n junction. Due to different doping levels, there is a displacement between the bands, as the doping level has to be the same. The energy-level diagram of a simple p–n junction is presented in Figure 6-60 (left), while Figure 6-60 (right) illustrates the energy-level diagram of a biased junction. Application of a low-voltage (≈ 1.5 V) forward bias diffuses electrons to the conduction band such that a population inversion can be achieved around the depletion zone of a p–n junction.

Electrons in the conduction band can then recombine with the holes, resulting in emission of photons with an energy near the bandgap energy. This reaction is described by

$$\text{electron} + \text{hole} = \text{photon} \tag{6.67}$$

With suitable energy, this process can also be stimulated by incoming photons.

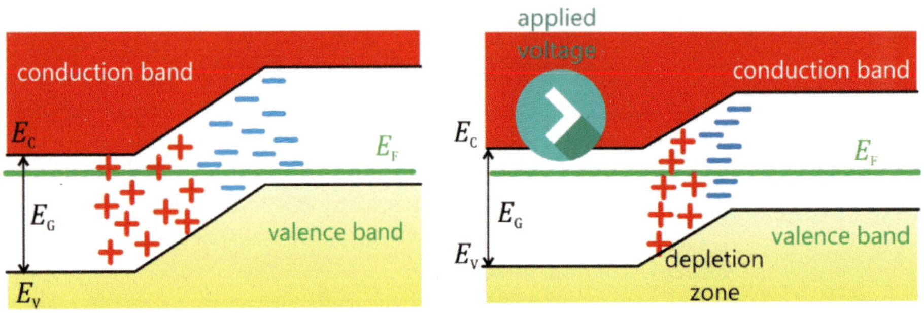


*Figure 6-60: Semiconductor laser p–n junction energy-level diagrams (left) in equilibrium and (right) in forward bias. Note that the depletion zone shortens when forward bias is applied.*

Forward bias provides the required energy, continuously feeding the depletion zone with electrons and holes. Other feeding options are used in the optically pumped semiconductor laser, where carriers are generated by the absorbed pump energy, and in quantum cascade lasers, where intra-band transitions are utilized.

The optical resonator is the crystal itself. Its surfaces are reflectors because the refractive index of the medium is very high. Thus, the surfaces of the crystal that are perpendicular to the desired direction of radiation are cleaved parallel to that direction, usually along the weaker crystalline planes. In order not to favor amplification along the other directions, the other surfaces are left unprocessed. For this reason, semiconductor lasers are called edge-emitting semiconductor lasers.

Within the edge-emitting laser structure, the laser beam is guided as if in a waveguide structure. Typically, the double-'sandwich' structure restricts the generated carriers to a narrow region while at the same time serving as a waveguide for the laser light. This arrangement leads to a low-threshold pump power and a high efficiency. Depending on the waveguide properties, particularly its transverse configuration, it is possible to obtain either a high-quality output that is compromised in power (a few hundreds of milliwatts), or (with a broad-area laser diode) a high output power (tens of watts or even > 100 W) that has a poor spatial intensity distribution.

The emitted radiation frequencies are provided as

$$E_G < h \cdot \nu < E_C - E_V \tag{6.68}$$

The maximum wavelength corresponds to a transition from the lower level inside the conduction band to a higher inside the valence band and is expressed by

$$\lambda_G = \frac{hc}{E_G} \approx \frac{1.24}{E_G\,[\text{eV}]} \tag{6.69}$$

Typical semiconducting materials are silicon (Si) and germanium (Ge), with $E_G$ 1.11 eV and 0.66 eV, respectively. These are elemental semiconductors. For emission in the visible, the required energy gap is approximately 2.5 eV. This may be achieved if the p–n junction is coupled to a pair of elements from the III and V columns of the periodic table [such as GaAs (gallium-arsenic) and AlSb (aluminum-antimony], or from the II and VI columns [for example, CdSe (cadmium-selenium)]. These are called **binary semiconductors**.

*Table 6-1: Energy gaps and their associated emission wavelengths in semiconductor lasers.*

| medium | energy $E_G$ [eV] | $\lambda_G$ [µm] |
|---|---|---|
| Ge | 0.66 | 1.88 |
| Si | 1.11 | 1.15 |
| AlP | 2.45 | 0.52 |
| AlAs | 2.16 | 0.57 |
| GaAs | 1.42 | 0.87 |
| IAs | 0.17 | 7.3 |

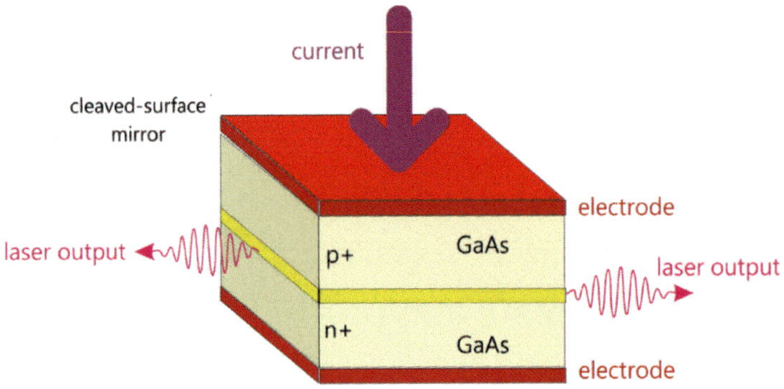


*Figure 6-61: Simplified illustration of a GaAs diode laser.*

The particular characteristics of the semiconductor laser are the following:

- Very small dimensions, low manufacturing cost, economical function, and mechanical stability due to the solid-state construction. This allows for their incorporation into integrated electron circuits.
- High efficiency and satisfactory power. At room temperature, semiconductor lasers have achieved power of 10 W in pulses of 100 ns duration.
- Direct function: The output can be amplitude- or frequency-modulated up to a gigahertz. The possibility of modulation is, in combination with circuit integration, the main reason that the solid-state laser finds so many applications in optical telecommunications.
- Large emission-radiation range. The wavelength depends on the value of the energy range and extends from 0.3 μm (blue diode) to 90 μm (the far-infrared), depending on the dopants.

*Table 6-2: Types, maximum power, operating wavelengths, and applications of semiconductor lasers.*

| Type | Maximum power | Operating wavelength | Applications |
|---|---|---|---|
| InGaN | 3 mW | 500 nm | optical memories<br>high-resolution displays<br>DVD players |
| GaAs | 5 mW | 840 nm | CD players |
| AlGaAs | 50 mW | 760 nm | laser printers |
| GaInAsP | 20 mW | 1300 nm | optical telecommunications |

# 6.5 LASER APPLICATIONS

> *'Initially, the laser was called an invention looking for a job.'*
>
> *George Harry Stine*

Because of their unique properties, lasers have a widespread and multitude of specialized scientific, technological, and medical applications. In many cases, such as interferometry, they are uniquely applied due to their high degree of coherence. In other cases, it is their high directionality and power that make them so useful. Moreover, their ability to produce pulses of very small temporal duration enables their application in many fast-developing effects. Today, their applications are so numerous that only a brief and select presentation is possible.

## 6.5.1 Applications in Physics and Chemistry

Lasers are present in every optics lab. They are employed in a broad spectrum of research and educational applications.

The very high degree of coherence is exceptionally suited to interferometry applications. A characteristic example is holography, one of the most interesting uses of lasers. Holography is involved in 3-D image storage, optical data storage (memory), and materials quality control and processing.

Lasers have extensive laser applications in spectroscopy, particularly in nonlinear, Raman, and Rayleigh spectroscopy, as well as in atmospheric remote sensing. We also mention laser-based light detection and ranging (LIDAR) technology, with applications in geology, remote sensing, and atmospheric physics. Lasers have been used aboard spacecraft such as the Cassini–Huygens mission, and as reference objects for telescopes that incorporate adaptive optics technologies.

## 6.5.2 Biomedical Applications

Lasers can be employed either as a diagnostic means, or as a fine surgical tool. A laser beam focused on a very small surface can achieve specific results. The focused laser interacts with tissue and transfers energy to tissue via absorption. **Coagulation** is the slow heating of tissue to destabilize proteins and other bio-molecules. A laser-heating coagulation of tissues above 50° C but below 100° C is called photocoagulation. The affected tissue shrinks due to water removal, and the tissue is burned and neutered. Thus, surgical lasers are applied for removing tumors and making masking incisions, as well as in other cosmetic treatments, such as removing age spots and tattoos. Surgical lasers are also employed in the treatment of various eye conditions, such as retinal disorders caused by diabetes and macular degeneration. Another application is **hemostatic laser surgery**, which is a bloodless incision–excision technique.

A pulsed laser with sufficiently high power density (> 100 W/cm$^2$) can quickly heat tissue to above 100° C. Water evaporates, and the affected tissue is removed in a process called photo-vaporization. Pulsed lasers are better-suited for such applications, since vaporization must not affect the surrounding tissue via thermal damage.

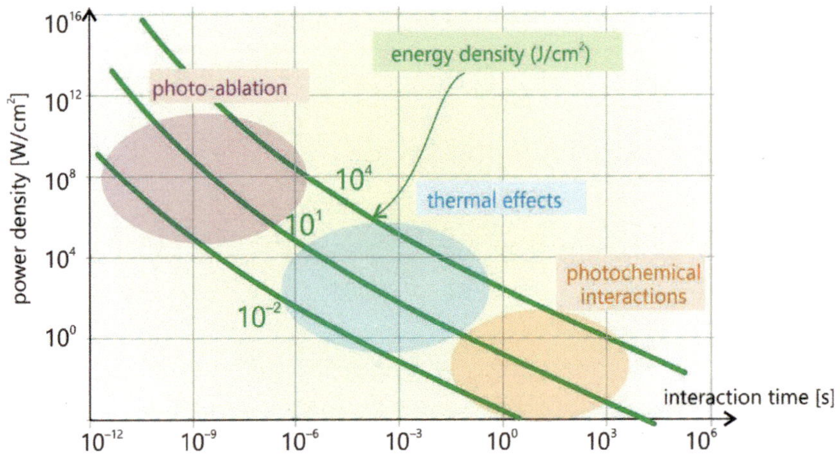


*Figure 6-62: Power density versus pulse length (interaction time) and the associated material interactions.*

When using high-power-density lasers in the ultraviolet, the photon energy is sufficient to cause molecular breakdown without localized heating. This process, which is called **photo-ablation**, produces 'clean' incisions. The thermal damage to the surrounding tissue is relatively small, as the extent of the thermal interaction is confined.

The best-known laser applications in medicine are in ophthalmology. These techniques are employed for management of retinal detachments, as well as for reshaping (sculpting) the cornea (the clear covering on the front of the eye that is also responsible for ⅔ of the total ocular power) by providing a modified anterior corneal curvature to compensate for refractive errors such as myopia (nearsightedness), hyperopia (farsightedness), and astigmatism.

The idea of using a laser to reshape the cornea to correct refractive errors dates back to 1983 with the collaboration of ophthalmologist Stephen Trokel and IBM Watson Research Center researchers Samuel Blum, Rangaswamy Srinivasan, and James J. Wynne. A paper published in the *American Journal of Ophthalmology* later that year introduced the field of refractive surgery[43] and was followed by years of experimentation and clinical trials. As a result, in 1995, the U.S. Food and Drug Administration (FDA) approved the first commercial excimer-laser-based refractive surgery system.

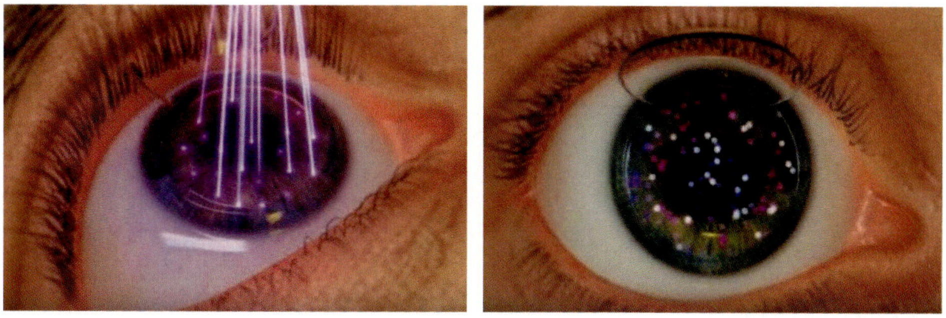

*Figure 6-63: Laser pulses on the corneal stroma in LASIK (laser-assisted* in situ *keratomileusis).*

The mechanism for corneal reshaping is photo-ablation. A laser beam of a select wavelength (UV from an ArF excimer laser), with an energy density greater than the ablation threshold (~ 50 mJ/cm$^2$ per pulse), is absorbed by the upper stromal tissue at a depth of a few microns. The resulting thermal absorption gives rise to local collagen molecular breakdown, developing a pressure gradient. Each laser pulse forms a mini 'crater' with a depth on the order of 0.3 μm and a cross-section corresponding to the cross-section of the focused spot, some tenths of a micron. Given the average corneal thickness (~ 500 μm), up to 100 μm of the stromal thickness is ablated, the precise amount depending on the desired correction (for example, 8.00 D of myopia) with proper superposition of a large number of such craters.

The part of the cornea that is reshaped is directly below the upper epithelial layer, the 'skin' of the cornea, which has an average thickness of 50 μm. This is the stroma, a collagen-layered structure that forms approximately 95% of the corneal thickness. Changes induced in the stroma are proven to be permanent.

There are a number of laser-vision-correction variations. The initially developed surface ablation technique, called **photorefractive keratectomy** (**PRK**) involves the mechanical removal of the corneal epithelium, and application of the excimer laser action on the exposed stroma. **LASIK** (laser-assisted *in situ* keratomileusis) is a further development that involves lifting the upper 100–150 μm of the cornea in the form of a flap, and applying excimer laser action on the exposed stromal 'bed.' Today LASIK is the most popular vision-correction surgery performed worldwide due to its rapid healing cycle and minimal pain. Millions of people around the world have had the procedure, and more than 90% of patients achieve 20/20 to 20/40 vision and are

[43] Trokel SL, Srinivasan R, Braren B. Excimer laser surgery of the cornea. Am J Ophthalmol. 1983; 96(6):710-5.

able to perform all or most of their daily activities without spectacle glasses or contact lenses. Modern improvements to the procedure enable many patients to achieve 20/15 vision.

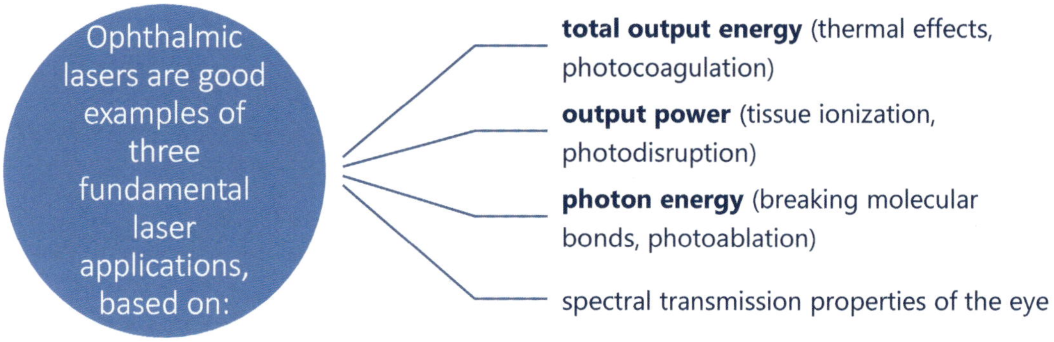


In LASIK, the flap was initially created with a mechanical blade, called a microkeratome. A fairly recent development is the femtosecond laser, which is a pulsed Nd:YAG laser in the near-infrared that can penetrate the cornea and focus inside it (the upper 100–120 µm), creating a lamellar cut. Thus, femtosecond-laser-assisted LASIK is the most advanced application.

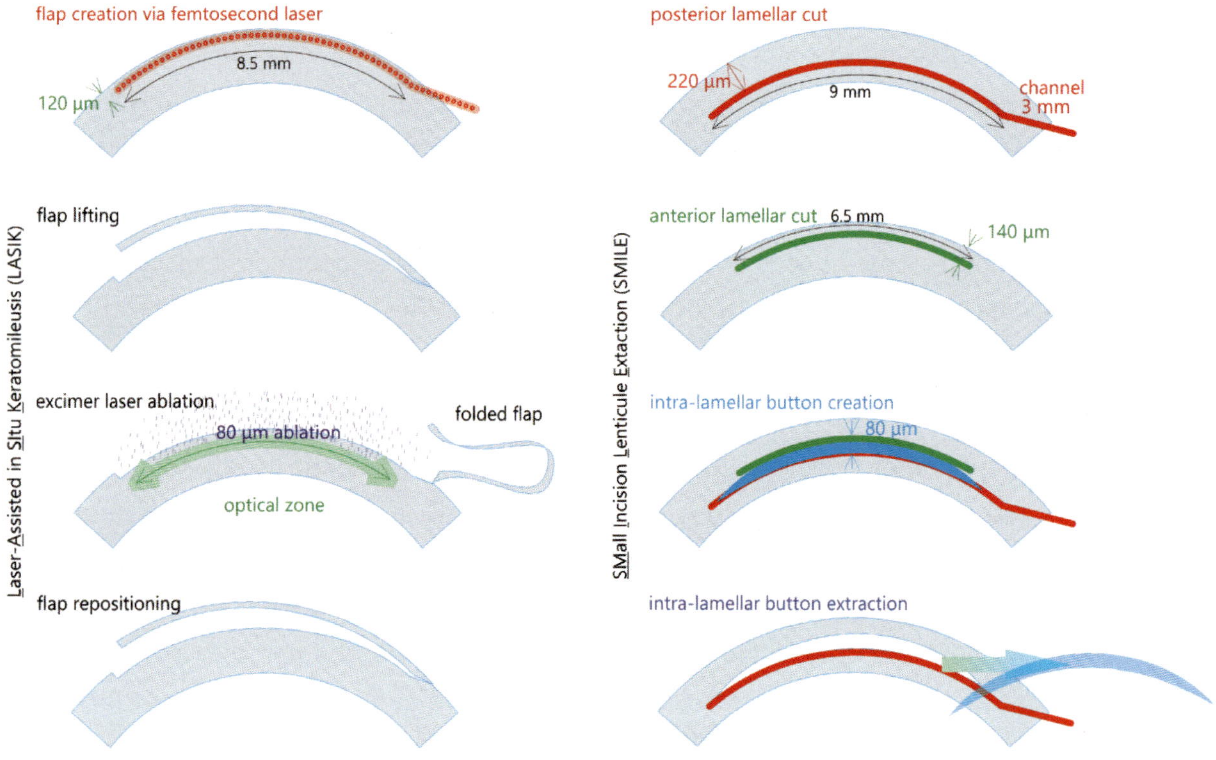


*Figure 6-64: (left) LASIK and (right) SMILE procedures for the correction of myopia.*

A new procedure called **small-incision lenticule extraction** (**SMILE**) involves only the use of a femtosecond laser to create two intrastromal lamellar cuts, forming a properly shaped corneal lenticule of the appropriate thickness, which is then simply extracted.

In addition to the laser unit, today's integrated refractive platforms consists of a proper laser-guidance optical system and a number of subsystems, such as eye tracking, focus control, and cyclotorsional compensation.

Historically, the first laser application in medicine was in dermatology, mainly facilitated by the ease of its external application. Not surprisingly, the first of such lasers was applied in the University Clinic of Harvard Medical School, Massachusetts General Hospital (Wellman Laboratories of Photomedicine)—it was precisely this hospital where the first modern surgical operation was performed more than a century ago. The Wellman Laboratories of Photomedicine holds yet another first. This is the lab that developed the use of a low-intensity infrared laser (actually a super-luminescent source) as a diagnostic application that enables the structure of biological tissue to be mapped.

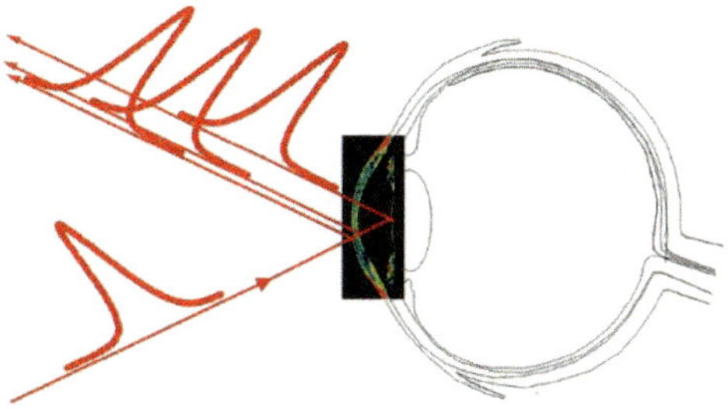

*Figure 6-65: The OCT technique obtains transverse optical cross-sections in biological tissue.*

The application is known as **optical coherence tomography** (**OCT**), in which a low-power pulse of a select wavelength (near-infrared) propagates through the tissue. Via interference with the reflection signal (backscattering), and applying laser-ranging interferometry techniques, the 'echo' of this signal is analyzed; thus, *in vivo* imaging of the tissue cross-section is obtained. The most significant application of OCT is in retinal imaging.

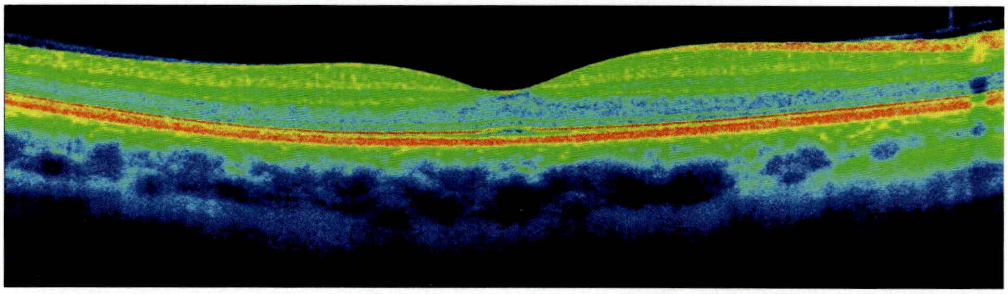

*Figure 6-66: OCT imaging of the central retinal section known as the macula.*

OCT is the optical analog of intravascular ultrasound (IV-US). The axial resolution corresponds to the wavelength. The higher the resolution, the more detail can be achieved. Today, OCT, and particularly frequency-domain OCT, presents the ultimate in resolution for many human tissues—not only for imaging in the eye, but also, if coupled to endoscopes with optical fibers, for imaging of the gastroenterological tract.

## 6.5.3 Materials Processing

Lasers are employed in metal processing such as cutting, welding, and precision boring. They are employed in integrated circuits and manufacturing of fine structures. The welding industry's 'darlings' are pulsed lasers (400 W Nd:YAG) and CW lasers (1000 W $CO_2$), which can focus the beam energy to a specific point to cause local melting. The high-precision beam directionality, the lack of wear on mechanical parts, and the possibility of remote operation, as well as the quickly produced and clean-cut features, are the most significant advantages of laser welding. The fuselage of the Airbus A318 aircraft was manufactured using this technique.

In addition to large-scale industrial laser applications, ultrashort-pulsed lasers are used in localized and specialized materials processing. A pulse (of a few nanoseconds or shorter) that is properly focused on a sample can cause initial heating (for surface density power up to $10^3$ W/mm$^2$), material melting ($10^3$ W/mm$^2$), and finally ablation, ionization, and plasma excitation.

The pulse energy therefore initially heats and evaporates a very small part of the sample; after that, the laser can ionize and excite the ablated material, creating a plasma. The atomic composition of the plasma is the same as the atomic composition of the sample. The excited plasma releases its energy in photonic emissions that correspond to the permissible atomic transitions; therefore, if the sample is properly analyzed spectroscopically, it is possible to conduct a qualitative and quantitative sample analysis. This technique is known as laser-induced breakdown spectroscopy (LIBS).

*Figure 6-67: The welder of the future! 1 kW Nd:YAG laser used in metal sheet boring.*

Figure 6-68 illustrates the plasma creation from a NaCl sample ionized with the third harmonic of a pulsed Nd:YAG laser. The plasma's intense yellow color is attributed to the sodium atomic line emission at 0.589 μm. The main chlorine emission lines are in the infrared (0.838 μm) and are detectable only using a photo-sensor with the proper spectral sensitivity.

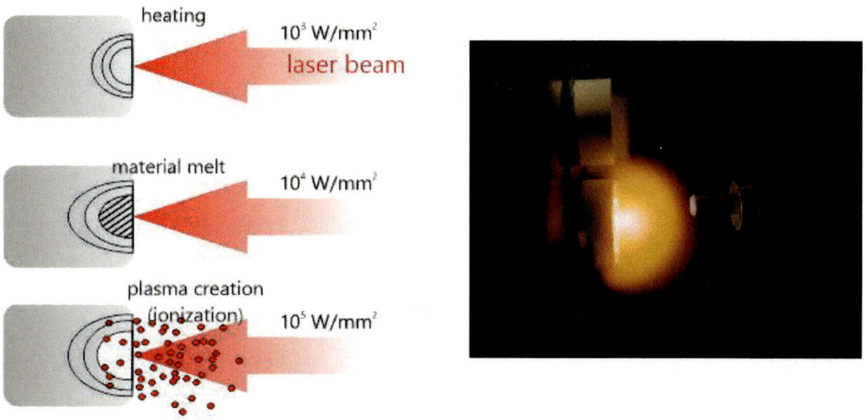


*Figure 6-68: Creation process of laser plasma.*

## 6.5.4 Optical Telecommunications

The very high directionality of a laser beam and its very large bandwidth are some of the significant advantages of laser applications in communications and information processing systems. Owing to their very high frequency and high degree of coherence, lasers are employed in satellite telecommunications with very low attenuation, minimum noise, and adaptability to a wide range applications—from the conventional microwave to optical telecommunication systems. Lasers are already used for transmitting high volumes of data over long distances.

The optical signal consists of a succession of pulses that correspond to binary information. These pulses propagate through optical waveguides or optical fibers. In solid-state and semiconductor lasers, the distribution is carried as beam amplitude or phase modulation.

A solid-state laser can emit a modulated signal in wavelengths (1.3 μm) for which the optical fiber displays the lowest dispersion, or at a wavelength (1.5 μm) for which it displays the lowest absorption. A solid-state laser can be integrated into semiconducting circuits due to its small size and low energy demands. With 50 million pulses per second, it can carry upwards of 5000 phone conversations through the same fiber optic!

Similar advantages are found in the use of lasers in information-processing systems. Such systems include optical storage devices, the simplest of which are the CD and DVD, either as read-only (CD-ROM) or as rewriteable CDs and DVDs. Information encoding is performed using a semiconductor GaAs laser operating at 840 nm, whose focused beam etches the encoded binary information onto the surface of a blank CD in a process of local mini-ablation. Reading is also performed with a diode laser. The encoded information may or may not, depending on whether the binary information is 0 or 1, produce a reflected signal, which is then detected by a photo-sensor.

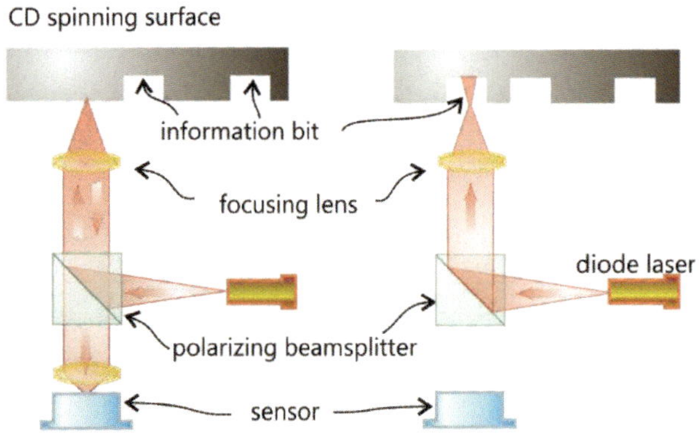


*Figure 6-69: Principle of CD operation.*

*Figure 6-70: Einstein tries to explain the optics inside a CD player to an interested youngster (© www.fiami.ch).*

# APPENDIX

## CONVENTIONS AND NOTATIONS

### Units (fundamental)

The International System of Units is followed.

| Unit name | Unit symbol | Physical Quantity |
|---|---|---|
| meter | m | length |
| kilogram | kg | mass |
| second | s | time |
| ampere | A | electric current |
| kelvin | K | thermodynamic temperature |
| mole | mol | amount of substance |
| candela | cd | luminous intensity |

### Decimal Marker and Grouping

The 2003 General Conference on Weights and Measures convention is followed.

- The period (.) is used as decimal marker symbol.
- A comma is used as grouping symbol in order to facilitate reading of numbers having more than four digits.
- The fundamental unit for length is the meter (m).

Example: 1.50 m means one meter and 50 centimeters.

| Length | Divisions | | |
|---|---|---|---|
| 1 m | 1 cm (centimeter) | $=10^{-2}$ m | 1 m = 100 cm |
| | 1 mm (millimeter) | $=10^{-3}$ m | 1 m = 1000 mm |
| | 1 μm (micrometer) | $=10^{-6}$ m | 1 m = 1,000,000 μm |
| | 1 nm (nanometer) | $=10^{-9}$ m | 1 m = 1,000,000,000 nm |

## Frequently Used Notation in Wave Optics

| Description | Notation | Value / Units |
|---|---|---|
| Wave disturbance | $y(z, t)$ | |
| Speed of light in vacuum | $c$ | $3 \times 10^8$ m·s$^{-1}$ |
| Speed of light in a medium | $u$ | less than $3 \times 10^8$ m·s$^{-1}$ |
| Electric field | $E$ | newton per coulomb: N·C$^{-1}$ = kg·m·s$^{-3}$·A$^{-1}$ |
| Magnetic field | $H$ | ampere per meter: A·m$^{-1}$ |
| Frequency | $\nu$ | hertz: Hz = s$^{-1}$ |
| Angular frequency | $\omega = 2\pi\nu$ | hertz |
| Wavelength | $\lambda$ | m (often µm/nm) |
| Wavevector | $k = 2\pi/\lambda$ | inverse length |
| Phase | $\varphi$ | 0 – 360° or 0 – $2\pi$ rad |
| Refractive index | $n$ | |
| Electric polarization | $P$ | C·m$^{-2}$ |

## Useful Notes

Angle units

| Angle | Divisions | | |
|---|---|---|---|
| 1° (degree) | 1′ (minute, arcmin) | =1°/60 | 1° = 60′ |
| | 1″ (second, arcsec) | =1′/60 = 1°/3600 | 1° = 3660 ″ |
| 1 rad (radian) | 1 rad ≈ 57.3° | | 1° = 0.0175 rad |

## Trigonometry

| Name | Noted as | Definition |
| --- | --- | --- |
| sine | sin | = opposite / hypotenuse |
| cosine | cos | = adjacent / hypotenuse |
| tangent | tan | = opposite / adjacent |
| cotangent | ctan | = adjacent / opposite |

## Useful Trigonometric Numbers

| angle (°) | angle (rad) | sin(ϑ) | cos(ϑ) | tan(ϑ) |
| --- | --- | --- | --- | --- |
| **0** | 0 | 0 | 1 | 0 |
| **30** | $\frac{\pi}{6}$ | $\frac{1}{2}$ | $\frac{\sqrt{3}}{2}$ | $\frac{\sqrt{3}}{3}$ |
| **45** | $\frac{\pi}{4}$ | $\frac{\sqrt{2}}{2}$ | $\frac{\sqrt{2}}{2}$ | 1 |
| **60** | $\frac{\pi}{3}$ | $\frac{\sqrt{3}}{2}$ | $\frac{1}{2}$ | $\sqrt{3}$ |
| **90** | $\frac{\pi}{2}$ | 1 | 2 | not defined |

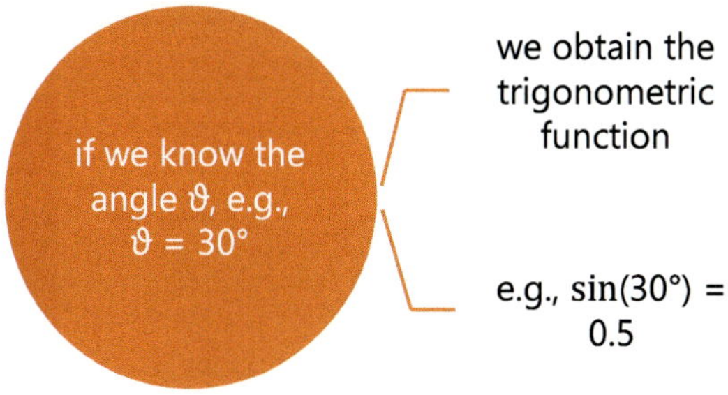


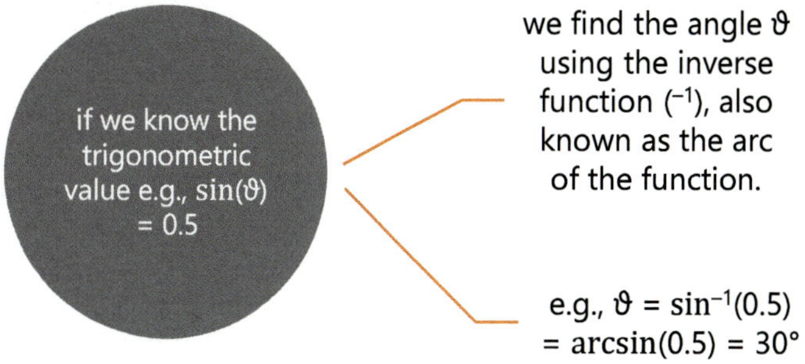

# ANSWERS TO QUIZ QUESTIONS

(Answers are provided to odd-numbered questions only.)

## Chapter 1 Light and Electromagnetism

1) c and e
3) a
5) a and e
7) c
9) d
11) d
13) a, b, c, and d
15) a
17) a and c
19) c
21) c
23) b
25) c
27) b
29) d
31) b
33) b
35) c

## Chapter 2 Polarization

1) d and e
3) d
5) b and c
7) b
9) c
11) b
13) c
15) f
17) a and f
19) a
21) c
23) c
25) b
27) a
29) b and c
31) c
33) b and e
35) b
37) c
39) b
41) d
43) d
45) c
47) b
49) b

## Chapter 3 Dispersion and Absorption

1) a
3) b
5) d
7) f
9) c
11) e
13) a
15) b and c
17) a and d
19) a

## Chapter 4 Interference

1) a
3) c
5) b
7) f
9) c
11) a
13) d
15) b and d
17) a
19) d
21) d
23) c
25) d
27) a and c
29) b
31) d
33) b
35) d
37) d
39) a, d and e
41) d
43) a and c
45) b
47) a
49) b
51) a and d
53) b
55) e
57) b
59) c

## Chapter 5 Diffraction

1) a
3) c and d
5) b
7) b and d
9) a
11) b
13) b
15) d
17) a and c
19) a and d
21) e
23) b
25) a
27) e
29) d
31) a
33) a
35) a and c
37) d
39) b
41) e
43) a and f
45) c
47) c
49) b
51) c
53) b
55) c
57) d
59) c

# INDEX

## A



## B



## C

## D

## E



## F

## G



## H



## I

## K



## L

## M



## N



## O



## P

## Q



## R

## S



## T



## U



## V



## W

## Y

**George Asimellis**, PhD, serves as Associate Professor of Optics and Research Director at the Kentucky College of Optometry, Pikeville, Kentucky, which he joined in 2015 as Founding Faculty. He oversees development and coordination of the Geometric Optics and Vision Science courses and development of the Laser Surgical Procedures course.

In the past, he served as head of Research at LaserVision.gr Institute, Athens, Greece, and as faculty in: the Physics Department, Aristotle University, Greece; Medical School, Democritus University, Greece; and the Electrical Engineering Department, George Mason University, Virginia.

His doctorate research involved advanced optical signal processing and pattern recognition techniques (PhD, Tufts University, Massachusetts), and optical coherence tomography (Fellowship, Harvard University, Massachusetts). He then worked on research and development of optoelectronic devices in a number in research centers in the USA. He has authored more than 75 peer-reviewed research publications, 8 scholarly books on optics and optical imaging, and a large number of presentations at international conferences and meetings.

He is on the Editorial Board of eight peer-reviewed journals, including the *Journal of Refractive Surgery*, for which he serves as Associate Editor. He received the 2017 Emerging Vision Scientist Award by the National Alliance for Eye and Vision Research (NAEVR).

His research interests include optoelectronic devices, anterior-segment (corneal and epithelial) imaging, keratoconus screening, ocular optics, and ophthalmological lasers. His recent contributions involve publications in clinical *in vivo* epithelial imaging and corneal cross-linking interventions.